Handbook of Applied Mathematics

Handbook of Applied Mathematics

Editor: Gregory Rago

NY RESEARCH PRESS
New York

Published by NY Research Press
118-35 Queens Blvd., Suite 400,
Forest Hills, NY 11375, USA
www.nyresearchpress.com

Handbook of Applied Mathematics
Edited by Gregory Rago

International Standard Book Number: 978-1-63238-879-7 (Hardback)

Cataloging-in-Publication Data

Handbook of applied mathematics / edited by Gregory Rago.
p. cm.
Includes bibliographical references and index.
ISBN 978-1-63238-879-7
1. Mathematics. 2. Mathematical models. I. Rago, Gregory.
QA36 .H36 2022
510--dc23

Contents

Preface

The main aim of this book is to educate learners and enhance their research focus by presenting diverse topics covering this vast field. This is an advanced book which compiles significant studies by distinguished experts in the area of analysis. This book addresses successive solutions to the challenges arising in the area of application, along with it; the book provides scope for future developments.

The discipline that is involved in developing and applying mathematically rigorous methods, computational algorithms and modeling insights to address real world problems is known as applied mathematics. It combines mathematical science and specialized knowledge for solving issues in fields like engineering, medicine and science. It is used in a wide range of situations such as predicting the behavior of financial markets, analyzing the flow of fluids around solids and constructing methods for multi criteria decision making. Some of the mathematical sciences which are closely associated with applied mathematics are scientific computing, statistics and actuarial science. This book elucidates new techniques and their applications in a multidisciplinary manner. While understanding the long-term perspectives of the subject, the book makes an effort in highlighting their impact as a modern tool for the growth of the discipline. This book, with its detailed analyses and data, will prove immensely beneficial to professionals and students involved in this area at various levels.

It was a great honour to edit this book, though there were challenges, as it involved a lot of communication and networking between me and the editorial team. However, the end result was this all-inclusive book covering diverse themes in the field.

Finally, it is important to acknowledge the efforts of the contributors for their excellent chapters, through which a wide variety of issues have been addressed. I would also like to thank my colleagues for their valuable feedback during the making of this book.

Editor

1

Viscous Dissipation Effects on the Motion of Casson Fluid over an Upper Horizontal Thermally Stratified Melting Surface of a Paraboloid of Revolution: Boundary Layer Analysis

T. M. Ajayi, A. J. Omowaye, and I. L. Animasaun

Department of Mathematical Sciences, Federal University of Technology, Akure, Ondo State, Nigeria

Correspondence should be addressed to T. M. Ajayi; stillmetunde@gmail.com

Academic Editor: Igor Andrianov

The problem of a non-Newtonian fluid flow past an upper surface of an object that is neither a perfect horizontal/vertical nor inclined/cone in which dissipation of energy is associated with temperature-dependent plastic dynamic viscosity is considered. An attempt has been made to focus on the case of two-dimensional Casson fluid flow over a horizontal melting surface embedded in a thermally stratified medium. Since the viscosity of the non-Newtonian fluid tends to take energy from the motion (kinetic energy) and transform it into internal energy, the viscous dissipation term is accommodated in the energy equation. Due to the existence of internal space-dependent heat source; plastic dynamic viscosity and thermal conductivity of the non-Newtonian fluid are assumed to vary linearly with temperature. Based on the boundary layer assumptions, suitable similarity variables are applied to nondimensionalized, parameterized and reduce the governing partial differential equations into a coupled ordinary differential equations. These equations along with the boundary conditions are solved numerically using the shooting method together with the Runge-Kutta technique. The effects of pertinent parameters are established. A significant increases in $\mathrm{Re}_x^{1/2}C_{fx}$ is guaranteed with S_t when magnitude of β is large. $\mathrm{Re}_x^{1/2}C_{fx}$ decreases with E_c and m.

1. Introduction

Within the last thirty years, the study of non-Newtonian fluid flow over a stretching surface has received significant attention due to its industrial applications. Such interest is fueled by its pertinent engineering applications in a number of fields as in spinning of filaments, continuous casting of metal, extrusion of polymers, crystal growing, glass fiber production, extrusion of plastic sheets, paper production, and process of condensation of metallic plates. Boundary layer analysis of fluid flow passing through a thick needle with variable diameter was investigated by Lee [1]. Historically, this can be referred to as the first report of flow adjacent to a surface with variable thickness where the effect of viscosity is highly significant. Thereafter, extensive studies were conducted on the boundary layer flows over a thin needle. Cebeci and Na [2] investigated the laminar free convection heat transfer from a needle. Ahmad et al. [3] examined the boundary layer flow over a moving thin needle with variable heat flux. The boundary layer flow over a stretching surface with variable thickness was analyzed by Fang et al. [4]. Recently, the flow of different fluids over an upper horizontal surface with variable thickness has been investigated extensively in [5–7]. This attracted Makinde and Animasaun [8, 9] to focus on the case of quartic autocatalysis kind of chemical reaction in the flow of an electrically conducting nanofluid containing gyrotactic-microorganism over an upper horizontal surface of a paraboloid of revolution in the presence and absence of thermophoresis and Brownian motion. In most cases, plastic dynamic viscosity of non-Newtonian Casson fluid tends to take energy away from the motion and transform it into internal energy. However, this area has been neglected.

In fluid mechanics, destruction of fluctuating velocity gradients due to viscous stresses is known as viscous dissipation. This partial irreversible processes is often referred to as transformation of kinetic energy into internal energy of the fluid (heating up the fluid due to viscosity since dissipation is high in the regions with large gradients). Pop [10] remarked that understanding the concept of energy dissipation and transport in nanoscale structures is of great importance for the design of energy-efficient circuits and energy-conversion systems. However, energy dissipation and transport of non-Newtonian fluid are also of importance to engineers and scientists. Motsumi and Makinde [11] examined the effects of viscous dissipation parameter (i.e., Eckert number), thermal diffusion, and thermal radiation on boundary layer flow of Cu-water and Al_2O_3-water nanofluids over a moving flat plate. In another theoretical study on combined effects of Newtonian heating and viscous dissipation parameter on boundary layer flow of copper and titania in water over stretchable wall, Makinde [12] reported an increase in the moving plate surface temperature and thermal boundary layer thickness. Depending on the admissible grouping of variables (parameterization), Eckert number and Brinkmann number ($Pr \times E_c$) may be used to quantify viscous dissipation. In addition, unsteady mixed convection in the flow of air over a semi-infinite stretching sheet taking into account the effect of viscous dissipation was carried out by Abd El-Aziz [13]. Both at steady stage ($A = 0$) and unsteady stage ($A = 1.5$), velocity and temperature of the flow increase with an increase in the magnitude of Eckert number. Recently, effects of viscous dissipation, Joule heating, and partial velocity slip on two-dimensional stagnation point flow were reported by Yasin et al. [14]. In another study conducted by Animasaun and Aluko [15], it is reported that when dynamic viscosity of air is assumed to vary linearly with temperature, normal negligible effect of Eckert number on velocity profiles will be noticed. Raju and Sandeep [16] focused on the motion of Casson fluid over a moving wedge with slip and observed a decrement in the temperature field with rising values of Eckert number. In the flow of non-Newtonian Casson fluid over an upper horizontal thermally stratified melting surface of a paraboloid of revolution, dissipation of smaller eddies due to molecular viscosity near the wall is significant. In the presence of constant magnetic field, electrically conducting Casson fluid flow over object that is neither a perfect horizontal/vertical nor inclined/cone is also an important issue.

The study of electrically conducting fluid flow is of considerable interest in modern metallurgical. This can be traced to the fact that most fluids in this sector are electrically conducting fluid. Historically, the first report on the motion of electrically conducting fluid in the presence of magnetic field was presented by Rossow [17]. Thereafter, Alfvén [18] reported that if a conducting liquid is placed in a constant magnetic field, every motion of the liquid generates a force called electromotive force (e.m.f.) which produces electric currents. One of the most significant importance of these contributions are its applications in engineering problems such as MHD generators, plasma studies, nuclear reactors, and geothermal energy extractions. Soundalgekar and Murty [19] investigated heat transfer in MHD flow with pressure gradient, suction, and injection. It was observed that an increase in the magnetic field parameter leads to an increase in fluid velocity, skin friction, rate of heat transfer, and a fall in temperature. Rajeswari et al. [20] observed that, due to the uniform magnetic field and suction at the wall of the surface, the concentration of the fluid decreases with the increase in chemical reaction parameter. Ghosh et al. [21] found that an increase in inclination of the applied magnetic field opposes primary flow and also reduces Grashof numbers. Das [22] concluded that increasing magnetic field and thermal radiation leads to deceleration of velocity but reverse is the effect for the melting parameter when the solid surface and the free stream move in the same direction. Motsa and Animasaun [23] presented the behavior of unsteady non-Darcian magnetohydrodynamic fluid flow past an impulsively using bivariate spectral local linearization analysis. Koriko et al. [24] illustrated the dynamics of two-dimensional magnetohydrodynamics (MHD) free convective flow of micropolar fluid along a vertical porous surface embedded in a thermally stratified medium. It was concluded that velocity profiles and microrotation profiles are strongly influenced by the magnetic field in the boundary layer, which decreases with an increase in the magnitude of magnetic parameter. Recently, theoretical investigation of MHD natural convection flow in vertical microchannel formed by two electrically nonconducting infinite vertical parallel plates and effects of MHD mixed convection on the flow through vertical pipe with time periodic boundary condition was presented explicitly by Jha and Aina [25, 26].

Several processes involving melting heating transfer in non-Newtonian fluids have promising applications in thermal engineering, such as melting permafrost, oil extraction, magma solidification, and thermal insulation. As such, a lot of experimental and theoretical work has been conducted in the kinetics of heat transfer accompanied with melting or solidification effect. The process of melting of ice placed in a hot stream of air at a steady state was first reported by Roberts [27]. Historically, this report can be referred to as the pioneering analysis of the melting phenomenon. Another novel report on melting phenomenon during forced convection heat transfer when an iceberg drifts in warm sea water was presented by Tien and Yen [28]. From their investigation, they observed that melting at the interface results in a decrease in the Nusselt number. Epstein and Cho [29] discussed the laminar film condensation on a vertical surface. Much later, melting heat transfer in a nanofluid boundary layer on a stretching circular cylinder was examined by Gorla et al. [30]. In the study of the effect of radiation on MHD mixed convection flow from a vertical plate embedded in a saturated porous media with melting, Adegbie et al. [31] reported that the Nusselt number decreases with increase in melting parameter. Adegbie et al. [31] stated that the temperature of UCM fluid flow over a melting surface is an increasing function of variable thermal conductivity parameter. Omowaye and Animasaun [32] investigated boundary layer analysis in the flow of upper convected Maxwell fluid flow. Due to the fact that temperature at the wall is zero, classical temperature-dependent viscosity and thermal conductivity linear models were modified to suit the case of both melting heat transfer and thermal stratification. In another study of micropolar fluid flow in the presence of temperature-dependent and space heat source, the analysis of the case where vortex viscosity is a constant function of temperature was reported in [33].

Within the past two decades, the effects of temperature-dependent viscosity on the fluid flow have become more important to engineers dealings with geothermal systems, crude oil extraction, and machinery lubrication. Due to friction and internal heat generated between two layers of fluid, viscosity and thermal conductivity of fluid substance may be affected by temperature; for more details see Batchelor [34], Lai and Kulacki [35], and Abd El-Aziz [36]. Proper consideration of this fact in the study on inherent irreversibility in a variable viscosity Couette flow by Makinde and Maserumule [37], numerical investigation of micropolar fluid flow over a nonlinear stretching sheet taking into account the effects of a temperature-dependent viscosity by Rahman et al. [38], effects of MHD on Casson fluid flow in the presence of Cattaneo-Christov heat flux by Malik et al. [39], fluid flow through a pipe with variation in viscosity by Makinde [40], Casson fluid flow within boundary layer over an exponentially stretching surface embedded in a thermally stratified medium by Animasaun [41], steady fully developed natural convection flow in a vertical annular microchannel having temperature-dependent viscosity in the presence of velocity slip and temperature jump at the annular microchannel surfaces by Jha et al. [42] have enhanced the body of knowledge on fluid flow, boundary layer analysis, and heat/mass transfer. In a recent experiment, Alam et al. [43] concluded that thermal boundary layer decreases with an increasing temperature-dependent viscosity. Hayat et al. [44] discussed the effect of variable thermal conductivity on the mixed convective flow over a porous medium stretching surface. In the article, the kinematics viscosity of Casson fluid was considered as a function depending on plastic dynamic viscosity, density, and Casson parameter and hence reported that increase in the magnitude of temperature-dependent viscosity parameter leads to an increase in fluid's velocity. The above literature review shows that there exists no published article on the effects of viscous dissipation in the flow of non-Newtonian Casson fluid over a upper horizontal thermally stratified melting surface of a paraboloid of revolution.

2. Formulation of the Problem

Consider a steady, incompressible, laminar flow of an electrically conducting non-Newtonian (Casson) fluid over a melting surface on upper horizontal paraboloid of revolution in the presence of viscous dissipation and thermal stratification. The x-axis is taken in the direction of motion and y-axis is normal to the flow as shown in Figures 1(a) and 1(b). A uniform magnetic field of strength B_o is applied normal to the flow. The induced magnetic field due to the motion of an electrically conducting Casson fluid is assumed to be so small; hence it is neglected. The stretching velocity is $U_w = U_o(x+b)^m$ and the wall is assumed impermeable. It is further assumed that the immediate fluid layer adjacent to the surface is specified as $y = A(x+b)^{(1-m)/2}$, where $m < 1$. Using Boussinesq approximation, the difference in inertia is negligible but gravity is sufficiently strong to make the specific weight appreciably different between any two layers of Casson fluid on the surface; hence the rate flow on this kind of surface is referred to as free convection. The body force term suitable to induce the flow over a surface which is neither a perfect horizontal/vertical nor inclined/cone is

$$-\frac{\partial p}{\partial x} + \rho g_x = g\gamma \frac{m+1}{2}(T_\infty - T) + 0. \tag{1}$$

Following the theory stated in Casson [45] and boundary layer assumptions, the rheological equation for an isotropic Casson fluid flow together with heat transfer is of the form

$$\frac{\partial u}{\partial x} + \frac{\partial v}{\partial y} = 0, \tag{2}$$

$$u\frac{\partial u}{\partial x} + v\frac{\partial u}{\partial y} = \frac{1}{\rho}\left(1+\frac{1}{\beta}\right)\frac{\partial}{\partial y}\left(\mu_b(T)\frac{\partial u}{\partial y}\right) + g\gamma \cdot \frac{m+1}{2}(T_\infty - T) - \frac{\sigma B_o^2}{\rho}u, \tag{3}$$

$$u\frac{\partial T}{\partial x} + v\frac{\partial T}{\partial y} = \frac{1}{\rho Cp}\frac{\partial}{\partial y}\left(\kappa_b(T)\frac{\partial T}{\partial y}\right) + \frac{\mu_b(T)}{\rho Cp}\left(1+\frac{1}{\beta}\right)\frac{\partial u}{\partial y}\frac{\partial u}{\partial y} + \frac{Q_o[T_w(x) - T_\infty]}{\rho C_p} \cdot \exp\left[-ny\sqrt{\frac{m+1}{2}}\sqrt{\frac{U_o}{\vartheta}}(x+b)^{(m-1)/2}\right]. \tag{4}$$

Suitable boundary conditions governing the flow along upper horizontal surface of a paraboloid of revolution are

$$u = U_o(x+b)^m, \quad \kappa_b(T)\frac{\partial T}{\partial y} = \rho[\lambda^* + c_s(T_m - T_o^*)]v(x,y), \quad T = T_m(x) \quad \text{at } y = A(x+b)^{(1-m)/2}, \tag{5}$$

$$u \longrightarrow 0, \quad T \longrightarrow T_\infty(x) \quad \text{as } y \longrightarrow \infty. \tag{6}$$

The formulation of the second term in boundary (5) states that the heat conducted to the melting surface on paraboloid of revolution is equal to the combination of heat of melting and the sensible heat required to raise the solid temperature T_o to its stratified melting temperature $T_m(x)$ (for details, see [29]). For lubricating fluids, heat generated by the internal friction and the corresponding increase in temperature affects the viscosity of the fluid; hence, it may not be realistic to constant function of plastic dynamic viscosity knowing fully well that space-dependent internal heat source is significant. In order to account for this variation, it is valid to consider the modified mathematical models of both the

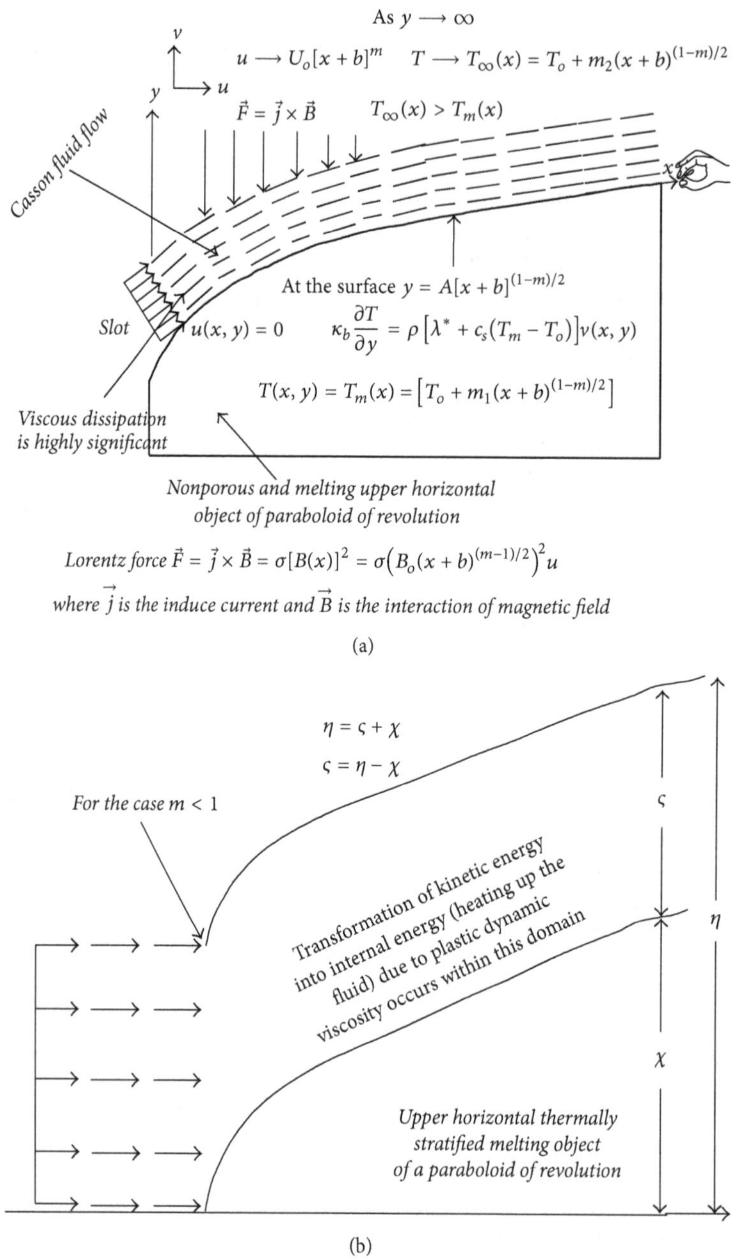

Figure 1: (a) The coordinate system of Casson fluid flow over upper horizontal thermally stratified melting surface of a paraboloid of revolution. (b) Graphical illustration of fluid domain and conversion of domain from $[\chi, \infty)$ to $[0, \infty)$.

temperature-dependent viscosity and thermal conductivity models proposed in [46] and adopted in [31, 32] as

$$\mu_b(T) = \mu_b^*\left[1 + b_1(T_\infty - T)\right],$$
$$\kappa_b(T) = \kappa_b^*\left[1 + b_2(T - T_m)\right], \quad (7)$$
$$\theta(\eta) = \frac{T - T_m}{T_\infty - T_o}.$$

Meanwhile, the models are still in good agreement with experimental data of Batchelor [34]. It is worth mentioning that the first and second terms in (7) are valid and reliable since $T_\infty(x) > T_m(x)$ in this study, whereas thermal stratification (T_m) at the melting wall and at the free stream is defined as

$$T_m(x) = T_o + m_1(x+b)^{(1-m)/2},$$
$$T_\infty(x) = T_o + m_2(x+b)^{(1-m)/2}. \quad (8)$$

Using similarity variable for temperature $\theta(\eta)$ in (7) to simplify $[1 - \theta(\eta)]$ we obtain suitable temperature difference in flow past thermally stratified horizontal melting surface of paraboloid of revolution as

$$T_\infty - T = (1-\theta)\left[T_\infty - T_o\right] - m_1(x+b)^{(1-m)/2}. \quad (9)$$

From thermal stratification models in (9), the following relations can be easily obtained

$$b_1(T_m - T_o) = b_1 m_1 (x+b)^{(1-m)/2},$$
$$b_1(T_\infty - T_o) = b_1 m_2 (x+b)^{(1-m)/2}, \tag{10}$$

where T_o is known as reference temperature. A significant difference between $(T_m - T_o)$ and $(T_\infty - T_o)$ can be easily obtained from (10). This can be traced to the fact that the linear stratification occurs at all points of "x" on the wall $(y = A(x+b)^{(1-m)/2})$ and at all points of "x" at the free stream as $(y \to \infty)$. In view of this, it is valid to define temperature-dependent viscous parameter ξ by considering the second term in (10) since $T_\infty(x) > T_m(x)$. Mathematically, the ratio of the first two terms in (10) can thus produce dimensionless thermal stratification parameter (S_t) as

$$\xi = b_1(T_\infty - T_o),$$
$$b_1(T_m - T_o) = \xi S_t, \tag{11}$$
$$S_t = \frac{m_1}{m_2}.$$

The stream function $\psi(x,y)$ and similarity variable η are of the form

$$u = \frac{\partial \psi}{\partial y},$$
$$v = -\frac{\partial \psi}{\partial x},$$
$$\eta = y\left(\frac{m+1}{2}\frac{U_o}{\vartheta}\right)^{1/2}(x+b)^{(m-1)/2}, \tag{12}$$
$$\psi = f\left(\frac{2}{m+1}\right)^{1/2}(\vartheta U_o)^{1/2}(x+b)^{(m+1)/2}.$$

It is important to note that the stream function $\psi(x,y)$ automatically satisfies continuity (2). The nonlinear partial differential equations (3) and (4) are reduced to the following nonlinear coupled ordinary differential equations

$$[1+\xi-\theta\xi-\xi S_t]\left(1+\frac{1}{\beta}\right)\frac{d^3 f}{d\eta^3} - \xi\left(1+\frac{1}{\beta}\right)\frac{d\theta}{d\eta}\frac{d^2 f}{d\eta^2} + f\frac{d^2 f}{d\eta^2} - \frac{2m}{m+1}\frac{df}{d\eta}\frac{df}{d\eta} + \frac{2H_a}{m+1}\frac{df}{d\eta} + G_t[(1-\theta)\xi - \xi S_t] = 0, \tag{13}$$

$$[1+\varepsilon\theta]\frac{d^2\theta}{d\eta^2} + \varepsilon\frac{d\theta}{d\eta}\frac{d\theta}{d\eta} - P_r S_t \frac{1-m}{m+1}\frac{df}{d\eta} + P_r f\frac{d\theta}{d\eta} - P_r\frac{1-m}{m+1}\theta\frac{df}{d\eta} + P_r E_c \frac{m+1}{2}[1+\xi-\theta\xi-\xi S_t]\left(1+\frac{1}{\beta}\right)\frac{d^2 f}{d\eta^2}\frac{d^2 f}{d\eta^2} + \frac{2P_r\Gamma}{m+1}e^{[-n\eta]} = 0. \tag{14}$$

In (13) and (14), melting parameter δ, Prandtl number P_r, magnetic parameter H_a, Eckert number E_c, temperature-dependent thermal conductivity parameter ε, space-dependent internal heat source parameter Γ, skin friction coefficient C_{fx}, Nusselt number Nu_x, and buoyancy parameter depending on volumetric-expansion coefficient due to temperature G_t are defined as

$$E_c = \frac{U_o^2 (x+b)^{(m+1)/2}}{C_p(T_w - T_\infty)},$$
$$\delta = \frac{C_p m_2 (x+b)^{(1-m)/2}}{\left[\lambda^* + c_s m_1 (x+b)^{(1-m)/2}\right]},$$
$$H_a = \frac{\sigma B_o^2}{\rho U_o},$$
$$\varepsilon = b_2(T_\infty - T_o), \tag{15}$$
$$\Gamma = \frac{Q_o}{\rho C_p U_o (x+b)^{m-1}},$$
$$G_t = \frac{g\gamma}{b_1 U_o^2 (x+b)^{2m-1}},$$
$$C_{fx} = \frac{\tau_w}{\rho\sqrt{(m+1)/2}\,(U_w)^2},$$
$$\mathrm{Nu}_x = \frac{(x+b)\,q_w}{\kappa_b[T_\infty - T_o]\sqrt{(m+1)/2}},$$

where τ_w is the shear stress (skin friction) between Casson fluid and upper surface of horizontal paraboloid of revolution and q_w is the heat flux at all points on the surface

$$\tau_w = \mu_b\left(1+\frac{1}{\beta}\right)\left.\frac{\partial u}{\partial y}\right|_{y=A(x+b)^{(1-m)/2}},$$
$$q_w = -\left(\kappa_b\left.\frac{\partial T}{\partial y}\right|_{y=A(x+b)^{(1-m)/2}}\right). \tag{16}$$

In order to nondimensionalize the boundary conditions (5) and (6), it is pertinent to note that the minimum value of y is not the starting point of the slot. This implies that all the conditions in (5) are not imposed at $y = 0$. As shown in Figures 1(a) and 1(b), it is obvious that it may not be realistic to say that $y = 0$ at all points on upper horizontal melting surface of a paraboloid of revolution. Hence, it is not valid to set $y = 0$ in similarity variable η. Upon using $y = A(x+b)^{(1-m)/2}$ the minimum value of y which accurately corresponds to minimum value of similarity variable

$$\eta = A\left(\frac{m+1}{2}\frac{U_o}{\vartheta}\right)^{1/2} = \chi. \tag{17}$$

This implies that, at the surface ($y = A(x+b)^{(1-m)/2}$), the boundary condition suitable to scale the boundary layer flow is $\eta = \chi$. The boundary condition becomes

$$\frac{df}{d\chi} = 0, \quad f(\chi) + \frac{\delta[1+\theta\varepsilon]}{P_r}\frac{d\theta}{d\chi} + \chi\frac{m-1}{m+1}\frac{df}{d\chi} = 0, \quad \theta(\chi) = 0 \quad \text{at } \chi = \eta. \tag{18}$$

$$\frac{df}{d\chi} \longrightarrow 1, \quad \theta(\chi) \longrightarrow 1 - S_t \quad \text{as } \chi \longrightarrow \infty. \tag{19}$$

Moreover, dimensionless governing equations (13) and (14) are depending on η while the boundary conditions (18) and (19) are functions and/or derivatives depending on χ. In order to transform the domain from $[\chi, \infty)$ to $[0, \infty)$ it is valid to adopt $F(\varsigma) = F(\eta - \chi) = f(\eta)$ and $\Theta(\varsigma) = \Theta(\eta - \chi) = \theta(\eta)$; for more details, see Figure 1(b). Considering the fact that Prandtl number is strongly dependent on plastic dynamic viscosity and thermal conductivity, and it is assumed that both properties vary linearly with temperature; hence for more accurate analysis of boundary layer as suggested in [47, 48], the P_r in (14) is

$$P_r = \frac{\mu_b C_p}{\kappa_b^*} = [1+\xi-\theta\xi-\xi S_t]\frac{\mu_b^* C_p}{\kappa_b^*} = P_{r\infty}[1+\xi-\theta\xi-\xi S_t] \tag{20}$$

Equation (18) reveals that Prandlt number at free stream is denoted as $P_{r\infty}$. The final dimensionless governing equation (coupled system of nonlinear ordinary differential equation) is

$$[1+\xi-\Theta\xi-\xi S_t]\left(1+\frac{1}{\beta}\right)\frac{d^3F}{d\varsigma^3} - \xi\left(1+\frac{1}{\beta}\right)\frac{d\Theta}{d\varsigma}\frac{d^2F}{d\varsigma^2} + F\frac{d^2F}{d\varsigma^2} - \frac{2m}{m+1}\frac{dF}{d\varsigma}\frac{dF}{d\varsigma} + \frac{2H_a}{m+1}\frac{dF}{d\varsigma} + G_t[(1-\Theta)\xi - \xi S_t] = 0, \tag{21}$$

$$[1+\varepsilon\Theta]\frac{d^2\Theta}{d\varsigma^2} + \varepsilon\frac{d\Theta}{d\varsigma}\frac{d\Theta}{d\varsigma} + [1+\xi-\Theta\xi-\xi S_t]\cdot\left(-P_{r\infty}S_t\frac{1-m}{m+1}\frac{dF}{d\varsigma} + P_{r\infty}F\frac{d\Theta}{d\varsigma} - P_{r\infty}\frac{1-m}{m+1}\Theta\frac{dF}{d\varsigma}\right) + P_{r\infty}E_c\frac{m+1}{2}[1+\xi-\Theta\xi - \xi S_t]^2\left(1+\frac{1}{\beta}\right)\frac{d^2F}{d\varsigma^2}\frac{d^2F}{d\varsigma^2} + P_{r\infty}[1+\xi-\Theta\xi - \xi S_t]\frac{2\Gamma}{m+1}e^{[-n\varsigma]} = 0. \tag{22}$$

Due to the fact that $\Theta(\varsigma) = 0$, the influence of temperature-dependent thermal conductivity on heat conduction during melting process diminishes. Dimensionless boundary condition reduces to

$$\frac{dF}{d\varsigma} = 0, \quad F(\varsigma) + \frac{\delta}{P_{r\infty}[1+\xi-\xi S_t]}\frac{d\Theta}{d\varsigma} + \chi\frac{m-1}{m+1}\frac{dF}{d\varsigma} = 0, \quad \Theta(\varsigma) = 0 \quad \text{at } \varsigma = 0. \tag{23}$$

$$\frac{dF}{d\varsigma} \longrightarrow 1, \quad \Theta(\varsigma) \longrightarrow 1 - S_t \quad \text{as } \varsigma \longrightarrow \infty. \tag{24}$$

Upon substituting the similarity variables (12) and models of physical quantities (i.e., C_{fx} and Nu_x) at the wall into (16) we obtain

$$\mathrm{Re}_x^{1/2}C_{fx} = \left(1+\frac{1}{\beta}\right)F''(0), \quad \mathrm{Nu}_x\mathrm{Re}_x^{-1/2} = -\Theta'(0). \tag{25}$$

3. Numerical Solution

Numerical solutions of the boundary valued problem (21)–(24) are obtained using classical Runge-Kutta method with shooting techniques and MATLAB package (bvp5c). The boundary value problem cannot be solved on an infinite interval and it would be impractical to solve it for even a very large finite interval; hence ς at infinity is 10. Using the method of superposition by Na [49], the boundary value problem of ODE has been reduced to a system of five simultaneous equations of first order (IVP) for five unknowns following the method of superposition. In order to integrate the corresponding IVP, the values of $F''(\varsigma = 0)$ and $\Theta'(\varsigma = 0)$ are required. However, such values do not exist after the nondimensionalization of the boundary conditions (5) and (6). The suitable guess values for $F''(\varsigma = 0)$ and $\Theta'(\varsigma = 0)$ are chosen and then integration is carried out. The calculated values of $F'(\varsigma)$ and $\Theta(\varsigma)$ at infinity ($\varsigma = 10$) are compared with the given boundary conditions in (24) and the estimated values $F''(\varsigma = 0)$ and $\Theta'(\varsigma = 0)$ are adjusted to give a better approximation for the solution. Series of values for $F''(\varsigma = 0)$ and $\Theta'(\varsigma = 0)$ are considered and applied with fourth-order classical Runge-Kutta method using step size $\Delta\varsigma = 0.01$.

TABLE 1: Validation of numerical technique: comparison between the solutions of classical Runge-Kutta together with shooting (RK4SM) and MATLAB solver bvp5c for the limiting case.

S_t	$F''(\varsigma = 0)$ (RK4SM)	$F''(\varsigma = 0)$ (bvp5c)	$\Theta'(\varsigma = 0)$ (RK4SM)	$\Theta'(\varsigma = 0)$ (bvp5c)
0	0.140452622100684	0.140452622451879	2.244340556008000	2.244340556781432
0.5	0.140381952095092	0.140381952471126	2.090185332523411	2.090185332413216
1	0.140582481552633	0.140582481451697	1.932324537463769	1.932324537412539

The above procedure is repeated until asymptotically converged results are obtained within a tolerance level of 10^{-6}. It is very important to remark that setting $\varsigma_\infty = 10$, all profiles are compatible with the boundary layer theory and asymptotically satisfies the conditions at free stream as suggested by Pantokratoras [50]. It is worth mentioning that there exist no related published articles that can be used to validate the accuracy of the numerical results. In view of this, (21)–(24) can easily be solved using ODE solvers such as MATLAB's bvp5c as explained in Kierzenka and Shampine [51] and Gökhan [52].

3.1. Verification of the Results. In order to verify the accuracy of the present analysis, the results of classical Runge-Kutta together with shooting have been compared with that of bvp5c solution for the limiting case when $\xi = 0.07$, $\varepsilon = 0.1$, $\beta = 0.1$, $E_c = 0.3$, $H_a = 0.25$, $\Gamma = 1$, $n = 0.09$, $P_{r\infty} = 1$, $m = 0.17$, $\delta = 0.2$, $\chi = 0.3$, and $G_t = 1$ at various values of S_t within the range $0 \le S_t \le 1$. As shown in Table 1, the comparison in the above case is found to be in good agreement. This good agreement is an encouragement for further study of the effects of other parameters.

4. Results and Discussion

The numerical computations have been carried out for various values of major parameters using the numerical scheme discussed in the previous section. This section presents the effect of different embedded physical parameters on the flow. Following Mustafa et al. [53], the ratio of momentum diffusivity to thermal diffusivity is considered to be unity (i.e., $P_r = 1$) due to the fact that Casson fluid flow under consideration possesses substantial yield stress.

4.1. Influence of Velocity Index Parameter m and Eckert Number E_c. At metalimnion level of thermal stratification ($S_t = 0.5$), when $\xi = 0.07$, $\varepsilon = 0.1$, $\beta = 0.1$, $E_c = 0.3$, $H_a = 0.25$, $\Gamma = 1$, $n = 0.09$, $P_{r\infty} = 1$, $\delta = 0.2$, $\chi = 0.3$, and $G_t = 1$, it is worth noting that both vertical and horizontal velocities decrease with m; see Figures 2 and 3. Combination of these practical meanings of velocity index parameter m justifies the decrease in both vertical and horizontal velocities profiles we obtained. Meanwhile, temperature distribution within the flow increases with m from a few distance after wall till free stream; see Figure 4. With an increase in the magnitude of m, Figure 5 reveals that temperature gradient profile $\Theta'(\varsigma)$ increases near the wall $0 \le \varsigma \le 3.1$ and decreases thereafter as $\varsigma \to 10$. It is very important to notice that velocity profile when $m = 0.15$ perfectly satisfies free stream condition asymptotically. This can be traced to the fact that

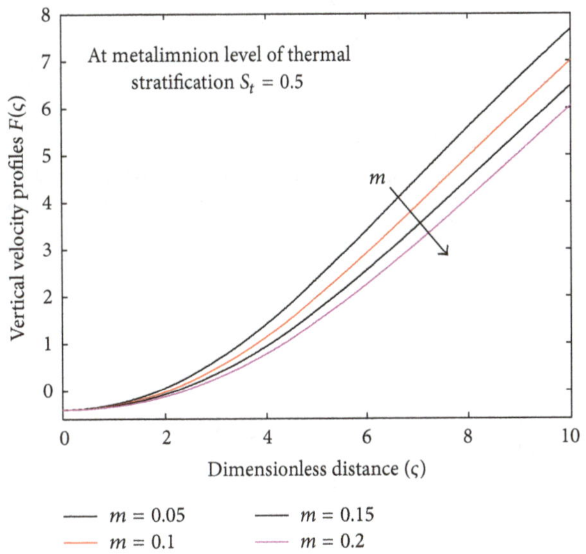

FIGURE 2: Effect of m on $F(\varsigma)$.

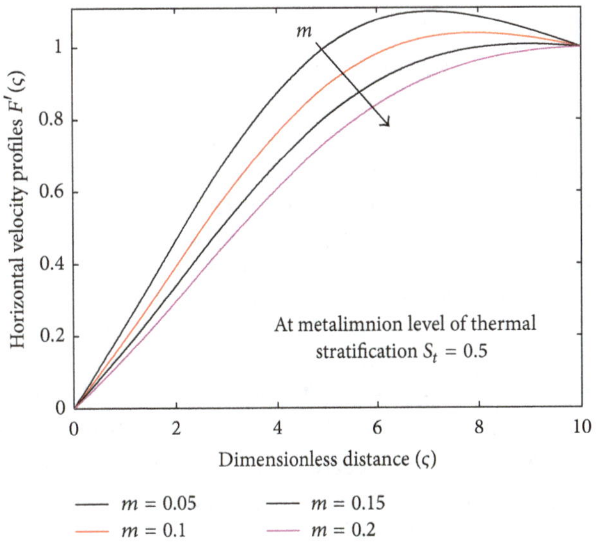

FIGURE 3: Effect of m on $F'(\varsigma)$.

$\Theta(0) = 0$ and as $\varsigma \to 10$, $\Theta(\varsigma) \to 1 - S_t$. In addition, graphical illustration of the function which describes the immediate layer of Casson fluid next to an upper horizontal surface of a paraboloid of revolution $y = A(x + b)^{(1-m)/2}$ against x (i.e., $A = 1$ and $b = 0.8$) at various values of m less than 1 and stretching velocity at the wall $U_w = U_o(x + b)^m$ when stretching rate (i.e., $U_o = 3$) are presented in

TABLE 2: Variations in Nusselt number $Nu_x Re_x^{-1/2}$ with Eckert number at various values of velocity index parameter when $\xi = 0.07$, $\varepsilon = 0.1$, $\beta = 0.1$, $E_c = 0.3$, $H_a = 0.25$, $\Gamma = 1$, $n = 0.09$, $P_{r\infty} = 1$, $S_t = 0.5$, $\delta = 0.2$, $\chi = 0.3$, and $G_t = 1$.

E_c	$Nu_x Re_x^{-1/2}$ When $m = 0.05$	$Nu_x Re_x^{-1/2}$ When $m = 0.15$	$Nu_x Re_x^{-1/2}$ When $m = 0.25$	$Nu_x Re_x^{-1/2}$ When $m = 0.35$
0	−1.9834	−2.0331	−2.0471	−2.0414
1	−2.2543	−2.2092	−2.1769	−2.1460
2	−2.4838	−2.3594	−2.2869	−2.2333
3	−2.6799	−2.4879	−2.3799	−2.3055
4	−2.8485	−2.5981	−2.4582	−2.3646
5	−2.9940	−2.6924	−2.5237	−2.4120

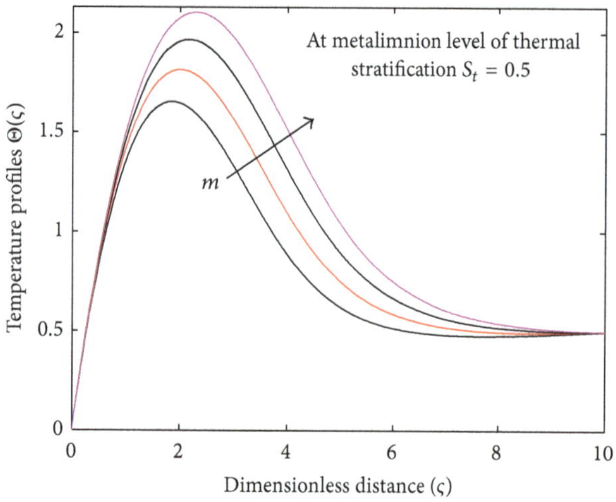

FIGURE 4: Effect of m on $\Theta(\varsigma)$.

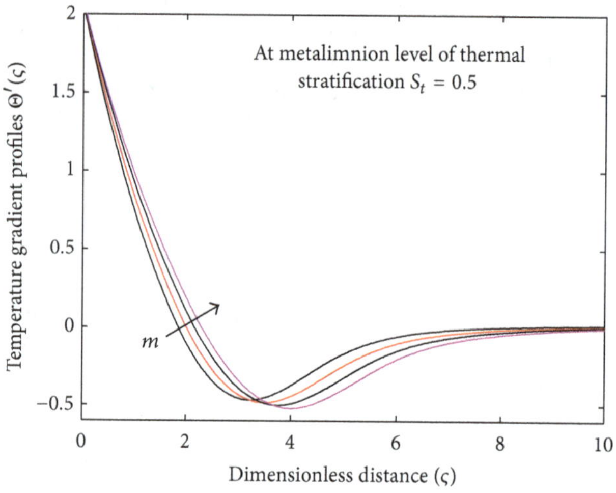

FIGURE 5: Effect of m on $\Theta'(\varsigma)$.

Figure 6. It is noticed that as m increases within $0.05 \leq m \leq 0.2$, the thickness of the paraboloid of revolution decreases but corresponding influence of stretching on the flow is an increasing function. The variation in local skin friction coefficient and local Nusselt number which is proportional to local heat transfer rate as stated in (25) as a function of viscous dissipation term and velocity index parameter is shown in Figure 7 and Table 2. Figure 7 shows that, at a fixed value of m, $Re_x^{1/2} C_{fx}$ decreases with Eckert number E_c. At constant value of E_c, unequal decrease in $Re_x^{1/2} C_{fx}$ with m is also observed. In addition, $Nu_x Re_x^{-1/2}$ decreases with E_c at various values of m. Table 2 shows that $Nu_x Re_x^{-1/2}$ increases highly significant when magnitude of E_c is large.

4.2. Influence of Non-Newtonian Casson Parameter β, Thermal Stratification Parameter S_t, and Eckert Number E_c. Figures 8–12 illustrate the effect of increasing the magnitude of non-Newtonian Casson parameter β on all the five profiles when $\xi = 0.07$, $\varepsilon = 0.1$, $E_c = 2$, $H_a = 0.25$, $\Gamma = 1$, $n = 0.09$, $P_{r\infty} = 1$, $m = 0.35$, $S_t = 0.5$, $\delta = 0.2$, $\chi = 0.3$, and $G_t = 1$. With an increase in the magnitude of β, it is observed that vertical velocity decreases, horizontal velocity decreases, temperature distribution increases only within the fluid domain ($2 \leq \varsigma \leq 8$), and the temperature gradient increases only near the wall. Figure 10 reveals that a distinct significant increase in $Re_x^{1/2} C_{fx}$ is guarantee with an initial increase in β from 0.2 to 0.25. Physically, increase in the magnitude of non-Newtonian Casson parameter ($\beta \to \infty$ implies sharp transition in the flow behavior from non-Newtonian fluid flow to Newtonian fluid flow. In view of this, resistance in the fluid flow is produced. It is worth mentioning that an increase in β implies a decrease in yield stress P_y of the Casson fluid and increase in the magnitude of plastic dynamic viscosity μ_b.

It is observed that the present study complements related studies on Casson fluid flow with temperature-dependent plastic dynamic viscosity on nonmelting surface; see Figures 8 and 9 in [41], Figures 3 and 4 in [54], and Figure 2 reported by Jasmine Benazir et al. [55]. The relationship between non-Newtonian Casson parameter β and Eckert number is sought for and illustrated graphically in Figures 13 and 14. Within $0 \leq E_c \leq 1.5$, there exists no significant difference in $Re_x^{1/2} C_{fx}$ with β. As shown in Figure 13, when the magnitude of $\beta = 0.25$, a distinct significant increase in $Re_x^{1/2} C_{fx}$ is observed due to an increase in the magnitude of viscous dissipation parameter. At small magnitude of β, local Nusselt number ($Nu_x Re_x^{-1/2}$) which is proportional to local heat transfer rate is found to be decreasing with E_c. At large value of β,

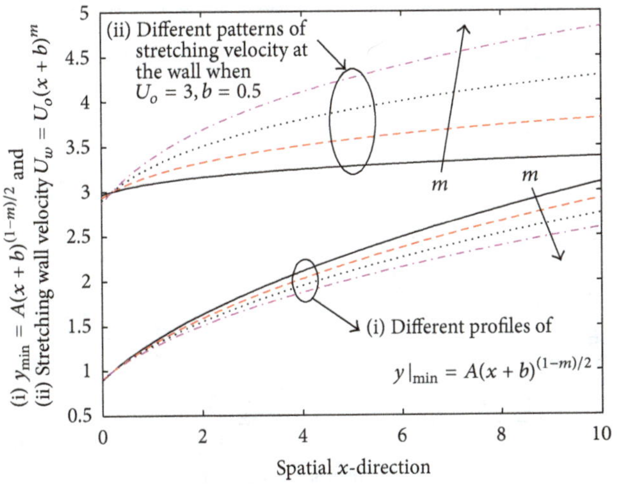

FIGURE 6: Graphical illustrations of (i) $y_{\min} = A(x+b)^{(1-m)/2}$; (ii) stretching wall velocity $U_w = U_o(x+b)^m$ at various values of m.

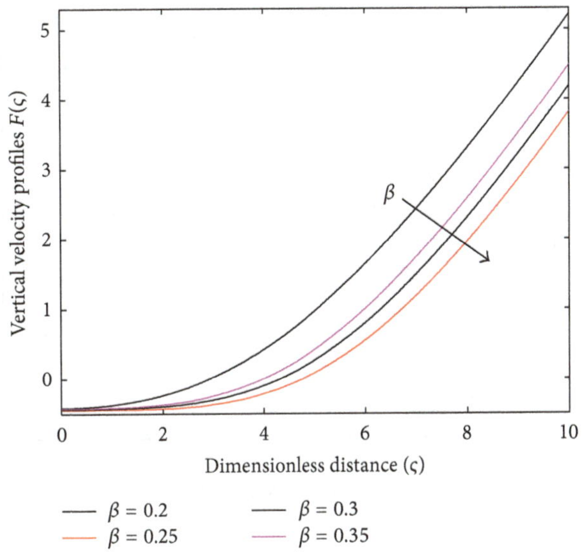

FIGURE 8: Effect of β on $F(\varsigma)$.

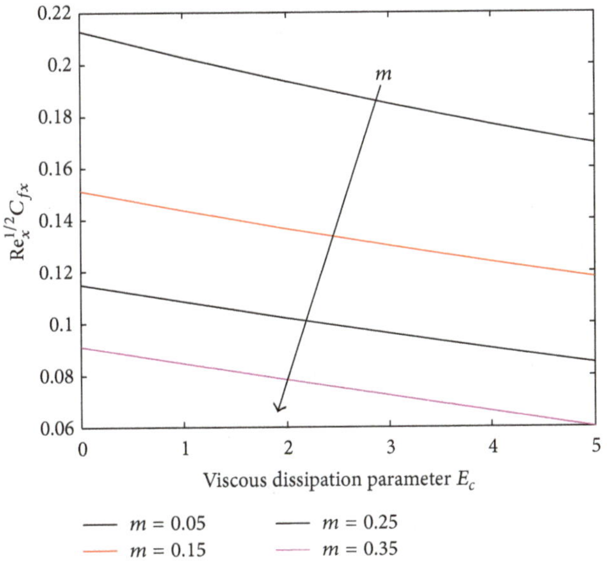

FIGURE 7: Variations in $\text{Re}_x^{1/2}C_{fx}$ with E_c at various values of m.

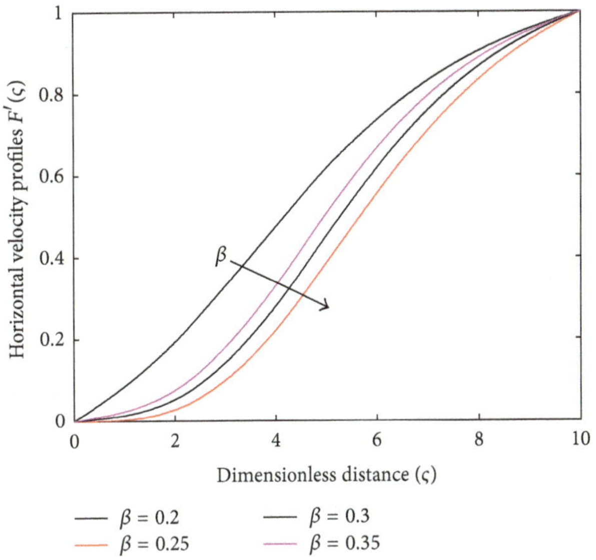

FIGURE 9: Effect of β on $F'(\varsigma)$.

$\text{Nu}_x\text{Re}_x^{-1/2}$ increases with E_c; see Figure 14. The simulation was further extended to unravels the relationship between non-Newtonian Casson parameter β, thermal stratification parameter S_t, and local skin friction coefficients when $\xi = 0.07$, $\varepsilon = 0.1$, $E_c = 2$, $H_a = 0.25$, $\Gamma = 1$, $n = 0.09$, $P_{r\infty} = 1$, $m = 0.17$, $\delta = 0.2$, $\chi = 0.3$, and $G_t = 1$. It is revealed in Figure 15 that $\text{Re}_x^{1/2}C_{fx}$ increases with β at epilimnion stage which is known as the highest and warmest layer ($S_t = 0$). In addition, a significant decrease in $\text{Re}_x^{1/2}C_{fx}$ is observed with an increase in β at hypolimnion stage which can be referred to as the coolest layer. Mathematically, when $S_t = 0$, this implies that $\Theta(\varsigma) = 1$ and maximum wall temperature at the wall explains the increase in $\text{Re}_x^{1/2}C_{fx}$ since increase in β corresponds to a decrease in yield stress P_y.

5. Conclusion

The boundary layer analysis of non-Newtonian Casson fluid flow over a horizontal melting surface embedded in a thermally stratified medium in the presence of viscous dissipation internal space heat source has been investigated numerically. The effects of the velocity power index, melting parameter, temperature-dependent viscous parameter, Eckert number, thermal conductivity, and magnetic interaction parameter were examined. Conclusions of the present analysis are as follows:

(1) Increase in the magnitude of velocity index parameter leads to a decrease in velocity and increase in temperature due to combine practical influence of the parameter.

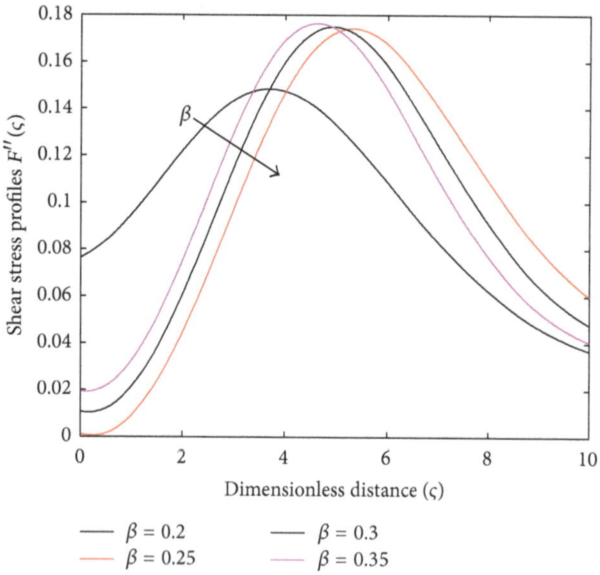

FIGURE 10: Effect of β on $F(\varsigma)$.

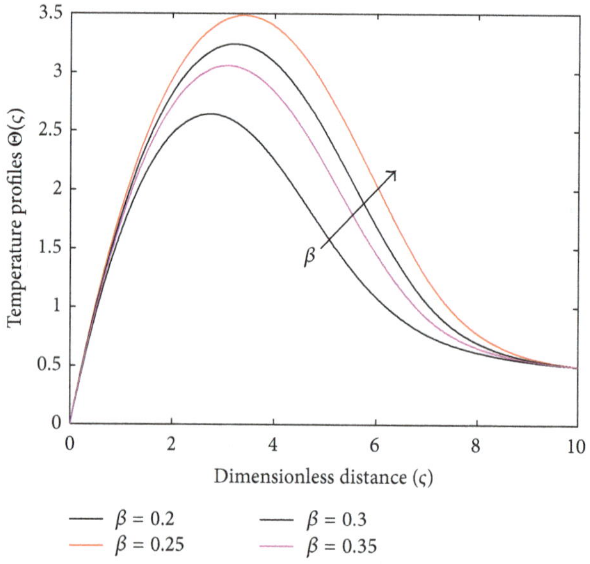

FIGURE 11: Effect of β on $F'(\varsigma)$.

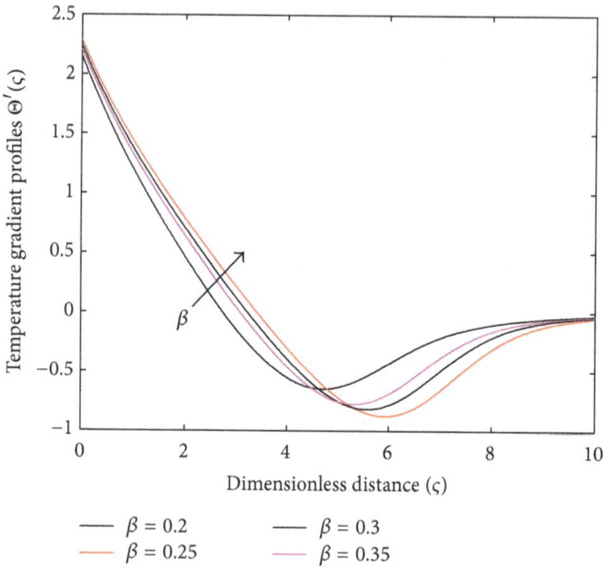

FIGURE 12: Effect of β on $F''(\varsigma)$.

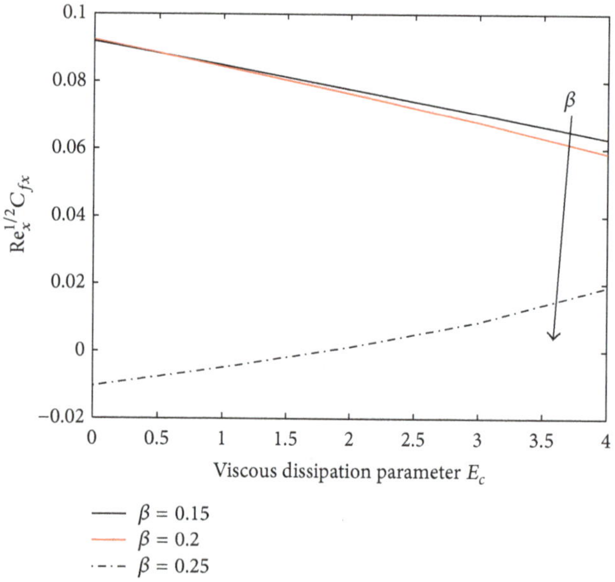

FIGURE 13: Variations in $\mathrm{Re}_x^{1/2} C_{fx}$ with E_c at various values of β.

(2) $\mathrm{Nu}_x \mathrm{Re}_x^{-1/2}$ decreases with E_c at various values of m within the interval $0.05 \leq m \leq 0.35$. $\mathrm{Re}_x^{1/2} C_{fx}$ decreases with E_c and m.

(3) At various values of viscous dissipation within $0 \leq E_c \leq 4$ and due to the nature of the flow past melting surface, a valid non-Newtonian parameter falls within the interval $0.000001 \leq \beta < 0.21$. Equivalent Newtonian fluid flow is guaranteed for $\beta \geq 0.23$.

(4) Local skin friction coefficient $\mathrm{Re}_x^{1/2} C_{fx}$ increases negligible with S_t when magnitude of β is small. A significant increases in $\mathrm{Re}_x^{1/2} C_{fx}$ is guaranteed with S_t when magnitude of β is large.

(5) In the case of Casson fluid flow over an upper horizontal thermally stratified melting surface of a paraboloid of revolution, decrease in horizontal velocity is guaranteed with an increase m and β.

(6) With an increase in the magnitude of m, the influence of stretching velocity at the wall $U_w = U_o(x+b)^m$ on horizontal and vertical velocities is stronger than that of $y = A(x+b)^{(1-m)/2}$ which describes the immediate fluid's layer next to upper horizontal surface of a paraboloid of revolution due to melting heat transfer.

An extension of the present study to the case of Williamson and Prandtl fluid flow over an upper horizontal thermally stratified melting surface of a paraboloid of revolution

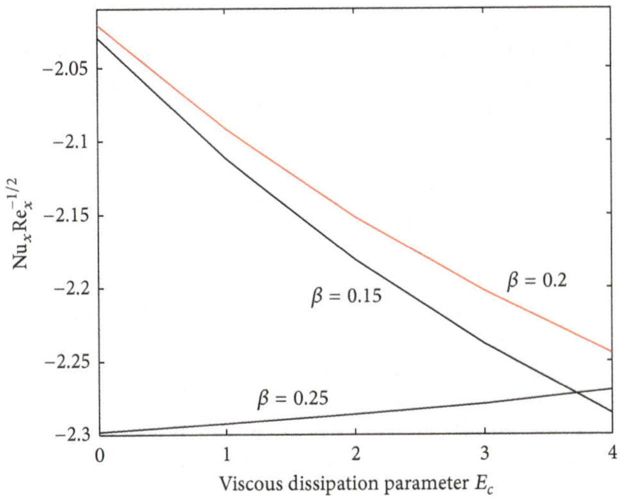

FIGURE 14: Variations in $Nu_x Re_x^{-1/2}$ with E_c at various values of β.

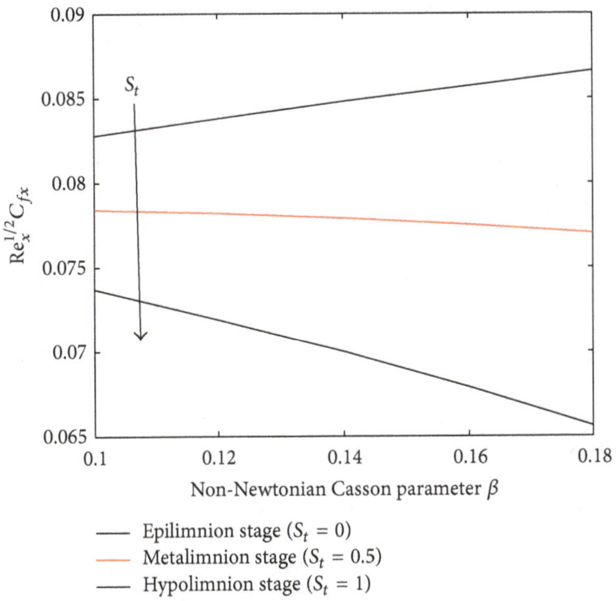

FIGURE 15: Variations in $Re_x^{1/2} C_{fx}$ with β at various values of S_t.

is hereby recommended. For suitable parametrization to achieve a comparative study, see [7].

Nomenclature

u: Velocity component in x direction
v: Velocity component in y direction
A: Coefficient related to stretching sheet
B_o: Magnetic field strength
C_f: Skin friction coefficient
C_p: Specific heat at constant pressure
x: Distance along the surface
y: Distance normal to the surface
b: Parameter related to stretching sheet
Nu_x: Local Nusselt number
P_r: Prandtl number
q_w: Heat transfer
T: Dimensional fluid temperature
U_w: Stretching velocity at the wall
b_1: Thermal property of the fluid
E_c: Eckert number
f, F: Dimensionless vertical velocity
κ_b: Thermal conductivity of the Casson fluid
m: Velocity power index
T_m: Melting temperature at wall
T_∞: Ambient Temperature
Re_x: Local Reynolds number
β: Non-Newtonian Casson parameter
ξ: Wall thickness parameter
μ_b: Plastic dynamic viscosity
λ^*: Latent heat of the fluid
δ: Melting parameter
θ, Θ: Dimensionless temperature
ϑ: Kinematic viscosity
ξ: Temperature-dependent viscosity parameter
ε: Temperature-dependent thermal conductivity parameter
ρ: Density of Casson fluid
σ: Electrical conductivity of the fluid
$\psi(x, y)$: Stream function.

Competing Interests

The authors declare that they have no competing interests.

References

[1] L. L. Lee, "Boundary layer over a thin needle," *Physics of Fluids*, vol. 10, no. 4, pp. 820–822, 1967.

[2] T. Cebeci and T. Y. Na, "Laminar free-convection heat transfer from a needle," *Physics of Fluids*, vol. 12, no. 2, pp. 463–465, 1969.

[3] S. Ahmad, R. Nazar, and L. Pop, "Mathematical modeling of boundary layer flow over a moving thin needle with variable heat flux," in *Advances in Numerical Methods*, vol. 11 of *Lecture Notes in Electrical Engineering*, pp. 43–54, Springer, Berlin, Germany, 2009.

[4] T. Fang, J. Zhang, and Y. Zhong, "Boundary layer flow over a stretching sheet with variable thickness," *Applied Mathematics and Computation*, vol. 218, no. 13, pp. 7241–7252, 2012.

[5] I. L. Animasaun, "47nm alumina-water nanofluid flow within boundary layer formed on upper horizontal surface of paraboloid of revolution in the presence of quartic autocatalysis chemical reaction," *Alexandria Engineering Journal*, vol. 55, no. 3, pp. 2375–2389, 2016.

[6] I. L. Animasaun and N. Sandeep, "Buoyancy induced model for the flow of 36 nm alumina-water nanofluid along upper horizontal surface of a paraboloid of revolution with variable thermal conductivity and viscosity," *Powder Technology*, vol. 301, pp. 858–867, 2016.

[7] O. Abegunrin, S. Okhuevbie, and I. Animasaun, "Comparison between the flow of two non-Newtonian fluids over an upper horizontal surface of paraboloid of revolution: boundary layer analysis," *Alexandria Engineering Journal*, vol. 55, no. 3, pp. 1915–1929, 2016.

[8] O. D. Makinde and I. L. Animasaun, "Bioconvection in MHD nanofluid flow with nonlinear thermal radiation and quartic autocatalysis chemical reaction past an upper surface of a paraboloid of revolution," *International Journal of Thermal Sciences*, vol. 109, pp. 159–171, 2016.

[9] O. Makinde and I. Animasaun, "Thermophoresis and Brownian motion effects on MHD bioconvection of nanofluid with nonlinear thermal radiation and quartic chemical reaction past an upper horizontal surface of a paraboloid of revolution," *Journal of Molecular Liquids*, vol. 221, pp. 733–743, 2016.

[10] E. Pop, "Energy dissipation and transport in nanoscale devices," *Nano Research*, vol. 3, no. 3, pp. 147–169, 2010.

[11] T. G. Motsumi and O. D. Makinde, "Effects of thermal radiation and viscous dissipation on boundary layer flow of nanofluids over a permeable moving flat plate," *Physica Scripta*, vol. 86, no. 4, Article ID 045003, 2012.

[12] O. D. Makinde, "Analysis of Sakiadis flow of nanofluids with viscous dissipation and Newtonian heating," *Applied Mathematics and Mechanics (English Edition)*, vol. 33, no. 12, pp. 1545–1554, 2012.

[13] M. Abd El-Aziz, "Unsteady mixed convection heat transfer along a vertical stretching surface with variable viscosity and viscous dissipation," *Journal of the Egyptian Mathematical Society*, vol. 22, no. 3, pp. 529–537, 2014.

[14] M. H. M. Yasin, A. Ishak, and I. Pop, "MHD stagnation-point flow and heat transfer with effects of viscous dissipation, Joule heating and partial velocity slip," *Scientific Reports*, vol. 5, Article ID 17848, 2015.

[15] I. L. Animasaun and O. B. Aluko, "Analysis on variable fluid viscosity of non-darcian flow over a moving vertical plate in a porous medium with suction and viscous dissipation," *IOSR Journal of Engineering*, vol. 4, no. 8, pp. 18–32, 2014.

[16] C. S. K. Raju and N. Sandeep, "MHD slip flow of a dissipative Casson fluid over a moving geometry with heat source/sink: a numerical study," *Acta Astronautica*, 2016.

[17] V. J. Rossow, "On the flow of electrically conducting fluid over a flat plate in the presence of a transverse magnetic field," NACA Technical Report 3071, 1957.

[18] H. Alfvén, "Existence of electromagnetic-hydrodynamic waves," *Nature*, vol. 150, no. 3805, pp. 405–406, 1942.

[19] V. M. Soundalgekar and T. V. R. Murty, "Heat transfer in MHD flow with pressure gradient, suction and injection," *Journal of Engineering Mathematics*, vol. 14, no. 2, pp. 155–159, 1980.

[20] R. Rajeswari, B. Jothiram, and V. K. Nelson, "Chemical reaction, heat and mass transfer on nonlinear MHD boundary layer flow through a vertical porous surface in the presence of suction," *Applied Mathematical Sciences*, vol. 3, no. 49–52, pp. 2469–2480, 2009.

[21] S. K. Ghosh, O. A. Bég, and A. Aziz, "A mathematical model for magnetohydrodynamic convection flow in a rotating horizontal channel with inclined magnetic field, magnetic induction and hall current effects," *World Journal of Mechanics*, vol. 1, no. 3, pp. 137–154, 2011.

[22] K. Das, "Radiation and melting effects on MHD boundary layer flow over a moving surface," *Ain Shams Engineering Journal*, vol. 5, no. 4, pp. 1207–1214, 2014.

[23] S. S. Motsa and I. L. Animasaun, "A new numerical investigation of some thermo-physical properties on unsteady mhd non-darcian flow past an impulsively started vertical surface," *Thermal Science*, vol. 19, supplement 1, pp. 249–258, 2015.

[24] O. K. Koriko, T. Oreyeni, A. J. Omowaye, and I. L. Animasaun, "Homotopy analysis of MHD free convective micropolar fluid flow along a vertical surface embedded in non-darcian thermally-stratified medium," *Open Journal of Fluid Dynamics*, vol. 6, no. 3, pp. 198–221, 2016.

[25] B. K. Jha and B. Aina, "Role of induced magnetic field on MHD natural convection flow in vertical microchannel formed by two electrically non-conducting infinite vertical parallel plates," *Alexandria Engineering Journal*, vol. 55, no. 3, pp. 2087–2097, 2016.

[26] B. K. Jha and B. Aina, "MHD mixed convection flow in a vertical pipe with time periodic boundary condition: steady periodic regime," *International Journal of Fluid Mechanics Research*, vol. 43, no. 4, pp. 350–367, 2016.

[27] L. Roberts, "On the melting of a semi-infinite body of ice placed in a hot stream of air," *Journal of Fluid Mechanics*, vol. 4, pp. 505–528, 1958.

[28] C. Tien and Y. Yen, "The effect of melting on forced convection heat transfer," *Journal of Applied Meteorology*, vol. 4, no. 4, pp. 523–527, 1965.

[29] M. Epstein and D. H. Cho, "Melting heat transfer in steady laminar flow over a flat plate," *Journal of Heat Transfer*, vol. 98, no. 3, pp. 531–533, 1976.

[30] R. S. Gorla, A. J. Chamkha, and E. Al-Meshaiei, "Melting heat transfer in a nanofluid boundary layer on a stretching circular cylinder," *Journal of Naval Architecture and Marine Engineering*, vol. 9, no. 1, pp. 1–10, 2012.

[31] K. S. Adegbie, A. J. Omowaye, A. B. Disu, and I. L. Animasaun, "Heat and mass transfer of upper convected maxwell fluid flow with variable thermo-physical properties over a horizontal melting surface," *Applied Mathematics*, vol. 6, no. 8, pp. 1362–1379, 2015.

[32] A. J. Omowaye and I. L. Animasaun, "Upper-convected maxwell fluid flow with variable thermo-physical properties over a melting surface situated in hot environment subject to thermal stratification," *Journal of Applied Fluid Mechanics*, vol. 9, no. 4, pp. 1777–1790, 2016.

[33] I. L. Animasaun, "Melting heat and mass transfer in stagnation point micropolar fluid flow of temperature dependent fluid viscosity and thermal conductivity at constant vortex viscosity," *Journal of the Egyptian Mathematical Society*, 2016.

[34] G. K. Batchelor, *An Introduction to Fluid Dynamics*, Cambridge University Press, Cambridge, UK, 1987.

[35] F. C. Lai and F. A. Kulacki, "The effect of variable viscosity on convective heat transfer along a vertical surface in a saturated porous medium," *International Journal of Heat and Mass Transfer*, vol. 33, no. 5, pp. 1028–1031, 1990.

[36] M. Abd El-Aziz, "Temperature dependent viscosity and thermal conductivity effects on combined heat and mass transfer in MHD three-dimensional flow over a stretching surface with Ohmic heating," *Meccanica*, vol. 42, article 375, 2007.

[37] O. D. Makinde and R. L. Maserumule, "Thermal criticality and entropy analysis for a variable viscosity Couette flow," *Physica Scripta*, vol. 78, no. 1, Article ID 015402, 2008.

[38] M. M. Rahman, M. A. Rahman, M. A. Samad, and M. S. Alam, "Heat transfer in a micropolar fluid along a non-linear stretching sheet with a temperature-dependent viscosity and variable surface temperature," *International Journal of Thermophysics*, vol. 30, no. 5, pp. 1649–1670, 2009.

[39] M. Y. Malik, M. Khan, T. Salahuddin, and I. Khan, "Variable viscosity and MHD flow in Casson fluid with Cattaneo-Christov

heat flux model: using Keller box method," *Engineering Science and Technology, an International Journal*, vol. 19, no. 4, pp. 1985–1992, 2016.

[40] O. D. Makinde, "Heat and mass transfer in a pipe with moving surface: effects of viscosity variation and energy dissipation," *Quaestiones Mathematicae*, vol. 24, no. 1, pp. 93–104, 2001.

[41] I. L. Animasaun, "Casson fluid flow with variable viscosity and thermal conductivity along exponentially stretching sheet embedded in a thermally stratified medium with exponentially heat generation," *Journal of Heat and Mass Transfer Research*, vol. 2, no. 2, pp. 63–78, 2015.

[42] B. K. Jha, B. Aina, and Z. Rilwanu, "Steady fully developed natural convection flow in a vertical annular microchannel having temperature dependent viscosity: an exact solution," *Alexandria Engineering Journal*, vol. 55, no. 2, pp. 951–958, 2016.

[43] M. S. Alam, M. M. Rahman, and M. A. Sattar, "Transient magnetohydrodynamic free convective heat and mass transfer flow with thermophoresis past a radiate inclined permeable plate in the presence of variable chemical reaction and temperature dependent viscosity," *Nonlinear Analysis: Modelling and Control*, vol. 14, no. 1, pp. 3–20, 2009.

[44] T. Hayat, S. A. Shehzad, M. Qasim, and A. Alsaedi, "Mixed convection flow by a porous sheet with variable thermal conductivityand convective boundary condition," *Brazilian Journal of Chemical Engineering*, vol. 31, no. 1, pp. 109–117, 2014.

[45] N. Casson, "A flow equation for pigment oil-suspensions of the printing ink type," in *Rheology of Disperse Systems*, C. C. Mill, Ed., p. 84, Pergamon Press, 1959.

[46] I. L. Animasaun, "Double diffusive unsteady convective micropolar flow past a vertical porous plate moving through binary mixture using modified Boussinesq approximation," *Ain Shams Engineering Journal*, vol. 7, no. 2, pp. 755–765, 2015.

[47] M. M. Rahman and K. M. Salahuddin, "Study of hydromagnetic heat and mass transfer flow over an inclined heated surface with variable viscosity and electric conductivity," *Communications in Nonlinear Science and Numerical Simulation*, vol. 15, no. 8, pp. 2073–2085, 2010.

[48] A. Pantokratoras, "Laminar free-convection over a vertical isothermal plate with uniform blowing or suction in water with variable physical properties," *International Journal of Heat and Mass Transfer*, vol. 45, no. 5, pp. 963–977, 2002.

[49] T. Y. Na, *Computational Methods in Engineering Boundary Value Problems*, Academic Press, New York, NY, USA, 2009.

[50] A. Pantokratoras, "A common error made in investigation of boundary layer flows," *Applied Mathematical Modelling*, vol. 33, no. 1, pp. 413–422, 2009.

[51] J. Kierzenka and L. F. Shampine, "A BVP solver based on residual control and the Maltab PSE," *ACM Transactions on Mathematical Software*, vol. 27, no. 3, pp. 299–316, 2001.

[52] F. S. Gökhan, *Effect of the Guess Function & Continuation Method on the Run Time of MATLAB BVP Solvers*, Edited by C. M. Ionescu, 2011.

[53] M. Mustafa, T. Hayat, I. Pop, and A. Aziz, "Unsteady boundary layer flow of a Casson fluid due to an impulsively started moving flat plate," *Heat Transfer—Asian Research*, vol. 40, no. 6, pp. 563–576, 2011.

[54] I. L. Animasaun, "Effects of thermophoresis, variable viscosity and thermal conductivity on free convective heat and mass transfer of non-darcian MHD dissipative Casson fluid flow with suction and nth order of chemical reaction," *Journal of the Nigerian Mathematical Society*, vol. 34, no. 1, pp. 11–31, 2015.

[55] A. Jasmine Benazir, R. Sivaraj, and M. M. Rashidi, "Comparison between casson fluid flow in the presence of heat and mass transfer from a vertical cone and flat plate," *Journal of Heat Transfer*, vol. 138, no. 11, Article ID 112005, 2016.

2

Green's Functions for Heat Conduction for Unbounded and Bounded Rectangular Spaces: Time and Frequency Domain Solutions

Inês Simões,[1,2] António Tadeu,[1,2] and Nuno Simões[1,2]

[1]*ITeCons, Rua Pedro Hispano, s/n, 3030-289 Coimbra, Portugal*
[2]*Department of Civil Engineering, University of Coimbra, Rua Luís Reis Santos, Pólo II, 3030-788 Coimbra, Portugal*

Correspondence should be addressed to Nuno Simões; nasimoes@dec.uc.pt

Academic Editor: Assunta Andreozzi

This paper presents a set of fully analytical solutions, together with explicit expressions, in the time and frequency domain for the heat conduction response of homogeneous unbounded and of bounded rectangular spaces (three-, two-, and one-dimensional spaces) subjected to point, line, and plane heat diffusion sources. Particular attention is given to the case of spatially sinusoidal, harmonic line sources. In the literature this problem is often referred to as the two-and-a-half-dimensional *fundamental solution* or 2.5D Green's functions. These equations are very useful for formulating three-dimensional thermodynamic problems by means of integral transforms methods and/or boundary elements. The image source technique is used to build up different geometries such as half-spaces, corners, rectangular pipes, and parallelepiped boxes. The final expressions are verified here by applying the equations to problems for which the solution is known analytically in the time domain.

1. Introduction

Problems in thermodynamics can often be solved with the aid of formulas or expressions known as Green's functions. These functions, or *fundamental solutions*, relate the field variables (heat fluxes and temperatures) at some location in a solid body caused by thermodynamic sources placed elsewhere in the medium.

The fundamental solutions most often used are point sources in a three-dimensional (3D), infinite homogeneous space; line sources acting within two-dimensional (2D) spaces; and plane sources heating one-dimensional (1D) spaces. The reason for these choices is that these three fundamental solutions are known in closed-form in time domain and have a relatively simple structure [1].

They are frequently combined to simulate heat conduction in the time domain or in a transform space defined by the Laplace transform, in half-spaces, infinite plates, rectangular 2D spaces, wedges, and rectangular 3D spaces [1–3]. Solutions have also been proposed to deal with multilayer systems, and they include the matrix method [1], the thermal quadrupole method [3], the thin layer method [4], and methods based on the definition of potentials [5–7]. Chen et al. have described the use of image method to solve 2D and 3D problems in unbounded and half-space domains containing circular or spherical shaped boundaries [8–10].

This paper compiles alternative fundamental solutions in explicit form, specifically Green's functions for harmonic 2D and 3D and harmonic (steady state) line sources whose amplitude varies sinusoidally in the third dimension. This last solution, which is often referred to in the literature as the 2.5D problem, can be of significant value when formulating 3D thermodynamics problems via boundary elements together with integral transforms. In addition, the proposed Green's functions are combined using an image source technique to model a half-space, a corner, a layer system, a laterally confined layer system, a solid rectangular column, a solid rectangular column with an end cross section, and a 3D parallelepiped inclusion. To the best of our knowledge, this is the first such derivation that promises to be efficient for formulating 3D thermodynamics problems using boundary elements and integral transforms.

Time domain solutions are obtained by applying inverse Fourier transforms, using complex frequencies to avoid aliasing phenomena. These solutions are validated by comparing computed responses with those obtained directly in the time domain.

2. Fundamental Solution

The transient heat transfer by conduction in an infinite, homogeneous space can be described by the diffusion equation in Cartesian coordinates:

$$\left(\frac{\partial^2}{\partial x^2}+\frac{\partial^2}{\partial y^2}+\frac{\partial^2}{\partial z^2}\right)T=\frac{1}{K}\frac{\partial T}{\partial t}, \quad (1)$$

in which t is time, $T(x,y,z,t)$ is the temperature at a point (x,y,z) in the domain, and K is the thermal diffusivity defined by $\lambda/(\rho c)$, where λ is the thermal conductivity, ρ is the density, and c is the specific heat of medium.

The solution of (1) can be obtained in the frequency domain after the application of a Fourier transform in the time domain, which leads to the following equation:

$$\left(\nabla^2+\left(\sqrt{\frac{-i\omega}{K}}\right)^2\right)T(x,y,z,\omega)=0, \quad (2)$$

where $i=\sqrt{-1}$, $k_1=\sqrt{-i\omega/K}$ and ω is the frequency.

Consider first an infinite, homogeneous space subjected at (x_0,y_0,z_0) to a harmonic *point heat source* of the form $\delta(x-x_0)\delta(y-y_0)\delta(z-z_0)e^{i\omega t}$. In this expression, $\delta(x-x_0)$, $\delta(y-y_0)$, and $\delta(z-z_0)$ are Dirac delta functions, and ω is the frequency of the source. The response of this heat source can be expressed by

$$\widehat{T}_f(x,y,z,\omega)=\frac{e^{-\sqrt{i\omega/K}\sqrt{r^2+(z-z_0)^2}}}{2\lambda\sqrt{r^2+(z-z_0)^2}}, \quad (3)$$

where $r=\sqrt{(x-x_0)^2+(y-y_0)^2}$.

Consider next an infinite, homogeneous space subjected to a spatially varying line heat source of the form $\delta(x-x_0)\delta(y-y_0)e^{i(\omega t-k_z z)}$, with k_z being the wavenumber in z. This source acts in one of the three coordinate directions, passes through (x_0,y_0), and varies sinusoidally in the z (i.e., third) dimension. This type of source is often referred to in the literature as a 2.5D source. The response to this source can be obtained by applying a spatial Fourier transform in the z direction to the equations for a point heat load.

Applying a Fourier transformation in the z direction leads to the solution

$$\widetilde{T}_f(x,y,k_z,\omega)=\frac{-i}{4\lambda}H_0(k_{1z}r), \quad (4)$$

where $k_{1z}=\sqrt{-i\omega/K-k_z^2}$, $H_0()$ are Hankel functions of the second kind and order 0.

The full 3D solution can then be achieved by applying an inverse Fourier transform in the k_z domain. This inverse Fourier transformation can be expressed as a discrete summation if we assume the existence of virtual sources, equally spaced at L_z along z, which enables the solution to be obtained by solving a limited number of 2D problems,

$$\widehat{T}_f(x,y,z,\omega)=\frac{2\pi}{L_z}\frac{-i}{4\lambda}\sum_{m_k=-M}^{M}H_0\left(\sqrt{-\frac{i\omega}{K}-k_{zm}^2}\,r\right)e^{-ik_{zm}z}, \quad (5)$$

with k_{zm} being the axial wavenumber given by $k_{zm}=(2\pi/L_z)m_k$. The distance L_z chosen must be big enough to prevent spatial contamination from the virtual sources.

Equation (4) can be further manipulated and written as a continuous superposition of heat plane phenomena,

$$\widetilde{T}_f(x,y,k_z,\omega)=\frac{-i}{4\pi\lambda}\int_{-\infty}^{+\infty}\left(\frac{e^{-ik_{1zx}|y-y_0|}}{k_{1zx}}\right)e^{-ik_x(x-x_0)}dk_x, \quad (6)$$

where $k_{1zx}=\sqrt{-i\omega/K-k_z^2-k_x^2}$ and $\operatorname{Im}(k_{1zx})\le 0$, and the integration is performed with respect to the horizontal wave number (k_x) in the x direction.

Assuming the existence of an infinite number of virtual sources, we can discretize these continuous integrals. The integral in the above equation can be transformed into a summation if an infinite number of such sources are distributed along the x direction, spaced at equal intervals L_x. The above equation can then be written as

$$\widetilde{T}_f(x,y,k_z,\omega)=E_0\sum_{n_x=-\infty}^{n=+\infty}\left(\frac{E}{k_{1zn}}\right)E_d, \quad (7)$$

where $E_0=-i/(2\lambda L_x)$, $E=e^{-ik_{1zn}|y|}$, $E_d=e^{-ik_n(x)}$, $k_{1zn}=\sqrt{-i\omega/K-k_z^2-k_n^2}$ and $\operatorname{Im}(k_{1zn})\le 0$, and $k_n=(2\pi/L_x)n_x$, which can in turn be approximated by a finite sum of equations (N). Note that $k_z=0$ is the 2D case, $\widetilde{T}_f(x,y,\omega)=E_0\sum_{n_x=-\infty}^{n=+\infty}(E/k_{1n})E_d$ with $k_{1n}=\sqrt{-i\omega/K-k_n^2}$.

Next, the above Green's functions are combined so as to define Green's functions for a half-space, a corner, a single layer system, a U system, a solid rectangular pipe, a solid open box, and a 3D parallelepiped box. Expressions in frequency and time solutions are provided. The time solutions obtained after the application of inverse spatial and frequency Fourier transforms are compared with those given by Green's functions defined directly in the time domain.

Green's functions for the different spaces are determined using the image source method. By this method a distribution of virtual sources and sinks are combined so as to give null temperatures (Dirichlet boundary conditions) or heat fluxes on the required boundaries (Neumann boundary conditions). Other boundary conditions, such as Robin, are not studied in this paper. In the case of solid bodies bounded by two parallel surfaces the number of sources, placed perpendicular to the surfaces, is theoretically infinite.

TABLE 1

	$r_{ijk}^{lmn} = \sqrt{(\bar{x}_i)^2 + (\bar{y}_j)^2 + (\bar{z}_k)^2}$	
$\bar{x}_0 = x - x_0$	$\bar{y}_0 = y - y_0$	$\bar{z}_0 = z - z_0$
$\bar{x}_1 = x + x_0 - 2d_1 l$	$\bar{y}_1 = y + y_0 - 2d_2 m$	$\bar{z}_1 = z + z_0 - 2d_3 n$
$\bar{x}_2 = x + x_0 + 2d_1(l-1)$	$\bar{y}_2 = y + y_0 + 2d_2(m-1)$	$\bar{z}_2 = z + z_0 + 2d_3(n-1)$
$\bar{x}_3 = x - x_0 - 2d_1 l$	$\bar{y}_3 = y - y_0 - 2d_2 m$	$\bar{z}_3 = z - z_0 - 2d_3 n$
$\bar{x}_4 = x - x_0 + 2d_1 l$	$\bar{y}_4 = y - y_0 + 2d_2 m$	$\bar{z}_4 = z - z_0 + 2d_3 n$

The superscripts l, m, and n identify the position of the virtual sources along the x, y, and z directions, respectively. The upper value of l, m, and n is defined by the convergence criteria. Each value of l, m, and n is associated with four possible source positions, which are identified by the subscripts i, j, and k for the x, y, and z directions, respectively. Thus, i, j, and k may take the values of 1, 2, 3, and 4.

The use of complex frequencies allows the contribution of the sources placed at greater distances to vanish and so to limit the number of the virtual sources. The use of complex frequencies with a small imaginary part, taking the form $\omega_c = \omega - i\eta$ (where $\omega = 0.7\Delta\omega$ and $\Delta\omega$ is the frequency increment), has the additional effect of avoiding the aliasing phenomena. This shift in the frequency domain is subsequently taken into account in the time domain by means of an exponential window, $e^{\eta t}$, applied to the response.

Green's functions are validated assuming that the medium is subject to a Dirac delta source. This type of source would require the solution to be computed in the frequency domain $[0.0, \infty]$ Hz. However, the response does not need to be computed for a very large number of frequencies since it decays very quickly as the frequency decays. Note that the static response for the frequency 0.0 Hz can be calculated thanks to the use of complex frequencies.

The number of virtual sources used depends directly on the predefined convergence criterion. As we move from one dimension to two dimensions and then to three dimensions, the number of sources grows significantly. Thus, although the method converges rapidly, the cost of computation grows significantly as we move from a one-dimensional to a three-dimensional problem.

3. Green's Functions

Green's functions in the time and frequency domain will be grouped for the following three cases:

(i) unbounded space, which includes Green's functions for 1D, 2D, and 3D sources;

(ii) two-dimensional space, which contains Green's functions for a half-space, a space bounded by two perpendicular planes, a single layer system, a U system, and a solid rectangular pipe, when subjected to 2D and 3D sources;

(iii) three-dimensional space, which compiles Green's functions for point sources placed in a solid open box and in a 3D parallelepiped box.

Special attention is given to the 2.5D solution in all cases since it enables the computation of the 3D heat field as a summation of 2D sources with varying spatial wavenumbers. Different boundary conditions are assumed and combined, namely, null temperatures or null heat fluxes. For each case, a scheme of the geometry is first illustrated (Figures 9–16) and then Green's functions are presented in the time and frequency domains. To verify the proposed solutions, responses in the time domain are included and computed directly in the time domain and in the frequency domain. A selection of results is presented in Figures 1–8 to illustrate the agreement between different solutions. Each figure includes a legend that indicates the comparisons performed.

All results showed a good agreement between the different formulations for all cases.

In the examples provided the Dirac delta source is positioned at the coordinate ($x_0 = 0.2$ m, $y_0 = 0.1$ m, $z_0 = 0.15$ m). This medium is homogeneous and is characterized by a thermal conductivity of $\lambda = 426.0\,\mathrm{W\cdot m^{-1}\cdot {}^\circ C^{-1}}$, a density $\rho = 10500.0\,\mathrm{kg\cdot m^{-3}}$, and a specific heat $c = 235.0\,\mathrm{J\cdot kg^{-1}\cdot {}^\circ C^{-1}}$. Computations are performed in the frequency range $[0.0, 40.96]$ Hz with a frequency increment of 0.01 Hz. The heat field is computed at the receiver point R ($x = 0.25$ m, $y = 0.15$ m, $z = 0.05$ m). The responses obtained with the time formulation (used to validate the frequency formulation) are determined for a time window $[0.0, 100.0]$ s with a time step of 0.0061 s. Depending on the type of geometry the medium may be bounded by planes placed at $x = 0.0$ m, $y = 0.0$ m, $z = 0.0$ m, $x = d_1 = 0.30$ m, $y = d_2 = 0.25$ m, or/and $z = d_3 = 0.20$ m.

The notation in Table 1 is used.

3.1. Unbounded Space. See Figure 9.

Equations for an Unbounded Space, Assuming 1D, 2D, and 3D Heat Sources. 1D source is as follows:

$$T(y,t) = \frac{e^{-|\bar{y}_0|^2/4K\tau}}{\rho c\,(4\pi K\tau)^{1/2}},$$
$$T(y,\omega) = \frac{-i}{2\lambda}\frac{e^{-ik_1|\bar{y}_0|}}{k_1}. \tag{8}$$

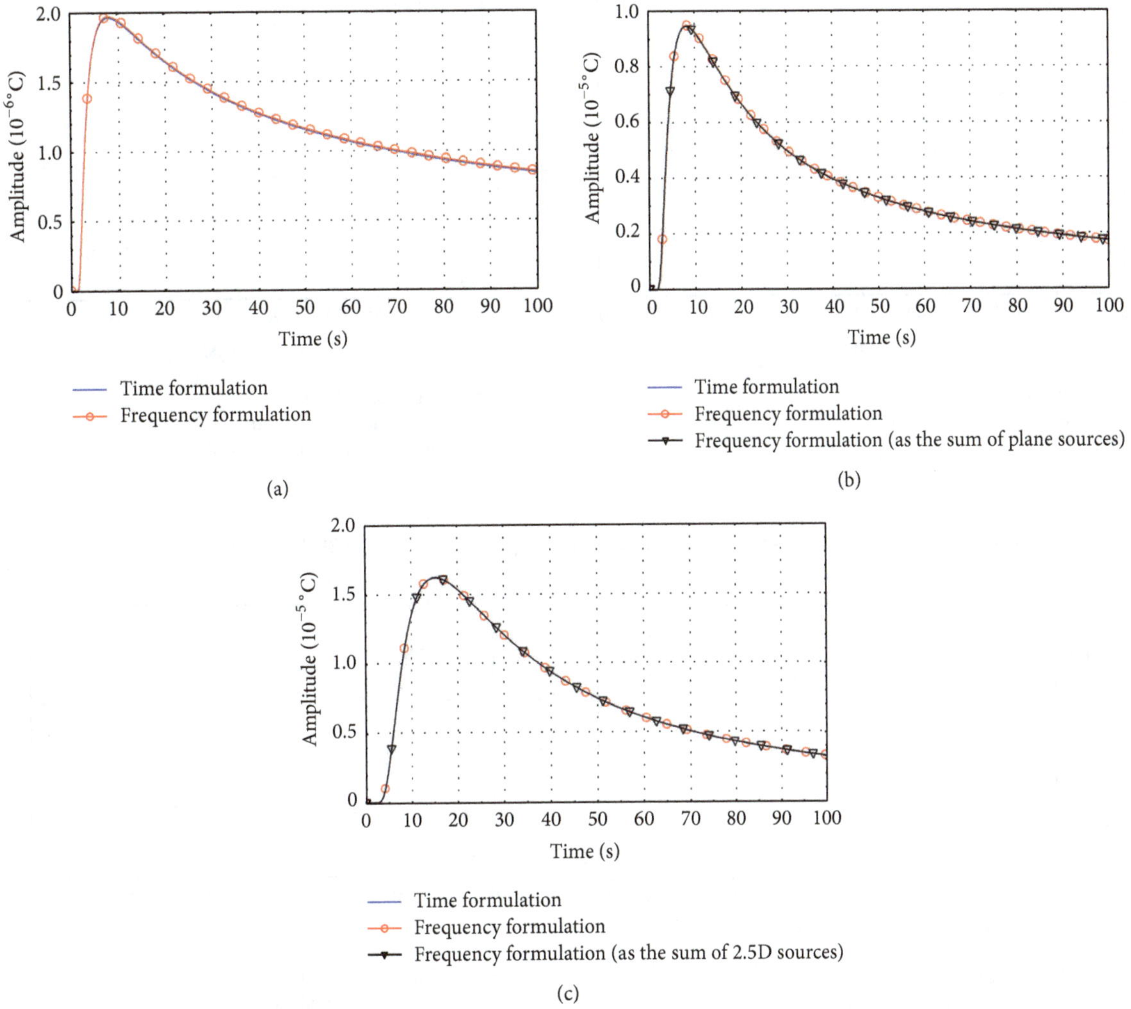

FIGURE 1: Responses for an unbounded space: (a) 1D heat source; (b) 2D heat source; (c) 3D heat source.

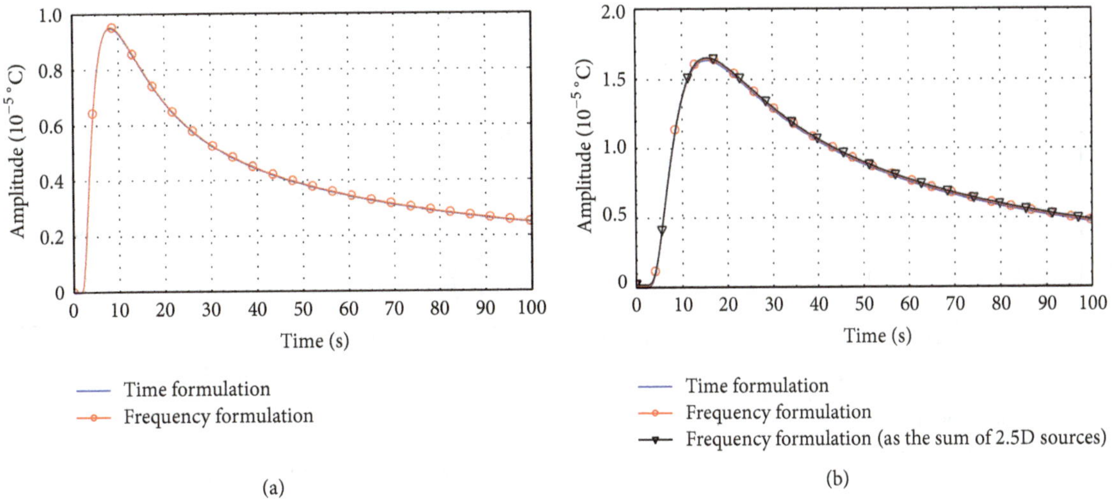

FIGURE 2: Response for the half-space (Case 1): (a) 2D heat source; (b) 3D heat source.

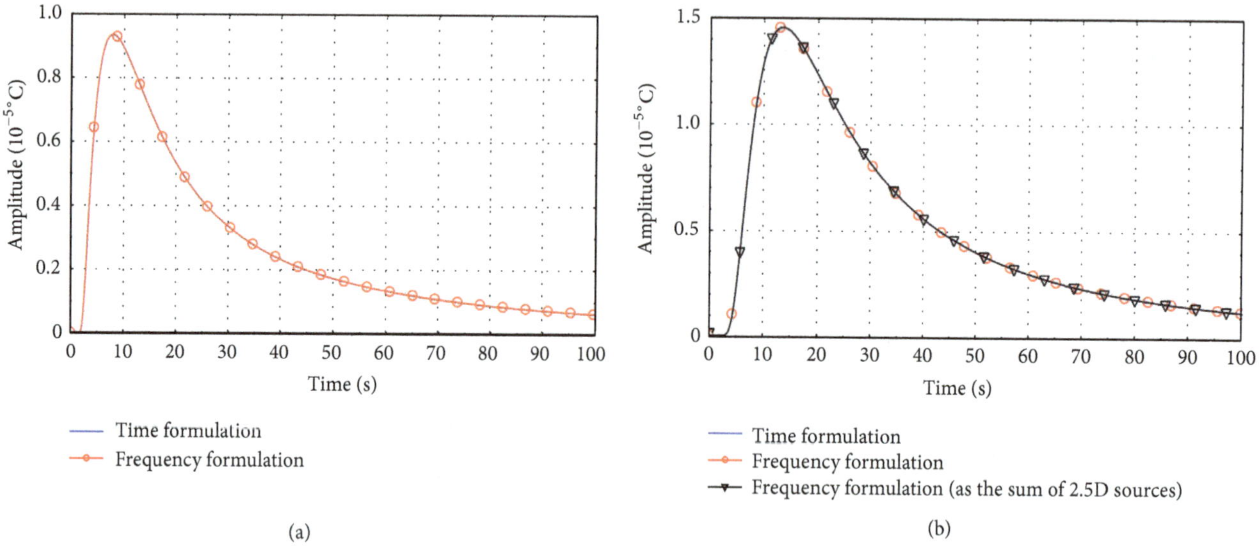

FIGURE 3: Response for the bounded space defined by $x \leq d_1$ and $y \geq 0$ (Case 3): (a) 2D heat source; (b) 3D heat source.

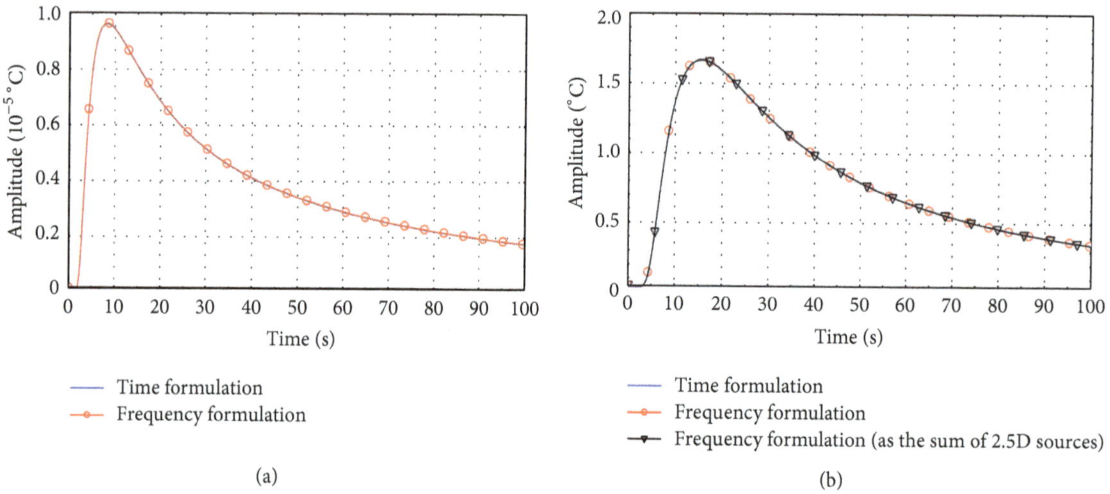

FIGURE 4: Response for the horizontal layer (Case 3): (a) 2D heat source; (b) 3D heat source.

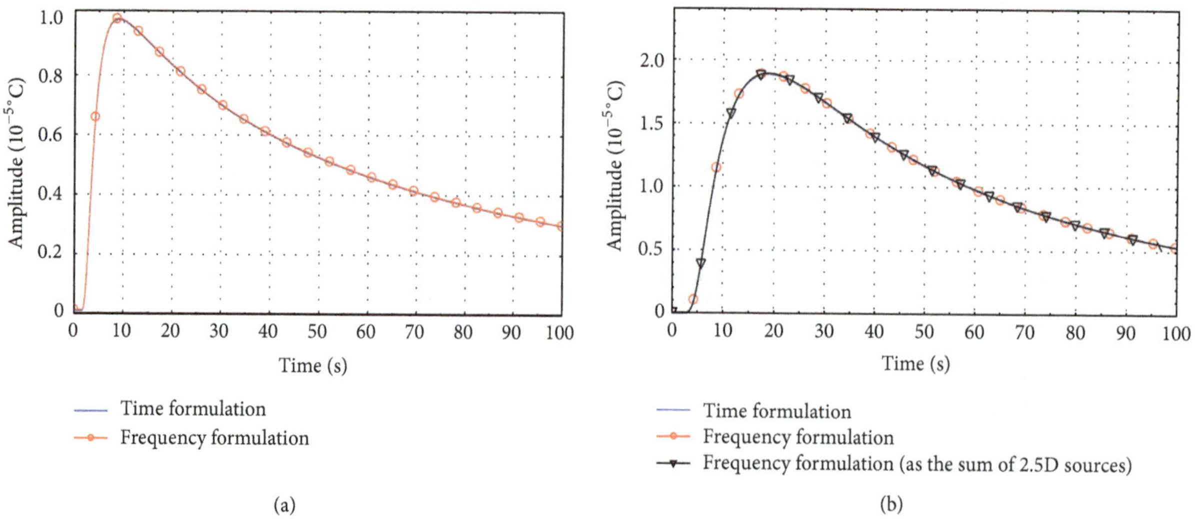

FIGURE 5: Response for the U system (Case 3): (a) 2D heat source; (b) 3D heat source.

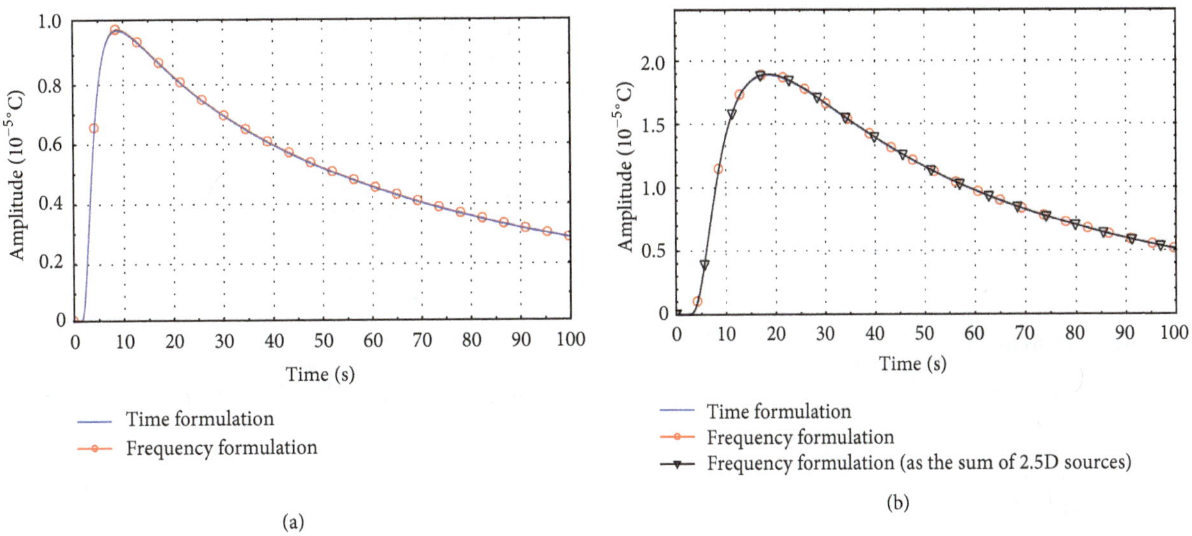

FIGURE 6: Response for the solid rectangular pipe (Case 3): (a) 2D heat source; (b) 3D heat source.

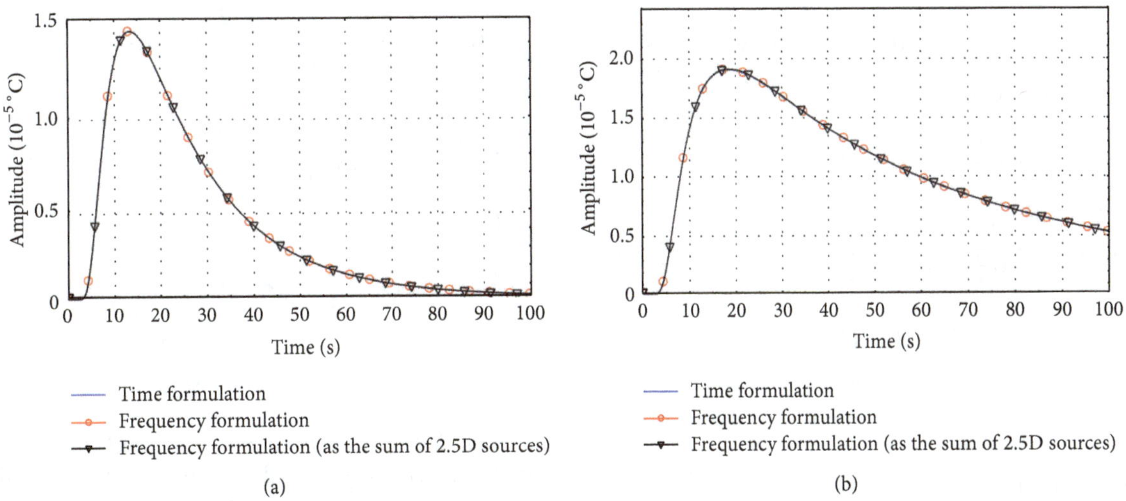

FIGURE 7: Response for the solid open box: (a) Case 2; (b) Case 3.

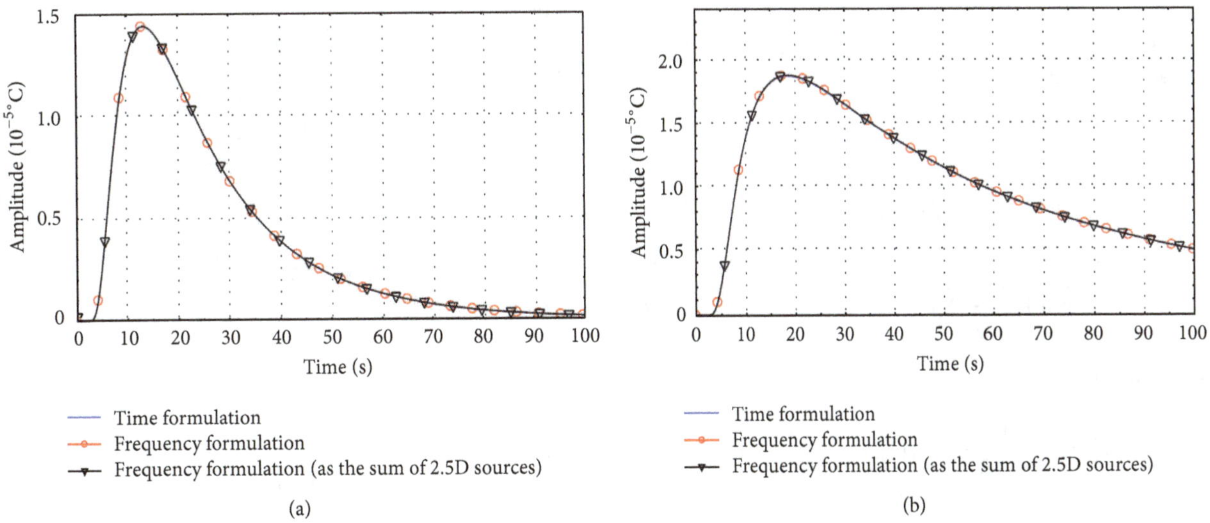

FIGURE 8: Responses for the 3D parallelepiped box: (a) Case 2. (b) Case 3.

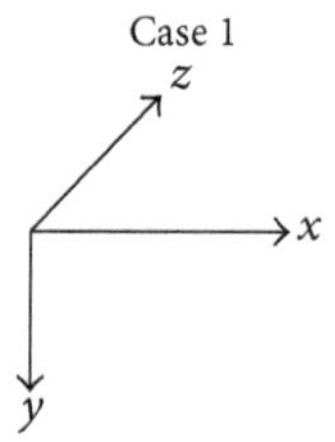

Figure 9: Unbounded space.

2D source is as follows:

$$T(x,y,t) = \frac{e^{-(r_{00}^{00})^2/4K\tau}}{\rho c(4\pi K\tau)},$$
$$T(x,y,\omega) = \frac{-i}{4\lambda} H_0\left(k_1 r_{00}^{00}\right), \tag{9}$$
$$T(x,y,\omega) = \frac{-i}{2\lambda L_x} \sum_{n_x=-\infty}^{\infty} \left[\frac{e^{-ik_{1n}|\bar{y}_0|}}{k_{1n}} e^{-ik_n(\bar{x}_0)}\right]$$

as the sum of plane sources.

3D source is as follows:

$$T(x,y,z,t) = \frac{e^{-(r_{000}^{000})^2/4K\tau}}{\rho c(4\pi K\tau)^{1.5}} \quad \text{with } \tau = t - t_0,$$
$$T(x,y,z,\omega) = \frac{1}{2\lambda r_{000}^{000}} e^{-\sqrt{i\omega/K} r_{000}^{000}}, \tag{10}$$
$$T(x,y,z,\omega) = \frac{-i\pi}{2\lambda L_z} \sum_{m_k=-\infty}^{\infty} \left[H_0\left(k_{1z} r_{00}^{00}\right) e^{-ik_z(\bar{z}_0)}\right]$$

as the sum of 2.5D sources.

3.2. Two-Dimensional Space

(a) Half-Space Defined by $y \geq 0$. Boundary conditions prescribed for the half-space (Cases 1 and 2) are shown in Figure 10.

Equations for the half-space (Cases 1 and 2), subjected to a 2D heat source, are as follows.

Case 1. Consider

$$T(x,y,t) = \frac{1}{\rho c(4\pi K\tau)} \sum_{j=0}^{1} e^{-(r_{0j}^{00})^2/4K\tau},$$
$$T(x,y,\omega) = \frac{-i}{4\lambda} \sum_{j=0}^{1} H_0\left(k_1 r_{0j}^{00}\right). \tag{11}$$

Case 2. Consider

$$T(x,y,t) = \frac{1}{\rho c(4\pi K\tau)} \sum_{j=0}^{1} \left[(-1)^j e^{-(r_{0j}^{00})^2/4K\tau}\right],$$
$$T(x,y,\omega) = \frac{-i}{4\lambda} \sum_{j=0}^{1} \left[(-1)^j H_0\left(k_1 r_{0j}^{00}\right)\right]. \tag{12}$$

Equations for the half-space (Cases 1 and 2), subjected to a 3D heat source, are as follows.

Case 1. Consider

$$T(x,y,z,t) = \frac{1}{\rho c(4\pi K\tau)^{1.5}} \sum_{j=0}^{1} e^{-(r_{0j0}^{000})^2/4K\tau},$$
$$T(x,y,z,\omega) = \frac{1}{2\lambda} \sum_{j=0}^{1} \frac{e^{-\sqrt{i\omega/K} r_{0j0}^{000}}}{r_{0j0}^{000}}, \tag{13}$$
$$T(x,y,z,\omega) = \frac{-i\pi}{2\lambda L_x} \sum_{j=0}^{1} \sum_{m_k=-\infty}^{\infty} \left[H_0\left(k_{1z} r_{0j}^{00}\right) e^{-ik_z(\bar{z}_0)}\right]$$

as the sum of 2.5D sources.

Case 2. Consider

$$T(x,y,z,t) = \frac{1}{\rho c(4\pi K\tau)^{1.5}} \sum_{j=0}^{1} \left[(-1)^j e^{-(r_{0j0}^{000})^2/4K\tau}\right],$$
$$T(x,y,z,\omega) = \frac{1}{2\lambda} \sum_{j=0}^{1} \left[(-1)^j \frac{e^{-\sqrt{i\omega/K} r_{0j0}^{000}}}{r_{0j0}^{000}}\right], \tag{14}$$
$$T(x,y,z,\omega) = \frac{-i\pi}{2\lambda L_x} \sum_{j=0}^{1} (-1)^j \sum_{m_k=-\infty}^{\infty} \left[H_0\left(k_{1z} r_{0j}^{00}\right) e^{-ik_z(\bar{z}_0)}\right]$$

as the sum of 2.5D sources.

(b) Bounded Space Defined by $x \leq d_1$ *and* $y \geq 0$. Boundary conditions prescribed for the bounded space defined by $x \leq d_1$ and $y \geq 0$ (Cases 1–4) are shown in Figure 11.

Equations for the bounded space defined by $x \leq d_1$ and $y \geq 0$ (Cases 1–4), subjected to a 2D heat source, are as follows.

Case 1. Consider

$$T(x,y,t) = \frac{1}{\rho c(4\pi K\tau)} \sum_{i=0}^{1} \sum_{j=0}^{1} \left[e^{-(r_{ij}^{i0})^2/4K\tau}\right],$$
$$T(x,y,\omega) = \frac{-i}{4\lambda} \sum_{i=0}^{1} \sum_{j=0}^{1} \left[H_0\left(k_1 r_{ij}^{i0}\right)\right]. \tag{15}$$

Case 2. Consider

$$T(x,y,t) = \frac{1}{\rho c(4\pi K\tau)} \sum_{i=0}^{1} \sum_{j=0}^{1} \left[(-1)^{(i+j)} e^{-(r_{ij}^{i0})^2/4K\tau}\right],$$
$$T(x,y,\omega) = \frac{-i}{4\lambda} \sum_{i=0}^{1} \sum_{j=0}^{1} \left[(-1)^{(i+j)} H_0\left(k_1 r_{ij}^{i0}\right)\right]. \tag{16}$$

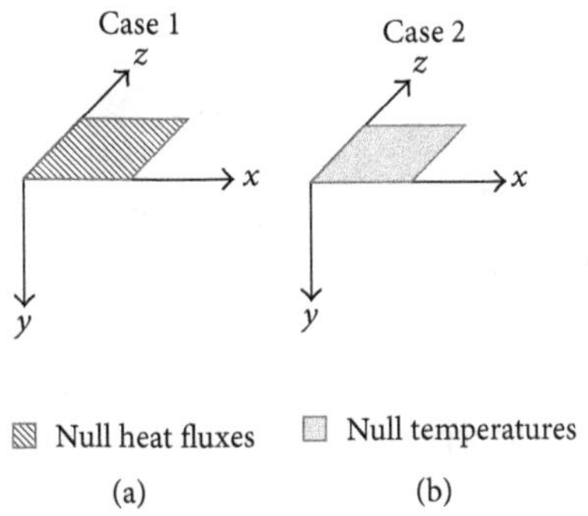

FIGURE 10: Boundary conditions prescribed for the half-space (Cases 1 and 2).

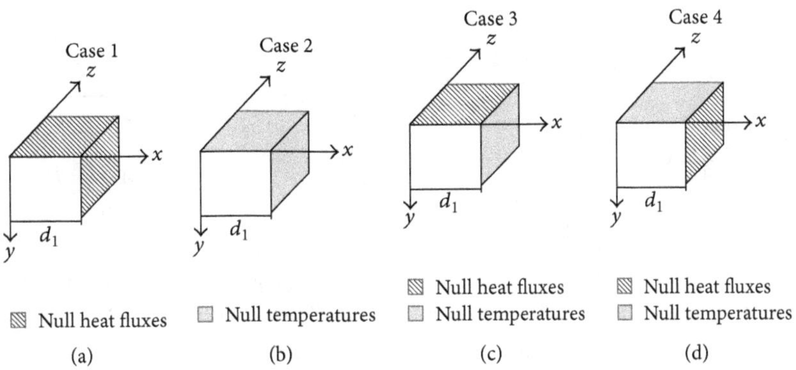

FIGURE 11: Boundary conditions prescribed for the bounded space defined by $x \leq d_1$ and $y \geq 0$ (Cases 1–4).

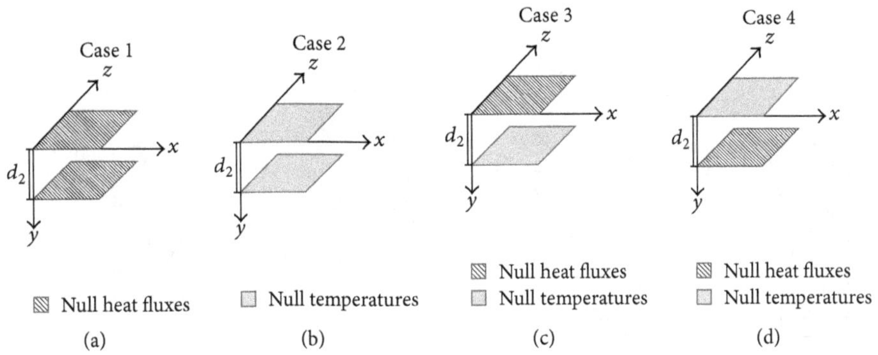

FIGURE 12: Boundary conditions prescribed for the horizontal layer (Cases 1–4).

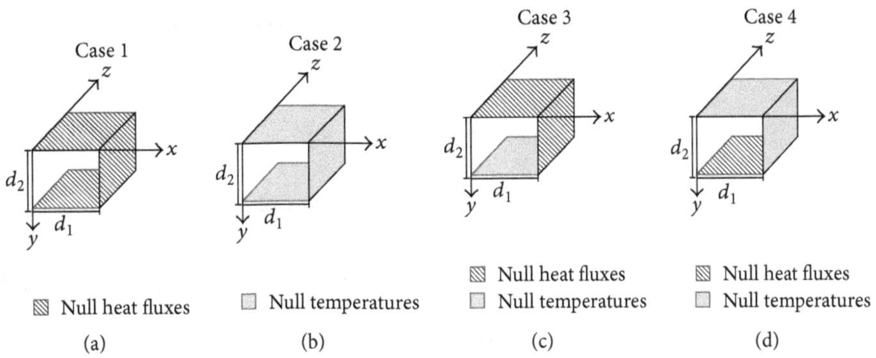

FIGURE 13: Boundary conditions prescribed for the U system (Cases 1–4).

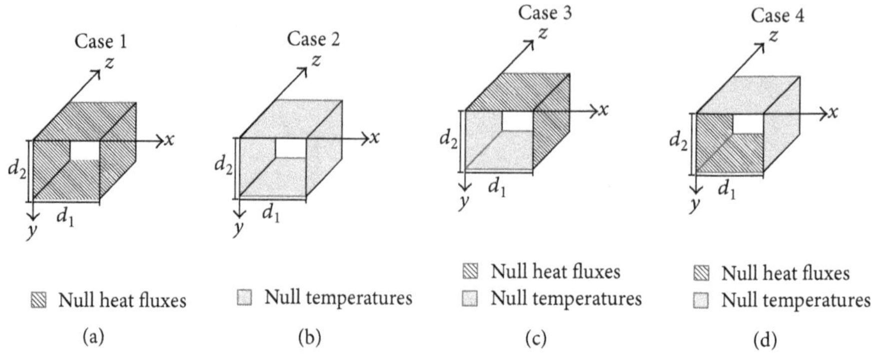

FIGURE 14: Boundary conditions prescribed for the solid rectangular pipe (Cases 1–4).

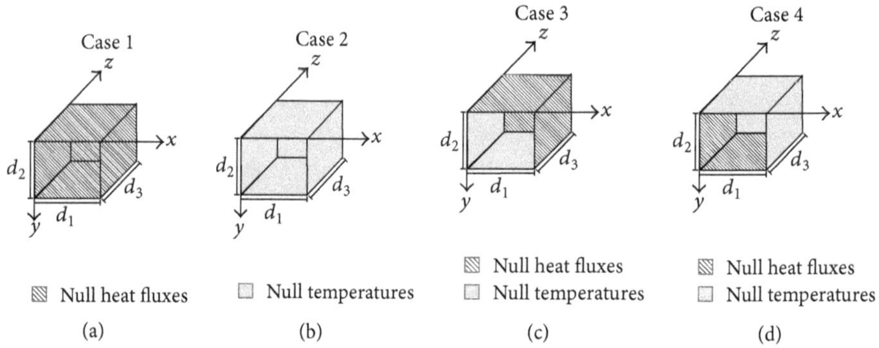

FIGURE 15: Boundary conditions prescribed for the solid open box (Cases 1–4).

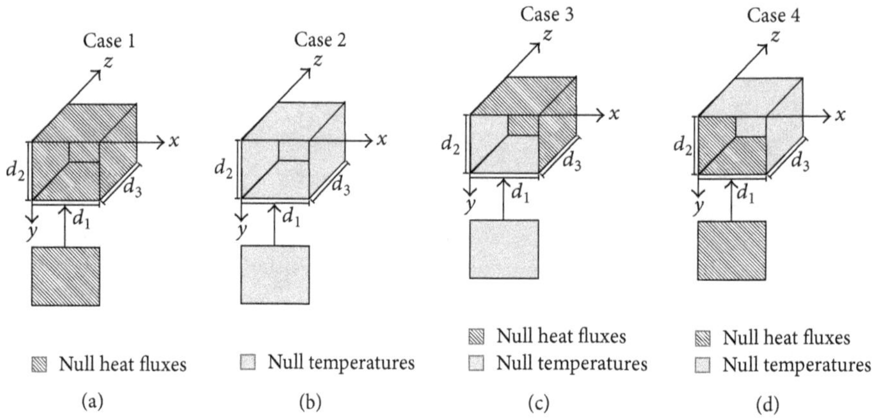

FIGURE 16: Boundary conditions for the 3D parallelepiped box (Cases 1–4).

Case 3. Consider

$$T(x,y,t) = \frac{1}{\rho c\,(4\pi K\tau)} \sum_{i=0}^{1}\sum_{j=0}^{1} \left[(-1)^i e^{-(r_{ij}^{i0})^2/4K\tau}\right],$$
$$T(x,y,\omega) = \frac{-i}{4\lambda} \sum_{i=0}^{1}\sum_{j=0}^{1} \left[(-1)^i H_0\left(k_1 r_{ij}^{i0}\right)\right]. \tag{17}$$

Case 4. Consider

$$T(x,y,t) = \frac{1}{\rho c\,(4\pi K\tau)} \sum_{i=0}^{1}\sum_{j=0}^{1} \left[(-1)^j e^{-(r_{ij}^{i0})^2/4K\tau}\right],$$
$$T(x,y,\omega) = \frac{-i}{4\lambda} \sum_{i=0}^{1}\sum_{j=0}^{1} \left[(-1)^j H_0\left(k_1 r_{ij}^{i0}\right)\right]. \tag{18}$$

Equations defined for the bounded space defined by $x \leq d_1$ and $y \geq 0$ (Cases 1–4) and subjected to a 3D heat source, are as follows.

Case 1. Consider

$$T(x,y,z,t) = \frac{1}{\rho c(4\pi K\tau)^{1.5}} \sum_{i=0}^{1}\sum_{j=0}^{1}\left[e^{-(r_{ij0}^{i00})^2/4K\tau}\right],$$

$$T(x,y,z,\omega) = \frac{1}{2\lambda}\sum_{i=0}^{1}\sum_{j=0}^{1}\left[(-1)^{(i+j)}\frac{e^{-\sqrt{i\omega/K}r_{ij0}^{i00}}}{r_{ij0}^{i00}}\right],$$

$$T(x,y,z,\omega) = \frac{-i\pi}{2\lambda L_x}\sum_{i=0}^{1}\sum_{j=0}^{1}\sum_{m_k=-\infty}^{\infty}\left[H_0\left(k_{1z}r_{ij}^{10}\right)e^{-ik_z(\overline{z}_0)}\right] \quad (19)$$

as the sum of 2.5D sources.

Case 2. Consider

$$T(x,y,z,t) = \frac{1}{\rho c(4\pi K\tau)^{1.5}} \sum_{i=0}^{1}\sum_{j=0}^{1}\left[(-1)^{(i+j)}e^{-(r_{ij0}^{i00})^2/4K\tau}\right],$$

$$T(x,y,z,\omega) = \frac{1}{2\lambda}\sum_{i=0}^{1}\sum_{j=0}^{1}\left[(-1)^{(i+j)}\frac{e^{-\sqrt{i\omega/K}r_{ij0}^{i00}}}{r_{ij0}^{i00}}\right],$$

$$T(x,y,z,\omega) = \frac{-i\pi}{2\lambda L_x}\sum_{i=0}^{1}\sum_{j=0}^{1}(-1)^{(i+j)}\sum_{m_k=-\infty}^{\infty}\left[H_0\left(k_{1z}r_{ij}^{i0}\right)e^{-ik_z(\overline{z}_0)}\right] \quad (20)$$

as the sum of 2.5D sources.

Case 3. Consider

$$T(x,y,z,t) = \frac{1}{\rho c(4\pi K\tau)^{1.5}} \sum_{i=0}^{1}\sum_{j=0}^{1}\left[(-1)^{i}e^{-(r_{ij0}^{i00})^2/4K\tau}\right],$$

$$T(x,y,z,\omega) = \frac{1}{2\lambda}\sum_{i=0}^{1}\sum_{j=0}^{1}\left[(-1)^{i}\frac{e^{-\sqrt{i\omega/K}r_{ij0}^{i00}}}{r_{ij0}^{i00}}\right],$$

$$T(x,y,z,\omega) = \frac{-i\pi}{2\lambda L_x}\sum_{i=0}^{1}\sum_{j=0}^{1}(-1)^{(i)}\sum_{m_k=-\infty}^{\infty}\left[H_0\left(k_{1z}r_{ij}^{i0}\right)e^{-ik_z(\overline{z}_0)}\right] \quad (21)$$

as the sum of 2.5D sources.

Case 4. Consider

$$T(x,y,z,t) = \frac{1}{\rho c(4\pi K\tau)^{1.5}} \sum_{i=0}^{1}\sum_{j=0}^{1}\left[(-1)^{j}e^{-(r_{ij0}^{i00})^2/4K\tau}\right],$$

$$T(x,y,z,\omega) = \frac{1}{2\lambda}\sum_{i=0}^{1}\sum_{j=0}^{1}\left[(-1)^{j}\frac{e^{-\sqrt{i\omega/K}r_{ij0}^{i00}}}{r_{ij0}^{i00}}\right],$$

$$T(x,y,z,\omega) = \frac{-i\pi}{2\lambda L_x}\sum_{i=0}^{1}\sum_{j=0}^{1}(-1)^{(j)}\sum_{m_k=-\infty}^{\infty}\left[H_0\left(k_{1z}r_{ij}^{i0}\right)e^{-ik_z(\overline{z}_0)}\right] \quad (22)$$

as the sum of 2.5D sources.

(c) Horizontal Layer Defined by $y = 0$ and $y = d_2$. Boundary conditions prescribed for the horizontal layer (Cases 1–4) are shown in Figure 12.

Equations for the horizontal layer (Cases 1–4), subjected to a 2D heat source, are as follows.

Case 1. Consider

$$T(x,y,t) = \frac{1}{\rho c(4\pi K\tau)}\left[e^{-(r_{00}^{00})^2/4K\tau} + \sum_{m=1}^{NS_y}\sum_{j=1}^{4}e^{-(r_{0j}^{0m})^2/4K\tau}\right],$$

$$T(x,y,\omega) = \frac{-i}{4\lambda}\left[H_0\left(k_1 r_{00}^{00}\right) + \sum_{m=1}^{NS_y}\sum_{j=1}^{4}H_0\left(k_1 r_{0j}^{0m}\right)\right]. \quad (23)$$

Case 2. Consider

$$T(x,y,t) = \frac{1}{\rho c(4\pi K\tau)}\left[e^{-(r_{00}^{00})^2/4K\tau} + \sum_{m=1}^{NS_y}\sum_{j=1}^{4}(-1)^{(j-1)!}e^{-(r_{0j}^{0m})^2/4K\tau}\right],$$

$$T(x,y,\omega) = \frac{-i}{4\lambda}\left[H_0\left(k_1 r_{00}^{00}\right) + \sum_{m=1}^{NS_y}\sum_{j=1}^{4}(-1)^{(j-1)!}H_0\left(k_1 r_{0j}^{0m}\right)\right]. \quad (24)$$

Case 3. Consider

$$T(x,y,t)=\frac{1}{\rho c(4\pi K\tau)}\left[e^{-(r_{00}^{00})^2/4K\tau}+\sum_{m=1}^{NS_y}(-1)^{(m-1)}\right.$$
$$\left.\cdot\left[\sum_{j=1}^{4}(-1)^{((j-1)!(j+1)+1)}e^{-(r_{0j}^{0m})^2/4K\tau}\right]\right],$$
$$T(x,y,\omega)=\frac{-i}{4\lambda}\left[H_0\left(k_1r_{00}^{00}\right)+\sum_{m=1}^{NS_y}(-1)^{(m-1)}\right.$$
$$\left.\cdot\left[\sum_{j=1}^{4}(-1)^{((j-1)!(j+1)+1)}H_0\left(k_1r_{0j}^{0m}\right)\right]\right]. \tag{25}$$

Case 4. Consider

$$T(x,y,t)=\frac{1}{\rho c(4\pi K\tau)}\left[e^{-(r_{00}^{00})^2/4K\tau}\right.$$
$$\left.+\sum_{m=1}^{NS_y}(-1)^{(m-1)}\left[\sum_{j=1}^{4}(-1)^{(j!+1)}e^{-(r_{0j}^{0m})^2/4K\tau}\right]\right],$$
$$T(x,y,\omega)=\frac{-i}{4\lambda}\left[H_0\left(k_1r_{00}^{00}\right)\right.$$
$$\left.+\sum_{m=1}^{NS_y}(-1)^{(m-1)}\left[\sum_{j=1}^{4}(-1)^{(j!+1)}H_0\left(k_1r_{0j}^{0m}\right)\right]\right]. \tag{26}$$

Equations for the horizontal layer (Cases 1–4), subjected to a 3D heat source, are as follows.

Case 1. Consider

$$T(x,y,z,t)=\frac{1}{\rho c(4\pi K\tau)^{1.5}}\left[e^{-(r_{000}^{000})^2/4K\tau}\right.$$
$$\left.+\sum_{m=1}^{NS_y}\sum_{j=1}^{4}e^{-(r_{0j0}^{0m0})^2/4K\tau}\right],$$
$$T(x,y,z,\omega)=\frac{1}{2\lambda}\left[\frac{e^{-\sqrt{i\omega/K}r_{000}^{000}}}{r_{000}^{000}}\right.$$
$$\left.+\sum_{m=1}^{NS_y}\sum_{j=1}^{4}\frac{e^{-\sqrt{i\omega/K}r_{0j0}^{0m0}}}{r_{0j0}^{0m0}}\right],$$
$$T(x,y,z,\omega)=\frac{-i\pi}{2\lambda L_x}\left[\sum_{m_k=-\infty}^{\infty}\left[H_0\left(k_{1z}r_{00}^{00}\right)e^{-ik_z(\overline{z}_0)}\right]\right.$$
$$\left.+\sum_{m=1}^{NS_y}\sum_{j=1}^{4}\sum_{m_k=-\infty}^{\infty}\left[H_0\left(k_{1z}r_{0j}^{0m}\right)e^{-ik_z(\overline{z}_0)}\right]\right]$$

as the sum of 2.5D sources. (27)

Case 2. Consider

$$T(x,y,z,t)=\frac{1}{\rho c(4\pi K\tau)^{1.5}}\left[e^{-(r_{000}^{000})^2/4K\tau}\right.$$
$$\left.+\sum_{m=1}^{NS_y}\sum_{j=1}^{4}(-1)^{(j-1)!}e^{-(r_{0j0}^{0m0})^2/4K\tau}\right],$$
$$T(x,y,z,\omega)=\frac{1}{2\lambda}\left[\frac{e^{-\sqrt{i\omega/K}r_{000}^{000}}}{r_{000}^{000}}\right.$$
$$\left.+\sum_{m=1}^{NS_y}\sum_{j=1}^{4}(-1)^{(j-1)!}\frac{e^{-\sqrt{i\omega/K}r_{0j0}^{0m0}}}{r_{0j0}^{0m0}}\right], \tag{28}$$
$$T(x,y,z,\omega)=\frac{-i\pi}{2\lambda L_x}\left[\sum_{m_k=-\infty}^{\infty}\left[H_0\left(k_{1z}r_{00}^{00}\right)e^{-ik_z(\overline{z}_0)}\right]\right.$$
$$\left.+\sum_{m=1}^{NS_y}\sum_{j=1}^{4}(-1)^{(j-1)!}\sum_{m_k=-\infty}^{\infty}\left[H_0\left(k_{1z}r_{0j}^{0m}\right)e^{-ik_z(\overline{z}_0)}\right]\right]$$

as the sum of 2.5D sources.

Case 3. Consider

$$T(x,y,z,t)=\frac{1}{\rho c(4\pi K\tau)^{1.5}}\left[e^{-(r_{000}^{000})^2/4K\tau}\right.$$
$$\left.+\sum_{m=1}^{NS_y}(-1)^{(m-1)}\sum_{j=1}^{4}(-1)^{((j-1)!(j+1)+1)}e^{-(r_{0j0}^{0m0})^2/4K\tau}\right],$$
$$T(x,y,z,\omega)=\frac{1}{2\lambda}\left[\frac{e^{-\sqrt{i\omega/K}r_{000}^{000}}}{r_{000}^{000}}+\sum_{m=1}^{NS_y}(-1)^{(m-1)}\right.$$
$$\left.\cdot\sum_{j=1}^{4}(-1)^{((j-1)!(j+1)+1)}\frac{e^{-\sqrt{i\omega/K}r_{0j0}^{0m0}}}{r_{0j0}^{0m0}}\right],$$

$$T(x,y,z,\omega)=\frac{-i\pi}{2\lambda L_x}\left[\sum_{m_k=-\infty}^{\infty}\left[H_0\left(k_{1z}r_{00}^{00}\right)e^{-ik_z(\overline{z}_0)}\right]\right.$$
$$+\sum_{m=1}^{NS_y}(-1)^{(m-1)}\sum_{j=1}^{4}(-1)^{(j-1)!}$$
$$\left.\cdot\sum_{m_k=-\infty}^{\infty}\left[H_0\left(k_{1z}r_{0j}^{0m}\right)e^{-ik_z(\overline{z}_0)}\right]\right]$$

as the sum of 2.5D sources. (29)

Case 4. Consider

$$T(x,y,z,t)=\frac{1}{\rho c(4\pi K\tau)^{1.5}}\left[e^{-(r_{000}^{000})^2/4K\tau}\right.$$
$$\left.+\sum_{m=1}^{NS_y}(-1)^{(m-1)}\sum_{j=1}^{4}(-1)^{(j!+1)}e^{-(r_{0j0}^{0m0})^2/4K\tau}\right],$$
$$T(x,y,z,\omega)=\frac{1}{2\lambda}\left[\frac{e^{-\sqrt{i\omega/K}r_{000}^{000}}}{r_{000}^{000}}+\sum_{m=1}^{NS_y}(-1)^{(m-1)}\right.$$
$$\left.\cdot\sum_{j=1}^{4}(-1)^{(j!+1)}\frac{e^{-\sqrt{i\omega/K}r_{0j0}^{0m0}}}{r_{0j0}^{0m0}}\right], \tag{30}$$

$$T(x,y,z,\omega)=\frac{-i\pi}{2\lambda L_x}\left[\sum_{m_k=-\infty}^{\infty}\left[H_0\left(k_{1z}r_{00}^{00}\right)e^{-ik_z(\overline{z}_0)}\right]\right.$$
$$+\sum_{m=1}^{NS_y}(-1)^{(m-1)}\sum_{j=1}^{4}(-1)^{(j!+1)}$$
$$\left.\cdot\sum_{m_k=-\infty}^{\infty}\left[H_0\left(k_{1z}r_{0j}^{0m}\right)e^{-ik_z(\overline{z}_0)}\right]\right]$$

as the sum of 2.5D sources.

(d) U System Bounded by $y\geq 0$, $y\leq d_2$, and $x\leq d_1$. Boundary conditions prescribed for the U system (Cases 1–4) are shown in Figure 13.

Equations for the U system (Cases 1–4), subjected to a 2D heat source, are as follows.

Case 1. Consider

$$T(x,y,t)=\frac{1}{\rho c(4\pi K\tau)}\sum_{i=0}^{1}\left[e^{-(r_{i0}^{i0})^2/4K\tau}+\sum_{m=1}^{NS_y}\sum_{j=1}^{4}e^{-(r_{ij}^{im})^2/4K\tau}\right],$$
$$T(x,y,\omega)=\frac{-i}{4\lambda}\sum_{i=0}^{1}\left[H_0\left(k_1r_{i0}^{i0}\right)+\sum_{m=1}^{NS_y}\sum_{j=1}^{4}H_0\left(k_1r_{ij}^{im}\right)\right]. \tag{31}$$

Case 2. Consider

$$T(x,y,t)=\frac{1}{\rho c(4\pi K\tau)}\sum_{i=0}^{1}(-1)^i$$
$$\cdot\left[e^{-(r_{i0}^{i0})^2/4K\tau}+\sum_{m=1}^{NS_y}\sum_{j=1}^{4}(-1)^{(j-1)!}e^{-(r_{ij}^{im})^2/4K\tau}\right],$$
$$T(x,y,\omega)=\frac{-i}{4\lambda}\sum_{i=0}^{1}(-1)^i$$
$$\cdot\left[H_0\left(k_1r_{i0}^{i0}\right)+\sum_{m=1}^{NS_y}\sum_{j=1}^{4}(-1)^{(j-1)!}H_0\left(k_1r_{ij}^{im}\right)\right]. \tag{32}$$

Case 3. Consider

$$T(x,y,t)=\frac{1}{\rho c(4\pi K\tau)}\sum_{i=0}^{1}\left[e^{-(r_{i0}^{i0})^2/4K\tau}\right.$$
$$\left.+\sum_{m=1}^{NS_y}(-1)^{(m-1)}\sum_{j=1}^{4}(-1)^{((j-1)!(j+1)+1)}e^{-(r_{ij}^{im})^2/4K\tau}\right],$$
$$T(x,y,\omega)=\frac{-i}{4\lambda}\sum_{i=0}^{1}\left[H_0\left(k_1r_{i0}^{i0}\right)\right.$$
$$\left.+\sum_{m=1}^{NS_y}(-1)^{(m-1)}\sum_{j=1}^{4}(-1)^{((j-1)!(j+1)+1)}H_0\left(k_1r_{ij}^{im}\right)\right]. \tag{33}$$

Case 4. Consider

$$T(x,y,t) = \frac{1}{\rho c(4\pi K\tau)}\sum_{i=0}^{1}(-1)^i\left[e^{-(r_{i0}^{i0})^2/4K\tau} + \sum_{m=1}^{NS_y}(-1)^{(m-1)}\sum_{j=1}^{4}(-1)^{(j!+1)}e^{-(r_{ij}^{im})^2/4K\tau}\right],$$

$$T(x,y,\omega) = \frac{-i}{4\lambda}\sum_{i=0}^{1}(-1)^i\left[H_0\left(k_1 r_{i0}^{i0}\right) + \sum_{m=1}^{NS_y}(-1)^{(m-1)}\sum_{j=1}^{4}(-1)^{(j!+1)}H_0\left(k_1 r_{ij}^{im}\right)\right]. \quad (34)$$

Equations for the U system (Cases 1–4), subjected to a 3D heat source, are as follows.

Case 1. Consider

$$T(x,y,z,t) = \frac{1}{\rho c(4\pi K\tau)^{1.5}}\sum_{i=0}^{1}\left[e^{-(r_{i00}^{i00})^2/4K\tau} + \sum_{m=1}^{NS_y}\sum_{j=1}^{4}e^{-(r_{ij0}^{im0})^2/4K\tau}\right],$$

$$T(x,y,z,\omega) = \frac{1}{2\lambda}\sum_{i=0}^{1}\left[\frac{e^{-\sqrt{i\omega/K}r_{i00}^{i00}}}{r_{i00}^{i00}} + \sum_{m=1}^{NS_y}\sum_{j=1}^{4}\frac{e^{-\sqrt{i\omega/K}r_{ij0}^{im0}}}{r_{ij0}^{im0}}\right],$$

$$T(x,y,z,\omega) = \frac{-i\pi}{2\lambda L_x}\cdot\sum_{i=0}^{1}\left[\sum_{m_k=-\infty}^{\infty}\left[H_0\left(k_{1z}r_{i0}^{i0}\right)e^{-ik_z(\overline{z}_0)}\right] + \sum_{m=1}^{NS_y}\sum_{j=1}^{4}\sum_{m_k=-\infty}^{\infty}\left[H_0\left(k_{1z}r_{ij}^{im}\right)e^{-ik_z(\overline{z}_0)}\right]\right] \quad (35)$$

as the sum of 2.5D sources.

Case 2. Consider

$$T(x,y,z,t) = \frac{1}{\rho c(4\pi K\tau)^{1.5}}\sum_{i=0}^{1}(-1)^i\left[e^{-(r_{i00}^{i00})^2/4K\tau} + \sum_{m=1}^{NS_y}\sum_{j=1}^{4}(-1)^{(j-1)!}e^{-(r_{ij0}^{im0})^2/4K\tau}\right],$$

$$T(x,y,z,\omega) = \frac{1}{2\lambda}\sum_{i=0}^{1}(-1)^i\left[\frac{e^{-\sqrt{i\omega/K}r_{i00}^{i00}}}{r_{i00}^{i00}} + \sum_{m=1}^{NS_y}\sum_{j=1}^{4}(-1)^{(j-1)!}\frac{e^{-\sqrt{i\omega/K}r_{ij0}^{im0}}}{r_{ij0}^{im0}}\right],$$

$$T(x,y,z,\omega) = \frac{-i\pi}{2\lambda L_x}\sum_{i=0}^{1}(-1)^i\cdot\left[\sum_{m_k=-\infty}^{\infty}\left[H_0\left(k_{1z}r_{i0}^{i0}\right)e^{-ik_z(\overline{z}_0)}\right] + \sum_{m=1}^{NS_y}\sum_{j=1}^{4}(-1)^{(j-1)!}\sum_{m_k=-\infty}^{\infty}\left[H_0\left(k_{1z}r_{ij}^{im}\right)e^{-ik_z(\overline{z}_0)}\right]\right]$$

as the sum of 2.5D sources. (36)

Case 3. Consider

$$T(x,y,z,t) = \frac{1}{\rho c(4\pi K\tau)^{1.5}}\sum_{i=0}^{1}(-1)^i\left[e^{-(r_{i00}^{i00})^2/4K\tau} + \sum_{m=1}^{NS_y}(-1)^{(m-1)}\sum_{j=1}^{4}(-1)^{((j-1)!(j+1)+1)}e^{-(r_{ij0}^{im0})^2/4K\tau}\right],$$

$$T(x,y,z,\omega) = \frac{1}{2\lambda}\sum_{i=0}^{1}(-1)^i\left[\frac{e^{-\sqrt{i\omega/K}r_{i00}^{i00}}}{r_{i00}^{i00}} + \sum_{m=1}^{NS_y}(-1)^{(m-1)}\sum_{j=1}^{4}(-1)^{((j-1)!(j+1)+1)}\frac{e^{-\sqrt{i\omega/K}r_{ij0}^{im0}}}{r_{ij0}^{im0}}\right], \quad (37)$$

$$T(x,y,z,\omega) = \frac{-i\pi}{2\lambda L_x}\sum_{i=0}^{1}(-1)^i\cdot\left[\sum_{m_k=-\infty}^{\infty}\left[H_0\left(k_{1z}r_{i0}^{i0}\right)e^{-ik_z(\overline{z}_0)}\right] + \sum_{m=1}^{NS_y}(-1)^{(m-1)}\cdot\sum_{j=1}^{4}(-1)^{((j-1)!(j+1)+1)}\cdot\sum_{m_k=-\infty}^{\infty}\left[H_0\left(k_{1z}r_{ij}^{im}\right)e^{-ik_z(\overline{z}_0)}\right]\right]$$

as the sum of 2.5D sources.

Case 4. Consider

$$T(x,y,z,t) = \frac{1}{\rho c (4\pi K\tau)^{1.5}} \sum_{i=0}^{1} (-1)^{i} \left[e^{-(r_{i00}^{i00})^2/4K\tau} \right.$$
$$\left. + \sum_{m=1}^{NS_y} (-1)^{(m-1)} \sum_{j=1}^{4} (-1)^{(j!+1)} e^{-(r_{ij0}^{im0})^2/4K\tau} \right],$$
$$T(x,y,z,\omega) = \frac{1}{2\lambda} \sum_{i=0}^{1} (-1)^{i} \left[\frac{e^{-\sqrt{i\omega/K} r_{i00}^{i00}}}{r_{i00}^{i00}} \right.$$
$$\left. + \sum_{m=1}^{NS_y} (-1)^{(m-1)} \sum_{j=1}^{4} (-1)^{(j!+1)} \frac{e^{-\sqrt{i\omega/K} r_{ij0}^{im0}}}{r_{ij0}^{im0}} \right], \quad (38)$$
$$T(x,y,z,\omega) = \frac{-i\pi}{2\lambda L_x} \sum_{i=0}^{1} (-1)^{i}$$
$$\cdot \left[\sum_{m_k=-\infty}^{\infty} \left[H_0\left(k_{1z} r_{i0}^{i0}\right) e^{-ik_z(\overline{z}_0)} \right] + \sum_{m=1}^{NS_y} (-1)^{(m-1)} \right.$$
$$\left. \cdot \sum_{j=1}^{4} (-1)^{(j!+1)} \sum_{m_k=-\infty}^{\infty} \left[H_0\left(k_{1z} r_{ij}^{im}\right) e^{-ik_z(\overline{z}_0)} \right] \right]$$

as the sum of 2.5D sources.

(e) Solid Rectangular Pipe Defined by $y \geq 0$, $y \leq d_2$, $x \geq 0$, and $x \leq d_1$. Boundary conditions prescribed for the solid rectangular pipe (Cases 1–4) are shown in Figure 14.

Equations for the solid rectangular pipe (Cases 1–4), subjected to a 2D heat source, are as follows.

Case 1. Consider

$$T(x,y,t) = \frac{1}{\rho c (4\pi K\tau)} \left[e^{-(r_{00}^{00})^2/4K\tau} \right.$$
$$+ \sum_{m=1}^{NS_y} \sum_{j=1}^{4} e^{-(r_{0j}^{0m})^2/4K\tau}$$
$$\left. + \sum_{l=1}^{NS_x} \sum_{i=1}^{4} \left[e^{-(r_{i0}^{l0})^2/4K\tau} + \sum_{m=1}^{NS_y} \sum_{j=1}^{4} e^{-(r_{ij}^{lm})^2/4K\tau} \right] \right], \quad (39)$$
$$T(x,y,\omega) = \frac{-i}{4\lambda} \left[H_0\left(k_1 r_{00}^{00}\right) + \sum_{m=1}^{NS_y} \sum_{j=1}^{4} H_0\left(k_1 r_{0j}^{0m}\right) \right.$$
$$\left. + \sum_{l=1}^{NS_x} \sum_{i=1}^{4} \left[H_0\left(k_1 r_{i0}^{l0}\right) + \sum_{m=1}^{NS_y} \sum_{j=1}^{4} H_0\left(k_1 r_{ij}^{lm}\right) \right] \right].$$

Case 2. Consider

$$T(x,y,t) = \frac{1}{\rho c (4\pi K\tau)} \left[e^{-(r_{00}^{00})^2/4K\tau} \right.$$
$$+ \sum_{m=1}^{NS_y} \sum_{j=1}^{4} (-1)^{(j-1)!} e^{-(r_{0j}^{0m})^2/4K\tau} + \sum_{l=1}^{NS_x} \sum_{i=1}^{4} (-1)^{(i-1)!}$$
$$\left. \cdot \left[e^{-(r_{i0}^{l0})^2/4K\tau} + \sum_{m=1}^{NS_y} \sum_{j=1}^{4} (-1)^{(j-1)!} e^{-(r_{ij}^{lm})^2/4K\tau} \right] \right],$$
$$(40)$$
$$T(x,y,\omega) = \frac{-i}{4\lambda} \left[H_0\left(k_1 r_{00}^{00}\right) \right.$$
$$+ \sum_{m=1}^{NS_y} \sum_{j=1}^{4} (-1)^{(j-1)!} H_0\left(k_1 r_{0j}^{0m}\right) + \sum_{l=1}^{NS_x} \sum_{i=1}^{4} (-1)^{(i-1)!}$$
$$\left. \cdot \left[H_0\left(k_1 r_{i0}^{l0}\right) + \sum_{m=1}^{NS_y} \sum_{j=1}^{4} (-1)^{(j-1)!} H_0\left(k_1 r_{ij}^{im}\right) \right] \right].$$

Case 3. Consider

$$T(x,y,t) = \frac{1}{\rho c (4\pi K\tau)} \left[e^{-(r_{00}^{00})^2/4K\tau} + \sum_{m=1}^{NS_y} (-1)^{(m-1)} \right.$$
$$\cdot \sum_{j=1}^{4} (-1)^{((j-1)!(j+1)+1)} e^{-(r_{0j}^{0m})^2/4K\tau} + \sum_{l=1}^{NS_x} (-1)^{(l-1)}$$
$$\cdot \sum_{i=1}^{4} (-1)^{(i!+1)} \left[e^{-(r_{i0}^{l0})^2/4K\tau} + \sum_{m=1}^{NS_y} (-1)^{(m-1)} \right.$$
$$\left. \left. \cdot \sum_{j=1}^{4} (-1)^{((j-1)!(j+1)+1)} e^{-(r_{ij}^{lm})^2/4K\tau} \right] \right],$$
$$(41)$$
$$T(x,y,\omega) = \frac{-i}{4\lambda} \left[H_0\left(k_1 r_{00}^{00}\right) + \sum_{m=1}^{NS_y} (-1)^{(m-1)} \right.$$
$$\cdot \sum_{j=1}^{4} (-1)^{((j-1)!(j+1)+1)} H_0\left(k_1 r_{0j}^{0m}\right) + \sum_{l=1}^{NS_x} (-1)^{(l-1)}$$
$$\cdot \sum_{i=1}^{4} (-1)^{(i!+1)} \left[H_0\left(k_1 r_{i0}^{l0}\right) + \sum_{m=1}^{NS_y} (-1)^{(m-1)} \right.$$
$$\left. \left. \cdot \sum_{j=1}^{4} (-1)^{((j-1)!(j+1)+1)} H_0\left(k_1 r_{ij}^{lm}\right) \right] \right].$$

Case 4. Consider

$$T(x,y,t) = \frac{1}{\rho c(4\pi K\tau)}\left[e^{-(r_{00}^{00})^2/4K\tau} + \sum_{m=1}^{NS_y}(-1)^{(m-1)} \cdot \sum_{j=1}^{4}(-1)^{(j!+1)} e^{-(r_{0j}^{0m})^2/4K\tau} + \sum_{l=1}^{NS_x}(-1)^{(l-1)} \cdot \sum_{i=1}^{4}(-1)^{((i-1)!(i+1)+1)}\left[e^{-(r_{i0}^{l0})^2/4K\tau} + \sum_{m=1}^{NS_y}(-1)^{(m-1)}\sum_{j=1}^{4}(-1)^{(j!+1)} e^{-(r_{ij}^{lm})^2/4K\tau}\right]\right],$$

$$T(x,y,\omega) = \frac{-i}{4\lambda}\left[H_0\left(k_1 r_{00}^{00}\right) + \sum_{m=1}^{NS_y}(-1)^{(m-1)} \cdot \sum_{j=1}^{4}(-1)^{(j!+1)} H_0\left(k_1 r_{0j}^{0m}\right) + \sum_{l=1}^{NS_x}(-1)^{(l-1)} \cdot \sum_{i=1}^{4}(-1)^{((i-1)!(i+1)+1)}\left[H_0\left(k_1 r_{i0}^{l0}\right) + \sum_{m=1}^{NS_y}(-1)^{(m-1)}\sum_{j=1}^{4}(-1)^{(j!+1)} H_0\left(k_1 r_{ij}^{lm}\right)\right]\right]. \quad (42)$$

Equations for the solid rectangular pipe (Cases 1–4), subjected to a 3D heat source, are as follows.

Case 1. Consider

$$T(x,y,z,t) = \frac{1}{\rho c(4\pi K\tau)^{1.5}}\left[e^{-(r_{000}^{000})^2/4K\tau} + \sum_{m=1}^{NS_y}\sum_{j=1}^{4} e^{-(r_{0j0}^{0m0})^2/4K\tau} + \sum_{l=1}^{NS_x}\sum_{i=1}^{4}\left[e^{-(r_{i00}^{l00})^2/4K\tau} + \sum_{m=1}^{NS_y}\sum_{j=1}^{4} e^{-(r_{ij0}^{lm0})^2/4K\tau}\right]\right],$$

$$T(x,y,z,\omega) = \frac{1}{2\lambda}\left[\frac{e^{-\sqrt{i\omega/K}r_{000}^{000}}}{r_{000}^{000}} + \sum_{m=1}^{NS_y}\sum_{j=1}^{4}\frac{e^{-\sqrt{i\omega/K}r_{0j0}^{0m0}}}{r_{0j0}^{0m0}} + \sum_{l=1}^{NS_x}\sum_{i=1}^{4}\left[\frac{e^{-\sqrt{i\omega/K}r_{i00}^{l00}}}{r_{i00}^{l00}} + \sum_{m=1}^{NS_y}\sum_{j=1}^{4}\frac{e^{-\sqrt{i\omega/K}r_{ij0}^{lm0}}}{r_{ij0}^{lm0}}\right]\right],$$

$$T(x,y,z,\omega) = \frac{-i\pi}{2\lambda L_x}\left[\sum_{m_k=-\infty}^{\infty}\left[H_0\left(k_{1z}r_{00}^{00}\right)e^{-ik_z(\overline{z}_0)}\right] + \sum_{m=1}^{NS_y}\sum_{j=1}^{4}\sum_{m_k=-\infty}^{\infty}\left[H_0\left(k_{1z}r_{0j}^{0m}\right)e^{-ik_z(\overline{z}_0)}\right] + \sum_{l=1}^{NS_x}\sum_{i=1}^{4}\left[\sum_{m_k=-\infty}^{\infty}\left[H_0\left(k_{1z}r_{i0}^{l0}\right)e^{-ik_z(\overline{z}_0)}\right] + \sum_{m=1}^{NS_y}\sum_{j=1}^{4}\sum_{m_k=-\infty}^{\infty}\left[H_0\left(k_{1z}r_{ij}^{lm}\right)e^{-ik_z(\overline{z}_0)}\right]\right]\right] \quad (43)$$

as the sum of 2.5D sources.

Case 2. Consider

$$T(x,y,z,t) = \frac{1}{\rho c(4\pi K\tau)^{1.5}}\left[e^{-(r_{000}^{000})^2/4K\tau} + \sum_{m=1}^{NS_y}\sum_{j=1}^{4}(-1)^{(j-1)!} e^{-(r_{0j0}^{0m0})^2/4K\tau} + \sum_{l=1}^{NS_x}\sum_{i=1}^{4}(-1)^{(i-1)!} \cdot \left[e^{-(r_{i00}^{l00})^2/4K\tau} + \sum_{m=1}^{NS_y}\sum_{j=1}^{4}(-1)^{(j-1)!} e^{-(r_{ij0}^{lm0})^2/4K\tau}\right]\right],$$

$$T(x,y,z,\omega) = \frac{1}{2\lambda}\left[\frac{e^{-\sqrt{i\omega/K}r_{000}^{000}}}{r_{000}^{000}} + \sum_{m=1}^{NS_y}\sum_{j=1}^{4}(-1)^{(j-1)!}\frac{e^{-\sqrt{i\omega/K}r_{0j0}^{0m0}}}{r_{0j0}^{0m0}} + \sum_{l=1}^{NS_x}\sum_{i=1}^{4}(-1)^{(i-1)!} \cdot \left[\frac{e^{-\sqrt{i\omega/K}r_{i00}^{l00}}}{r_{i00}^{l00}} + \sum_{m=1}^{NS_y}\sum_{j=1}^{4}(-1)^{(j-1)!}\frac{e^{-\sqrt{i\omega/K}r_{ij0}^{lm0}}}{r_{ij0}^{lm0}}\right]\right],$$

$$T(x,y,z,\omega) = \frac{-i\pi}{2\lambda L_x}\left[\sum_{m_k=-\infty}^{\infty}\left[H_0\left(k_{1z}r_{00}^{00}\right)e^{-ik_z(\overline{z}_0)}\right] + \sum_{m=1}^{NS_y}\sum_{j=1}^{4}(-1)^{(j-1)!}\sum_{m_k=-\infty}^{\infty}\left[H_0\left(k_{1z}r_{0j}^{0m}\right)e^{-ik_z(\overline{z}_0)}\right] + \sum_{l=1}^{NS_x}\sum_{i=1}^{4}(-1)^{(i-1)!}\left[\sum_{m_k=-\infty}^{\infty}\left[H_0\left(k_{1z}r_{i0}^{l0}\right)e^{-ik_z(\overline{z}_0)}\right]\right.$$

$$+\sum_{m=1}^{NS_y}\sum_{j=1}^{4}(-1)^{(j-1)!} \cdot \sum_{m_k=-\infty}^{\infty}\left[H_0\left(k_{1z}r_{ij}^{lm}\right)e^{-ik_z(\overline{z}_0)}\right]\Bigg]\Bigg]$$

as the sum of 2.5D sources. (44)

Case 3. Consider

$$T(x,y,z,t)=\frac{1}{\rho c(4\pi K\tau)^{1.5}}\Bigg[e^{-(r_{000}^{000})^2/4K\tau} + \sum_{m=1}^{NS_y}(-1)^{(m-1)}\sum_{j=1}^{4}(-1)^{((j-1)!(j+1)+1)}e^{-(r_{0j0}^{0m0})^2/4K\tau} + \sum_{l=1}^{NS_x}(-1)^{(l-1)}\sum_{i=1}^{4}(-1)^{(i!+1)}\Bigg[e^{-(r_{i00}^{l00})^2/4K\tau} + \sum_{m=1}^{NS_y}(-1)^{(m-1)} \cdot \sum_{j=1}^{4}(-1)^{((j-1)!(j+1)+1)}e^{-(r_{ij0}^{lm0})^2/4K\tau}\Bigg]\Bigg],$$

$$T(x,y,z,\omega)=\frac{1}{2\lambda}\Bigg[\frac{e^{-\sqrt{i\omega/K}r_{000}^{000}}}{r_{000}^{000}} + \sum_{m=1}^{NS_y}(-1)^{(m-1)} \cdot \sum_{j=1}^{4}(-1)^{((j-1)!(j+1)+1)}\frac{e^{-\sqrt{i\omega/K}r_{0j0}^{0m0}}}{r_{0j0}^{0m0}} + \sum_{l=1}^{NS_x}(-1)^{(l-1)} \cdot \sum_{i=1}^{4}(-1)^{(i!-1)}\Bigg[\frac{e^{-\sqrt{i\omega/K}r_{i00}^{l00}}}{r_{i00}^{l00}} + \sum_{m=1}^{NS_y}(-1)^{(m-1)} \cdot \sum_{j=1}^{4}(-1)^{((j-1)!(j+1)+1)}\frac{e^{-\sqrt{i\omega/K}r_{ij0}^{lm0}}}{r_{ij0}^{lm0}}\Bigg]\Bigg],$$

$$T(x,y,z,\omega)=\frac{-i\pi}{2\lambda L_x}\Bigg[\sum_{m_k=-\infty}^{\infty}\left[H_0\left(k_{1z}r_{00}^{00}\right)e^{-ik_z(\overline{z}_0)}\right] + \sum_{m=1}^{NS_y}(-1)^{(m-1)}\sum_{j=1}^{4}(-1)^{((j-1)!(j+1)+1)} \cdot \sum_{m_k=-\infty}^{\infty}\left[H_0\left(k_{1z}r_{0j}^{0m}\right)e^{-ik_z(\overline{z}_0)}\right] + \sum_{l=1}^{NS_x}(-1)^{(l-1)}$$

$$\cdot\sum_{i=1}^{4}(-1)^{(i!-1)}\Bigg[\sum_{m_k=-\infty}^{\infty}\left[H_0\left(k_{1z}r_{i0}^{l0}\right)e^{-ik_z(\overline{z}_0)}\right] + \sum_{m=1}^{NS_y}(-1)^{(m-1)}\sum_{j=1}^{4}(-1)^{((j-1)!(j+1)+1)} \cdot \sum_{m_k=-\infty}^{\infty}\left[H_0\left(k_{1z}r_{ij}^{lm}\right)e^{-ik_z(\overline{z}_0)}\right]\Bigg]\Bigg]$$

as the sum of 2.5D sources. (45)

Case 4. Consider

$$T(x,y,z,t)=\frac{1}{\rho c(4\pi K\tau)^{1.5}}\Bigg[e^{-(r_{000}^{000})^2/4K\tau} + \sum_{m=1}^{NS_y}(-1)^{(m-1)}\sum_{j=1}^{4}(-1)^{(j!+1)}e^{-(r_{0j0}^{0m0})^2/4K\tau} + \sum_{l=1}^{NS_x}(-1)^{(l-1)}\sum_{i=1}^{4}(-1)^{((i-1)!(i+1)+1)}\Bigg[e^{-(r_{i00}^{l00})^2/4K\tau} + \sum_{m=1}^{NS_y}(-1)^{(m-1)}\sum_{j=1}^{4}(-1)^{(j!+1)}e^{-(r_{ij0}^{lm0})^2/4K\tau}\Bigg]\Bigg],$$

$$T(x,y,z,\omega)=\frac{1}{2\lambda}\Bigg[\frac{e^{-\sqrt{i\omega/K}r_{000}^{000}}}{r_{000}^{000}} + \sum_{m=1}^{NS_y}\left[(-1)^{(m-1)}\sum_{j=1}^{4}(-1)^{(j!+1)}\frac{e^{-\sqrt{i\omega/K}r_{0j0}^{0m0}}}{r_{0j0}^{0m0}}\right] + \sum_{l=1}^{NS_x}(-1)^{(l-1)}\sum_{i=1}^{4}(-1)^{((j-1)!(j+1)+1)}\Bigg[\frac{e^{-\sqrt{i\omega/K}r_{i00}^{l00}}}{r_{i00}^{l00}} + \sum_{m=1}^{NS_y}(-1)^{(m-1)}\sum_{j=1}^{4}(-1)^{(j!+1)}\frac{e^{-\sqrt{i\omega/K}r_{ij0}^{lm0}}}{r_{ij0}^{lm0}}\Bigg]\Bigg],$$

$$T(x,y,z,\omega)=\frac{-i\pi}{2\lambda L_x}\Bigg[\sum_{m_k=-\infty}^{\infty}\left[H_0\left(k_{1z}r_{00}^{00}\right)e^{-ik_z(\overline{z}_0)}\right] + \sum_{m=1}^{NS_y}(-1)^{(m-1)}\sum_{j=1}^{4}(-1)^{(j!+1)} \cdot \sum_{m_k=-\infty}^{\infty}\left[H_0\left(k_{1z}r_{0j}^{0m}\right)e^{-ik_z(\overline{z}_0)}\right] + \sum_{l=1}^{NS_x}(-1)^{(l-1)}$$

$$\cdot \sum_{i=1}^{4} (-1)^{((i-1)!(j+1)+1)} \left[\sum_{m_k=-\infty}^{\infty} \left[H_0\left(k_{1z} r_{i0}^{l0}\right) e^{-ik_z(\overline{z}_0)}\right] \right.$$

$$+ \sum_{m=1}^{NS_y} (-1)^{(m-1)} \sum_{j=1}^{4} (-1)^{(j!+1)}$$

$$\left. \left. \cdot \sum_{m_k=-\infty}^{\infty} \left[H_0\left(k_{1z} r_{ij}^{lm}\right) e^{-ik_z(\overline{z}_0)}\right] \right] \right]$$

as the sum of 2.5D sources.

(46)

3.3. Three-Dimensional Space

(a) Solid Open Box Defined by $y \geq 0$, $y \leq d_2$, $x \geq 0$, $x \leq d_1$, *and* $z \leq d_3$. Boundary conditions prescribed for the solid open box (Cases 1–4) are shown in Figure 15.

Equations for the solid open box (Cases 1–4), subjected to a 3D heat source, are as follows.

Case 1. Consider

$$T(x,y,z,t) = \frac{1}{\rho c (4\pi K \tau)^{1.5}} \sum_{n=0}^{1} \left[e^{-(r_{00n}^{00n})^2/4K\tau} + \sum_{m=1}^{NS_y} \sum_{j=1}^{4} e^{-(r_{0jn}^{0mn})^2/4K\tau} + \sum_{l=1}^{NS_x} \sum_{i=1}^{4} \left[e^{-(r_{i0n}^{l0n})^2/4K\tau} + \sum_{m=1}^{NS_y} \sum_{j=1}^{4} e^{-(r_{ijn}^{lmn})^2/4K\tau} \right] \right],$$

$$T(x,y,z,\omega) = \frac{1}{2\lambda} \sum_{n=0}^{1} \left[\frac{e^{-\sqrt{i\omega/K} r_{00n}^{00n}}}{r_{00n}^{00n}} + \sum_{m=1}^{NS_y} \sum_{j=1}^{4} \frac{e^{-\sqrt{i\omega/K} r_{0jn}^{0mn}}}{r_{0jn}^{0mn}} + \sum_{l=1}^{NS_x} \sum_{i=1}^{4} \left[\frac{e^{-\sqrt{i\omega/K} r_{i0n}^{l0n}}}{r_{i0n}^{l0n}} + \sum_{m=1}^{NS_y} \sum_{j=1}^{4} \frac{e^{-\sqrt{i\omega/K} r_{ijn}^{lmn}}}{r_{ijn}^{lmn}} \right] \right],$$

$$T(x,y,z,\omega) = \frac{-i\pi}{2\lambda L_x} \sum_{n=0}^{1} \left[\sum_{m_k=-\infty}^{\infty} \left[H_0\left(k_{1z} r_{00}^{00}\right) e^{-ik_z(\overline{z}_n)}\right] + \sum_{m=1}^{NS_y} \sum_{j=1}^{4} \sum_{m_k=-\infty}^{\infty} \left[H_0\left(k_{1z} r_{0j}^{0m}\right) e^{-ik_z(\overline{z}_n)}\right] \right. \qquad (47)$$

$$\left. + \sum_{l=1}^{NS_x} \sum_{i=1}^{4} \left[\sum_{m_k=-\infty}^{\infty} \left[H_0\left(k_{1z} r_{i0}^{l0}\right) e^{-ik_z(\overline{z}_n)}\right] + \sum_{m=1}^{NS_y} \sum_{j=1}^{4} \sum_{m_k=-\infty}^{\infty} \left[H_0\left(k_{1z} r_{ij}^{lm}\right) e^{-ik_z(\overline{z}_n)}\right] \right] \right]$$

as the sum of 2.5D sources.

Case 2. Consider

$$T(x,y,z,t) = \frac{1}{\rho c (4\pi K \tau)^{1.5}} \sum_{n=0}^{1} (-1)^n \left[e^{-(r_{00n}^{00n})^2/4K\tau} \right.$$

$$+ \sum_{m=1}^{NS_y} \sum_{j=1}^{4} (-1)^{(j-1)!} e^{-(r_{0jn}^{0mn})^2/4K\tau} + \sum_{l=1}^{NS_x} \sum_{i=1}^{4} (-1)^{(i-1)!}$$

$$\left. \cdot \left[e^{-(r_{i0n}^{l0n})^2/4K\tau} + \sum_{m=1}^{NS_y} \sum_{j=1}^{4} (-1)^{(j-1)!} e^{-(r_{ijn}^{lmn})^2/4K\tau} \right] \right],$$

$$T(x,y,z,\omega) = \frac{1}{2\lambda} \sum_{n=0}^{1} (-1)^n \left[\frac{e^{-\sqrt{i\omega/K} r_{00n}^{00n}}}{r_{00n}^{00n}} \right.$$

$$+ \sum_{m=1}^{NS_y} \sum_{j=1}^{4} (-1)^{(j-1)!} \frac{e^{-\sqrt{i\omega/K} r_{0jn}^{0mn}}}{r_{0jn}^{0mn}} + \sum_{l=1}^{NS_x} \sum_{i=1}^{4} (-1)^{(i-1)!}$$

$$\left. \cdot \left[\frac{e^{-\sqrt{i\omega/K} r_{i0n}^{l0n}}}{r_{i0n}^{l0n}} + \sum_{m=1}^{NS_y} \sum_{j=1}^{4} (-1)^{(j-1)!} \frac{e^{-\sqrt{i\omega/K} r_{ijn}^{lmn}}}{r_{ijn}^{lmn}} \right] \right],$$

$$T(x,y,z,\omega) = \frac{-i\pi}{2\lambda L_x} \sum_{n=0}^{1} (-1)^n$$

$$\cdot \left[\sum_{m_k=-\infty}^{\infty} \left[H_0\left(k_{1z} r_{00}^{00}\right) e^{-ik_z(\overline{z}_n)}\right] \right.$$

$$+ \sum_{m=1}^{NS_y} \sum_{j=1}^{4} (-1)^{(j-1)!} \sum_{m_k=-\infty}^{\infty} \left[H_0\left(k_{1z} r_{0j}^{0m}\right) e^{-ik_z(\overline{z}_n)}\right]$$

$$+ \sum_{l=1}^{NS_x} \sum_{i=1}^{4} (-1)^{(i-1)!} \left[\sum_{m_k=-\infty}^{\infty} \left[H_0\left(k_{1z} r_{i0}^{l0}\right) e^{-ik_z(\overline{z}_n)}\right] \right.$$

$$+ \sum_{m=1}^{NS_y} \sum_{j=1}^{4} (-1)^{(j-1)!}$$

$$\left. \left. \cdot \sum_{m_k=-\infty}^{\infty} \left[H_0\left(k_{1z} r_{ij}^{lm}\right) e^{-ik_z(\overline{z}_n)}\right] \right] \right]$$

as the sum of 2.5D sources.

(48)

Case 3. Consider

$$
\begin{aligned}
T(x,y,z,t) &= \frac{1}{\rho c (4\pi K\tau)^{1.5}} \sum_{n=0}^{1} \Bigg[e^{-(r_{00n}^{00n})^2/4K\tau} \\
&+ \sum_{m=1}^{NS_y} (-1)^{(m-1)} \sum_{j=1}^{4} (-1)^{((j-1)!(j+1)+1)} e^{-(r_{0jn}^{0mn})^2/4K\tau} \\
&+ \sum_{l=1}^{NS_x} (-1)^{(l-1)} \sum_{i=1}^{4} (-1)^{(i!+1)} \Bigg[e^{-(r_{i0n}^{l0n})^2/4K\tau} \\
&+ \sum_{m=1}^{NS_y} (-1)^{(m-1)} \\
&\cdot \sum_{j=1}^{4} (-1)^{((j-1)!(j+1)+1)} e^{-(r_{ijn}^{lmn})^2/4K\tau} \Bigg]\Bigg], \\
T(x,y,z,\omega) &= \frac{1}{2\lambda} \sum_{n=0}^{1} \Bigg[\frac{e^{-\sqrt{i\omega/K} r_{00n}^{00n}}}{r_{000}^{000}} + \sum_{m=1}^{NS_y} (-1)^{(m-1)} \\
&\cdot \sum_{j=1}^{4} (-1)^{((j-1)!(j+1)+1)} \frac{e^{-\sqrt{i\omega/K} r_{0jn}^{0mn}}}{r_{0jn}^{0mn}} + \sum_{l=1}^{NS_x} (-1)^{(l-1)} \\
&\cdot \sum_{i=1}^{4} (-1)^{(i!-1)} \Bigg[\frac{e^{-\sqrt{i\omega/K} r_{i0n}^{l0n}}}{r_{i0n}^{l0n}} + \sum_{m=1}^{NS_y} (-1)^{(m-1)} \\
&\cdot \sum_{j=1}^{4} (-1)^{((j-1)!(j+1)+1)} \frac{e^{-\sqrt{i\omega/K} r_{ij0}^{lmn}}}{r_{ijn}^{lmn}} \Bigg]\Bigg], \\
T(x,y,z,\omega) &= \frac{-i\pi}{2\lambda L_x} \Bigg[\sum_{m_k=-\infty}^{\infty} \left[H_0 \left(k_{1z} r_{00}^{00} \right) e^{-ik_z(\overline{z}_n)} \right] \\
&+ \sum_{m=1}^{NS_y} (-1)^{(m-1)} \sum_{j=1}^{4} (-1)^{((j-1)!(j+1)+1)} \\
&\cdot \sum_{m_k=-\infty}^{\infty} \left[H_0 \left(k_{1z} r_{0j}^{0m} \right) e^{-ik_z(\overline{z}_n)} \right] + \sum_{l=1}^{NS_x} (-1)^{(l-1)} \\
&\cdot \sum_{i=1}^{4} (-1)^{(i!-1)} \Bigg[\sum_{m_k=-\infty}^{\infty} \left[H_0 \left(k_{1z} r_{i0}^{l0} \right) e^{-ik_z(\overline{z}_n)} \right] \\
&+ \sum_{m=1}^{NS_y} (-1)^{(m-1)} \sum_{j=1}^{4} (-1)^{((j-1)!(j+1)+1)} \\
&\cdot \sum_{m_k=-\infty}^{\infty} \left[H_0 \left(k_{1z} r_{ij}^{lm} \right) e^{-ik_z(\overline{z}_n)} \right] \Bigg]\Bigg]
\end{aligned}
\tag{49}
$$

as the sum of 2.5D.

Case 4. Consider

$$
\begin{aligned}
T(x,y,z,t) &= \frac{1}{\rho c (4\pi K\tau)^{1.5}} \sum_{n=0}^{1} (-1)^n \Bigg[e^{-(r_{00n}^{00n})^2/4K\tau} \\
&+ \sum_{m=1}^{NS_y} (-1)^{(m-1)} \sum_{j=1}^{4} (-1)^{(j!+1)} e^{-(r_{0jn}^{0mn})^2/4K\tau} \\
&+ \sum_{l=1}^{NS_x} (-1)^{(l-1)} \sum_{i=1}^{4} (-1)^{((i-1)!(i+1)+1)} \Bigg[e^{-(r_{i0n}^{l0n})^2/4K\tau} \\
&+ \sum_{m=1}^{NS_y} (-1)^{(m-1)} \sum_{j=1}^{4} (-1)^{(j!+1)} e^{-(r_{ijn}^{lmn})^2/4K\tau} \Bigg]\Bigg], \\
T(x,y,z,\omega) &= \frac{1}{2\lambda} \sum_{n=0}^{1} (-1)^n \Bigg[\frac{e^{-\sqrt{i\omega/K} r_{00n}^{00n}}}{r_{00n}^{00n}} \\
&+ \sum_{m=1}^{NS_y} (-1)^{(m-1)} \sum_{j=1}^{4} (-1)^{(j!+1)} \frac{e^{-\sqrt{i\omega/K} r_{0jn}^{0mn}}}{r_{0jn}^{0mn}} \\
&+ \sum_{l=1}^{NS_x} (-1)^{(l-1)} \sum_{i=1}^{4} (-1)^{((i-1)!(i+1)+1)} \Bigg[\frac{e^{-\sqrt{i\omega/K} r_{i0n}^{l0n}}}{r_{i0n}^{l0n}} \\
&+ \sum_{m=1}^{NS_y} (-1)^{(m-1)} \sum_{j=1}^{4} (-1)^{(j!+1)} \frac{e^{-\sqrt{i\omega/K} r_{ijn}^{lmn}}}{r_{ijn}^{lmn}} \Bigg]\Bigg], \\
T(x,y,z,\omega) &= \frac{-i\pi}{2\lambda L_x} \sum_{n=0}^{1} (-1)^n \\
&\cdot \Bigg[\sum_{m_k=-\infty}^{\infty} \left[H_0 \left(k_{1z} r_{00}^{00} \right) e^{-ik_z(\overline{z}_n)} \right] + \sum_{m=1}^{NS_y} (-1)^{(m-1)} \\
&\cdot \sum_{j=1}^{4} (-1)^{(j!+1)} \sum_{m_k=-\infty}^{\infty} \left[H_0 \left(k_{1z} r_{0j}^{0m} \right) e^{-ik_z(\overline{z}_n)} \right] \\
&+ \sum_{l=1}^{NS_x} (-1)^{(l-1)} \sum_{i=1}^{4} (-1)^{((i-1)!(i+1)+1)} \\
&\cdot \Bigg[\sum_{m_k=-\infty}^{\infty} \left[H_0 \left(k_{1z} r_{i0}^{l0} \right) e^{-ik_z(\overline{z}_n)} \right] + \sum_{m=1}^{NS_y} (-1)^{(m-1)!} \\
&\cdot \sum_{j=1}^{4} (-1)^{(j!+1)} \sum_{m_k=-\infty}^{\infty} \left[H_0 \left(k_{1z} r_{ij}^{lm} \right) e^{-ik_z(\overline{z}_n)} \right] \Bigg]\Bigg]
\end{aligned}
\tag{50}
$$

as the sum of 2.5D sorces.

(b) 3D Parallelepiped Box Defined by $y \geq 0$, $y \leq d_2$, $x \geq 0$, $x \leq d_1$, $z \geq 0$, and $z \leq d_3$. Boundary conditions for the 3D parallelepiped box (Cases 1–4) are shown in Figure 16.

Equations for the 3D parallelepiped box (Cases 1–4), subjected to a 3D heat source, are as follows.

Case 1. Consider

$$T(x,y,z,t)=\frac{1}{\rho c(4\pi K\tau)^{1.5}}\left[e^{-(r_{000}^{000})^2/4K\tau}+\sum_{m=1}^{NS_y}\sum_{j=1}^{4}e^{-(r_{0j0}^{0m0})^2/4K\tau}+\sum_{l=1}^{NS_x}\sum_{i=1}^{4}\left[e^{-(r_{i00}^{l00})^2/4K\tau}+\sum_{m=1}^{NS_y}\sum_{j=1}^{4}e^{-(r_{ij0}^{lm0})^2/4K\tau}\right]\right.$$
$$\left.+\sum_{n=1}^{NS_z}\sum_{k=1}^{4}\left[e^{-(r_{00k}^{00n})^2/4K\tau}+\sum_{m=1}^{NS_y}\sum_{j=1}^{4}e^{-(r_{0jk}^{0mn})^2/4K\tau}+\sum_{l=1}^{NS_x}\sum_{i=1}^{4}\left[e^{-(r_{i0k}^{l0n})^2/4K\tau}+\sum_{m=1}^{NS_y}\sum_{j=1}^{4}e^{-(r_{ijk}^{lmn})^2/4K\tau}\right]\right]\right],$$

$$T(x,y,z,\omega)=\frac{1}{2\lambda}\left[\frac{e^{-\sqrt{i\omega/K}r_{000}^{000}}}{r_{000}^{000}}+\sum_{m=1}^{NS_y}\sum_{j=1}^{4}\frac{e^{-\sqrt{i\omega/K}r_{0j0}^{0m0}}}{r_{0j0}^{0m0}}+\sum_{l=1}^{NS_x}\sum_{i=1}^{4}\left[\frac{e^{-\sqrt{i\omega/K}r_{i00}^{l00}}}{r_{i00}^{l00}}+\sum_{m=1}^{NS_y}\sum_{j=1}^{4}\frac{e^{-\sqrt{i\omega/K}r_{ij0}^{lm0}}}{r_{ij0}^{lm0}}\right]\right.$$
$$\left.+\sum_{n=1}^{NS_z}\sum_{k=1}^{4}\left[\frac{e^{-\sqrt{i\omega/K}r_{00k}^{00n}}}{r_{00k}^{00n}}+\sum_{m=1}^{NS_y}\sum_{j=1}^{4}\frac{e^{-\sqrt{i\omega/K}r_{0jk}^{0mn}}}{r_{0jk}^{0mn}}+\sum_{l=1}^{NS_x}\sum_{i=1}^{4}\left[\frac{e^{-\sqrt{i\omega/K}r_{i0k}^{l0n}}}{r_{i0k}^{l0n}}+\sum_{m=1}^{NS_y}\sum_{j=1}^{4}\frac{e^{-\sqrt{i\omega/K}r_{ijk}^{lmn}}}{r_{ijk}^{lmn}}\right]\right]\right], \quad (51)$$

$$T(x,y,z,\omega)=\frac{-i\pi}{2\lambda L_x}\left[\sum_{m_k=-\infty}^{\infty}\left[H_0\left(k_{1z}r_{00}^{00}\right)e^{-ik_z(\overline{z}_0)}\right]+\sum_{m=1}^{NS_y}\sum_{j=1}^{4}\sum_{m_k=-\infty}^{\infty}\left[H_0\left(k_{1z}r_{0j}^{0m}\right)e^{-ik_z(\overline{z}_0)}\right]\right.$$
$$+\sum_{l=1}^{NS_x}\sum_{i=1}^{4}\left[\sum_{m_k=-\infty}^{\infty}\left[H_0\left(k_{1z}r_{i0}^{l0}\right)e^{-ik_z(\overline{z}_0)}\right]+\sum_{m=1}^{NS_y}\sum_{j=1}^{4}\sum_{m_k=-\infty}^{\infty}\left[H_0\left(k_{1z}r_{ij}^{lm}\right)e^{-ik_z(\overline{z}_0)}\right]\right]$$
$$\left.+\sum_{n=1}^{NS_z}\sum_{k=1}^{4}\left[\sum_{m_k=-\infty}^{\infty}\left[H_0\left(k_{1z}r_{00}^{00}\right)e^{-ik_z(\overline{z}_n)}\right]+\sum_{m=1}^{NS_y}\sum_{j=1}^{4}\sum_{m_k=-\infty}^{\infty}\left[H_0\left(k_{1z}r_{0j}^{0m}\right)e^{-ik_z(\overline{z}_n)}\right]+\sum_{l=1}^{NS_x}\sum_{i=1}^{4}\left[\sum_{m_k=-\infty}^{\infty}\left[H_0\left(k_{1z}r_{i0}^{l0}\right)e^{-ik_z(\overline{z}_n)}\right]+\sum_{m=1}^{NS_y}\sum_{j=1}^{4}\sum_{m_k=-\infty}^{\infty}\left[H_0\left(k_{1z}r_{ij}^{lm}\right)e^{-ik_z(\overline{z}_n)}\right]\right]\right]\right]$$

as the sum of 2.5D sources.

Case 2. Consider

$$T(x,y,z,t)=\frac{1}{\rho c(4\pi K\tau)^{1.5}}\left[e^{-(r_{000}^{000})^2/4K\tau}\right.$$
$$+\sum_{m=1}^{NS_y}\sum_{j=1}^{4}(-1)^{(j-1)!}e^{-(r_{0j0}^{0m0})^2/4K\tau}+\sum_{l=1}^{NS_x}\sum_{i=1}^{4}(-1)^{(i-1)!}$$
$$\cdot\left[e^{-(r_{i00}^{l00})^2/4K\tau}+\sum_{m=1}^{NS_y}\sum_{j=1}^{4}(-1)^{(j-1)!}e^{-(r_{ij0}^{lm0})^2/4K\tau}\right]$$
$$+\sum_{n=1}^{NS_z}\sum_{k=1}^{4}(-1)^{(k-1)!}\left[e^{-(r_{00k}^{00n})^2/4K\tau}\right.$$
$$+\sum_{m=1}^{NS_y}\sum_{j=1}^{4}(-1)^{(j-1)!}e^{-(r_{0jk}^{0mn})^2/4K\tau}+\sum_{l=1}^{NS_x}\sum_{i=1}^{4}(-1)^{(i-1)!}$$
$$\left.\left.\cdot\left[e^{-(r_{i0k}^{l0n})^2/4K\tau}+\sum_{m=1}^{NS_y}\sum_{j=1}^{4}(-1)^{(j-1)!}e^{-(r_{ijk}^{lmn})^2/4K\tau}\right]\right]\right],$$

$$T(x,y,z,\omega)=\frac{1}{2\lambda}\left[\frac{e^{-\sqrt{i\omega/K}r_{000}^{000}}}{r_{000}^{000}}\right.$$
$$+\sum_{m=1}^{NS_y}\sum_{j=1}^{4}(-1)^{(j-1)!}\frac{e^{-\sqrt{i\omega/K}r_{0j0}^{0m0}}}{r_{0j0}^{0m0}}+\sum_{l=1}^{NS_x}\sum_{i=1}^{4}(-1)^{(i-1)!}$$
$$\cdot\left[\frac{e^{-\sqrt{i\omega/K}r_{i00}^{l00}}}{r_{i00}^{l00}}+\sum_{m=1}^{NS_y}\sum_{j=1}^{4}(-1)^{(j-1)!}\frac{e^{-\sqrt{i\omega/K}r_{ij0}^{lm0}}}{r_{ij0}^{lm0}}\right]$$
$$+\sum_{n=1}^{NS_z}\sum_{k=1}^{4}(-1)^{(k-1)!}\left[\frac{e^{-\sqrt{i\omega/K}r_{00k}^{00n}}}{r_{00k}^{00n}}\right.$$
$$+\sum_{m=1}^{NS_y}\sum_{j=1}^{4}(-1)^{(j-1)!}\frac{e^{-\sqrt{i\omega/K}r_{0jk}^{0mn}}}{r_{0jk}^{0mn}}+\sum_{l=1}^{NS_x}\sum_{i=1}^{4}(-1)^{(i-1)!}$$
$$\left.\left.\cdot\left[\frac{e^{-\sqrt{i\omega/K}r_{i0k}^{l0n}}}{r_{i0k}^{l0n}}+\sum_{m=1}^{NS_y}\sum_{j=1}^{4}(-1)^{(j-1)!}\frac{e^{-\sqrt{i\omega/K}r_{ijk}^{lmn}}}{r_{ijk}^{lmn}}\right]\right]\right],$$

$$T(x,y,z,\omega)=\frac{-i\pi}{2\lambda L_x}\left[\sum_{m_k=-\infty}^{\infty}\left[H_0\left(k_{1z}r_{00}^{00}\right)e^{-ik_z(\overline{z}_0)}\right]\right.$$
$$+\sum_{m=1}^{NS_y}\sum_{j=1}^{4}(-1)^{(j-1)!}\sum_{m_k=-\infty}^{\infty}\left[H_0\left(k_{1z}r_{0j}^{0m}\right)e^{-ik_z(\overline{z}_0)}\right]$$
$$+\sum_{l=1}^{NS_x}\sum_{i=1}^{4}(-1)^{(i-1)!}\left[\sum_{m_k=-\infty}^{\infty}\left[H_0\left(k_{1z}r_{i0}^{l0}\right)e^{-ik_z(\overline{z}_0)}\right]\right.$$
$$\left.+\sum_{m=1}^{NS_y}\sum_{j=1}^{4}(-1)^{(j-1)!}\sum_{m_k=-\infty}^{\infty}\left[H_0\left(k_{1z}r_{ij}^{lm}\right)e^{-ik_z(\overline{z}_0)}\right]\right]$$

$$+\sum_{n=1}^{NS_z}\sum_{k=1}^{4}(-1)^{(k-1)!}\left[\sum_{m_k=-\infty}^{\infty}\left[H_0\left(k_{1z}r_{00}^{00}\right)e^{-ik_z(\overline{z}_n)}\right]\right.$$

$$+\sum_{m=1}^{NS_y}\sum_{j=1}^{4}(-1)^{(j-1)!}\sum_{m_k=-\infty}^{\infty}\left[H_0\left(k_{1z}r_{0j}^{0m}\right)e^{-ik_z(\overline{z}_n)}\right]$$

$$+\sum_{l=1}^{NS_x}\sum_{i=1}^{4}(-1)^{(i-1)!}\left[\sum_{m_k=-\infty}^{\infty}\left[H_0\left(k_{1z}r_{i0}^{l0}\right)e^{-ik_z(\overline{z}_n)}\right]\right.$$

$$\left.\left.\left.+\sum_{m=1}^{NS_y}\sum_{j=1}^{4}(-1)^{(j-1)!}\sum_{m_k=-\infty}^{\infty}\left[H_0\left(k_{1z}r_{ij}^{lm}\right)e^{-ik_z(\overline{z}_n)}\right]\right]\right]\right] \tag{52}$$

as the sum of 2.5D sources.

Case 3. Consider

$$T(x,y,z,t)=\frac{1}{\rho c(4\pi K\tau)^{1.5}}\left[e^{-(r_{000}^{000})^2/4K\tau}+\sum_{m=1}^{NS_y}(-1)^{(m-1)}\sum_{j=1}^{4}(-1)^{((j-1)!(j+1)+1)}e^{-(r_{0j0}^{0m0})^2/4K\tau}+\sum_{l=1}^{NS_x}(-1)^{(l-1)}\sum_{i=1}^{4}(-1)^{(i!+1)}\right.$$

$$\cdot\left[e^{-(r_{i00}^{l00})^2/4K\tau}+\sum_{m=1}^{NS_y}(-1)^{(m-1)}\sum_{j=1}^{4}(-1)^{((j-1)!(j+1)+1)}e^{-(r_{ij0}^{lm0})^2/4K\tau}\right]+\sum_{n=1}^{NS_z}(-1)^{(n-1)}\sum_{k=1}^{4}(-1)^{(k!+1)}\left[e^{-(r_{00k}^{00n})^2/4K\tau}\right.$$

$$+\sum_{m=1}^{NS_y}(-1)^{(m-1)}\sum_{j=1}^{4}(-1)^{((j-1)!(j+1)+1)}e^{-(r_{0jk}^{0mn})^2/4K\tau}+\sum_{l=1}^{NS_x}(-1)^{(l-1)}\sum_{i=1}^{4}(-1)^{(i!+1)}$$

$$\left.\left.\cdot\left[e^{-(r_{i0k}^{l0n})^2/4K\tau}+\sum_{m=1}^{NS_y}(-1)^{(m-1)}\sum_{j=1}^{4}(-1)^{((j-1)!(j+1)+1)}e^{-(r_{ijk}^{lmn})^2/4K\tau}\right]\right]\right],$$

$$T(x,y,z,\omega)=\frac{1}{2\lambda}\left[\frac{e^{-\sqrt{i\omega/K}r_{000}^{000}}}{r_{000}^{000}}+\sum_{m=1}^{NS_y}\left[(-1)^{(m-1)}\sum_{j=1}^{4}(-1)^{((j-1)!(j+1)+1)}\frac{e^{-\sqrt{i\omega/K}r_{0j0}^{0m0}}}{r_{0j0}^{0m0}}\right]+\sum_{l=1}^{NS_x}(-1)^{(l-1)}\sum_{i=1}^{4}(-1)^{(i!+1)}\right.$$

$$\cdot\left[\frac{e^{-\sqrt{i\omega/K}r_{i00}^{l00}}}{r_{i00}^{l00}}+\sum_{m=1}^{NS_y}(-1)^{(m-1)}\sum_{j=1}^{4}(-1)^{((j-1)!(j+1)+1)}\frac{e^{-\sqrt{i\omega/K}r_{ij0}^{lm0}}}{r_{ij0}^{lm0}}\right]+\sum_{n=1}^{NS_z}(-1)^{(n-1)}\sum_{k=1}^{4}(-1)^{(k!+1)}\left[\frac{e^{-\sqrt{i\omega/K}r_{00k}^{00n}}}{r_{00k}^{00n}}\right.$$

$$+\sum_{m=1}^{NS_y}\left[(-1)^{(m-1)}\sum_{j=1}^{4}(-1)^{((j-1)!(j+1)+1)}\frac{e^{-\sqrt{i\omega/K}r_{0jk}^{0mn}}}{r_{0jk}^{0mn}}\right]+\sum_{l=1}^{NS_x}(-1)^{(l-1)}\sum_{i=1}^{4}(-1)^{(i!+1)}$$

$$\left.\left.\cdot\left[\frac{e^{-\sqrt{i\omega/K}r_{i0k}^{l0n}}}{r_{i0k}^{l0n}}+\sum_{m=1}^{NS_y}(-1)^{(m-1)}\sum_{j=1}^{4}(-1)^{((j-1)!(j+1)+1)}\frac{e^{-\sqrt{i\omega/K}r_{ijk}^{lmn}}}{r_{ijk}^{lmn}}\right]\right]\right],$$

$$T(x,y,z,\omega)=\frac{-i\pi}{2\lambda L_x}\left[\sum_{m_k=-\infty}^{\infty}\left[H_0\left(k_{1z}r_{00}^{00}\right)e^{-ik_z(\overline{z}_0)}\right]+\sum_{m=1}^{NS_y}(-1)^{(m-1)}\sum_{j=1}^{4}(-1)^{((j-1)!(j+1)+1)}\sum_{m_k=-\infty}^{\infty}\left[H_0\left(k_{1z}r_{0j}^{0m}\right)e^{-ik_z(\overline{z}_0)}\right]\right.$$

$$+\sum_{l=1}^{NS_x}(-1)^{(l-1)}\sum_{i=1}^{4}(-1)^{(i!+1)}\left[\sum_{m_k=-\infty}^{\infty}\left[H_0\left(k_{1z}r_{i0}^{l0}\right)e^{-ik_z(\overline{z}_0)}\right]+\sum_{m=1}^{NS_y}(-1)^{(m-1)}\sum_{j=1}^{4}(-1)^{((j-1)!(j+1)+1)}\right.$$

$$\left.\cdot\sum_{m_k=-\infty}^{\infty}\left[H_0\left(k_{1z}r_{ij}^{lm}\right)e^{-ik_z(\overline{z}_0)}\right]\right]+\sum_{n=1}^{NS_z}(-1)^{(n-1)}\sum_{k=1}^{4}(-1)^{(k!+1)}\left[\sum_{m_k=-\infty}^{\infty}\left[H_0\left(k_{1z}r_{00}^{00}\right)e^{-ik_z(\overline{z}_n)}\right]+\sum_{m=1}^{NS_y}(-1)^{(m-1)}\right.$$

$$\cdot \sum_{j=1}^{4} (-1)^{((j-1)!(j+1)+1)} \sum_{m_k=-\infty}^{\infty} \left[H_0 \left(k_{1z} r_{0j}^{0m} \right) e^{-ik_z(\overline{z}_n)} \right] + \sum_{l=1}^{NS_x} (-1)^{(l-1)}$$

$$\cdot \sum_{i=1}^{4} \left[\sum_{m_k=-\infty}^{\infty} \left[H_0 \left(k_{1z} r_{i0}^{l0} \right) e^{-ik_z(\overline{z}_n)} \right] + \sum_{m=1}^{NS_y} (-1)^{(m-1)} \sum_{j=1}^{4} (-1)^{((j-1)!(j+1)+1)} \sum_{m_k=-\infty}^{\infty} \left[H_0 \left(k_{1z} r_{ij}^{lm} \right) e^{-ik_z(\overline{z}_n)} \right] \right] \right] \right]$$

as the sum of 2.5D sources. (53)

Case 4. Consider

$$T(x,y,z,t) = \frac{1}{\rho c (4\pi K\tau)^{1.5}} \left[e^{-(r_{000}^{000})^2/4K\tau} \right.$$

$$+ \sum_{m=1}^{NS_y} (-1)^{(m-1)} \sum_{j=1}^{4} (-1)^{(j!+1)} e^{-(r_{0j0}^{0m0})^2/4K\tau}$$

$$+ \sum_{l=1}^{NS_x} (-1)^{(l-1)} \sum_{i=1}^{4} (-1)^{((i-1)!(i+1)+1)} \left[e^{-(r_{i00}^{l00})^2/4K\tau} \right.$$

$$\left. + \sum_{m=1}^{NS_y} (-1)^{(m-1)} \sum_{j=1}^{4} (-1)^{(j!+1)} e^{-(r_{ij0}^{lm0})^2/4K\tau} \right]$$

$$+ \sum_{n=1}^{NS_z} (-1)^{(n-1)} \sum_{k=1}^{4} (-1)^{((k-1)!(k+1)+1)} \left[e^{-(r_{00k}^{00n})^2/4K\tau} \right.$$

$$+ \sum_{m=1}^{NS_y} (-1)^{(m-1)} \sum_{j=1}^{4} (-1)^{(j!+1)} e^{-(r_{0jk}^{0mn})^2/4K\tau}$$

$$+ \sum_{l=1}^{NS_x} (-1)^{(l-1)} \sum_{i=1}^{4} (-1)^{((i-1)!(i+1)+1)} \left[e^{-(r_{i0k}^{l0n})^2/4K\tau} \right.$$

$$\left. \left. \left. + \sum_{m=1}^{NS_y} (-1)^{(m-1)} \sum_{j=1}^{4} (-1)^{(j!+1)} e^{-(r_{ijk}^{lmn})^2/4K\tau} \right] \right] \right],$$

$$T(x,y,z,\omega) = \frac{1}{2\lambda} \left[\frac{e^{-\sqrt{i\omega/K} r_{000}^{000}}}{r_{000}^{000}} + \sum_{m=1}^{NS_y} \left[(-1)^{(m-1)} \right. \right.$$

$$\left. \cdot \sum_{j=1}^{4} (-1)^{(j!+1)} \frac{e^{-\sqrt{i\omega/K} r_{0j0}^{0m0}}}{r_{0j0}^{0m0}} \right] + \sum_{l=1}^{NS_x} (-1)^{(l-1)}$$

$$\cdot \sum_{i=1}^{4} (-1)^{((i-1)!(i+1)+1)} \left[\frac{e^{-\sqrt{i\omega/K} r_{i00}^{l00}}}{r_{i00}^{l00}} + \sum_{m=1}^{NS_y} (-1)^{(m-1)} \right.$$

$$\left. \cdot \sum_{j=1}^{4} (-1)^{(j!+1)} \frac{e^{-\sqrt{i\omega/K} r_{ij0}^{lm0}}}{r_{ij0}^{lm0}} \right] + \sum_{n=1}^{NS_z} (-1)^{(n-1)}$$

$$\cdot \sum_{k=1}^{4} (-1)^{((k-1)!(k+1)+1)} \left[\frac{e^{-\sqrt{i\omega/K} r_{00k}^{00n}}}{r_{00k}^{00n}} \right.$$

$$+ \sum_{m=1}^{NS_y} \left[(-1)^{(m-1)} \sum_{j=1}^{4} (-1)^{(j!+1)} \frac{e^{-\sqrt{i\omega/K} r_{0jk}^{0mn}}}{r_{0jk}^{0mn}} \right]$$

$$+ \sum_{l=1}^{NS_x} (-1)^{(l-1)} \sum_{i=1}^{4} (-1)^{((i-1)!(i+1)+1)} \left[\frac{e^{-\sqrt{i\omega/K} r_{i0k}^{l0n}}}{r_{i0k}^{l0n}} \right.$$

$$\left. \left. \left. + \sum_{m=1}^{NS_y} (-1)^{(m-1)} \sum_{j=1}^{4} (-1)^{(j!+1)} \frac{e^{-\sqrt{i\omega/K} r_{ijk}^{lmn}}}{r_{ijk}^{lmn}} \right] \right] \right],$$

$$T(x,y,z,\omega) = \frac{-i\pi}{2\lambda L_x} \left[\sum_{m_k=-\infty}^{\infty} \left[H_0 \left(k_{1z} r_{00}^{00} \right) e^{-ik_z(\overline{z}_0)} \right] \right.$$

$$+ \sum_{m=1}^{NS_y} (-1)^{(m-1)} \sum_{j=1}^{4} (-1)^{(j!+1)} \sum_{m_k=-\infty}^{\infty} \left[H_0 \left(k_{1z} r_{0j}^{0m} \right) \right.$$

$$\left. \cdot e^{-ik_z(\overline{z}_0)} \right] + \sum_{l=1}^{NS_x} (-1)^{(l-1)} \sum_{i=1}^{4} (-1)^{((i-1)!(i+1)+1)}$$

$$\cdot \left[\sum_{m_k=-\infty}^{\infty} \left[H_0 \left(k_{1z} r_{i0}^{l0} \right) e^{-ik_z(\overline{z}_0)} \right] + \sum_{m=1}^{NS_y} (-1)^{(m-1)} \right.$$

$$\left. \cdot \sum_{j=1}^{4} (-1)^{(j!+1)} \sum_{m_k=-\infty}^{\infty} \left[H_0 \left(k_{1z} r_{ij}^{lm} \right) e^{-ik_z(\overline{z}_0)} \right] \right]$$

$$+ \sum_{n=1}^{NS_z} (-1)^{(n-1)} \sum_{k=1}^{4} (-1)^{((k-1)!(k+1)+1)}$$

$$\cdot \left[\sum_{m_k=-\infty}^{\infty} \left[H_0 \left(k_{1z} r_{00}^{00} \right) e^{-ik_z(\overline{z}_n)} \right] + \sum_{m=1}^{NS_y} (-1)^{(m-1)} \right.$$

$$\cdot \sum_{j=1}^{4} (-1)^{(j!+1)} \sum_{m_k=-\infty}^{\infty} \left[H_0 \left(k_{1z} r_{0j}^{0m} \right) e^{-ik_z(\overline{z}_n)} \right]$$
$$+ \sum_{l=1}^{NS_x} (-1)^{(l-1)} \sum_{i=1}^{4} (-1)^{((i-1)!(i+1)+1)}$$
$$\cdot \left[\sum_{m_k=-\infty}^{\infty} \left[H_0 \left(k_{1z} r_{i0}^{l0} \right) e^{-ik_z(\overline{z}_n)} \right] + \sum_{m=1}^{NS_y} (-1)^{(m-1)} \right.$$
$$\left. \cdot \sum_{j=1}^{4} (-1)^{(j!+1)} \sum_{m_k=-\infty}^{\infty} \left[H_0 \left(k_{1z} r_{ij}^{lm} \right) e^{-ik_z(\overline{z}_n)} \right] \right] \Bigg] \Bigg]$$

as the sum of 2.5D sources.

(54)

4. Conclusions

Fully analytical solutions for heat conduction for unbounded and rectangular spaces subjected to point, line, and plane sources have been presented. Two boundary conditions were assumed, namely, the Dirichlet and Neumann boundary conditions. Particular attention was given to the two-and-a-half-dimensional fundamental solution or 2.5D Green's functions defined for spatially sinusoidal, harmonic line sources. The final expressions were validated by applying the equations to the problem of a Dirac delta source, for which the solutions in the time domain are known in analytical form. Excellent agreement was found between the numerical solutions given by Fourier synthesis and the exact solutions.

Conflict of Interests

The authors declare that there is no conflict of interests regarding the publication of this paper.

Acknowledgments

This work has been framed under the Initiative Energy for Sustainability of the University of Coimbra and supported by the Energy and Mobility for Sustainable Regions (EMSURE) Project (CENTRO-07-0224-FEDER-002004) and was supported in part by the POCI-01-0247-FEDER-003179 (Revi Clean Facade) Project funded by Portugal 2020 through the Operational Programme for Competitiveness Factors (COMPETE 2020).

References

[1] H. S. Carslaw and J. C. Jaeger, *Conduction of Heat in Solids*, Oxford University Press, 2nd edition, 1959.

[2] J. V. Beck, K. D. Cole, A. Haji-Sheikh, and B. Litkouhi, *Heat Conduction using Green's Function*, Hemisphere Publishing, New York, NY, USA, 2nd edition, 2010.

[3] J. V. Beck, "Green's function solution for transient heat conduction problems," *International Journal of Heat and Mass Transfer*, vol. 27, no. 8, pp. 1235–1244, 1984.

[4] A. E. Joanni and E. Kausel, "Heat diffusion in layered media via the thin-layer method," *International Journal for Numerical Methods in Engineering*, vol. 78, no. 6, pp. 692–712, 2009.

[5] A. Tadeu, I. Simões, N. Simões, and J. Prata, "Simulation of dynamic linear thermal bridges using a boundary element method model in the frequency domain," *Energy and Buildings*, vol. 43, no. 12, pp. 3685–3695, 2011.

[6] A. Tadeu and N. Simões, "Three-dimensional fundamental solutions for transient heat transfer by conduction in an unbounded medium, half-space, slab and layered media," *Engineering Analysis with Boundary Elements*, vol. 30, no. 5, pp. 338–349, 2006.

[7] A. Tadeu, I. Simões, J. António, and L. Sousa, "Simulation of the 3D sound pressure level inside closed absorbing acoustic rooms bounded by non-parallel floor and ceiling surfaces, and parallel sidewalls," *Acta Acustica united with Acustica*, vol. 98, no. 6, pp. 894–906, 2012.

[8] J.-T. Chen, H.-C. Shieh, J.-J. Tsai, and J.-W. Lee, "A study on the method of fundamental solutions using an image concept," *Applied Mathematical Modelling*, vol. 34, no. 12, pp. 4253–4266, 2010.

[9] J.-T. Chen, H.-C. Shieh, Y.-T. Lee, and J.-W. Lee, "Image solutions for boundary value problems without sources," *Applied Mathematics and Computation*, vol. 216, no. 5, pp. 1453–1468, 2010.

[10] J.-T. Chen, Y.-T. Lee, S.-R. Yu, and S.-C. Shieh, "Equivalence between the Trefftz method and the method of fundamental solution for the annular Green's function using the addition theorem and image concept," *Engineering Analysis with Boundary Elements*, vol. 33, no. 5, pp. 678–688, 2009.

3

Uniqueness of Solutions to a Nonlinear Elliptic Hessian Equation

Siyuan Li

School of Mathematics and Applied Statistics, Faculty of Engineering and Information Sciences, University of Wollongong, Wollongong, NSW 2522, Australia

Correspondence should be addressed to Siyuan Li; sl296@uowmail.edu.au

Academic Editor: Carlos Conca

Through an Alexandrov-Fenchel inequality, we establish the general Brunn-Minkowski inequality. Then we obtain the uniqueness of solutions to a nonlinear elliptic Hessian equation on $\mathbb{S}^n$.

1. Introduction

According to a general Brunn-Minkowski inequality, we obtain a proof of the uniqueness of solutions to the following fully nonlinear elliptic Hessian equation:

$$\sigma_k\left(u_{ij}+u\delta_{ij}\right)=fu^{p-1} \quad \text{on } \mathbb{S}^n, \tag{1}$$

where u is the support function of convex bodies, u_{ij} are the second-order covariant derivations of u with respect to any orthonormal frame $\{e_1,e_2,\ldots,e_n\}$ on $\mathbb{S}^n$, δ_{ij} is the standard Kronecker symbol, $\mathbb{S}^n$ is the unit sphere of n-dimension, f is a positive function defined on $\mathbb{S}^n$, $k\in\{1,2,\ldots,n\}$, $p>1$, and σ_k is the *kth elementary symmetric function* defined as follows: for $\lambda=(\lambda_1,\lambda_2,\ldots,\lambda_n)\in\mathbb{R}^n$,

$$\sigma_k(\lambda)=\sum_{1\leqslant i_1<i_2<\cdots<i_k\leqslant n}\lambda_{i_1}\lambda_{i_2}\cdots\lambda_{i_k}. \tag{2}$$

The definition can be extended to any symmetric matrix $W\in\mathbb{R}^{n\times n}$ by $\sigma_k(W)=\sigma_k(\lambda(W))$, where $\lambda(W)=(\lambda_1(W),\lambda_2(W),\ldots,\lambda_n(W))$ is the eigenvalue vector of W.

Equation (1) arrives from the geometry of convex bodies. A compact convex subset of Euclidean $(n+1)$-space $\mathbb{R}^{n+1}$ with nonempty interiors is called a *convex body*. An important concept related to a convex body Q is its support function.

Definition 1. Let M (the boundary of a convex body Q) be a smooth, closed, uniformly convex hypersurface enclosing the origin in $\mathbb{R}^{n+1}$. Assume that M is parameterized by its inverse Gauss map $X:\mathbb{S}^n\to M\subset\mathbb{R}^{n+1}$; the *support function* u of M (or Q) is defined by

$$u(x)=\langle x,X(x)\rangle,\quad \forall x\in\mathbb{S}^n, \tag{3}$$

where $\langle\cdot,\cdot\rangle$ denotes the standard inner product in $\mathbb{R}^{n+1}$.

u is convex after being extended as a function of homogeneous degree 1 in $\mathbb{R}^{n+1}$. Conversely, any continuous convex function u of homogeneous degree 1 determines a convex body as follows:

$$Q=\left\{y\in\mathbb{R}^{n+1}: y\cdot x\leqslant u(x),\ \forall x\in\mathbb{S}^n\right\}. \tag{4}$$

From some basic concepts to support function, Minkowski sum [see Definition 4], and mixed volumes [see Definition 5], Minkowski developed a set of theories related to convex bodies. If $k=n$ and $p=1$, (1) is the Monge-Ampère equation corresponding to the classical Minkowski problem

$$\det\left(u_{ij}+u\delta_{ij}\right)=f \quad \text{on } \mathbb{S}^n, \tag{5}$$

which has been solved by Nirenberg [1], Pogorelov [2, 3], Cheng and Yau [4], and many others. When $p=1$, (1) is the classical Christoffel-Minkowski problem:

$$\sigma_k\left(u_{ij}+u\delta_{ij}\right)=f \quad \text{on } \mathbb{S}^n. \tag{6}$$

A necessary condition [3] for (6) to have a solution is

$$\int_{\mathbb{S}^n} x_i f(x)\, ds = 0, \quad \forall i = 1, 2, \ldots, n+1, \tag{7}$$

where ds is the standard area form on $\mathbb{S}^n$. Guan et al. [5] obtained that (7) is sufficient for (6) to have an admissible solution [see Definition 6].

Firey [6] generalized the Minkowski sum to p-sum [see Definition 4] from $p = 1$ to $p \geqslant 1$ in 1962. Later, Lutwak [7] extended the classical surface area measure to the p-sum cases. Also in [7], Lutwak first introduced the general Minkowski problem, which is called L_p-Minkowski problem thereafter. In the smooth category, L_p-Minkowski problem is equivalent to considering the following Monge-Ampère equation:

$$\det\left(u_{ij} + u\delta_{ij}\right) = fu^{p-1} \quad \text{on } \mathbb{S}^n. \tag{8}$$

The uniqueness of L_p-Minkowski problem for $p > 1$ and $p \neq n+1$ (the uniqueness holds up to a dilation if $p = n+1$) has been solved in [7]. However, the uniqueness for $p < 1$ is difficult and still open. In [8], Jian et al. obtained that, for any $-n-1 < p < 0$, there exists a positive function $f \in C^{\infty}(\mathbb{S}^n)$ to guarantee that (8) has two different solutions, which means that we need more conditions to consider the uniqueness.

When considering cases $1 \leqslant k < n$, attention is paid to the generalized Christoffel-Minkowski problem. In the smooth category, we need to study the k-Hessian equation (1).

For (1), Hu et al. [9] got the existence and uniqueness of solutions to (1) when $1 \leqslant k < n$ and $p > k+1$ under appropriate conditions. However, the uniqueness of (1) when $p < 1$ has not been solved well. In this paper, we study the uniqueness of (1) for $p > 1$.

Our main result is the following.

Theorem 2. *Suppose u is a positive admissible solution of*

$$\sigma_k\left(u_{ij} + u\delta_{ij}\right) = fu^{p_0} \quad \text{on } \mathbb{S}^n, \tag{9}$$

where $1 \leqslant k < n$, $k \in \mathbb{Z}$, $p_0 \in \mathbb{R}^+ \setminus \{k\}$, and f is a positive function defined on the unit sphere $\mathbb{S}^n$ and then the uniqueness holds. If $p_0 = k$, the uniqueness holds up to a dilation, which means that if u solves (9), then $\{au : \forall a \in \mathbb{R}^+\}$ are the whole solutions of (9).

Remark 3. Here, we rewrite (1) by (9), where $p_0 = p - 1$.

The organization of this paper is as follows. In Section 2, we show some basic concepts and lemmas which have been obtained by Guan et al. in [10]. In Section 3, we prove two useful propositions according to the methods in [11]. In the last section, we prove the main theorem.

2. Preliminaries

Definition 4. Given two convex bodies Q_1 and Q_2 in $\mathbb{R}^{n+1}$ with respective support functions u_1, u_2, and $\lambda, \mu \geqslant 0$ ($\lambda + \mu > 0$), the *Minkowski sum* $\lambda Q_1 + \mu Q_2 \subset \mathbb{R}^{n+1}$ is defined by the convex body whose support function is $\lambda u_1 + \mu u_2$.

For $p \geqslant 1$, let Q_1 and Q_2 be two convex bodies containing the origin in $\mathbb{R}^{n+1}$ in their interiors, and $\lambda, \mu \geqslant 0$ ($\lambda + \mu > 0$). The convex body $\lambda \circ Q_1 +_p \mu \circ Q_2$, whose support function is given by $(\lambda u_1^p + \mu u_2^p)^{1/p}$, is called *Firey's p-sum* of Q_1 and Q_2, where "$+_p$" means the p-summation and "$\circ$" means Firey's multiplication.

Definition 5. Let $Q_1, Q_2, \ldots, Q_r$ be convex bodies in $\mathbb{R}^{n+1}$ and the volume of their Minkowski sum

$$Q = \lambda_1 Q_1 + \lambda_2 Q_2 + \cdots + \lambda_r Q_r, \quad \lambda_i \geqslant 0, \tag{10}$$

is an $(n+1)$th degree homogeneous polynomial of the family $\lambda_1, \lambda_2, \ldots, \lambda_r$. Specially, the volume of Q is

$$\begin{aligned} \operatorname{Vol}(Q) &= \operatorname{Vol}\left(\lambda_1 Q_1 + \lambda_2 Q_2 + \cdots + \lambda_r Q_r\right) \\ &= \sum_{i_1, i_2, \ldots, i_{n+1}=1}^{r} \lambda_{i_1}\lambda_{i_2}\cdots\lambda_{i_{n+1}} V\left(Q_{i_1}, Q_{i_2}, \ldots, Q_{i_{n+1}}\right), \end{aligned} \tag{11}$$

where the functions V are symmetric. Then $V(Q_1, Q_2, \ldots, Q_{n+1})$ is called the *Minkowski mixed volume* of $Q_1, Q_2, \ldots, Q_{n+1}$.

Definition 6. For $k \in \{1, 2, \ldots, n\}$, let Γ_k be the *convex cone* in $\mathbb{R}^n$ which is determined by

$$\Gamma_k = \left\{\lambda \in \mathbb{R}^n : \sigma_1(\lambda) > 0, \sigma_2(\lambda) > 0, \ldots, \sigma_k(\lambda) > 0\right\}. \tag{12}$$

A function $u \in C^2(\mathbb{S}^n)$ is called *k-convex* if

$$W(x) = \left\{u_{ij}(x) + u(x)\delta_{ij}\right\} \in \Gamma_k, \quad \forall x \in \mathbb{S}^n, \tag{13}$$

and u is called an *admissible solution* to (1) if u is k-convex and satisfies (1).

Definition 7. Let $A_1, A_2, \ldots, A_m$ be symmetric real $k \times k$ matrices, $\lambda_1, \lambda_2, \ldots, \lambda_m \in \mathbb{R}$; the determinant of $\lambda_1 A_1 + \cdots + \lambda_m A_m$ is a homogeneous polynomial of degree k in $\lambda_1, \lambda_2, \ldots, \lambda_m$. Namely,

$$\begin{aligned} &\det\left(\lambda_1 A_1 + \cdots + \lambda_m A_m\right) \\ &\quad = \sum_{i_1, \ldots, i_k=1}^{m} \lambda_{i_1}\cdots\lambda_{i_k} D_k\left(A_{i_1}, \ldots, A_{i_k}\right). \end{aligned} \tag{14}$$

In fact, the coefficient $\lambda_{i_1}\cdots\lambda_{i_k}$ depends only on $A_{i_1}, \ldots, A_{i_k}$; then they are uniquely determined. $D_k(A_1, \ldots, A_k)$ is called the *mixed discriminant* of $A_1, \ldots, A_k$.

For later applications, we collect some results here which have been proved in [10].

Lemma 8. *Let $u_1, u_2, \ldots, u_{n+1}$ be the support function of convex bodies $Q_1, Q_2, \ldots, Q_{n+1}$, respectively. Denoting Minkowski mixed volume $V(Q_1, Q_2, \ldots, Q_{n+1})$ by $V(u_1, u_2, \ldots, u_{n+1})$ and*

$$W_m = \left\{(u_m)_{ij} + u_m\delta_{ij}\right\}, \quad m = 1, 2, \ldots, n+1, \tag{15}$$

then

$$V(u_1,u_2,\ldots,u_{n+1}) = \int_{\mathbb{S}^n} u_1 D_n(W_2,W_3,\ldots,W_{n+1})\,ds, \tag{16}$$

where $D_n(W_2,W_3,\ldots,W_{n+1})$ *is the mixed discriminant [see Definition 7] of* $W_2,W_3,\ldots,W_{n+1}$.

Remark 9. For all $1 \leqslant k \leqslant n$, setting $u_{k+2} = \cdots = u_{n+1} = 1$, then

$$V(u_1,\ldots,u_{k+1},1,\ldots,1) := V_{k+1}(u_1,u_2,\ldots,u_{k+1}) = \int_{\mathbb{S}^n} u_1 D_k(W_2,W_3,\ldots,W_{k+1})\,ds, \tag{17}$$

where $D_k(W_2,W_3,\ldots,W_{k+1})$ is the mixed discriminant of $W_2,W_3,\ldots,W_{k+1}$. Furthermore, if $u_1 = u_2 = \cdots = u_{n+1} = u$, denote $V(u_1,u_2,\ldots,u_{n+1}) := V(u)$ and $V_{k+1}(u_1,u_2,\ldots,u_{k+1}) := V_{k+1}(u)$; then

$$V(u) = \int_{\mathbb{S}^n} u \det\left(u_{ij} + u\delta_{ij}\right) ds, \qquad V_{k+1}(u) = \int_{\mathbb{S}^n} u\sigma_k\left(u_{ij} + u\delta_{ij}\right) ds. \tag{18}$$

Lemma 10. *V is a symmetric multilinear form on* $(C^2(\mathbb{S}^n))^{n+1}$.

Lemma 11. *For any function* $u \in C^2(\mathbb{S}^n)$, $W = \{u_{ij} + u\delta_{ij}\}$, $1 \leqslant k < n$, *we have the Minkowski type integral formula,*

$$\int_{\mathbb{S}^n} u\sigma_k(W)\,ds = \int_{\mathbb{S}^n} \sigma_{k+1}(W)\,ds, \tag{19}$$

where ds is the standard area element on $\mathbb{S}^n$.

The following is a form of Alexandrov-Fenchel inequality for positive k-convex functions which comes from [10].

Lemma 12 (Alexandrov-Fenchel inequality). *If* $u_1,u_2,\ldots,u_k$ *are k-convex,* u_1 *is positive, and there exists* $l \in \{2,3,\ldots,k\}$ *such that* $u_l \geqslant 0$ *on* $\mathbb{S}^n$, *then, for any* $v \in C^2(\mathbb{S}^n)$,

$$V_{k+1}^2(v,u_1,u_2,\ldots,u_k) \geqslant V_{k+1}(u_1,u_1,u_2,\ldots,u_k)\,V_{k+1}(v,v,u_2,\ldots,u_k), \tag{20}$$

with equality if and only if $v = au_1 + \sum_{i=1}^{n+1} a_i x_i$ *for some constants* $a,a_1,\ldots,a_{n+1}$.

3. Two Important Propositions

Now we prove two important propositions. The methods we use are from [11].

Proposition 13. *Suppose* $u_0,u_1 > 0$ *are k-convex; then*

$$V_{k+1}^{1/(k+1)}((1-t)u_0 + tu_1) \geqslant (1-t)V_{k+1}^{1/(k+1)}(u_0) + tV_{k+1}^{1/(k+1)}(u_1), \quad \forall t \in [0,1], \tag{21}$$

with equality if and only if $u_0 = au_1 + \sum_{i=1}^{n+1} a_i x_i$ *for some constants* $a,a_1,\ldots,a_{n+1}$.

Proof. We only need to prove that

$$F(t) = V_{k+1}^{1/(k+1)}((1-t)u_0 + tu_1) \tag{22}$$

is concave on $[0,1]$. Setting $u_t = (1-t)u_0 + tu_1$, $t \in [0,1]$, we have

$$F(t) = V_{k+1}^{1/(k+1)}\Big(\overbrace{u_t,u_t,\ldots,u_t}^{k+1}\Big). \tag{23}$$

By the symmetric multilinear property of V, it is obvious that

$$F'(t) = V_{k+1}^{1/(k+1)-1}\Big(\overbrace{u_t,\ldots,u_t}^{k+1}\Big)V_{k+1}\Big(-u_0 + u_1,\overbrace{u_t,\ldots,u_t}^{k}\Big), \tag{24}$$

$$\begin{aligned}F''(t) = {}& kV_{k+1}^{1/(k+1)-2}\Big(\overbrace{u_t,\ldots,u_t}^{k+1}\Big) \cdot \Big[V_{k+1}\Big(\overbrace{u_t,\ldots,u_t}^{k+1}\Big) \cdot V_{k+1}\Big(-u_0 + u_1,-u_0 + u_1,\overbrace{u_t,\ldots,u_t}^{k-1}\Big) \\ & - V_{k+1}^2\Big(-u_0 + u_1,\overbrace{u_t,\ldots,u_t}^{k}\Big)\Big] \leqslant 0,\end{aligned} \tag{25}$$

where the last inequality uses (20); thus F is a concave function on $[0,1]$. The equality condition is checked easily. ☐

Proposition 14 (general Brunn-Minkowski inequality). *Supposing* $u_0,u_1 > 0$ *are k-convex, then*

$$\int_{\mathbb{S}^n} u_1\sigma_k\left((u_0)_{ij} + u_0\delta_{ij}\right) ds \geqslant V_{k+1}^{1/(k+1)}(u_1)\,V_{k+1}^{1-1/(k+1)}(u_0), \tag{26}$$

with equality if and only if $u_0 = au_1 + \sum_{i=1}^{n+1} a_i x_i$ *for some constants* $a,a_1,\ldots,a_{n+1}$.

Proof. Setting

$$F(t) = V_{k+1}^{1/(k+1)}((1-t)u_0 + tu_1) - (1-t)V_{k+1}^{1/(k+1)}(u_0) - tV_{k+1}^{1/(k+1)}(u_1), \tag{27}$$

then $F(0) = F(1) = 0$. By (21), $F(t) \geqslant 0$; thus $F'(0) \geqslant 0$; namely,

$$V_{k+1}^{1/(k+1)-1}(u_0)\,V_{k+1}\Big(-u_0 + u_1,\overbrace{u_0,\ldots,u_0}^{k}\Big) + V_{k+1}^{1/(k+1)}(u_0) - V_{k+1}^{1/(k+1)}(u_1) \geqslant 0. \tag{28}$$

Then

$$V_{k+1}^{1/(k+1)-1}(u_0)\int_{\mathbb{S}^n}(-u_0+u_1)\sigma_k\left((u_0)_{ij}+u_0\delta_{ij}\right)ds + V_{k+1}^{1/(k+1)}(u_0) \geqslant V_{k+1}^{1/(k+1)}(u_1). \quad (29)$$

By (19),

$$V_{k+1}^{1/(k+1)-1}(u_0)\int_{\mathbb{S}^n}u_1\sigma_k\left((u_0)_{ij}+u_0\delta_{ij}\right)ds \geqslant V_{k+1}^{1/(k+1)}(u_1), \quad (30)$$

and then

$$\int_{\mathbb{S}^n}u_1\sigma_k\left((u_0)_{ij}+u_0\delta_{ij}\right)ds \geqslant V_{k+1}^{1/(k+1)}(u_1)\,V_{k+1}^{1-1/(k+1)}(u_0). \quad (31)$$

□

4. Proof of Theorem 2

Now we prove Theorem 2. The main methods are from [7, 12].

Proof. Assuming that (9) has two solutions u and v, then we consider the equation in the following three cases.

Case 1 ($p_0 > k$). Supposing x_0 is the maximum value point of $G = u/v$, then at x_0, we have

$$\begin{aligned} 0 &= \nabla \ln G = \frac{\nabla u}{u} - \frac{\nabla v}{v}, \\ 0 &\geqslant \nabla^2 \ln G = \left(\frac{\nabla^2 u}{u} - \frac{(\nabla u)^2}{u^2}\right) - \left(\frac{\nabla^2 v}{v} - \frac{(\nabla v)^2}{v^2}\right) \\ &= \frac{\nabla^2 u}{u} - \frac{\nabla^2 v}{v}; \end{aligned} \quad (32)$$

that is,

$$\frac{\nabla^2 u}{u} \leqslant \frac{\nabla^2 v}{v}. \quad (33)$$

Hence

$$\begin{aligned} fu^{p_0}(x_0) &= u^k(x_0)\,\sigma_k\left(\frac{u_{ij}}{u}+\delta_{ij}\right)(x_0) \\ &\leqslant u^k(x_0)\,\sigma_k\left(\frac{v_{ij}}{v}+\delta_{ij}\right)(x_0) \\ &= \frac{u^k(x_0)}{v^k(x_0)} f v^{p_0}(x_0); \end{aligned} \quad (34)$$

therefore

$$u^{p_0-k}(x_0) \leqslant v^{p_0-k}(x_0) \Longrightarrow G(x_0) = \frac{u(x_0)}{v(x_0)} \leqslant 1; \quad (35)$$

then

$$\frac{u}{v} \leqslant 1. \quad (36)$$

Similarly, we have $v/u \leqslant 1$. Thus $u \equiv v$.

Case 2 ($0 < p_0 < k$). We have

$$u^{-p_0}\sigma_k\left(u_{ij}+u\delta_{ij}\right) = v^{-p_0}\sigma_k\left(v_{ij}+v\delta_{ij}\right); \quad (37)$$

then

$$\begin{aligned} V_{k+1}(u) &= \int_{\mathbb{S}^n} u\sigma_k\left(u_{ij}+u\delta_{ij}\right)ds \\ &= \int_{\mathbb{S}^n}\left(\frac{u}{v}\right)^{p_0+1} v\sigma_k\left(v_{ij}+v\delta_{ij}\right)ds \\ &\geqslant \left[\int_{\mathbb{S}^n} u\sigma_k\left(v_{ij}+v\delta_{ij}\right)ds\right]^{p_0+1} \\ &\cdot \left[\int_{\mathbb{S}^n} v\sigma_k\left(v_{ij}+v\delta_{ij}\right)ds\right]^{-p_0} \geqslant V_{k+1}^{(p_0+1)/(k+1)}(u) \\ &\cdot V_{k+1}^{(kp_0+k)/(k+1)}(v)\,V_{k+1}^{-p_0}(v) = V_{k+1}^{(p_0+1)/(k+1)}(u) \\ &\cdot V_{k+1}^{1-(p_0+1)/(k+1)}(v), \end{aligned} \quad (38)$$

where we have used Hölder inequality in the first inequality and used (26) in the second one. Hence $V_{k+1}(u) = V_{k+1}(v)$, which forces both the equalities to hold. By the equality condition, there exists a constant $a \in \mathbb{R}$ such that $v = au$. By (9), we know $a = 1$. Therefore, $u \equiv v$.

Case 3 ($p_0 = k$). According to Case 2, when $p_0 = k$, we have

$$\begin{aligned} V_{k+1}(u) &= \int_{\mathbb{S}^n} u\sigma_k\left(u_{ij}+u\delta_{ij}\right)ds = \int_{\mathbb{S}^n}\left(\frac{u}{v}\right)^{k+1} \\ &\cdot v\sigma_k\left(v_{ij}+v\delta_{ij}\right)ds \geqslant \left[\int_{\mathbb{S}^n} u\sigma_k\left(v_{ij}+v\delta_{ij}\right)ds\right]^{k+1} \\ &\cdot \left[\int_{\mathbb{S}^n} v\sigma_k\left(v_{ij}+v\delta_{ij}\right)ds\right]^{-k} \geqslant V_{k+1}(u)\,V_{k+1}^{k}(v) \\ &\cdot V_{k+1}^{-k}(v) = V_{k+1}(u); \end{aligned} \quad (39)$$

then all the equalities hold. Thus there exists $a \in \mathbb{R}$, such that $v = au$. Therefore $\{au : \forall a \in \mathbb{R}^+\}$ are the whole solutions of (9).

Now we complete the proof of Theorem 2. □

Competing Interests

The author declares no competing interests.

References

[1] L. Nirenberg, "The Weyl and Minkowski problems in differential geometry in the large," *Communications on Pure and Applied Mathematics*, vol. 6, no. 3, pp. 337–394, 1953.

[2] A. V. Pogorelov, "On existence of a convex surface with a given sum principal radii of curvature," *Uspekhi Matematicheskikh Nauk*, vol. 8, no. 3, pp. 127–130, 1953 (Russian).

[3] A. V. Pogorelov, *The Multidimensional Minkowski Problem*, Winston, Washington, DC, USA, 1978.

[4] S. Y. Cheng and S. T. Yau, "On the regularity of the solution of the n-dimensional Minkowski problem," *Communications on Pure and Applied Mathematics*, vol. 29, no. 5, pp. 495–516, 1976.

[5] P. Guan, X.-N. Ma, and F. Zhou, "The Christofel-Minkowski problem. III. Existence and convexity of admissible solutions," *Communications on Pure and Applied Mathematics*, vol. 59, no. 9, pp. 1352–1376, 2006.

[6] W. J. Firey, "p-means of convex bodies," *Mathematica Scandinavica*, vol. 10, pp. 17–24, 1962.

[7] E. Lutwak, "The Brunn-Minkowski-Firey theory I: mixed volumes and the Minkowski problem," *Journal of Differential Geometry*, vol. 38, no. 1, pp. 131–150, 1993.

[8] H. Jian, J. Lu, and X.-J. Wang, "Nonuniqueness of solutions to the L_p-Minkowski problem," *Advances in Mathematics*, vol. 281, pp. 845–856, 2015.

[9] C. Q. Hu, X.-N. Ma, and C. Shen, "On the Christoffel-Minkowski problem of Firey's p-sum," *Calculus of Variations and Partial Differential Equations*, vol. 21, no. 2, pp. 137–155, 2004.

[10] P. F. Guan, X.-N. Ma, N. Trudinger, and X. Zhu, "A form of Alexandrov-Fenchel inequality," *Pure and Applied Mathematics Quarterly*, vol. 6, no. 4, pp. 999–1012, 2010.

[11] R. Schneider, *Convex Bodies: the Brunn-Minkowski Theory*, Cambridge University Press, Cambridge, UK, 2013.

[12] K.-S. Chou and X.-J. Wang, "The L_p-Minkowski problem and the Minkowski problem in centroaffine geometry," *Advances in Mathematics*, vol. 205, no. 1, pp. 33–83, 2006.

4

Some Properties of the Strong Primitivity of Nonnegative Tensors

Lihua You, Yafei Chen, and Pingzhi Yuan

School of Mathematical Sciences, South China Normal University, Guangzhou 510631, China

Correspondence should be addressed to Lihua You; ylhua@scnu.edu.cn and Pingzhi Yuan; yuanpz@scnu.edu.cn

Academic Editor: Ali R. Ashrafi

We show that an order m dimension 2 tensor is primitive if and only if its majorization matrix is primitive, and then we obtain the characterization of order m dimension 2 strongly primitive tensors and the bound of the strongly primitive degree. Furthermore, we study the properties of strongly primitive tensors with $n \geq 3$ and propose some problems for further research.

1. Introduction

In recent years, the study of tensors and the spectra of tensors (and hypergraphs) with their various applications has attracted extensive attention and interest, since the work of L. Qi ([1]) and L.H. Lim ([2]) in 2005.

As is in [1], an order m dimension n tensor $\mathbb{A} = (a_{i_1i_2\ldots i_m})_{1\leq i_j\leq n\ (j=1,\ldots,m)}$ over the complex field $\mathbb{C}$ is a multidimensional array with all entries

$$a_{i_1i_2\cdots i_m} \in \mathbb{C}\left(i_1,\ldots,i_m \in [n] = \{1,\ldots,n\}\right). \tag{1}$$

A tensor $\mathbb{A} = (a_{i_1i_2\ldots i_m})$ is called a nonnegative tensor if all of its entries $a_{i_1i_2\ldots i_m}$ are nonnegative. Clearly, adjacency tensors and signless Laplacian tensors are nonnegative.

In the theory of nonnegative matrices, the notion of primitivity plays an important role in the convergence of the Collatz method. For a nonnegative matrix A, the following are equivalent [3]:

(1) Let $\rho(A)$ be the spectral radius of A. Then A is irreducible and $\rho(A)$ is greater than any other eigenvalue in modulus.

(2) The only A-invariant nonempty subset of the boundary of the positive cone is $\{0\}$.

(3) There exists a natural number r such that A^r is positive.

Matrices which satisfy any of the above conditions are called primitive. The least such r such that A^r is positive is called the primitive exponent (or simply, exponent) of A and is denoted by $\exp(A)$.

In [4], K.C. Chang et al. defined the primitivity of nonnegative tensors (as Definition 1), extended the theory of nonnegative matrices to nonnegative tensors, and proved the convergence of the NQZ method which is an extension of the Collatz method and can be used to find the largest eigenvalue of any nonnegative irreducible tensor.

Definition 1 (see [4]). Let $\mathbb{A}$ be a nonnegative tensor with order m and dimension n, $x = (x_1, x_2, \ldots, x_n)^T \in \mathbb{R}^n$ a vector, and $x^{[r]} = (x_1^r, x_2^r, \ldots, x_n^r)^T$. Define the map $T_{\mathbb{A}}$ from $\mathbb{R}^n$ to $\mathbb{R}^n$ as $T_{\mathbb{A}}(x) = (\mathbb{A}x^{m-1})^{[1/(m-1)]}$. If there exists some positive integer r such that $T_{\mathbb{A}}^r(x) > 0$ for all nonnegative nonzero vectors $x \in \mathbb{R}^n$, then $\mathbb{A}$ is called primitive and the smallest such integer r is called the primitive degree of $\mathbb{A}$, denoted by $\gamma(\mathbb{A})$.

As in [1], let $\mathbb{A} = (a_{i_1i_2\ldots i_m})_{1\leq i_j\leq n\ (j=1,\ldots,m)}$ be an order m dimension n tensor over the complex field $\mathbb{C}$, $x = (x_1, \ldots, x_n)^T \in \mathbb{C}_n$, $x^{[r]} = (x_1^r, x_2^r, \ldots, x_n^r)$, and $\mathbb{A}x^{m-1}$ be a vector $\in \mathbb{C}_n$ whose ith component is defined as follows:

$$\left(\mathbb{A}x^{m-1}\right)_i = \sum_{i_2,\ldots,i_m=1}^{n} a_{ii_2\cdots i_m} x_{i_2} \cdots x_{i_m}. \tag{2}$$

Then a number $\lambda \in \mathbb{C}$ is called an eigenvalue of $\mathbb{A}$ if there exists a nonzero vector $x \in \mathbb{C}_n$ such that

$$\mathbb{A}x^{m-1} = \lambda x^{[m-1]}. \tag{3}$$

Recently, Shao [5] defined the general product of two n-dimensional tensors as follows, and one of the applications of the tensor product is that $\mathbb{A}x^{m-1}$ can be simply written as $\mathbb{A} \cdot x$.

Definition 2 (see [5]). Let $\mathbb{A}$ (and $\mathbb{B}$) be an order $m \geq 2$ (and $k \geq 1$), dimension n tensor, respectively. Define the general product $\mathbb{A} \cdot \mathbb{B}$ (sometimes simplified as $\mathbb{AB}$), to be the following tensor $\mathbb{D}$ of order $(m-1)(k-1)+1$ and dimension n:

$$d_{i\alpha_1\cdots\alpha_{m-1}} = \sum_{i_2,\dots,i_m=1}^{n} a_{ii_2\cdots i_m} b_{i_2\alpha_1} \cdots b_{i_m\alpha_{m-1}} \quad (4)$$

$$\left(i \in [n],\ \alpha_1, \dots, \alpha_{m-1} \in [n]^{k-1}\right).$$

The tensor product is a generalization of the usual matrix product and satisfies a very useful property: the associative law ([5], Theorem 1.1). By the associative law, we can define $\mathbb{A}^k$ as the product of k many tensors $\mathbb{A}$.

With the general product, when $k = 1$ and $\mathbb{B} = x = (x_1, \dots, x_n)^T \in \mathbb{C}^n$ is a vector of dimension n, then $\mathbb{A} \cdot \mathbb{B} = \mathbb{A} \cdot x$ is still a vector of dimension n, and for any $i \in [n]$, $(\mathbb{A} \cdot \mathbb{B})_i = (\mathbb{A} \cdot x)_i = \sum_{i_2,\dots,i_m=1}^{n} a_{ii_2\dots i_m} x_{i_2} \cdots x_{i_m} = (\mathbb{A}x^{m-1})_i$, where $\mathbb{A}x^{m-1}$ is defined in (2). Thus we have $\mathbb{A}x^{m-1} = \mathbb{A} \cdot x$.

In order to study eigenvalue, Pearson defined "essentially positive" tensors as Definition 3. By the general product of tensors, Shao obtained Proposition 4 and Definition 5 which is equivalent to Definition 3.

Definition 3 (see [6], Definition 3.1). A nonnegative tensor $\mathbb{A}$ is called essentially positive, if, for any nonnegative nonzero vector $x \in \mathbb{R}^n$, $\mathbb{A} \cdot x > 0$ holds.

Proposition 4 (see [5], Proposition 4.1). *Let $\mathbb{A}$ be an order m and dimension n nonnegative tensor. Then the following three conditions are equivalent:*

(1) For any $i, j \in [n]$, $a_{ij\dots j} > 0$ holds.

(2) For any $j \in [n]$, $\mathbb{A} \cdot e_j > 0$ holds (where e_j is the j-th column of the identity matrix I_n).

(3) For any nonnegative nonzero vector $x \in \mathbb{R}^n$, $\mathbb{A} \cdot x > 0$ holds.

Definition 5 (see [5], Definition 4.1). A nonnegative tensor $\mathbb{A}$ is called essentially positive, if it satisfies one of the three conditions in Proposition 4.

Based on the above arguments and the zero patterns defined by Shao in [5], Shao showed a characterization of primitive tensors and defined the primitive degree as follows.

Proposition 6 (see [5], Theorem 4.1). *A nonnegative tensor $\mathbb{A}$ is primitive if and only if there exists some positive integer r such that $\mathbb{A}^r$ is essentially positive. Furthermore, the smallest such r is the primitive degree of $\mathbb{A}$, $\gamma(\mathbb{A})$.*

The concept of the majorization matrix of a tensor introduced by Pearson is very useful.

Definition 7 (see [6], Definition 2.1). The majorization matrix $M(\mathbb{A})$ of the tensor $\mathbb{A}$ is defined as $(M(\mathbb{A}))_{ij} = a_{ij\dots j}$ for $i, j \in [n]$.

By Definition 5, Proposition 6, and Definition 7, the following characterization of the primitive tensors was easily obtained.

Proposition 8 (see [7], Remark 2.6). *Let $\mathbb{A}$ be a nonnegative tensor with order m and dimension n. Then $\mathbb{A}$ is primitive if and only if there exists some positive integer r such that $M(\mathbb{A}^r) > 0$. Furthermore, the smallest such r is the primitive degree of $\mathbb{A}$, $\gamma(\mathbb{A})$.*

On the primitive degree $\gamma(\mathbb{A})$, Shao proposed the following conjecture for further research.

Conjecture 9 (see [5], Conjecture 1). *When m is fixed, then there exists some polynomial $f(n)$ on n such that $\gamma(\mathbb{A}) \leq f(n)$ for all nonnegative primitive tensors of order m and dimension n.*

In the case of $m = 2$ ($\mathbb{A}$ is a matrix), the well-known Wielandt upper bound tells us that we can take $f(n) = (n-1)^2 + 1$. Recently, the authors [7] confirmed Conjecture 9 by proving Theorem 10.

Theorem 10 (see [7], Theorem 1.2). *Let $\mathbb{A}$ be a nonnegative primitive tensor with order m and dimension n. Then its primitive degree $\gamma(\mathbb{A}) \leq (n-1)^2 + 1$, and the upper bound is tight.*

They also showed that there are no gaps in the tensor case in [8], which implies that the result of the case $m \geq 3$ is totally different from the case $m = 2$ ($\mathbb{A}$ is a matrix). In [5], Shao also proposed the concept of strongly primitive tensor for further research.

Definition 11 (see [5], Definition 4.3). Let $\mathbb{A}$ be a nonnegative tensor with order m and dimension n. If there exists some positive integer k such that $\mathbb{A}^k > 0$ is a positive tensor, then $\mathbb{A}$ is called strongly primitive, and the smallest such k is called the strongly primitive degree of $\mathbb{A}$.

Let $\mathbb{A} = (a_{i_1 i_2 \dots i_m})$ be a nonnegative tensor with order m and dimension n. It is clear that if $\mathbb{A}$ is strongly primitive, then $\mathbb{A}$ is primitive. For convenience, let $\eta(\mathbb{A})$ be the strongly primitive degree of $\mathbb{A}$. Clearly, $\gamma(\mathbb{A}) \leq \eta(\mathbb{A})$. In fact, it is obvious that, in the matrix case ($m = 2$), a nonnegative matrix A is primitive if and only if A is strongly primitive, and $\gamma(A) = \eta(A) = \exp(A)$. But in the case $m \geq 3$ Shao gave an example to show that these two concepts are not equivalent. In [8], the authors proposed the following question.

Question 12 ([8], Question 4.18). Can we define and study the strongly primitive degree, the strongly primitive degree set, the j-strongly primitive degree of strongly primitive tensors and so on?

Based on Question 12, we study primitive tensors and strongly primitive tensors in this paper, show that an order m dimension 2 tensor is primitive if and only if its majorization matrix is primitive, and obtain the characterization of order m dimension 2 strongly primitive tensors and the bound of the strongly primitive degree. Furthermore, we study the properties of strongly primitive tensors with $n \geq 3$ and propose some problems for further research.

2. Preliminaries

In [8], the authors obtained the following Proposition 13 and gave Example 15 by computing the strongly primitive degree.

Proposition 13 ([8], Proposition 4.16). *Let $\mathbb{A} = (a_{i_1 i_2 \ldots i_m})$ be a nonnegative strongly primitive tensor with order m and dimension n. Then for any $\alpha \in [n]^{m-1}$, there exists some $i \in [n]$ such that $a_{i\alpha} > 0$.*

Let $k(\geq 0)$, $n(\geq 2)$, $q(\geq 0)$, $r(\geq 1)$ be integers and $k = (n-1)q + r$ with $1 \leq r \leq n-1$ when $k \geq 1$. In [7–9], the authors defined some nonnegative tensors with order m and dimension n as follows:

$$\mathbb{A}_k = \left(a^{[k]}_{i_1 i_2 \cdots i_m}\right)_{1 \leq i_j \leq n\ (j=1,\ldots,m)}, \tag{5}$$

where one has the following:

(1)

$$M(\mathbb{A}_k) = M_1 = \begin{pmatrix} 0 & 0 & \cdots & 0 & 1 & 1 \\ 1 & 0 & \cdots & 0 & 0 & 0 \\ 0 & 1 & \cdots & 0 & 0 & 0 \\ \vdots & \vdots & \ddots & \vdots & \vdots & \vdots \\ 0 & 0 & \cdots & 1 & 0 & 0 \\ 0 & 0 & \cdots & 0 & 1 & 0 \end{pmatrix}. \tag{6}$$

(2) $a^{[0]}_{ii_2 \ldots i_m} = 0$, if $i_2 \ldots i_m \neq i_2 \ldots i_2$ for any $i \in [n]$.

(3) $a^{[k]}_{i\alpha} = 1$, if $i \in [n] \backslash \{r-q, r-q+1, \ldots, r, r+1\} \pmod n$ and $\alpha = i_2 \ldots i_m \in [n]^{m-1}$ with $\{i_2, \ldots, i_m\} = \{r-q-1, r\} \pmod n$.

(4) $a^{[k]}_{i_1 i_2 \ldots i_m} = 0$, except for (1) and (3).

The authors showed the tensors $\mathbb{A}_k$ $(k \geq 0)$ are primitive, the primitive degree $\gamma(\mathbb{A}_0) = (n-1)^2 + 1$ ([7]) and $\gamma(\mathbb{A}_k) = k+n$ ([8], Theorem 3.3) for $1 \leq k \leq n^2 - 3n + 2$.

Remark 14. It is clear that, for any $\mathbb{A}_k$ $(0 \leq k \leq n^2 - 3n + 2)$, there exists some $\alpha \in [n]^{m-1}$, for any $i \in [n]$, $a^{[k]}_{i\alpha} = 0$. Thus, for each $0 \leq k \leq n^2 - 3n + 2$, $\mathbb{A}_k$ is not a strongly primitive tensor by Proposition 13.

Example 15 ([8], Example 4.17). Let $m = n = 3$, and let $\mathbb{A} = (a_{i_1 i_2 i_3})$ be a nonnegative tensor with order m and dimension n, where $a_{111} = a_{222} = a_{333} = a_{233} = a_{311} = 0$ and other $a_{i_1 i_2 i_3} = 1$. Then $\eta(\mathbb{A}) = 4$.

Remark 16. In fact, we can obtain $\gamma(\mathbb{A}) = \eta(\mathbb{A}) = 4$ because of $a^{(3)}_{233333333} = 0$, where $\mathbb{A}^3 = (a^{(3)}_{i_1 i_2 \ldots i_9})$.

In the computation of Example 15, we note that the following equation is useful and will be used repeatedly. It is not difficult to obtain the equation by the general product of two n-dimensional tensors which is defined in Definition 1.2 in [5].

Let $\mathbb{A}$ be a nonnegative primitive tensor with order m and dimension n, $\alpha_2, \ldots, \alpha_m \in [n]^{(m-1)^{k-1}}$. Then we have

$$\left(\mathbb{A}^k\right)_{i\alpha_2 \cdots \alpha_m} = \sum_{i_2, i_3, \ldots, i_m = 1}^{n} a_{ii_2 i_3 \cdots i_m} \left(\mathbb{A}^{k-1}\right)_{i_2 \alpha_2} \cdots \left(\mathbb{A}^{k-1}\right)_{i_m \alpha_m}. \tag{7}$$

Proposition 17 (see [7], Proposition 2.7). *Let $\mathbb{A}$ be a nonnegative primitive tensor with order m and dimension n and $M(\mathbb{A})$ be the majorization matrix of $\mathbb{A}$. Then we have the following:*

(1) For each $j \in [n]$, there exists an integer $i \in [n] \backslash \{j\}$ such that $(M(\mathbb{A}))_{ij} > 0$.

(2) There exist some $j \in [n]$ and integers u, v with $1 \leq u < v \leq n$ such that $(M(\mathbb{A}))_{uj} > 0$ and $(M(\mathbb{A}))_{vj} > 0$.

Let $\alpha = jj \ldots j \in [n]^{m-1}$; then $M(\mathbb{A})_{ij} = a_{i\alpha}$. We can see that Proposition 13 is the generalization of result (1) of Proposition 17 from a primitive tensor to a strongly primitive tensor. We note that Proposition 17 played an important role in [7], and if $\mathbb{A}$ is a nonnegative strongly primitive tensor, then $\mathbb{A}$ must be a nonnegative primitive tensor; thus result (2) of Proposition 17 also holds for nonnegative strongly primitive tensors.

Proposition 18. *Let $\mathbb{A} = (a_{i_1 i_2 \ldots i_m})$ be a nonnegative strongly primitive tensor with order m and dimension n. Then there exists at least one $j \in [n]$ and integers u, v with $1 \leq u < v \leq n$ such that $(M(\mathbb{A}))_{uj} > 0$ and $(M(\mathbb{A}))_{vj} > 0$.*

Proposition 19. *Let $\mathbb{A} = (a_{i_1 i_2 \ldots i_m})_{1 \leq i_j \leq n (j=1,\ldots,m)}$ be a nonnegative tensor with order m and dimension n and $\mathbb{A} \neq \mathbb{J}$. For given $i \in [n]$, if $a_{i\alpha} = a_{jii\ldots i} = 1$ for any $\alpha \in [n]^{m-1}$ and any $j \in [n] \backslash \{i\}$, then $\mathbb{A}$ is strongly primitive with $\eta(\mathbb{A}) = 2$.*

Proof. By (7), for any $k \in [n]$ and $\alpha_2, \ldots, \alpha_m \in [n]^{m-1}$, we have $(\mathbb{A}^2)_{k\alpha_2 \ldots \alpha_m} = \sum_{k_2, k_3, \ldots, k_m = 1}^{n} a_{kk_2 \ldots k_m} a_{k_2 \alpha_2} \cdots a_{k_m \alpha_m} \geq a_{kii\ldots i} a_{i\alpha_2} \cdots a_{i\alpha_m} = 1$, which implies $\mathbb{A}$ is strongly primitive and $\eta(\mathbb{A}) = 2$. □

Remark 20. From Proposition 19, we can see the following:

(1) There exist at least $n(2^{(n-1)(n^{m-1}-1)} - 1)$ strongly primitive tensors such that their strongly primitive degree is equal to 2.

(2) We cannot improve the result of Proposition 13 any more by the fact that there exists $i \in [n]$ such that $a_{i\alpha} = 1 > 0$ for any $\alpha \in [n]^{m-1}$ and there is exactly one i such that $a_{i\alpha} > 0$ for any $\alpha \neq ii \ldots i$.

(3) Similarly, we cannot improve the result of Proposition 18 any more by the fact that there is exactly one $i \in [n]$ such that $(M(\mathbb{A}))_{ui} > 0$ for any $u \in [n]$ and for any other $j \in [n] \backslash \{i\}$, there exists only $i \in [n]$ such that $(M(\mathbb{A}))_{ij} > 0$.

(4) What is more, combining the above arguments, we know whether a nonnegative tensor is a nonnegative strongly

primitive tensor or not, and the values of the strongly primitive degree of a nonnegative strongly primitive tensor do not depend on the number of nonzero entries but the positions of the nonzero entries.

Proposition 21. *Let $\mathbb{A} = (a_{i_1 i_2 \dots i_m})$ be a nonnegative strongly primitive tensor with order m and dimension n. Then for any $i \in [n]$, there exists some $\alpha \in [n]^{m-1}$ such that $a_{i\alpha} > 0$.*

Proof. Since $\mathbb{A}$ is strongly primitive, there exists some $k > 0$ such that $\mathbb{A}^k > 0$ by Definition 1. Assume that there exists some $i \in [n]$ such that $a_{i\alpha} = 0$ for any $\alpha \in [n]^{m-1}$. Then by (7), we have

$$\left(\mathbb{A}^k\right)_{i\alpha_2 \cdots \alpha_m} = \sum_{i_2, i_3, \dots, i_m = 1}^{n} a_{i i_2 i_3 \cdots i_m} \left(\mathbb{A}^{k-1}\right)_{i_2 \alpha_2} \cdots \left(\mathbb{A}^{k-1}\right)_{i_m \alpha_m} = 0, \tag{8}$$

which leads to a contraction. □

Remark 22. Let $\mathbb{A} = (a_{i_1 i_2 \dots i_m})_{1 \le i_j \le n (j = 1, \dots, m)}$ be a nonnegative tensor with order m and dimension n. For given $i \in [n]$, we take $a_{i\alpha} = a_{jii \dots i} = 1$ for any $\alpha \in [n]^{m-1}$ and any $j \in [n] \backslash \{i\}$ and any other entry $a_{i_1 i_2 \dots i_m} = 0$. Then $\mathbb{A}$ is strongly primitive with $\eta(\mathbb{A}) = 2$ by Proposition 19. This implies that we cannot improve the result of Proposition 21 anymore, and it indicates the importance of the positions of the nonzero entries again.

Proposition 23. *Let $\mathbb{A}$ be a nonnegative strongly primitive tensor and $k = \eta(\mathbb{A})$. Then, for any integer $t > k > 0$, we have $\mathbb{A}^t > 0$.*

Proof. It is clear that $\mathbb{A}^k > 0$ by $k = \eta(\mathbb{A})$. We only need to show $\mathbb{A}^{k+1} > 0$; say, for any $i \in [n]$ and any $\alpha_2, \dots, \alpha_m \in [n]^{(m-1)^k}$, we show $(\mathbb{A}^{k+1})_{i\alpha_2 \dots \alpha_m} > 0$.

By Proposition 21, there exists some $\alpha = j_2 j_3 \cdots j_m \in [n]^{m-1}$ such that $a_{i\alpha} = a_{ij_2 \dots j_m} > 0$. By $\mathbb{A}^k > 0$ we have $(\mathbb{A}^k)_{j_2 \alpha_2} > 0, \dots, (\mathbb{A}^k)_{j_m \alpha_m} > 0$; then by (7), we have

$$\begin{aligned} \left(\mathbb{A}^{k+1}\right)_{i\alpha_2 \cdots \alpha_m} &= \sum_{i_2, i_3, \dots, i_m = 1}^{n} a_{i i_2 i_3 \cdots i_m} \left(\mathbb{A}^{k}\right)_{i_2 \alpha_2} \left(\mathbb{A}^{k}\right)_{i_3 \alpha_3} \cdots \left(\mathbb{A}^{k}\right)_{i_m \alpha_m} \\ &\ge a_{i j_2 j_3 \cdots j_m} \left(\mathbb{A}^{k}\right)_{j_2 \alpha_2} \cdots \left(\mathbb{A}^{k}\right)_{j_m \alpha_m} > 0. \end{aligned} \tag{9}$$

□

Proposition 24. *Let $\mathbb{A}$ be a nonnegative tensor with order m and dimension n and t be a positive integer. Then $\mathbb{A}$ is strongly primitive if and only if $\mathbb{A}^t$ is strongly primitive.*

Proof. Firstly, the sufficiency is obvious. Now we show the necessity. Let $k = \eta(\mathbb{A})$. Then $\mathbb{A}^k > 0$ by $\mathbb{A}$ is strongly primitive. Let s be a positive integer such that $st \ge k$; then $\mathbb{A}^{st} > 0$ by Proposition 23. Thus $(\mathbb{A}^t)^s = \mathbb{A}^{st} > 0$, which implies $\mathbb{A}^t$ is strongly primitive. □

3. A Characterization of the (Strongly) Primitive Tensor with Order m and Dimension 2

In this section, we study primitive tensors and strongly primitive tensors in this paper, show that an order m dimension 2 tensor is primitive if and only if its majorization matrix is primitive, and obtain the characterization of order m dimension 2 strongly primitive tensors and the bound of the strongly primitive degree.

Lemma 25 (see [5], Corollary 4.1). *Let $\mathbb{A}$ be a nonnegative tensor with order m and dimension n. If $M(\mathbb{A})$ is primitive, then $\mathbb{A}$ is also primitive and in this case we have $\gamma(\mathbb{A}) \le \gamma(M(\mathbb{A})) \le (n-1)^2 + 1$.*

Theorem 26. *Let $\mathbb{A}$ be a nonnegative tensor with order m and dimension $n = 2$. Then $\mathbb{A}$ is primitive if and only if $M(\mathbb{A})$ is primitive.*

Proof. Firstly, the sufficiency is obvious by Lemma 25. Now we only show the necessity. Clearly, all primitive (0,1) matrices of order 2 are listed as follows:

$$\begin{pmatrix} 1 & 1 \\ 1 & 0 \end{pmatrix}, \quad \begin{pmatrix} 0 & 1 \\ 1 & 1 \end{pmatrix}, \quad \begin{pmatrix} 1 & 1 \\ 1 & 1 \end{pmatrix}. \tag{10}$$

Let $\mathbb{A}$ be primitive. Then $\gamma(\mathbb{A}) \le 2$ by Theorem 10 and $M(\mathbb{A}^2) > 0$ by Proposition 8. Now we assume that $M(\mathbb{A})$ is not primitive; we will show $\mathbb{A}$ is also not primitive.

It is not difficult to find that

$$\left(\mathbb{A}^2\right)_{ijj \cdots j} = \sum_{i_2, i_3, \dots, i_m = 1}^{2} a_{i i_2 i_3 \cdots i_m} a_{i_2 jj \cdots j} \cdots a_{i_m jj \cdots j} \tag{11}$$

$$= a_{i22 \cdots 2} \left(a_{2jj \cdots j}\right)^{m-1} + \sum_{i_2, i_3, \dots, i_m = 1, i_2 i_3 \cdots i_m \ne 22 \cdots 2}^{2} a_{i i_2 \cdots i_m} a_{i_2 jj \cdots j} \cdots a_{i_m jj \cdots j} \tag{12}$$

$$= a_{i11 \cdots 1} \left(a_{1jj \cdots j}\right)^{m-1} + \sum_{i_2, i_3, \dots, i_m = 1, i_2 i_3 \cdots i_m \ne 11 \cdots 1}^{2} a_{i i_2 \cdots i_m} a_{i_2 jj \cdots j} \cdots a_{i_m jj \cdots j}. \tag{13}$$

In (12), we note that $i_2 i_3 \dots i_m \ne 22 \dots 2$, which implies that there exists at least one entry, say, $i_s = 1$, where $2 \le s \le m$; then $a_{1jj \dots j} \in \{a_{i_2 jj \dots j}, \dots, a_{i_m jj \dots j}\}$.

Similarly, in (13), we note that $i_2 i_3 \dots i_m \ne 11 \dots 1$, which implies that there exists at least one entry, say, $i_s = 2$, where $2 \le s \le m$; then $a_{2jj \dots j} \in \{a_{i_2 jj \dots j}, \dots, a_{i_m jj \dots j}\}$.

Thus, by (12), (13), and the above arguments, we have

$$M\left(\mathbb{A}^2\right)_{ij}=\left(\mathbb{A}^2\right)_{ijj\cdots j}=a_{i22\cdots 2}\left(a_{2jj\cdots j}\right)^{m-1}+a_{1jj\cdots j}P \quad (14)$$

$$=a_{i11\cdots 1}\left(a_{1jj\cdots j}\right)^{m-1}+a_{2jj\cdots j}Q. \quad (15)$$

Since $M(\mathbb{A})$ is not primitive; by (10), we can complete the proof by the following two cases.

Case 1. $M(\mathbb{A})=\left(\begin{smallmatrix} * & 0\\ * & *\end{smallmatrix}\right)$ or $\left(\begin{smallmatrix} * & *\\ 0 & *\end{smallmatrix}\right)$.

Subcase 1.1. $M(\mathbb{A})_{12}=0$.
Then $a_{12\ldots 2}=0$. By (14), we have $M(\mathbb{A}^2)_{12}=(\mathbb{A}^2)_{12\ldots 2}=0$, which implies $\mathbb{A}^2$ is not essentially positive.

Subcase 1.2. $M(\mathbb{A})_{21}=0$.
Then $a_{21\ldots 1}=0$. By (15), we have $M(\mathbb{A}^2)_{21}=(\mathbb{A}^2)_{21\ldots 1}=0$, which implies $\mathbb{A}^2$ is not essentially positive.

Case 2. $M(\mathbb{A})=\left(\begin{smallmatrix} 0 & 1\\ 1 & 0\end{smallmatrix}\right)$.
Then we have $M(\mathbb{A})_{11}=M(\mathbb{A})_{22}=0$; that is, $a_{11\ldots 1}=a_{22\ldots 2}=0$; by (15) we have $M(\mathbb{A}^2)_{12}=(\mathbb{A}^2)_{12\ldots 2}=0$, which implies $\mathbb{A}^2$ is not essentially positive.
Based on the above two cases and Proposition 6, we complete the proof of the necessity. □

A nature question is whether the result of Theorem 26 is true for $n\geq 3$ or not. The following Example 27 shows that the necessity of Theorem 26 is false with $n\geq 3$.

Example 27. Let $\mathbb{A}=(a_{i_1i_2\ldots i_m})$ be a nonnegative tensor of order m and dimension $n\geq 3$, where

$$a_{i_1i_2\cdots i_m}=\begin{cases}0, & \text{if } i_1=1,\ i_2=i_3=\cdots=i_m\neq 1;\\ 1, & \text{otherwise.}\end{cases} \quad (16)$$

Then $\mathbb{A}$ is (strongly) primitive, but $M(\mathbb{A})$ is not primitive.

Proof. By direct calculation and Definition 2, we know that $\mathbb{A}^2$ is the tensor of order $(m-1)^2+1$ and dimension n, and for any $1\leq i\leq n$, we have

$$\begin{aligned}\left(\mathbb{A}^2\right)_{i\alpha_2\alpha_3\cdots\alpha_m}&=\sum_{i_2,i_3,\ldots,i_m=1}^{2}a_{ii_2i_3\cdots i_m}a_{i_2\alpha_2}\cdots a_{i_m\alpha_m}\\ &\geq\begin{cases}a_{12\cdots m}a_{2\alpha_2}\cdots a_{m\alpha_m}=1, & \text{if } i=1;\\ a_{ii\cdots i}a_{i\alpha_2}\cdots a_{i\alpha_m}=1, & \text{otherwise.}\end{cases}\end{aligned} \quad (17)$$

Obviously, $\mathbb{A}^2$ is positive; then $\mathbb{A}$ is strongly primitive with $\eta(\mathbb{A})=2$ and thus $\mathbb{A}$ is primitive with $\gamma(\mathbb{A})=2$.
On the other hand, by the definition of $\mathbb{A}$, we have

$$M\left(\mathbb{A}\right)=\begin{pmatrix}1 & 0 & \cdots & 0\\ 1 & 1 & \cdots & 1\\ & & \vdots & \\ 1 & 1 & \cdots & 1\end{pmatrix}. \quad (18)$$

Since the associated digraph of $M(\mathbb{A})$ is not strongly connected, thus $M(\mathbb{A})$ is not primitive. □

Next, we will study the strongly primitive degree of order m and dimension 2 tensors. Firstly, we discuss an example with order $m=5$ and dimension $n=2$ tensor as follows.

Definition 28 (see [10]). Let $\mathbb{A}$ be a tensor with order m and dimension n. The i-th slice of $\mathbb{A}$, denoted by $\mathbb{A}[i]$, is the subtensor of $\mathbb{A}$ with order $m-1$ and dimension n such that $(\mathbb{A}[i])_{i_2\ldots i_m}=a_{ii_2\ldots i_m}$.

Example 29. Let $\mathbb{A}=(a_{i_1i_2i_3i_4i_5})_{1\leq i_j\leq 2(j=1,\ldots,5)}$ be a nonnegative tensor with order $m=5$ and dimension $n=2$, where $a_{12122}=a_{21121}=0$ and other $a_{i_1i_2i_3i_4i_5}=1$. Then there exists at least one zero element in each slice of $\mathbb{A}^2$.

Proof. Let $\alpha_1=2122$, $\alpha_2=1121$, and denote $\beta_2=\beta_4=\beta_5=\alpha_1$, $\beta_3=\alpha_2$. Then we have

$$\begin{aligned}&\left(A^2\right)_{1\beta_2\beta_3\beta_4\beta_5}\\ &=\!=\sum_{i_2,i_3,i_4,i_5=1}^{2}a_{1i_2i_3i_4i_5}a_{i_2\beta_2}a_{i_3\beta_3}a_{i_4\beta_4}a_{i_5\beta_5}\\ &=\!=\sum_{i_3,i_4,i_5=1}^{2}a_{11i_3i_4i_5}a_{1\beta_2}a_{i_3\beta_3}a_{i_4\beta_4}a_{i_5\beta_5}\\ &\quad+\sum_{i_3,i_4,i_5=1}^{2}a_{12i_3i_4i_5}a_{2\beta_2}a_{i_3\beta_3}a_{i_4\beta_4}a_{i_5\beta_5}\\ &\overset{a_{1\beta_2}=0}{=\!=}\sum_{i_3,i_4,i_5=1}^{2}a_{12i_3i_4i_5}a_{2\beta_2}a_{i_3\beta_3}a_{i_4\beta_4}a_{i_5\beta_5}\\ &=\!=\sum_{i_4,i_5=1}^{2}a_{121i_4i_5}a_{2\beta_2}a_{1\beta_3}a_{i_4\beta_4}a_{i_5\beta_5}\\ &\quad+\sum_{i_4,i_5=1}^{2}a_{122i_4i_5}a_{2\beta_2}a_{2\beta_3}a_{i_4\beta_4}a_{i_5\beta_5}\\ &\overset{a_{2\beta_3}=0}{=\!=}\sum_{i_4,i_5=1}^{2}a_{121i_4i_5}a_{2\beta_2}a_{1\beta_3}a_{i_4\beta_4}a_{i_5\beta_5}\\ &=\!=\sum_{i_5=1}^{2}a_{1211i_5}a_{2\beta_2}a_{1\beta_3}a_{1\beta_4}a_{i_5\beta_5}\\ &\quad+\sum_{i_5=1}^{2}a_{1212i_5}a_{2\beta_2}a_{1\beta_3}a_{2\beta_4}a_{i_5\beta_5}\\ &\overset{a_{1\beta_4}=0}{=\!=}\sum_{i_5=1}^{2}a_{1212i_5}a_{2\beta_2}a_{1\beta_3}a_{2\beta_4}a_{i_5\beta_5}\\ &=\!=a_{12121}a_{2\beta_2}a_{1\beta_3}a_{2\beta_4}a_{1\beta_5}\end{aligned}$$

$$+ a_{12122}a_{2\beta_2}a_{1\beta_3}a_{2\beta_4}a_{2\beta_5}$$

$$\xlongequal{a_{1\beta_5}=0} a_{12122}a_{2\beta_2}a_{1\beta_3}a_{2\beta_4}a_{2\beta_5} \xlongequal{a_{12122}=0} 0. \tag{19}$$

Similarly, we let $\gamma_2 = \gamma_3 = \gamma_5 = \alpha_2$ and $\gamma_4 = \alpha_1$, we can show $(\mathbb{A}^2)_{2\gamma_2\gamma_3\gamma_4\gamma_5} = 0$, and we omit it.

Combining the above arguments, we know there exists at least one zero element in each slice of $\mathbb{A}^2$ by $(A^2)_{1\beta_2\beta_3\beta_4\beta_5} = (\mathbb{A}^2)_{2\gamma_2\gamma_3\gamma_4\gamma_5} = 0$. □

Similarly, the result of Example 29 can be generalized to any nonnegative tensor with order m and dimension $n = 2$.

Lemma 30. *Let $\mathbb{A} = (a_{i_1i_2\dots i_m})_{1\le i_j\le 2(j=1,\dots,m)}$ be a nonnegative tensor with order m and dimension $n = 2$. If there exist $\alpha_1 = j_2j_3\cdots j_m \in [2]^{m-1}$ and $\alpha_2 = k_2k_3\dots k_m \in [2]^{m-1}$ such that $a_{1\alpha_1} = a_{2\alpha_2} = 0$, for any $2 \le t \le m$, let $\beta_t = \begin{cases}\alpha_2, & \text{if } j_t=1;\\ \alpha_1, & \text{if } j_t=2,\end{cases}$ and $\gamma_t = \begin{cases}\alpha_2, & \text{if } k_t=1;\\ \alpha_1, & \text{if } k_t=2.\end{cases}$ Then*

$$\begin{aligned}(\mathbb{A}^2)_{1\beta_2\cdots\beta_m} &= 0,\\ (\mathbb{A}^2)_{2\gamma_2\cdots\gamma_m} &= 0.\end{aligned} \tag{20}$$

Proof. We first show $(\mathbb{A}^2)_{1\beta_2\dots\beta_m} = 0$. For any $2 \le t \le m$, $j_t \in \{1,2\}$, we denote $\overline{j_t} \in \{1,2\}\backslash\{j_t\}$. Then we have $a_{\overline{j_t}\beta_t} = 0$ by $a_{1\alpha_1} = a_{2\alpha_2} = 0$, and

$$\begin{aligned}(\mathbb{A}^2)_{1\beta_2\cdots\beta_m} &= \sum_{i_2,i_3,\dots,i_m=1}^{2} a_{1i_2i_3\cdots i_m}a_{i_2\beta_2}\cdots a_{i_m\beta_m}\\
&= \sum_{i_3,\dots,i_m=1}^{2} a_{11i_3\cdots i_m}a_{1\beta_2}a_{i_3\beta_3}\cdots a_{i_m\beta_m}\\
&\quad + \sum_{i_3,\dots,i_m=1}^{2} a_{12i_3\cdots i_m}a_{2\beta_2}a_{i_3\beta_3}\cdots a_{i_m\beta_m}\\
&\xlongequal{a_{\overline{j_2}\beta_2}=0} \sum_{i_3,\dots,i_m=1}^{2} a_{1j_2i_3\cdots i_m}a_{j_2\beta_{j_2}}a_{i_3\beta_3}\cdots a_{i_m\beta_m}\\
&= \sum_{i_4,\dots,i_m=1}^{2} a_{1j_21i_4\cdots i_m}a_{j_2\beta_{j_2}}a_{1\beta_3}a_{i_4\beta_4}\cdots a_{i_m\beta_m}\\
&\quad + \sum_{i_4,\dots,i_m=1}^{2} a_{1j_22i_4\cdots i_m}a_{j_2\beta_{j_2}}a_{2\beta_3}a_{i_4\beta_4}\cdots a_{i_m\beta_m}\\
&\xlongequal{a_{\overline{j_3}\beta_3}=0} \sum_{i_4,\dots,i_m=1}^{2} a_{1j_2j_3i_4\cdots i_m}a_{j_2\beta_{j_2}}a_{j_3\beta_{j_3}}a_{i_4\beta_4}\cdots a_{i_m\beta_m}\\
&= \cdots\cdots\\
&= \sum_{i_m=1}^{2} a_{1j_2j_3\cdots j_{m-1}i_m}a_{j_2\beta_{j_2}}a_{j_3\beta_{j_3}}\cdots a_{j_{m-1}\beta_{j_{m-1}}}a_{i_m\beta_m}\\
&= a_{1j_2\cdots j_{m-1}1}a_{j_2\beta_{j_2}}\cdots a_{j_{m-1}\beta_{j_{m-1}}}a_{1\beta_m}\\
&\quad + a_{1j_2\cdots j_{m-1}2}a_{j_2\beta_{j_2}}\cdots a_{j_{m-1}\beta_{j_{m-1}}}a_{2\beta_m}\\
&\xlongequal{a_{\overline{j_m}\beta_m}=0} a_{1j_2\cdots j_m}a_{j_2\beta_{j_2}}\cdots a_{j_m\beta_{j_m}} \xlongequal{a_{1\alpha_1}=0} 0.\end{aligned} \tag{21}$$

Similarly, for any $2 \le t \le m$, $k_t \in \{1,2\}$, we denote $\overline{k_t} \in \{1,2\}\backslash\{k_t\}$. Then we have $a_{\overline{k_t}\gamma_t} = 0$ and we can show $(\mathbb{A}^2)_{2\gamma_2\dots\gamma_m} = 0$ by

$$(\mathbb{A}^2)_{2\gamma_2\cdots\gamma_m} = \sum_{i_2,i_3,\dots,i_m=1}^{2} a_{2i_2i_3\cdots i_m}a_{i_2\gamma_2}\cdots a_{i_m\gamma_m} \tag{22}$$

and similar process of the above arguments. Thus we complete the proof of (20). □

Theorem 31. *Let $\mathbb{A} = (a_{i_1i_2\dots i_m})_{1\le i_j\le 2(j=1,\dots,m)}$ be a nonnegative tensor with order m and dimension $n = 2$. If there exist $\alpha_1 = j_2j_3\cdots j_m \in [2]^{m-1}$ and $\alpha_2 = k_2k_3\dots k_m \in [2]^{m-1}$ such that $a_{1\alpha_1} = a_{2\alpha_2} = 0$, then $\mathbb{A}$ is not strongly primitive.*

Proof. Now we show that there exists at least one zero element in each slice of $\mathbb{A}^r$ by induction on $r(\ge 2)$.

Firstly, by Lemma 30, we know there exists at least one zero element in each slice of $\mathbb{A}^2$. Now we assume that there exists at least one zero element in each slice of $\mathbb{A}^{r-1}$; say, there exist $\delta_1, \delta_2 \in [2]^{(m-1)^{r-1}}$ such that $(\mathbb{A}^{r-1})_{1\delta_1} = (\mathbb{A}^{r-1})_{2\delta_2} = 0$. For any $2 \le t \le m$, let $\beta_t = \begin{cases}\delta_2 & \text{if } j_t=1\\ \delta_1 & \text{if } j_t=2\end{cases}$ and $\gamma_t = \begin{cases}\delta_2 & \text{if } k_t=1\\ \delta_1 & \text{if } k_t=2.\end{cases}$ Then by (7) and the similar proof of Lemma 30, we have

$$\begin{aligned}&(\mathbb{A}^r)_{1\beta_2\cdots\beta_m}\\ &= \sum_{i_2,i_3,\dots,i_m=1}^{n} a_{1i_2i_3\cdots i_m}(\mathbb{A}^{r-1})_{i_2\beta_2}\cdots(\mathbb{A}^{r-1})_{i_m\beta_m} = 0,\end{aligned} \tag{23}$$

and

$$\begin{aligned}&(\mathbb{A}^r)_{2\gamma_2\cdots\gamma_m}\\ &= \sum_{i_2,i_3,\dots,i_m=1}^{n} a_{2i_2i_3\cdots i_m}(\mathbb{A}^{r-1})_{i_2\gamma_2}\cdots(\mathbb{A}^{r-1})_{i_m\gamma_m} = 0.\end{aligned} \tag{24}$$

By (23) and (24), we obtain that there exists at least one zero element in each slice of $\mathbb{A}^r$, and thus we complete the proof. □

Now we give the characterization of the strongly primitive tensor with order m and dimension 2.

Theorem 32. *Let $\mathbb{A} = (a_{i_1i_2\dots i_m})_{1\le i_j\le 2(j=1,\dots,m)}$ be a nonnegative tensor with order m and dimension $n = 2$. Then*

(1) *$\mathbb{A}$ is strongly primitive if and only if one of the following holds:*

(a) $\mathbb{A} = \mathbb{J}$;

(b) $\mathbb{A} \neq \mathbb{J}$ *and* $a_{1i_2\ldots i_m} = a_{211\ldots1} = 1 (1 \leq i_j \leq 2, j = 2,\ldots,m)$;

(c) $\mathbb{A} \neq \mathbb{J}$ *and* $a_{2i_2\ldots i_m} = a_{122\ldots2} = 1 (1 \leq i_j \leq 2, j = 2,\ldots,m)$;

(2) *if* $\mathbb{A}$ *is strongly primitive, then* $\eta(\mathbb{A}) \leq 2$.

Proof. Firstly, we show that (1) is sufficient. It is easy to see that $\mathbb{A} = \mathbb{J}$ is strongly primitive with $\eta(\mathbb{J}) = 1$, and if $\mathbb{A}$ satisfies (b) or (c), $\mathbb{A}$ is strongly primitive with $\eta(\mathbb{A}) = 2$ by Proposition 19 immediately.

Now we show the necessity of (1); that is, if $\mathbb{A}$ does not satisfy the conditions of (a), (b), or (c), then we will show that $\mathbb{A}$ is not strongly primitive. We complete the proof by the following three cases.

Case 1. $a_{1\alpha} = 1$ for any $\alpha \in [2]^{m-1}$ and $a_{211\ldots1} = 0$.

It is not difficult to find that $M(\mathbb{A}) = \left(\begin{smallmatrix} 1 & 1 \\ 0 & * \end{smallmatrix}\right)$. Then $\mathbb{A}$ is not primitive by Theorem 26, and thus $\mathbb{A}$ is not strongly primitive.

Case 2. $a_{2\alpha} = 1$ for any $\alpha \in [2]^{m-1}$ and $a_{122\ldots2} = 0$.

Similarly, we can find that $M(\mathbb{A}) = \left(\begin{smallmatrix} * & 0 \\ 1 & 1 \end{smallmatrix}\right)$. Then $\mathbb{A}$ is not primitive by Theorem 26, and thus $\mathbb{A}$ is not strongly primitive.

Case 3. There is at least one zero element in each slice of $\mathbb{A}$.

Then there exist $\alpha_1 = j_2 j_3 \cdots j_m \in [2]^{m-1}$ and $\alpha_2 = k_2 k_3 \ldots k_m \in [2]^{m-1}$ such that $a_{1\alpha_1} = a_{2\alpha_2} = 0$. Thus $\mathbb{A}$ is not strongly primitive by Theorem 31.

(2) If $\mathbb{A}$ is strongly primitive, by Definition 11 and the proof of (1), we obtain $\eta(\mathbb{A}) \leq 2$ immediately. □

Remark 33. By Theorem 32, we can see that the strongly primitive degree $\eta(\mathbb{A})$ of an nonnegative tensor with order m and dimension $n = 2$ is irrelevant to its order m.

4. Some Properties and Problems of Order m Dimension $n(\geq 3)$ Strongly Primitive Tensors

In this section, we will study some properties of the strongly primitive tensors with order m and dimension $n \geq 3$ and propose some questions for further research.

Proposition 34. *Let* $\mathbb{A} = (a_{i_1 i_2\ldots i_m})_{1\leq i_j\leq n (j=1,\ldots,m)}$ *be a nonnegative tensor with order m and dimension n. Let* $s \in [n]$, $2 \leq t \leq m$, $j_t^{(s)} \in [n]$, *and* $\alpha_s = j_2^{(s)} j_3^{(s)} \ldots j_m^{(s)}$. *If* $a_{i\alpha_s} = 0$ *for any* $i \in [n]$ *and any* $s \in [n]\backslash\{i\}$, *then there exist* $\gamma_1, \gamma_2, \ldots, \gamma_n \in [n]^{(m-1)^2}$ *such that* $(\mathbb{A}^2)_{i\gamma_k} = 0$ *for any* $i \in [n]$ *and any* $k \in [n]\backslash\{i\}$.

Proof. For each $i \in [n]$, let $\beta_t^{(k)} = \alpha_{j_t^{(k)}} \in [n]^{m-1}$ for any $k \in [n]$ and $2 \leq t \leq m$; then $a_{\overline{j_t^{(k)}}\beta_t^{(k)}} = 0$ for any $\overline{j_t^{(k)}} \in [n]\backslash\{j_t^{(k)}\}$ by $a_{i\alpha_s} = 0$ for any $i \in [n]$ and any $s \in [n]\backslash\{i\}$. Let $\gamma_k = \beta_2^{(k)}\beta_3^{(k)}\ldots\beta_m^{(k)} \in [n]^{(m-1)^2}$ for any $k \in [n]$. Now we show $(\mathbb{A}^2)_{i\gamma_k} = 0$ for any $k \in [n]\backslash\{i\}$.

$$\begin{aligned}
\left(\mathbb{A}^2\right)_{i\gamma_k} &= \left(\mathbb{A}^2\right)_{i\beta_2^{(k)}\beta_3^{(k)}\cdots\beta_m^{(k)}} \\
&= \sum_{i_2,i_3,\ldots,i_m=1}^{n} a_{ii_2i_3\cdots i_m} a_{i_2\beta_2^{(k)}} \cdots a_{i_m\beta_m^{(k)}} \\
&= \sum_{i_3,\ldots,i_m=1}^{n} a_{i1i_3\cdots i_m} a_{1\beta_2^{(k)}} a_{i_3\beta_3^{(k)}} \cdots a_{i_m\beta_m^{(k)}} + \cdots \\
&\quad + \sum_{i_3,\ldots,i_m=1}^{n} a_{ini_3\cdots i_m} a_{n\beta_2^{(k)}} a_{i_3\beta_3^{(k)}} \cdots a_{i_m\beta_m^{(k)}} \\
&\overset{a_{\overline{j_2^{(k)}}\beta_2^{(k)}}=0}{=} \sum_{i_3,\ldots,i_m=1}^{n} a_{ij_2^{(k)}i_3\cdots i_m} a_{j_2^{(k)}\beta_2^{(k)}} a_{i_3\beta_3^{(k)}} \cdots a_{i_m\beta_m^{(k)}} \\
&= \sum_{i_4,\ldots,i_m=1}^{n} a_{ij_2^{(k)}1i_4\cdots i_m} a_{j_2^{(k)}\beta_2^{(k)}} a_{1\beta_3^{(k)}} a_{i_4\beta_4^{(k)}} \\
&\quad \cdots a_{i_m\beta_m^{(k)}} + \cdots + \sum_{i_4,\ldots,i_m=1}^{n} a_{ij_2^{(k)}ni_4\cdots i_m} a_{j_2^{(k)}\beta_2^{(k)}} a_{n\beta_3^{(k)}} a_{i_4\beta_4^{(k)}} \\
&\quad \cdots a_{i_m\beta_m^{(k)}} \\
&\overset{a_{\overline{j_3^{(k)}}\beta_3^{(k)}}=0}{=} \sum_{i_4,\ldots,i_m=1}^{n} a_{ij_2^{(k)}j_3^{(k)}i_4\cdots i_m} a_{j_2^{(k)}\beta_2^{(k)}} a_{j_3^{(k)}\beta_3^{(k)}} a_{i_4\beta_4^{(k)}} \\
&\quad \cdots a_{i_m\beta_m^{(k)}} = \cdots \\
&= \sum_{i_m=1}^{n} a_{ij_2^{(k)}j_3^{(k)}\cdots j_{m-1}^{(k)}i_m} a_{j_2^{(k)}\beta_2^{(k)}} a_{j_3^{(k)}\beta_3^{(k)}} \\
&\quad \cdots a_{j_{m-1}^{(k)}\beta_{m-1}^{(k)}} a_{i_m\beta_m^{(k)}} \\
&= a_{ij_2^{(k)}j_3^{(k)}\cdots j_{m-1}^{(k)}1} a_{j_2^{(k)}\beta_2^{(k)}} a_{j_3^{(k)}\beta_3^{(k)}} \\
&\quad \cdots a_{j_{m-1}^{(k)}\beta_{m-1}^{(k)}} a_{1\beta_m^{(k)}} + \cdots + a_{ij_2^{(k)}j_3^{(k)}\cdots j_{m-1}^{(k)}n} a_{j_2^{(k)}\beta_2^{(k)}} a_{j_3^{(k)}\beta_3^{(k)}} \\
&\quad \cdots a_{j_{m-1}^{(k)}\beta_{m-1}^{(k)}} a_{n\beta_m^{(k)}} \\
&\overset{a_{\overline{j_m^{(k)}}\beta_m^{(k)}}=0}{=} a_{ij_2^{(k)}j_3^{(k)}\cdots j_{m-1}^{(k)}j_m^{(k)}} a_{j_2^{(k)}\beta_2^{(k)}} a_{j_3^{(k)}\beta_3^{(k)}} \\
&\quad \cdots a_{j_{m-1}^{(k)}\beta_{m-1}^{(k)}} a_{j_m^{(k)}\beta_m^{(k)}} = a_{i\alpha_k} a_{j_2^{(k)}\beta_2^{(k)}} a_{j_3^{(k)}\beta_3^{(k)}} \\
&\quad \cdots a_{j_{m-1}^{(k)}\beta_{m-1}^{(k)}} a_{j_m^{(k)}\beta_m^{(k)}} \overset{a_{i\alpha_k}=0}{=} 0.
\end{aligned} \tag{25}$$

We note that $k \in [n]\backslash\{i\}$ which means there are $n-1$ zero elements in i-th slice of $\mathbb{A}^2$; thus we complete the proof by $i \in [n]$. □

We note that Proposition 34 is the generalization of Lemma 30; now we will obtain the generalization of Theorem 31.

Theorem 35. *Let $\mathbb{A} = (a_{i_1 i_2 \ldots i_m})_{1\le i_j\le n(j=1,\ldots,m)}$ be a nonnegative tensor with order m and dimension n. Let $s \in [n]$, $2 \le t \le m$, $j_t^{(s)} \in [n]$, and $\alpha_s = j_2^{(s)} j_3^{(s)} \ldots j_m^{(s)}$. If $a_{i\alpha_s} = 0$ for any $i \in [n]$ and any $s \in [n]\backslash\{i\}$, then $\mathbb{A}$ is not strongly primitive.*

Proof. Now we show that there exist $\varepsilon_1, \varepsilon_2, \ldots, \varepsilon_n \in [n]^{(m-1)^r}$ such that $(\mathbb{A}^r)_{i\varepsilon_k} = 0$ for any $i \in [n]$ and any $k \in [n]\backslash\{i\}$ by induction on $r(\ge 2)$; say, there exist at least $n-1$ zero elements in each slice of $\mathbb{A}^r$ and thus $\mathbb{A}$ is not strongly primitive.

Firstly, by Proposition 34, we know there exist $\gamma_1, \gamma_2, \ldots, \gamma_n \in [n]^{(m-1)^2}$ such that $(\mathbb{A}^2)_{i\gamma_k} = 0$ for any $i \in [n]$ and any $k \in [n]\backslash\{i\}$; say, there exist at least $n-1$ zero elements in each slice of $\mathbb{A}^2$. Now we assume that there exist $\delta_1, \delta_2, \ldots, \delta_n \in [n]^{(m-1)^{r-1}}$ such that $(\mathbb{A}^{r-1})_{i\delta_k} = 0$ for any $i \in [n]$ and any $k \in [n]\backslash\{i\}$; say, there exist at least $n-1$ zero elements in each slice of $\mathbb{A}^{r-1}$.

Let $\eta_t^{(s)} = \delta_{j_t^{(s)}}$ for any $s \in [n]$ and $2 \le t \le m$; then $(\mathbb{A}^{r-1})_{\overline{j_t^{(s)}}\eta_t^{(s)}} = 0$ for any $\overline{j_t^{(s)}} \in [n]\backslash\{j_t^{(s)}\}$ by $(\mathbb{A}^{r-1})_{i\delta_k} = 0$ for any $i \in [n]$ and any $k \in [n]\backslash\{i\}$. Let $\varepsilon_k = \eta_2^{(k)}\eta_3^{(k)}\ldots\eta_m^{(k)} \in [n]^{(m-1)^r}$ for any $k \in [n]$. Now we show $(\mathbb{A}^r)_{i\varepsilon_k} = 0$ for any $i \in [n]$ and any $k \in [n]\backslash\{i\}$.

By (7) and similar proof of Proposition 34, we have

$$\begin{aligned}(\mathbb{A}^r)_{i\varepsilon_k} &= (\mathbb{A}^r)_{i\eta_2^{(k)}\eta_3^{(k)}\cdots\eta_m^{(k)}} \\ &= \sum_{i_2,i_3,\ldots,i_m=1}^{n} a_{ii_2i_3\cdots i_m}\left(\mathbb{A}^{r-1}\right)_{i_2\eta_2^{(k)}}\cdots\left(\mathbb{A}^{r-1}\right)_{i_m\eta_m^{(k)}} \\ &= 0,\end{aligned} \tag{26}$$

and then we complete the proof. □

Proposition 36. *Let $\mathbb{A} = (a_{i_1 i_2 \cdots i_m})_{1\le i_j\le n(j=1,\cdots,m)}$ be a nonnegative tensor with order m and dimension n, and $M(\mathbb{A})$ be the majorization matrix of $\mathbb{A}$. If there exist $i, j \in [n]$, such that $(M(\mathbb{A}))_{ij} > 0, (M(\mathbb{A}))_{uj} = 0$ for any $u \in [n]\backslash\{i\}$ and $(M(\mathbb{A}))_{ji} > 0, (M(\mathbb{A}))_{vi} = 0$ for any $v \in [n]\backslash\{j\}$, then $\mathbb{A}$ is not primitive, and thus $\mathbb{A}$ is not strongly primitive.*

Proof. Firstly, we show the following assertion:

If k is odd, then $(M(\mathbb{A}^k))_{ij} > 0$, $(M(\mathbb{A}^k))_{ji} > 0$, $(M(\mathbb{A}^k))_{uj} = 0$ for any $u \in [n]\backslash\{i\}$, $(M(\mathbb{A}^k))_{vi} = 0$ for any $v \in [n]\backslash\{j\}$.

If k is even, then $(M(\mathbb{A}^k))_{ii} > 0$, $(M(\mathbb{A}^k))_{jj} > 0$, $(M(\mathbb{A}^k))_{ui} = 0$ for any $u \in [n]\backslash\{i\}$, $(M(\mathbb{A}^k))_{vj} = 0$ for any $v \in [n]\backslash\{j\}$.

When $k = 1$, the above result holds which is obvious. When $k = 2$, by Definition 7 and (7), we have

$$\begin{aligned}\left(M\left(\mathbb{A}^2\right)\right)_{ii} &= \left(\mathbb{A}^2\right)_{ii\cdots i} \\ &= \sum_{i_2,i_3,\ldots,i_m=1}^{n} a_{ii_2i_3\cdots i_m}a_{i_2i\cdots i}\cdots a_{i_mi\cdots i} \\ &= a_{ij\cdots j}\left(a_{ji\cdots i}\right)^{m-1} \\ &= (M(\mathbb{A}))_{ij}\left((M(\mathbb{A}))_{ji}\right)^{m-1} > 0,\end{aligned} \tag{27}$$

and for any $u \in [n]\backslash\{i\}$, we have

$$\begin{aligned}\left(M\left(\mathbb{A}^2\right)\right)_{ui} &= \left(\mathbb{A}^2\right)_{ui\cdots i} \\ &= \sum_{i_2,i_3,\ldots,i_m=1}^{n} a_{ui_2i_3\cdots i_m}a_{i_2i\cdots i}\cdots a_{i_mi\cdots i} \\ &= a_{uj\cdots j}\left(a_{ji\cdots i}\right)^{m-1} \\ &= (M(\mathbb{A}))_{uj}\left((M(\mathbb{A}))_{ji}\right)^{m-1} = 0.\end{aligned} \tag{28}$$

Similarly, we can show $(M(\mathbb{A}^2))_{jj} > 0$ and $(M(\mathbb{A}^2))_{vj} = 0$ for any $v \in [n]\backslash\{j\}$.

Now we assume that, for any k, the above assertion holds. Then, for $k+1$, we consider the following two cases.

Case 1. k is odd.

Then, by (7), we have

$$\begin{aligned}\left(M\left(\mathbb{A}^{k+1}\right)\right)_{ii} &= \left(\mathbb{A}^{k+1}\right)_{ii\cdots i} \\ &= \sum_{i_2,i_3,\ldots,i_m=1}^{n} a_{ii_2i_3\cdots i_m}\left(\mathbb{A}^k\right)_{i_2i\cdots i}\cdots\left(\mathbb{A}^k\right)_{i_mi\cdots i} \\ &= a_{ij\cdots j}\left(\left(\mathbb{A}^k\right)_{ji\cdots i}\right)^{m-1} \\ &= (M(\mathbb{A}))_{ij}\left(\left(M\left(\mathbb{A}^k\right)\right)_{ji}\right)^{m-1} > 0,\end{aligned} \tag{29}$$

and, for any $u \in [n]\backslash\{i\}$, we have

$$\begin{aligned}\left(M\left(\mathbb{A}^{k+1}\right)\right)_{ui} &= \left(\mathbb{A}^{k+1}\right)_{ui\cdots i} \\ &= \sum_{i_2,i_3,\ldots,i_m=1}^{n} a_{ui_2i_3\cdots i_m}\left(\mathbb{A}^k\right)_{i_2i\cdots i}\cdots\left(\mathbb{A}^k\right)_{i_mi\cdots i} \\ &= a_{uj\cdots j}\left(\left(\mathbb{A}^k\right)_{ji\cdots i}\right)^{m-1} \\ &= (M(\mathbb{A}))_{uj}\left(\left(M\left(\mathbb{A}^k\right)\right)_{ji}\right)^{m-1} = 0.\end{aligned} \tag{30}$$

Similarly, we can show $(M(\mathbb{A}^{k+1}))_{jj} > 0$ and $(M(\mathbb{A}^{k+1}))_{vj} = 0$ for any $v \in [n]\backslash\{j\}$.

Case 2. k is even.

By (7) and similar proof of Case 1, we can show $(M(\mathbb{A}^{k+1}))_{ij} > 0$, $(M(\mathbb{A}^{k+1}))_{ji} > 0$, $(M(\mathbb{A}^{k+1}))_{uj} = 0$ for any $u \in [n]\backslash\{i\}$, and $(M(\mathbb{A}^{k+1}))_{vi} = 0$ for any $v \in [n]\backslash\{j\}$.

By Proposition 8 and the above assertion, we know $\mathbb{A}$ is not primitive, and thus $\mathbb{A}$ is not strongly primitive. □

Let $\mathbb{A} = (a_{i_1 i_2 \cdots i_m})_{1\le i_j\le n(j=1,\cdots,m)}$ be a nonnegative strongly primitive tensor with order m and dimension n. When $n = 2$, we know $\eta(\mathbb{A}) \le 2$ by Theorem 32. When $n \ge 3$, we do not know the value or bound of $\eta(\mathbb{A})$. Even $n = 3$, we do not find out all strongly primitive tensors. Thus we think it is not easy to obtain the value or bound of $\eta(\mathbb{A})$. Based on the computation of the case $n = 3$, we propose the following problem for further research.

Question 37. Let $n \geq 3$, $\mathbb{A} = (a_{i_1 i_2 \cdots i_m})_{1 \leq i_j \leq n (j=1,\cdots,m)}$ be a nonnegative strongly primitive tensor with order m and dimension n. Then $\eta(\mathbb{A}) < (n-1)^2 + 1$.

In [7, 9], the authors gave some algebraic characterizations of a nonnegative primitive tensor, and in [11] the authors showed that a nonnegative tensor is primitive if and only if the greatest common divisor of all the cycles in the associated directed hypergraph is equal to 1. It is natural for us to consider the following.

Question 38. Study the algebraic or graphic characterization of a nonnegative strongly primitive tensor.

We are sure the above two questions are interesting and not easy.

Conflicts of Interest

The authors declare that they have no conflicts of interest.

Acknowledgments

L. You's research is supported by the National Natural Science Foundation of China (Grant no. 11571123) and the Guangdong Provincial Natural Science Foundation (Grant no. 2015A030313377). P. Yuan's research is supported by the NSF of China (Grant no. 11671153). Y. Chen's research is supported by the Scientific Research Foundation of Graduate School of South China Normal University (Grant no. 2015lkxm19).

References

[1] L. Qi, "Eigenvalues of a real supersymmetric tensor," *Journal of Symbolic Computation*, vol. 40, no. 6, pp. 1302–1324, 2005.

[2] L.-H. Lim, "Singular values and eigenvalues of tensors: a variational approach," in *Proceedings of the 1st IEEE International Workshop on Computational Advances in Multi-Sensor Adaptive Processing (CAMSAP '05)*, pp. 129–132, Puerto Vallarta, Mexico, December 2005.

[3] N. J. Pullman, *Matrix theory and its applications, Selected topics, Pure and Applied Mathematics 35*, Marcel Dekker, Inc., New York, USA, 1976.

[4] K.-C. Chang, K. J. Pearson, and T. Zhang, "Primitivity, the convergence of the NQZ method, and the largest eigenvalue for nonnegative tensors," *SIAM Journal on Matrix Analysis and Applications*, vol. 32, no. 3, pp. 806–819, 2011.

[5] J.-Y. Shao, "A general product of tensors with applications," *Linear Algebra and Its Applications*, vol. 439, no. 8, pp. 2350–2366, 2013.

[6] K. Pearson, "Essentially positive tensors," *International Journal of Algebra*, vol. 4, pp. 421–427, 2010.

[7] P. Yuan, Z. He, and L. You, "A conjecture on the primitive degree of tensors," *Linear Algebra and Its Applications*, vol. 450, pp. 175–185, 2014.

[8] P. Yuan, Z. He, and L. You, "Further results and some open problems on the primitive degree of nonnegative tensors," *Linear Algebra and Its Applications*, vol. 480, pp. 72–92, 2015.

[9] Z. He, P. Yuan, and L. You, "On the exponent set of nonnegative primitive tensors," *Linear Algebra and Its Applications*, vol. 465, pp. 376–390, 2015.

[10] J.-Y. Shao, H.-Y. Shan, and L. Zhang, "On some properties of the determinants of tensors," *Linear Algebra and Its Applications*, vol. 439, no. 10, pp. 3057–3069, 2013.

[11] L.-B. Cui, W. Li, and M. K. Ng, "Primitive tensors and directed hypergraphs," *Linear Algebra and Its Applications*, vol. 471, pp. 96–108, 2015.

5

On a Bivariate Spectral Homotopy Analysis Method for Unsteady Mixed Convection Boundary Layer Flow, Heat, and Mass Transfer due to a Stretching Surface in a Rotating Fluid

Sandile S. Motsa and Zodwa G. Makukula

University of Swaziland, Private Bag 4, Kwaluseni, Matsapha M201, Swaziland

Correspondence should be addressed to Zodwa G. Makukula; zmzmakukula@gmail.com

Academic Editor: Saeid Abbasbandy

A bivariate spectral homotopy analysis method (BSHAM) is extended to solutions of systems of nonlinear coupled partial differential equations (PDEs). The method has been used successfully to solve a nonlinear PDE and is now tested with systems. The method is based on a new idea of finding solutions that obey a rule of solution expression that is defined in terms of the bivariate Lagrange interpolation polynomials. The BSHAM is used to solve a system of coupled nonlinear partial differential equations modeling the unsteady mixed convection boundary layer flow, heat, and mass transfer due to a stretching surface in a rotating fluid, taking into consideration the effect of buoyancy forces. Convergence of the numerical solutions was monitored using the residual error of the PDEs. The effects of the flow parameters on the local skin-friction coefficient, the Nusselt number, and the Sherwood number were presented in graphs.

1. Introduction

Fluid flow dynamics due to a stretching surface find applications in the production of sheeting material including metal and polymer sheets. Examples include the cooling of an infinite metallic plate in a cooling bath, paper production, the boundary layer along material handling conveyers, the aerodynamic extrusion of plastic sheets, glass blowing, metal spinning, and drawing plastic films, to name just a few of these applications [1]. The study of a two-dimensional stretching sheet flow was pioneered by Crane [2], who presented an exact analytical solution for the steady two-dimensional stretching sheet problem in a quiescent fluid. Due the important applications of the flow, many researchers have developed Crane's model to study different aspects of boundary layer flow due to a stretching surface.

The two-dimensional flow in the boundary layer induced by a rotating fluid past a stretching surface was first studied by Wang [3]. Such flows have applications in geophysics and astrophysics, in solar physics involved in the sunspot development, and in self-cooled liquid metal blankets in fusion reactors when the container is being rotated [4]. Rajeswari and Nath [5] extended Wang's model to study unsteady flow. Other reported studies include that of Nazar et al. [6], who carried out a numerical investigation of the induced unsteady flow due to a stretching surface in a rotating fluid, where the unsteadiness is caused by the suddenly stretched surface. The resulting equations were solved using the Keller-box method.

The homotopy analysis method was used by Tan and Liao [7] to find accurate analytic approximations of the partial differential equations modeling the unsteady viscous flows due to a suddenly stretching surface in a rotating fluid. Abbas et al. [4] carried out a numerical study to investigate the unsteady magnetohydrodynamic (MHD) boundary layer flow and heat transfer in an incompressible rotating viscous fluid over a stretching continuous sheet. The resulting governing partial differential equations were solved using the Keller-box method. The spectral relaxation method was used to solve a coupled highly nonlinear system of partial differential equations due to an unsteady flow over a stretching surface in an incompressible rotating viscous fluid by Awad et al. [8]. The effects of binary chemical reaction and Arrhenius activation energy were taken into consideration.

Govardhan et al. [9] considered the unsteady flow due to a stretching surface in a rotating fluid in the presence of porous media, where the unsteadiness is caused by the suddenly stretched surface. The transformed governing partial differential equations were solved numerically by using the Adams predictor-corrector method for some values of the physically governing parameters. The Keller-box method was used by Hafidzuddin and Nazar [10], to carry out a numerical study for the steady magnetohydrodynamics rotating boundary layer flow and heat transfer of a viscous fluid over a permeable shrinking sheet.

Motivated by the first success of the bivariate spectral homotopy analysis method (BSHAM) in finding the solution of a nonlinear PDE reported in [11], the aim of the study is to extend the application of the BSHAM to solutions of systems of coupled nonlinear PDEs. The test problem is that of a three-dimensional unsteady laminar viscous flow with heat and mass transfer due to a stretching sheet in a rotating fluid. The model is an extension of that considered by Nazar et al. [6] and Tan and Liao [7], to include heat, mass transfer, and buoyancy effects on the flow. Using similarity transformations that were first used by Williams and Rhyne [12], the unsteady Navier-Stokes equations were reduced into a system of coupled nonlinear partial differential equations. We would like to acknowledge the fact that the problem can be solved using other methods like the original homotopy analysis method (HAM) as in the works of Liao [13, 14], but not with the type of linear operators used with the BSHAM.

The BSHAM is based on finding solutions that obey a rule of solution expression that is expressed in terms of bivariate Lagrange interpolation polynomials. The efficiency and accuracy of the method will be validated using residual errors of the differential equations. Simulations showing important flow characteristics such as the skin-friction coefficient, heat, and mass transfer rates will be carried out.

2. Governing Equations

We consider the unsteady boundary layer flows with heat and mass transfer, caused by a stretching surface at $z = 0$ in a rotating fluid. Let u, v, w be the velocity components in the direction of Cartesian axes x, y, z, respectively, with the axes rotating at an angular velocity Ω in the z direction. When $t < 0$, the surface rotates at an angular velocity Ω in the z direction so that the fluid is at rest relative to the surface. Initially (for $t = 0$), both fluid and plate are stationary and have constant temperature T_∞ and concentration C_∞. At time $t = 0$, the surface at $z = 0$ is impulsively stretched in the x-direction. For this time, the surface temperature and concentration vary from T_∞ to T_w ($T_w > T_\infty$) and C_∞ to C_w ($C_w > C_\infty$), respectively. Due to the Coriolis force, the fluid motion is three-dimensional and the corresponding governing equations are given as [6, 7].

$$\frac{\partial u}{\partial x} + \frac{\partial v}{\partial y} + \frac{\partial w}{\partial z} = 0, \tag{1}$$

$$\begin{aligned} \frac{\partial u}{\partial t} + u\frac{\partial u}{\partial x} + v\frac{\partial u}{\partial y} + w\frac{\partial u}{\partial z} - 2\Omega v \\ = -\frac{1}{\rho}\frac{\partial p}{\partial x} + \nu\nabla^2 u \\ + g\{\beta_T(T - T_\infty) + \beta_C(C - C_\infty)\}, \end{aligned} \tag{2}$$

$$\frac{\partial v}{\partial t} + u\frac{\partial v}{\partial x} + v\frac{\partial v}{\partial y} + w\frac{\partial v}{\partial z} + 2\Omega u = -\frac{1}{\rho}\frac{\partial p}{\partial y} + \nu\nabla^2 v, \tag{3}$$

$$\frac{\partial w}{\partial t} + u\frac{\partial w}{\partial x} + v\frac{\partial w}{\partial y} + w\frac{\partial w}{\partial z} = -\frac{1}{\rho}\frac{\partial p}{\partial z} + \nu\nabla^2 w, \tag{4}$$

$$\frac{\partial T}{\partial t} + u\frac{\partial T}{\partial x} + v\frac{\partial T}{\partial y} + w\frac{\partial T}{\partial z} = \alpha\frac{\partial^2 T}{\partial z^2}, \tag{5}$$

$$\frac{\partial C}{\partial t} + u\frac{\partial C}{\partial x} + v\frac{\partial C}{\partial y} + w\frac{\partial C}{\partial z} = D\frac{\partial^2 C}{\partial z^2}, \tag{6}$$

where p denotes the pressure, ρ is the density, ν is the kinematic viscosity, ∇^2 is the three-dimensional Laplacian, T and C are the fluid temperature and concentration, respectively, g is the acceleration due to gravity, β_T is the thermal expansion coefficient, β_C is the concentration expansion coefficient, α is the thermal diffusivity, and D is the diffusion coefficient. The initial and boundary conditions are given as follows:

$t < 0$:

$$\begin{aligned} & u = v = w = 0, \\ & T = T_\infty, \\ & C = C_\infty, \\ & \qquad \text{for any } x, y, z. \end{aligned} \tag{7}$$

$t \geq 0$:

$$\begin{aligned} & u = ax, \\ & v = w = 0, \\ & T = T_w, \\ & C = C_w, \\ & \qquad \text{at } z = 0, \end{aligned} \tag{8}$$

$$\begin{aligned} & u \longrightarrow 0, \\ & v \longrightarrow 0, \\ & w \longrightarrow 0, \\ & T \longrightarrow 0, \\ & C \longrightarrow 0, \\ & \qquad \text{as } z \longrightarrow \infty. \end{aligned} \tag{9}$$

Introducing the following similarity transformations [12] in (2)–(6),

$$\begin{aligned} \xi &= 1 - e^{-\tau}, \quad \tau = at, \\ \eta &= \left(\frac{a}{\nu}\right)^{1/2} \xi^{-1/2} z, \\ u &= axf'(\xi,\eta), \\ v &= axg(\xi,\eta), \\ w &= -(a\nu)^{1/2}\xi^{1/2} f(\xi,\eta), \\ T &= T_\infty + (T_w - T_\infty)\theta(\xi,\eta), \\ C &= C_\infty + (C_w - C_\infty)\phi(\xi,\eta), \end{aligned} \tag{10}$$

gives rise to the coupled system of nonlinear PDEs

$$\begin{aligned} f''' + \frac{\eta}{2}(1-\xi) f'' + \xi\left(ff'' - f'^2 + 2\lambda g + \gamma\theta + \sigma\phi\right) \\ = \xi(1-\xi)\frac{\partial f'}{\partial \xi}, \end{aligned} \tag{11}$$

$$\begin{aligned} g'' + \frac{\eta}{2}(1-\xi) g' + \xi\left(fg' - f'g - 2\lambda f'\right) \\ = \xi(1-\xi)\frac{\partial g}{\partial \xi}, \end{aligned} \tag{12}$$

$$\frac{1}{\Pr}\theta'' + \frac{\eta}{2}(1-\xi)\theta' + \xi f\theta' = \xi(1-\xi)\frac{\partial \theta}{\partial \xi}, \tag{13}$$

$$\frac{1}{\mathrm{Sc}}\phi'' + \frac{\eta}{2}(1-\xi)\phi' + \xi f\phi' = \xi(1-\xi)\frac{\partial \phi}{\partial \xi}, \tag{14}$$

with the corresponding boundary conditions

$$\begin{aligned} f(\xi,0) &= 0, \\ f'(\xi,0) &= 1, \\ f'(\xi,\infty) &= 0, \\ g(\xi,0) &= 0, \\ g(\xi,\infty) &= 0, \\ \theta(\xi,0) &= 1, \\ \theta(\xi,\infty) &= 0, \\ \phi(\xi,0) &= 1, \\ \phi(\xi,\infty) &= 0. \end{aligned} \tag{15}$$

In the above, λ is an important dimensionless parameter signifying the relative importance of rotation rate to stretching rate, γ is the buoyancy parameter, σ is the concentration buoyancy parameter, Pr is the Prandtl number, and Sc the Schmidt number given by

$$\begin{aligned} \lambda &= \frac{\Omega}{a}, \\ \gamma &= \frac{g\beta_T(T_w - T_\infty)}{a^2 x}, \\ \sigma &= \frac{g\beta_C(C_w - C_\infty)}{a^2 x}, \\ \Pr &= \frac{\nu}{\alpha}, \\ \mathrm{Sc} &= \frac{\nu}{D}. \end{aligned} \tag{16}$$

The important physical parameters for the type of boundary layer flow under consideration are the skin-friction coefficient, Nusselt number, and Sherwood number. From the velocity field, the shearing stress at the surface can be obtained, which in the nondimensional form (skin-friction coefficient) in the x- and y-directions using (10) is given by

$$\begin{aligned} C_f^x &= (\xi \mathrm{Re}_x)^{-1/2} f''(\xi,0), \\ C_f^y &= (\xi \mathrm{Re}_x)^{-1/2} g'(\xi,0), \end{aligned} \tag{17}$$

where $\mathrm{Re}_x = ax^2/\nu$ is the local Reynolds number.

Using the temperature field, the heat transfer coefficient at the surface can be obtained, which, in the nondimensional form, in terms of the Nusselt number, is given by

$$\mathrm{Nu}_r = \mathrm{Nu}_x(\xi \mathrm{Re}_x)^{-1/2} = -\theta'(\xi,0). \tag{18}$$

From the concentration field, we are able to get the mass transfer coefficient at the plate, which, in the nondimensional form, in terms of the Sherwood number, is given by

$$\mathrm{Sh}_r = \mathrm{Sh}(\xi \mathrm{Re}_x)^{-1/2} = -\phi'(\xi,0). \tag{19}$$

3. The Bivariate Spectral Homotopy Analysis Method (BSHAM)

In this section, we give a description of the method of solution used to solve the governing equations (11)–(14), the bivariate spectral homotopy analysis method (BSHAM). The method is a recent innovation presented by Motsa [11]. It is an extension of the spectral homotopy analysis method (SHAM), whereby base functions used are defined in terms of the bivariate Lagrange interpolation polynomials. The first step of the solution algorithm is to reduce the nonlinear system of PDEs into a system of nonlinear ordinary differential equation (ODEs). This is achieved by applying the spectral collocation in the ξ direction. The resulting equations are then integrated using the spectral collocation method. Following Motsa [11], we approximate the exact solutions of $f(\eta,\xi)$, $g(\eta,\xi)$, $\theta(\eta,\xi)$, and $\phi(\eta,\xi)$ by Lagrange interpolation polynomials that interpolate the functions at the so-called Gauss-Lobatto collocation points defined as

$$\zeta_i = \cos\frac{i\pi}{N_t}, \quad i = 0, 1, 2, \ldots, N_t, \tag{20}$$

where $\zeta \in [-1, 1]$ and $\zeta = 2\xi - 1$, so that the unknown functions $f(\eta, \xi)$, $g(\eta, \xi)$, $\theta(\eta, \xi)$, and $\phi(\eta, \xi)$ are approximated as

$$\begin{aligned} f(\eta,\xi) &\approx \sum_{j=0}^{N_t} f(\eta,\xi_j) L_j(\xi) = \sum_{j=0}^{N_t} f_j(\eta) L_j(\xi), \\ g(\eta,\xi) &\approx \sum_{j=0}^{N_t} g(\eta,\xi_j) L_j(\xi) = \sum_{j=0}^{N_t} g_j(\eta) L_j(\xi), \\ \theta(\eta,\xi) &\approx \sum_{j=0}^{N_t} \theta(\eta,\xi_j) L_j(\xi) = \sum_{j=0}^{N_t} \theta_j(\eta) L_j(\xi), \\ \phi(\eta,\xi) &\approx \sum_{j=0}^{N_t} \phi(\eta,\xi_j) L_j(\xi) = \sum_{j=0}^{N_t} \phi_j(\eta) L_j(\xi), \end{aligned} \tag{21}$$

with $y_j(\eta) = y(\eta, \xi_j)$, $y = f, g, \theta$ or ϕ and $L_j(\xi)$ are the characteristic Lagrange cardinal polynomials given by

$$\begin{aligned} L_j(\xi) &= \prod_{\substack{j=0 \\ j\neq k}}^{M} \frac{\xi - \xi_k}{\xi_j - \xi_k}, \\ L_j(\xi_k) &= \delta_{jk} = \begin{cases} 0 & \text{if } j \neq k \\ 1 & \text{if } j = k. \end{cases} \end{aligned} \tag{22}$$

Equations (21) are then substituted in the governing nonlinear system of PDEs (11)–(14) and collocation is applied; that is, the equations are required to be satisfied exactly at the collocation points ξ_i, $i = 0, 1, 2, \ldots, N_t$. An important step in the substitution process is the evaluation of the derivatives of $L_j(\xi)$ with respect to ξ. Several formulas exist for computing the derivatives if the collocation points are chosen to be Chebyshev-Gauss-Lobatto points. The values of the derivatives at the Chebyshev-Gauss-Lobatto points are computed as

$$\begin{aligned} \left.\frac{\partial f'}{\partial \xi}\right|_{\xi=\xi_i} &= 2\sum_{j=0}^{N_t} d_{i,j} f_j'(\eta), \\ \left.\frac{\partial g}{\partial \xi}\right|_{\xi=\xi_i} &= 2\sum_{j=0}^{N_t} d_{i,j} g_j(\eta), \\ \left.\frac{\partial \theta}{\partial \xi}\right|_{\xi=\xi_i} &= 2\sum_{j=0}^{N_t} d_{i,j} \theta_j(\eta), \\ \left.\frac{\partial \phi}{\partial \xi}\right|_{\xi=\xi_i} &= 2\sum_{j=0}^{N_t} d_{i,j} \phi_j(\eta), \\ & i = 0, 1, 2, \ldots, N_t, \end{aligned} \tag{23}$$

where $d_{i,j}$ $(i, j = 0, 1, \ldots, N_t)$ are entries of the standard Chebyshev differentiation matrix $\mathbf{d} = [d_{i,j}]$ of size $(N_t + 1) \times (N_t + 1)$ (see, e.g., [15, 16]).

Consequently, substituting (23) in (11)–(14) gives

$$\begin{aligned} & f_i''' + \frac{1}{2}(1-\xi_i)\eta f_i'' - 2\xi_i(1-\xi_i)\sum_{j=0}^{N_t} d_{i,j} f_j' \\ & \quad + \xi_i\left(f_i f_i'' - (f_i')^2 + 2\lambda g_i + \gamma\theta_i + \sigma\phi_i\right) = 0, \\ & g_i'' + \frac{1}{2}(1-\xi_i)\eta g_i' - 2\xi_i(1-\xi_i)\sum_{j=0}^{N_t} d_{i,j} g_j \\ & \quad + \xi_i\left(f_i g_i' - f_i' g_i - 2\lambda f_i'\right) = 0, \\ & \frac{1}{\Pr}\theta_i'' + \frac{1}{2}(1-\xi_i)\eta\theta_i' - 2\xi_i(1-\xi_i)\sum_{j=0}^{N_t} d_{i,j}\theta_j \\ & \quad + \xi_i f_i \theta_i' = 0, \\ & \frac{1}{\mathrm{Sc}}\phi_i'' + \frac{1}{2}(1-\xi_i)\eta\phi_i' - 2\xi_i(1-\xi_i)\sum_{j=0}^{N_t} d_{i,j}\phi_j \\ & \quad + \xi_i f_i \phi_i' = 0, \end{aligned} \tag{24}$$

subject to the boundary conditions

$$\begin{aligned} f_i(0) &= 0, \\ f_i'(0) &= 1, \\ f_i'(\infty) &= 0, \\ g_i(0) &= 0, \\ g_i(\infty) &= 0, \\ \theta_i(0) &= 1, \\ \theta_i(\infty) &= 0, \\ \phi_i(0) &= 1, \\ \phi_i(\infty) &= 0, \\ & i = 0, 1, \ldots, N_t. \end{aligned} \tag{25}$$

With the initial conditions at $\xi = 0$ (which corresponds to $\zeta = \zeta_{N_t}$) known, note that the system (24) forms a system of N_t nonlinear coupled ODEs to be solved for f_i, g_i, θ_i, and ϕ_i, $i = 0, 1, \ldots, N_t - 1$ subject to the corresponding boundary conditions. The next step is to apply the HAM basic principle of decoupling the system of equations into a sequence of linear systems. We refer interested readers about the details on Liao's HAM to [17, 18]. The decoupling process is made possible through the formation of the so-called zeroth-order deformation equation, defined as

$$\begin{aligned}
&(1-q)\mathscr{L}_{f_i}\left[\left\{F_0(\eta;q),\ldots,F_{N_t-1}(\eta;q)\right\}-f_{i,0}(\eta)\right]\\
&\quad = q\hbar\left\{\mathscr{N}_{f_i}\left[F_0(\eta;q),\ldots,F_{N_t-1}(\eta;q)\right]\right.\\
&\qquad \left.-2\xi_i(1-\xi_i)\,d_{i,M}f'_{N_t}\right\},\\
&(1-q)\mathscr{L}_{g_i}\left[\left\{G_0(\eta;q),\ldots,G_{N_t-1}(\eta;q)\right\}-g_{i,0}(\eta)\right]\\
&\quad = q\hbar\left\{\mathscr{N}_{g_i}\left[G_0(\eta;q),\ldots,G_{N_t-1}(\eta;q)\right]\right.\\
&\qquad \left.-2\xi_i(1-\xi_i)\,d_{i,M}g_{N_t}\right\},\\
&(1-q)\mathscr{L}_{\theta_i}\left[\left\{\Theta_0(\eta;q),\ldots,\Theta_{N_t-1}(\eta;q)\right\}-\theta_{i,0}(\eta)\right]\\
&\quad = q\hbar\left\{\mathscr{N}_{\theta_i}\left[\Theta_0(\eta;q),\ldots,\Theta_{N_t-1}(\eta;q)\right]\right.\\
&\qquad \left.-2\xi_i(1-\xi_i)\,d_{i,M}\theta_{N_t}\right\},\\
&(1-q)\mathscr{L}_{\phi_i}\left[\left\{\Phi_0(\eta;q),\ldots,\Phi_{N_t-1}(\eta;q)\right\}-\phi_{i,0}(\eta)\right]\\
&\quad = q\hbar\left\{\mathscr{N}_{\phi_i}\left[\Phi_0(\eta;q),\ldots,\Phi_{N_t-1}(\eta;q)\right]\right.\\
&\qquad \left.-2\xi_i(1-\xi_i)\,d_{i,M}\phi_{N_t}\right\},
\end{aligned}\tag{26}$$

with corresponding boundary conditions

$$\begin{aligned}
F_i(0;q)&=0,\\
F'_i(0;q)&=0,\\
F_i(\infty;q)&=0,\\
G_i(0;q)&=0,\\
G_i(\infty;q)&=0,\\
\Theta_i(0;q)&=0,\\
\Theta_i(\infty;q)&=0,\\
\Phi_i(0;q)&=0,\\
\Phi_i(\infty;q)&=0.
\end{aligned}\tag{27}$$

In the above equations, $q \in [0,1]$ is an embedding parameter and $\hbar$ represents a nonzero convergence controlling auxiliary parameter. The initial approximate solutions of $f_i(\eta)$, $g_i(\eta)$, $\theta_i(\eta)$, and $\phi_i(\eta)$ are given as $f_{i,0}(\eta)$, $g_{i,0}(\eta)$, $\theta_{i,0}(\eta)$, and $\phi_{i,0}(\eta)$, respectively, for $i = 0,1,2,\ldots,N_t-1$. The components $\mathscr{L}_p$ and $\mathscr{N}_p$, $p = f_i, g_i, \theta_i$ or ϕ_i, are the linear and nonlinear operators of the unknown functions defined by

$$\begin{aligned}
\mathscr{L}_{f_i} &= f_i''' + \frac{1}{2}(1-\xi_i)\eta f_i'' + \xi_i\left(f_i''-f_i'\right)\\
&\quad - 2\xi_i(1-\xi_i)\sum_{j=0}^{N_t} d_{i,j}f_j',\\
\mathscr{N}_{f_i} &= f_i''' + \frac{1}{2}(1-\xi_i)\eta f_i'' - 2\xi_i(1-\xi_i)\sum_{j=0}^{N_t} d_{i,j}f_j'\\
&\quad + \xi_i\left(f_if_i'' - \left(f_i'\right)^2 + 2\lambda g_i + \gamma\theta_i + \sigma\phi_i\right),\\
\mathscr{L}_{g_i} &= g_i'' + \frac{1}{2}(1-\xi_i)\eta g_i' - 2\xi_i(1-\xi_i)\sum_{j=0}^{N_t} d_{i,j}g_j,\\
\mathscr{N}_{g_i} &= g_i'' + \frac{1}{2}(1-\xi_i)\eta g_i' - 2\xi_i(1-\xi_i)\sum_{j=0}^{N_t} d_{i,j}g_j\\
&\quad + \xi_i\left(f_ig_i' - f_i'g_i - 2\lambda f_i'\right),\\
\mathscr{L}_{\theta_i} &= \frac{1}{\Pr}\theta_i'' + \frac{1}{2}(1-\xi_i)\eta\theta_i' - 2\xi_i(1-\xi_i)\sum_{j=0}^{N_t} d_{i,j}\theta_j,\\
\mathscr{N}_{\theta_i} &= \frac{1}{\Pr}\theta_i'' + \frac{1}{2}(1-\xi_i)\eta\theta_i' - 2\xi_i(1-\xi_i)\sum_{j=0}^{N_t} d_{i,j}\theta_j\\
&\quad + \xi_if_i\theta_i',\\
\mathscr{L}_{\phi_i} &= \frac{1}{\mathrm{Sc}}\phi_i'' + \frac{1}{2}(1-\xi_i)\eta\phi_i' - 2\xi_i(1-\xi_i)\sum_{j=0}^{N_t} d_{i,j}\phi_j,\\
\mathscr{N}_{\phi_i} &= \frac{1}{\mathrm{Sc}}\phi_i'' + \frac{1}{2}(1-\xi_i)\eta\phi_i' - 2\xi_i(1-\xi_i)\sum_{j=0}^{N_t} d_{i,j}\phi_j\\
&\quad + \xi_if_i\phi_i'.
\end{aligned}\tag{28}$$

The initial approximate solutions are chosen to ensure that

$$\begin{aligned}
&\mathscr{L}_{f_i}\left[f_{i,0}\right] = 2\xi_i(1-\xi_i)\,d_{i,M}f'_{N_t},\\
&\qquad f_{i,0}(0)=0,\;\; f'_{i,0}=1,\;\; f'_{i,0}(\infty)=0,\\
&\mathscr{L}_{g_i}\left[g_{i,0}\right] = 2\xi_i(1-\xi_i)\,d_{i,M}g_{N_t},\\
&\qquad g_{i,0}(0)=0,\;\; g_{i,0}(\infty)=0,\\
&\mathscr{L}_{\theta_i}\left[\theta_{i,0}\right] = 2\xi_i(1-\xi_i)\,d_{i,M}\theta_{N_t},\\
&\qquad \theta_{i,0}(0)=0,\;\; \theta_{i,0}(\infty)=0,\\
&\mathscr{L}_{\phi_i}\left[\phi_{i,0}\right] = 2\xi_i(1-\xi_i)\,d_{i,M}\phi_{N_t},\\
&\qquad \phi_{i,0}(0)=0,\;\; \phi_{i,0}(\infty)=0.
\end{aligned}\tag{29}$$

From the zeroth-order equations (26), it can be shown that as q varies from 0 to 1, the unknown variables $F_i(\eta;q)$,

$G_i(\eta;q)$, $\Theta_i(\eta;q)$, and $\Phi_i(\eta;q)$ vary from their initial approximate solutions $f_{i,0}(\eta)$, $g_{i,0}(\eta)$, $\theta_{i,0}(\eta)$, and $\phi_{i,0}(\eta)$ to the corresponding solutions $f_i(\eta)$, $g_i(\eta)$, $\theta_i(\eta)$, and $\phi_i(\eta)$, respectively, of the governing equations (24). Expanding $F_i(\eta;q)$, $G_i(\eta;q)$, $\Theta_i(\eta;q)$, and $\Phi_i(\eta;q)$ using Taylor series about q gives

$$\begin{aligned}
F_i(\eta;q) &= F_i(\eta;0) + \sum_{n=1}^{+\infty} f_{i,n}(\eta)\, q^n, \\
G_i(\eta;q) &= G_i(\eta;0) + \sum_{n=1}^{+\infty} g_{i,n}(\eta)\, q^n, \\
\Theta_i(\eta;q) &= \Theta_i(\eta;0) + \sum_{n=1}^{+\infty} \theta_{i,n}(\eta)\, q^n, \\
\Phi_i(\eta;q) &= \Phi_i(\eta;0) + \sum_{n=1}^{+\infty} \phi_{i,n}(\eta)\, q^n,
\end{aligned} \tag{30}$$

where

$$\begin{aligned}
f_{i,n}(\eta) &= \frac{1}{n!}\frac{\partial^n F_i(\eta;q)}{\partial q^n}\bigg|_{q=0}, \\
g_{i,n}(\eta) &= \frac{1}{n!}\frac{\partial^n G_i(\eta;q)}{\partial q^n}\bigg|_{q=0}, \\
\theta_{i,n}(\eta) &= \frac{1}{n!}\frac{\partial^n \Theta_i(\eta;q)}{\partial q^n}\bigg|_{q=0}, \\
\phi_{i,n}(\eta) &= \frac{1}{n!}\frac{\partial^n \Phi_i(\eta;q)}{\partial q^n}\bigg|_{q=0}.
\end{aligned} \tag{31}$$

Therefore, since $F_i(\eta;1) = f_i(\eta)$, $G_i(\eta;1) = g_i(\eta)$, $\Theta_i(\eta;1) = \theta_i(\eta)$, $\Phi_i(\eta;1) = \phi_i(\eta)$ and $F_i(\eta;0) = f_{i,0}(\eta,\xi)$, $G_i(\eta;0) = g_{i,0}(\eta,\xi)$, $\Theta_i(\eta;0) = \theta_{i,0}(\eta,\xi)$, $\Phi_i(\eta;0) = \phi_{i,0}(\eta,\xi)$, we obtain

$$\begin{aligned}
f_i(\eta) &= f_{i,0}(\eta) + \sum_{m=1}^{+\infty} f_{i,m}(\eta), \\
g_i(\eta) &= g_{i,0}(\eta) + \sum_{m=1}^{+\infty} g_{i,m}(\eta), \\
\theta_i(\eta) &= \theta_{i,0}(\eta) + \sum_{m=1}^{+\infty} \theta_{i,m}(\eta), \\
\phi_i(\eta) &= \phi_{i,0}(\eta) + \sum_{m=1}^{+\infty} \phi_{i,m}(\eta).
\end{aligned} \tag{32}$$

Convergence of the series (32) depends on a proper choice of the auxiliary parameter $\hbar$. Next, higher order deformation equations are formed, where we are able to get the terms of the series from their solution, $f_{i,m}$, $g_{i,m}$, $\theta_{i,m}$, and $\phi_{i,m}$, $m \geq 1$. The higher order deformation equations are constructed by differentiating the zeroth-order deformation equations (26) m times with respect to q, then dividing by $m!$, and finally setting $q = 0$ given as follows:

$$\begin{aligned}
\mathscr{L}_{f_i}\left[f_{i,m}(\eta) - \chi_m f_{i,m-1}(\eta)\right] &= \hbar R^{f_i}_{i,m-1}, \\
\mathscr{L}_{g_i}\left[g_{i,m}(\eta) - \chi_m g_{i,m-1}(\eta)\right] &= \hbar R^{g_i}_{i,m-1}, \\
\mathscr{L}_{\theta_i}\left[\theta_{i,m}(\eta) - \chi_m \theta_{i,m-1}(\eta)\right] &= \hbar R^{\theta_i}_{i,m-1}, \\
\mathscr{L}_{\phi_i}\left[\phi_{i,m}(\eta) - \chi_m \phi_{i,m-1}(\eta)\right] &= \hbar R^{\phi_i}_{i,m-1},
\end{aligned} \tag{33}$$

with boundary conditions

$$\begin{aligned}
f_{i,m}(0) &= 0, \\
f'_{i,m}(0) &= 0, \\
f'_{i,m}(\infty) &= 0, \\
g_{i,m}(0) &= 0, \\
g_{i,m}(\infty) &= 0, \\
\theta_{i,m}(0) &= 0, \\
\theta_{i,m}(\infty) &= 0, \\
\phi_{i,m}(0) &= 0, \\
\phi_{i,m}(\infty) &= 0.
\end{aligned} \tag{34}$$

From (33),

$$\begin{aligned}
R^{f_i}_{i,m-1} &= \mathscr{L}_{f_i}\left[f_{i,m-1}\right] - 2(1-\chi_m)\xi_i(1-\xi_i)\, d_{i,M} f'_{N_t} \\
&\quad + \xi_i\left(2\lambda g_{i,m-1} + \gamma\theta_{i,m-1}\right) \\
&\quad + \xi_i\left(\sigma\phi_{i,m-1} + \sum_{n=0}^{m-1}\left[f_{i,m-1-n} f''_{i,n} - f'_{i,m-1-n} f'_{i,n}\right]\right), \\
R^{g_i}_{i,m-1} &= \mathscr{L}_{g_i}\left[g_{i,m-1}\right] - 2(1-\chi_m)\xi_i(1-\xi_i)\, d_{i,M} g_{N_t} \\
&\quad - 2\xi\lambda f'_{i,m-1} + \xi_i\sum_{n=0}^{m-1}\left[f_{i,m-1-n} g'_{i,n} - f'_{i,m-1-n} g_{i,n}\right], \\
R^{\theta_i}_{i,m-1} &= \mathscr{L}_{\theta_i}\left[\theta_{i,m-1}\right] - 2(1-\chi_m)\xi_i(1-\xi_i)\, d_{i,M}\theta_{N_t} \\
&\quad + \xi_i\sum_{n=0}^{m-1} f_{i,m-1-n}\theta'_{i,n}, \\
R^{\phi_i}_{i,m-1} &= \mathscr{L}_{\phi_i}\left[\phi_{i,m-1}\right] - 2(1-\chi_m)\xi_i(1-\xi_i)\, d_{i,M}\phi_{N_t} \\
&\quad + \xi_i\sum_{n=0}^{m-1} f_{i,m-1-n}\phi'_{i,n}.
\end{aligned} \tag{35}$$

At this stage, the Chebyshev spectral collocation method is applied independently in the η direction to solve for the initial approximate solutions $f_{i,0}$, $g_{i,0}$, $\theta_{i,0}$, and $\phi_{i,0}$ and the higher order deformation equations giving rise to $f_{i,m}$, $g_{i,m}$, $\theta_{i,m}$, and $\phi_{i,m}$. In the spectral domain, the derivatives of the variables with respect to η are defined in terms of the Chebyshev differentiation matrix; for example, the derivative in $f_{i,m}$ will be given as

$$\left.\frac{d^p f_{i,m}}{d\eta^p}\right|_{\eta=\eta_r} = \left(\frac{2}{\eta_e}\right)^p \sum_{s=0}^{N} D^p_{r,s} f_{i,m}(\overline{\eta}_s) = \mathbf{D}^p \mathbf{F}_{i,m}, \quad r = 0, 1, 2, \ldots, N, \tag{36}$$

similar to the other variables. In (36), p is the order of the derivative, $\mathbf{D} = (2/\eta_e)[D_{r,s}]$ $(r, s = 0, 1, 2, \ldots, N_x)$ with $[D_{r,s}]$ being an $(N_x+1)\times(N_x+1)$ Chebyshev derivative matrix, and the vector $\mathbf{F}_{i,m}$ is defined as

$$\mathbf{F}_{i,m} = \left[f_{i,m}(\overline{\eta}_0), f_{i,m}(\overline{\eta}_1), \ldots, f_{i,m}(\overline{\eta}_{N_x})\right]^t. \tag{37}$$

The derivative formula (36) is a result of a linear transformation used to convert the interval $\eta \in [0, \eta_e]$ to $\overline{\eta} \in [-1, 1]$ where η_e is a sufficiently large finite value chosen to approximate the conditions at ∞. This gives rise to the matrix equations:

$$\begin{aligned} A^{f_i}_{i,m-1}\mathbf{F}_{i,m} &= \hbar \mathbf{R}^{f_i}_{i,m-1}, \\ A^{g_i}_{i,m-1}\mathbf{G}_{i,m} &= \hbar \mathbf{R}^{g_i}_{i,m-1}, \\ A^{\theta_i}_{i,m-1}\mathbf{\Theta}_{i,m} &= \hbar \mathbf{R}^{\theta_i}_{i,m-1}, \\ A^{\phi_i}_{i,m-1}\mathbf{\Phi}_{i,m} &= \hbar \mathbf{R}^{\phi_i}_{i,m-1}, \end{aligned} \tag{38}$$

where

$$\begin{aligned}
A^{f_i}_{i,i} &= \mathbf{D}^3 + \mathbf{a}_0\mathbf{D}^2 + \mathbf{a}_1\mathbf{D} - 2\xi_i(1-\xi_i)\, d_{i,i}\mathbf{D}, \\
A^{f_i}_{i,j} &= -2\xi_i(1-\xi_i)\, d_{i,j}\mathbf{D}, \\
\mathbf{R}^{f_i}_{i,m-1} &= (\hbar + \chi_m) \\
&\quad \cdot \left(\mathbf{D}^3 + \mathbf{a}_0\mathbf{D}^2 + \mathbf{a}_1\mathbf{D} - 2\xi_i(1-\xi_i)\sum_{j=0}^{Nt-1} d_{i,j}\mathbf{D}\right) \\
&\quad \cdot \mathbf{F}_{i,m-1} + \xi_i\left(\gamma\mathbf{\Theta}_{i,m-1} + \sigma\mathbf{\Phi}_{i,m-1} + 2\lambda\mathbf{G}_{i,m-1}\right) \\
&\quad - 2\hbar\xi_i(1-\xi_i)\, d_{i,Nt}\mathbf{D}\mathbf{F}_{i,Nt} \\
&\quad + \xi_i\hbar\sum_{n=0}^{m-1}\left[\mathbf{F}_{i,m-1-n}\mathbf{D}^2\mathbf{F}_{i,n} - \mathbf{D}\mathbf{F}_{i,m-1-n}\mathbf{D}\mathbf{F}_{i,n}\right], \\
A^{g_i}_{i,i} &= \mathbf{D}^2 + \mathbf{b}_0\mathbf{D} + -2\xi_i(1-\xi_i)\, d_{i,i}\mathbf{I}, \\
A^{g_i}_{i,j} &= -2\xi_i(1-\xi_i)\, d_{i,j}\mathbf{I}, \\
\mathbf{R}^{g_i}_{i,m-1} &= (\hbar + \chi_m) \\
&\quad \cdot \left(\mathbf{D}^2 + \mathbf{b}_0\mathbf{D} - 2\xi_i(1-\xi_i)\sum_{j=0}^{Nt-1} d_{i,j}\mathbf{I}\right)\mathbf{G}_{i,m-1} \\
&\quad - 2\hbar\lambda\mathbf{D}\mathbf{F}_{i,m-1} - 2\hbar\xi_i(1-\xi_i)\, d_{i,Nt}\mathbf{G}_{i,Nt} \\
&\quad + \xi_i\hbar\sum_{n=0}^{m-1}\left[\mathbf{F}_{i,m-1-n}\mathbf{D}\mathbf{G}_{i,n} - \mathbf{D}\mathbf{F}_{i,m-1-n}\mathbf{G}_{i,n}\right]. \\
A^{\theta_i}_{i,i} &= \mathbf{D}^2 + \mathbf{c}_0\mathbf{D} + -2\xi_i(1-\xi_i)\, d_{i,i}\mathbf{I}, \\
A^{\theta_i}_{i,j} &= -2\xi_i(1-\xi_i)\, d_{i,j}\mathbf{I}, \\
\mathbf{R}^{\theta_i}_{i,m-1} &= (\hbar + \chi_m) \\
&\quad \cdot \left(\mathbf{D}^2 + \mathbf{c}_0\mathbf{D} - 2\xi_i(1-\xi_i)\sum_{j=0}^{Nt-1} d_{i,j}\mathbf{I}\right)\mathbf{\Theta}_{i,m-1} \\
&\quad - 2\hbar\xi_i(1-\xi_i)\, d_{i,Nt}\mathbf{\Theta}_{i,Nt} + \xi_i\hbar\sum_{n=0}^{m-1}\mathbf{F}_{i,m-1-n}\mathbf{D}\mathbf{\Theta}_{i,n}, \\
A^{\phi_i}_{i,i} &= \mathbf{D}^2 + \mathbf{d}_0\mathbf{D} + -2\xi_i(1-\xi_i)\, d_{i,i}\mathbf{I}, \\
A^{\phi_i}_{i,j} &= -2\xi_i(1-\xi_i)\, d_{i,j}\mathbf{I}, \\
\mathbf{R}^{\phi_i}_{i,m-1} &= (\hbar + \chi_m) \\
&\quad \cdot \left(\mathbf{D}^2 + \mathbf{d}_0\mathbf{D} - 2\xi_i(1-\xi_i)\sum_{j=0}^{Nt-1} d_{i,j}\mathbf{I}\right)\mathbf{\Phi}_{i,m-1} \\
&\quad - 2\hbar\xi_i(1-\xi_i)\, d_{i,Nt}\mathbf{\Phi}_{i,Nt} + \xi_i\hbar\sum_{n=0}^{m-1}\mathbf{F}_{i,m-1-n}\mathbf{D}\mathbf{\Phi}_{i,n}.
\end{aligned} \tag{39}$$

4. Results and Discussion

In this section, numerical results of the governing system of partial differential equations (11)–(14) obtained using the bivariate spectral homotopy analysis method are presented. The results show convergence of the series solutions against the convergence controlling parameter, $\hbar$, residual error curves and variation of the skin-friction coefficient, and Nusselt and Sherwood numbers with the different flow parameters. The number of collocation points used was 60 and 14 in space (η) and time (ξ), respectively, if not stated.

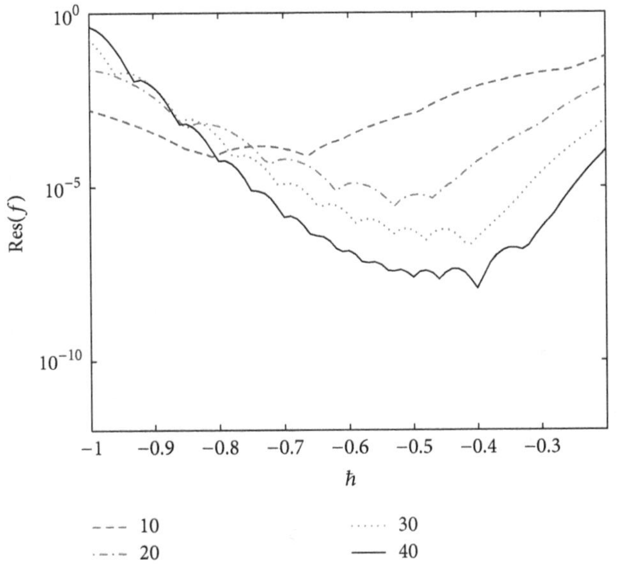

FIGURE 1: Residual error in $f(\eta, \xi)$ against $\hbar$ at different iteration levels ($\lambda = 1, = \gamma = 0.5$, Pr = 3, Sc = 2, $\xi = 0.45$).

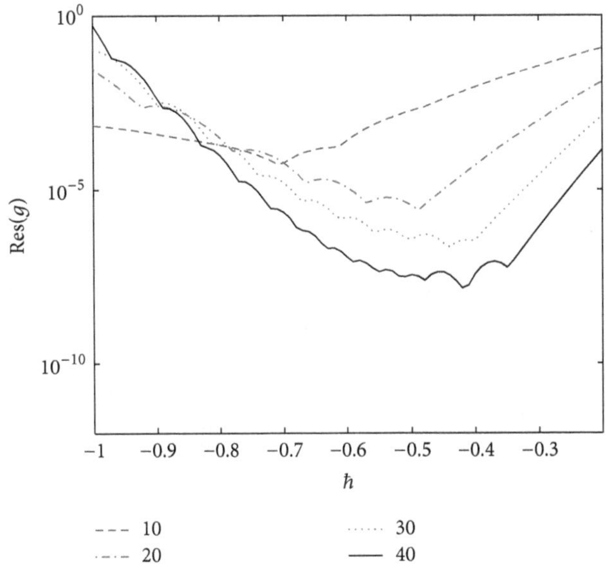

FIGURE 2: Residual error in $g(\eta, \xi)$ against $\hbar$ at different iteration levels ($\lambda = 1$, $\sigma = \gamma = 0.5$, Pr = 3, Sc = 2, $\xi = 0.45$).

The residual error measures the extent to which the numerical solutions approximate the true solution of the governing differential equations (11)–(14). It is used in this study to monitor the accuracy of the numerical scheme. For each variable function, the residual error is defined by

$$\begin{aligned}
\text{Res}(f) &= \left\|\Delta_f\left[\mathbf{F}_i, \mathbf{G}_i, \mathbf{\Theta}_i, \mathbf{\Phi}_i\right]\right\|_\infty, \\
\text{Res}(g) &= \left\|\Delta_g\left[\mathbf{F}_i, \mathbf{G}_i\right]\right\|_\infty, \\
\text{Res}(\theta) &= \left\|\Delta_\theta\left[\mathbf{F}_i, \mathbf{\Theta}_i\right]\right\|_\infty, \\
\text{Res}(\phi) &= \left\|\Delta_\phi\left[\mathbf{F}_i, \mathbf{\Phi}_i\right]\right\|_\infty,
\end{aligned} \tag{40}$$

where Δ_f, Δ_g, Δ_θ, and Δ_ϕ represent the governing nonlinear PDEs (11), (12), (13), and (14), respectively, and $\mathbf{F}_i$, $\mathbf{G}_i$, $\mathbf{\Theta}_i$, and $\mathbf{\Phi}_i$ are the BSHAM approximate solutions at the time collocation points ξ_i.

Figures 1–4 present the residual error curves against the convergence controlling parameter $\hbar$, at various iteration levels. The residual errors are seen to decrease with increase in iterations which shows that the numerical scheme converges. From Figures 1–4, we are also able to pick the value of $\hbar$ that will give optimal results.

Variation of the residual error with time ξ is presented in Figures 5–8. Again the accuracy of the results is seen to improve with improvement in iterations proving convergence of the scheme. The error is seen to be uniform for values of time.

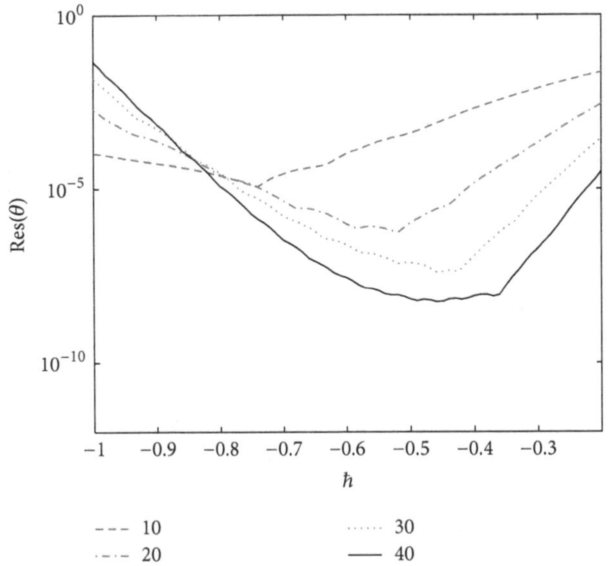

FIGURE 3: Residual error in $\theta(\eta, \xi)$ against $\hbar$ at different iteration levels ($\lambda = 1$, $\sigma = \gamma = 0.5$, Pr = 3, Sc = 2, $\xi = 0.45$).

The effect of the different flow parameters on the local skin friction along the x-direction, $f''(0, \xi)$, is shown in Figures 9–13. Figure 9 shows the effect of the rotating ratio parameter λ and it can be seen that the local friction decreases with increase in λ. These findings are in agreement with those of [8, 9]. The effect of γ is shown to increase the local skin-friction coefficient in Figure 10. This is caused by the fact that γ is positive, thus assisting the flow enhancing the fluid velocity.

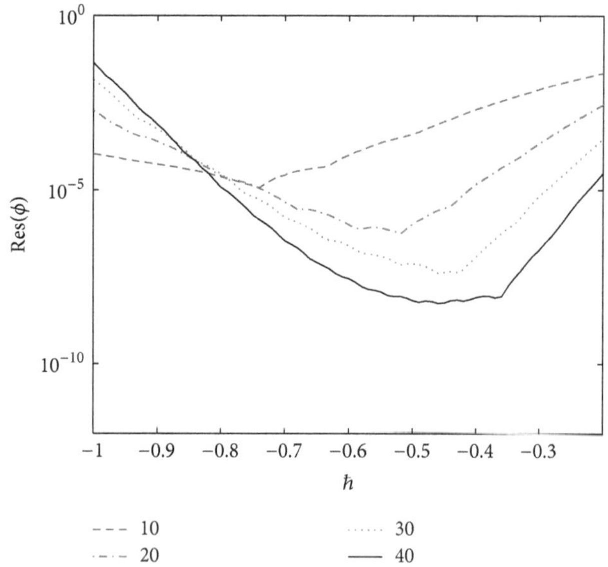

FIGURE 4: Residual error in $\phi(\eta, \xi)$ against $\hbar$ at different iteration levels ($\lambda = 1$, $\sigma = \gamma = 0.5$, $\Pr = 3$, $\mathrm{Sc} = 2$, $\xi = 0.45$).

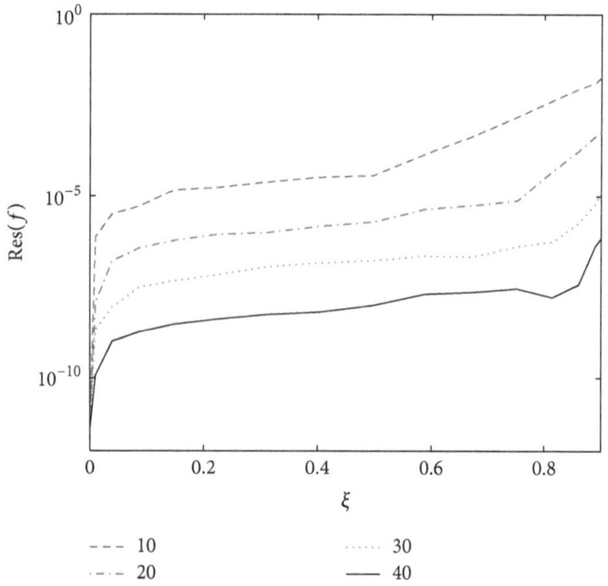

FIGURE 5: Residual error in $f(\eta, \xi)$ against ξ at different iteration levels ($\lambda = 1$, $\sigma = \gamma = 0.5$, $\Pr = 3$, $\mathrm{Sc} = 2$, $\hbar = -0.5$).

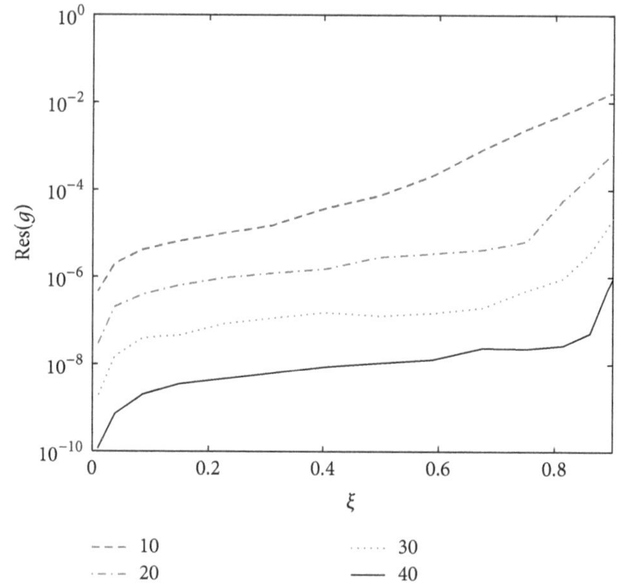

FIGURE 6: Residual error in $g(\eta, \xi)$ against ξ at different iteration levels ($\lambda = 1$, $\sigma = \gamma = 0.5$, $\Pr = 3$, $\mathrm{Sc} = 2$, $\hbar = -0.5$).

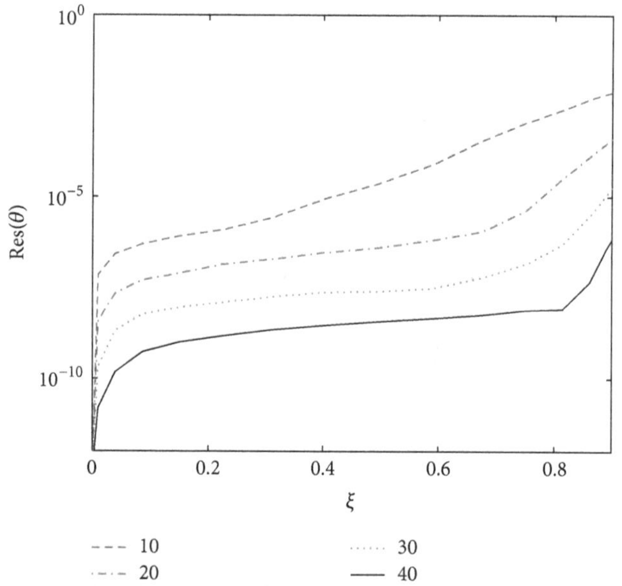

FIGURE 7: Residual error in $\theta(\eta, \xi)$ against ξ at different iteration levels ($\lambda = 1$, $\sigma = \gamma = 0.5$, $\Pr = 3$, $\mathrm{Sc} = 2$, $\hbar = -0.5$).

A similar effect is observed in Figure 11 with σ because positive values of σ also enhance the fluid velocity, in turn increasing the local skin friction. The effects of the Prandtl number and the Schmidt number are to decrease the local skin-friction coefficient $f''(0, \xi)$ as shown in Figures 12 and 13, respectively. An increase in the Prandtl number implies an increase in kinematic viscosity, in turn decreasing the velocity of the fluid flow. The skin friction will decrease with decrease in the velocity profile. The same effect is experienced with Sc; that is, increasing Sc implies an increase in the kinematic viscosity.

Figures 14–18 depict the effect of the flow parameters on the local skin-friction coefficient along the y-direction, $g'(0, \xi)$. As expected, the skin friction along y-direction will also decrease with increase in the rotating ratio parameter λ (presented in Figure 14) as observed earlier with $f''(0, \xi)$ in Figure 9.

The effects of γ and σ on $g'(0, \xi)$ are depicted in Figures 15 and 16, respectively. Increasing either γ or σ enhances the flow along the x-direction, thus decreasing that along

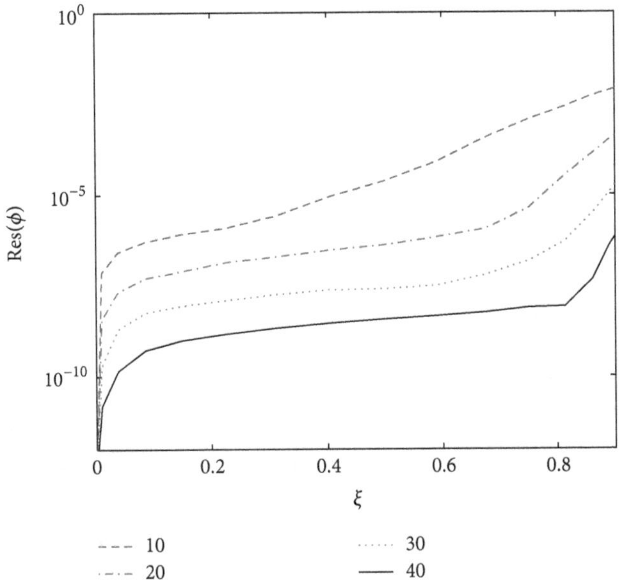

FIGURE 8: Residual error in $\phi(\eta, \xi)$ against ξ at different iteration levels ($\lambda = 1,\ \sigma = \gamma = 0.5,\ \text{Pr} = 3,\ \text{Sc} = 2,\ \hbar = -0.5$).

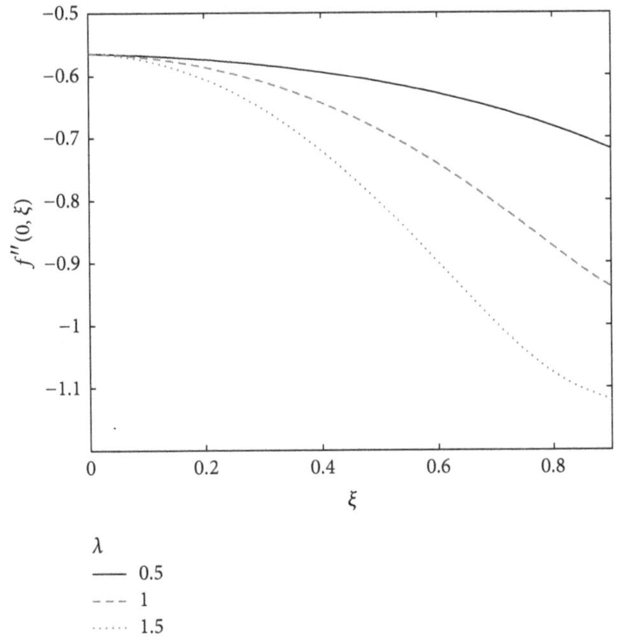

FIGURE 9: Effect of λ on $f''(0, \xi)$ ($\sigma = \gamma = 0.5,\ \text{Pr} = 3,\ \text{Sc} = 2,\ \hbar = -0.5$).

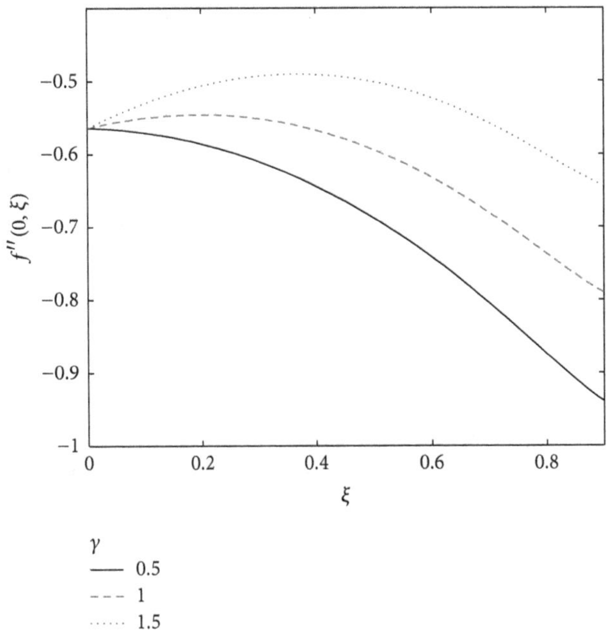

FIGURE 10: Effect of γ on $f''(0, \xi)$ ($\sigma = 0.5\ \lambda = 0.5,\ \text{Pr} = 3,\ \text{Sc} = 2,\ \hbar = -0.5$).

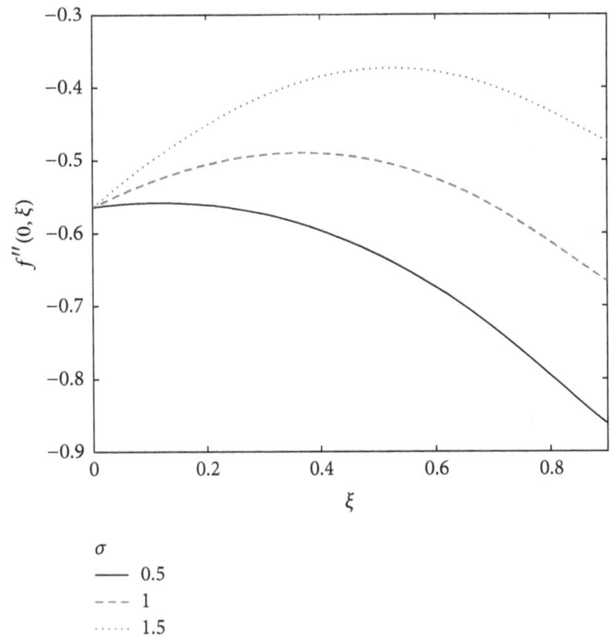

FIGURE 11: Effect of σ on $f''(0, \xi)$ ($\lambda = 1,\ \gamma = 0.5,\ \text{Pr} = 3,\ \text{Sc} = 2,\ \hbar = -0.5$).

the y-direction. The momentum boundary layer along y-direction decreases, in turn decreasing the local skin-friction coefficient along y-direction, $g'(0, \xi)$.

Figures 17 and 18 display the effect of the Prandtl and Schmidt number on $g'(0, \xi)$, respectively. For both flow parameters, the effect is insignificant for the unsteady-state flow. A slight increase in $g'(0, \xi)$ with increase in both Pr and Sc is reported towards the steady-state flow.

Figure 19 presents the effect of the rotating parameter λ on the local Nusselt number. The local Nusselt number decreases with increase in λ with the effect being more momentous towards the steady-state flow. Increasing λ decreases the thermal boundary layer thickness, as reported in [10]. A thinner thermal boundary layer consecutively results in a decrease of the heat transfer rate at the surface.

The effects of γ and σ on the local Nusselt number are presented in Figures 20 and 21, respectively. Both buoyancy forces are seen to enhance the temperature field and the

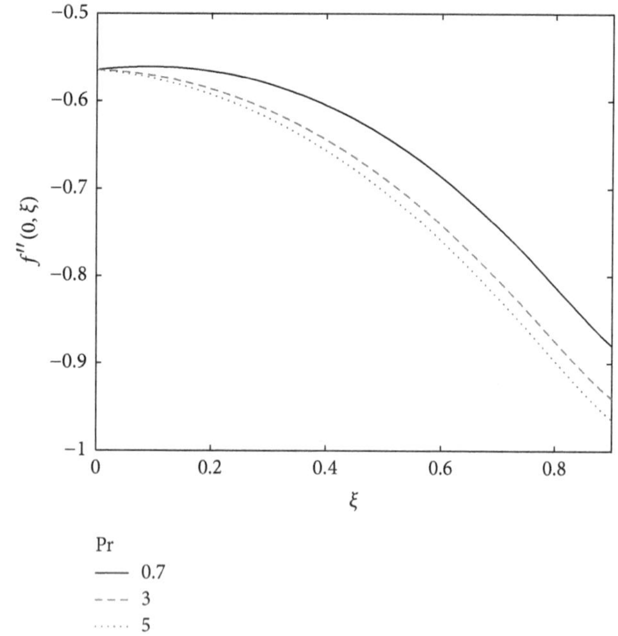

FIGURE 12: Effect of Pr on $f''(0, \xi)$ ($\sigma = \gamma = 0.5,\ \lambda = 1,\ \text{Sc} = 2,\ \hbar = -0.5$).

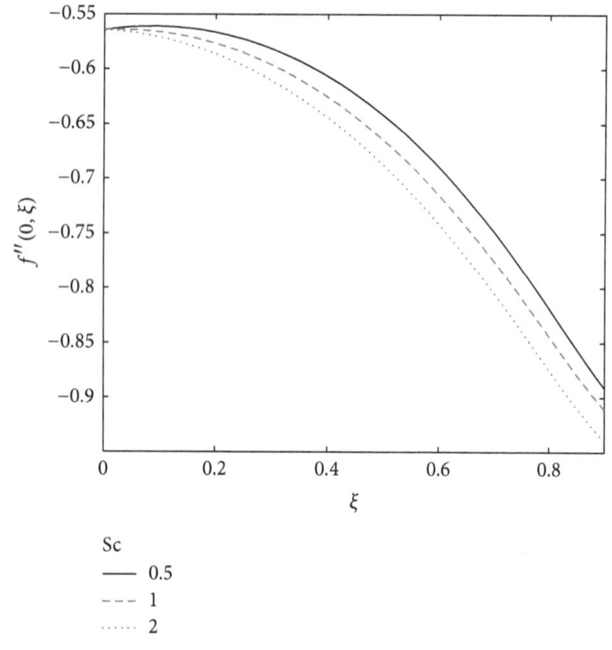

FIGURE 13: Effect of Sc on $f''(0, \xi)$ ($\sigma = \gamma = 0.5,\ \text{Pr} = 3,\ \lambda = 1,\ \hbar = -0.5$).

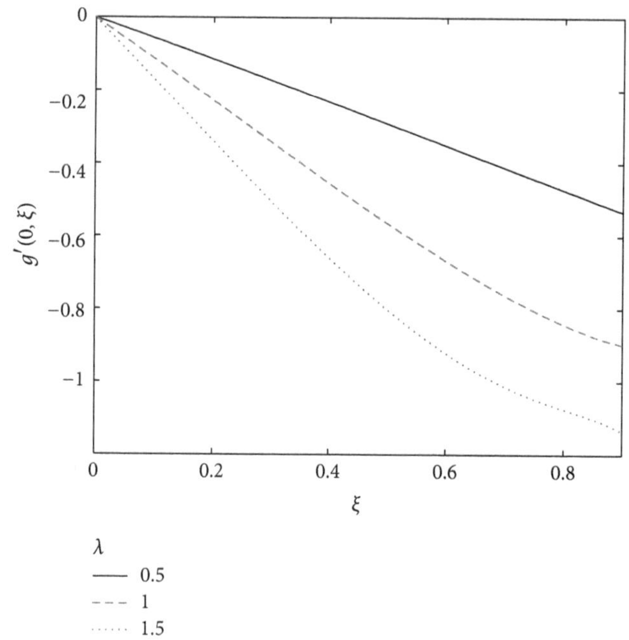

FIGURE 14: Effect of λ on $g'(0, \xi)$. ($\sigma = \gamma = 0.5,\ \text{Pr} = 3,\ \text{Sc} = 2,\ \hbar = -0.5$).

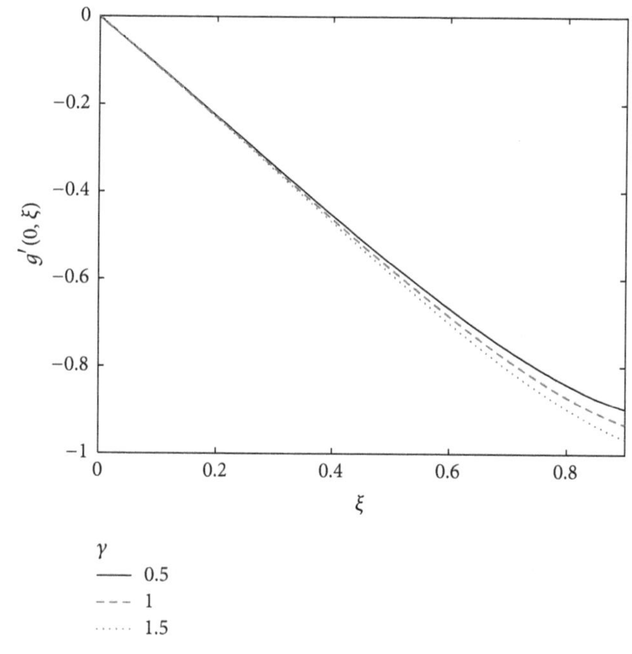

FIGURE 15: Effect of γ on $g'(0, \xi)$ ($\sigma = 0.5,\ \lambda = 1,\ \text{Pr} = 3,\ \text{Sc} = 2,\ \hbar = -0.5$).

thermal boundary layer. This explains the increase in the heat transfer rate with increase in both γ and σ.

The effect of the Prandtl number on the local Nusselt number is presented in Figure 22. The effect of Pr is to increase the heat transfer rate. Physically, this is caused by the fact that increasing Pr reduces the temperature field and the thermal boundary layer thickness, in turn increasing the heat transfer rate. The increase with ξ is significantly pronounced for all values of ξ.

Variation of $-\theta'(0, \xi)$ with the Schmidt number is depicted in Figure 23. The heat transfer rate decreases with increase in the Schmidt number. The findings are in agreement with those of Awad et al. [8].

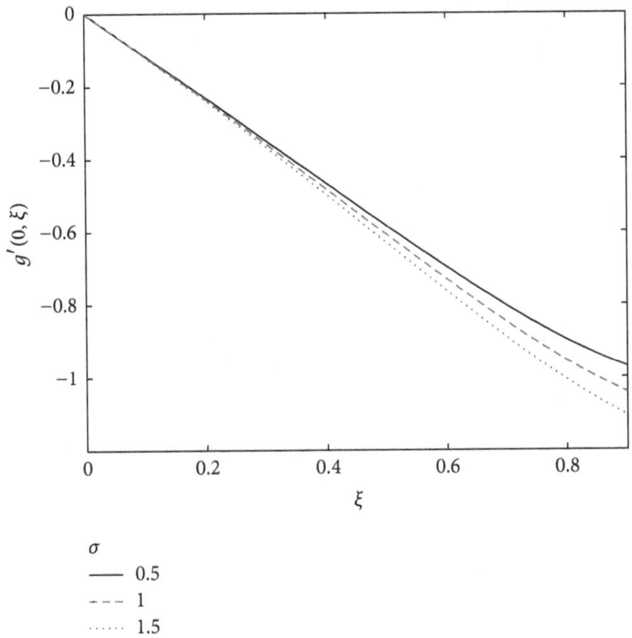

FIGURE 16: Effect of σ on $g'(0,\xi)$ ($\lambda = \gamma = 0.5$, $\Pr = 3$, $Sc = 2$, $\hbar = -0.5$).

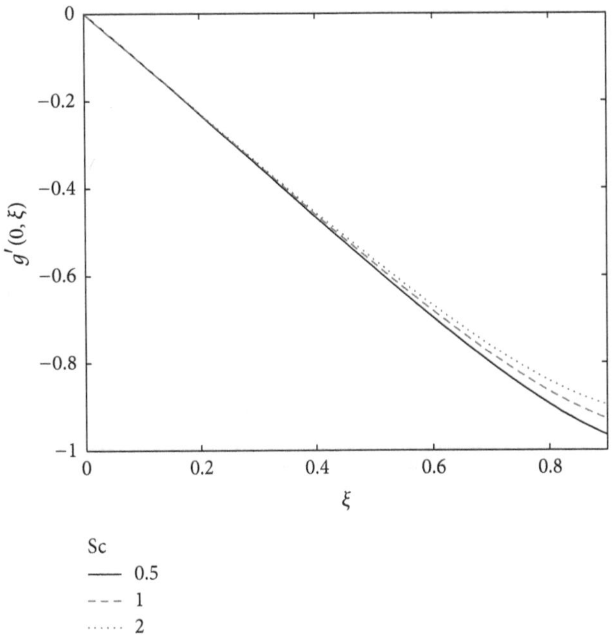

FIGURE 18: Effect of Sc on $g'(0,\xi)$ ($\sigma = \gamma = 0.5$, $\Pr = 3$, $\lambda = 1$, $\hbar = -0.5$).

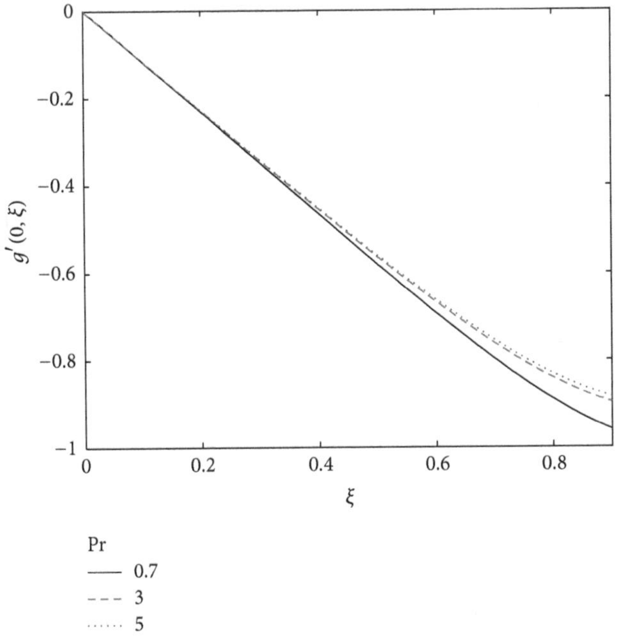

FIGURE 17: Effect of Pr on $g'(0,\xi)$. ($\sigma = \gamma = 0.5$, $\lambda = 1$, $Sc = 2$, $\hbar = -0.5$).

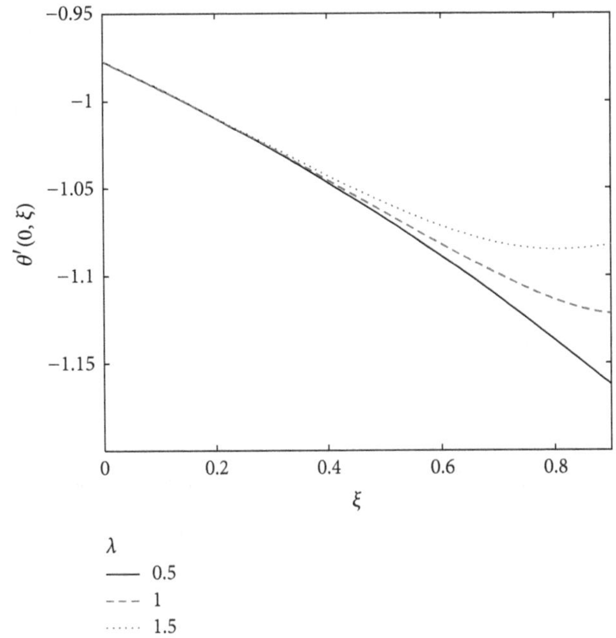

FIGURE 19: Effect of λ on $\theta'(0,\xi)$ ($\sigma = \gamma = 0.5$, $\Pr = 3$, $Sc = 2$, $\hbar = -0.5$).

For a fixed value of ξ, the local Nusselt number increases with Pr as displayed in Figure 22. These findings are consistent with those of Awad et al. [8], whereby it is shown that the thermal boundary layer decreases with increase in Pr, in turn reducing the heat transfer rate at the surface. The effect of the Schmidt number on the local Nusselt number is insignificant for the unsteady-state flow as displayed in Figure 23. A slight increase in $-\theta'(0,\xi)$ is observed towards the steady-state flow.

The effects of the flow parameters on the mass transfer rate are presented in Figures 24–28. The Sherwood number is shown to decrease with increase in λ and Pr in Figures 24 and 27, respectively. The effects of γ, σ, and Sc are to enhance mass transfer rate, presented in Figures 25, 26, and 28, respectively. The results in the present study are consistent with those of a similar study carried out by [8].

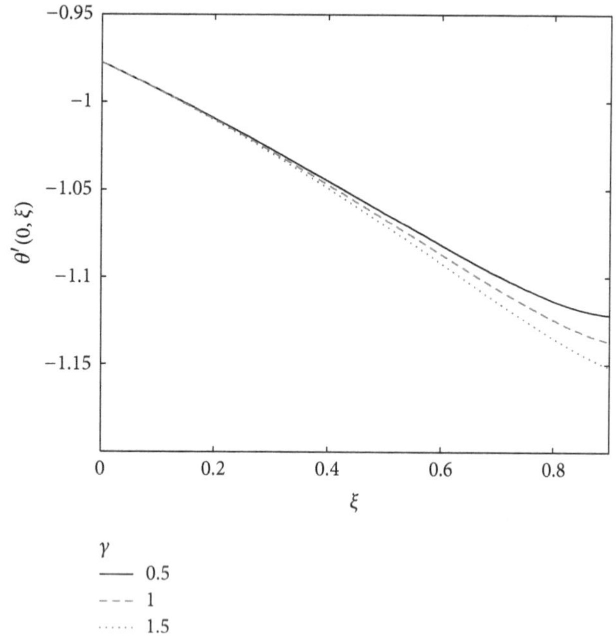

FIGURE 20: Effect of γ on $\theta'(0,\xi)$ ($\sigma = 0.5$, $\lambda = 1$, $\mathrm{Pr} = 3$, $\mathrm{Sc} = 2$, $\hbar = -0.5$).

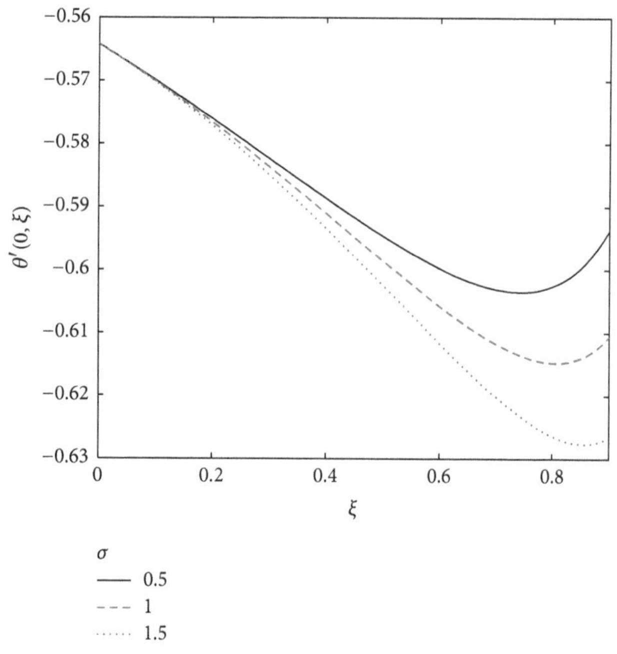

FIGURE 21: Effect of σ on $\theta'(0,\xi)$ ($\lambda = 1$, $\gamma = 0.5$, $\mathrm{Pr} = 3$, $\mathrm{Sc} = 2$, $\hbar = -0.5$).

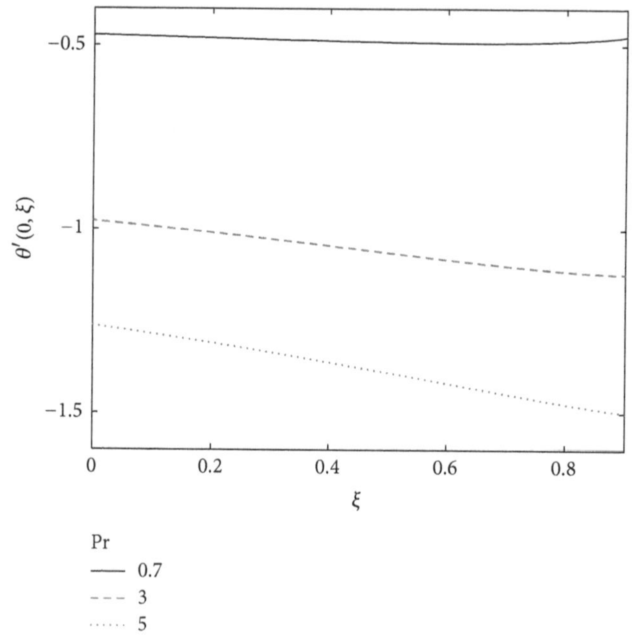

FIGURE 22: Effect of Pr on $\theta'(0,\xi)$ ($\sigma = \gamma = 0.5$, $\lambda = 1$, $\mathrm{Sc} = 2$, $\hbar = -0.5$).

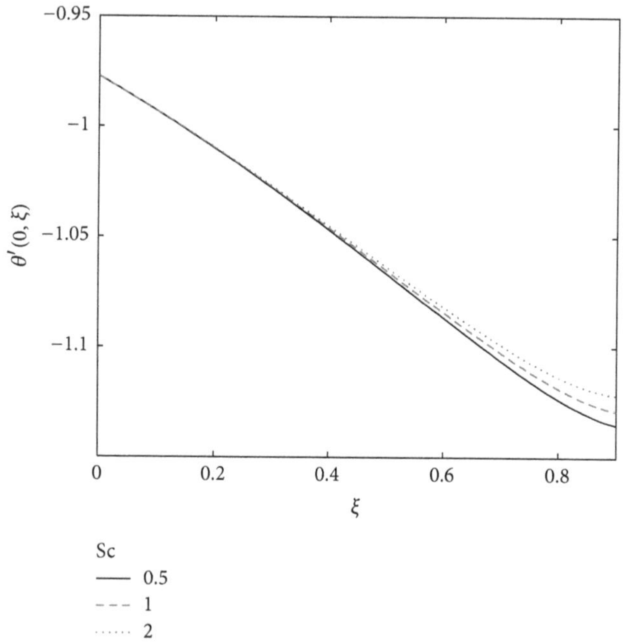

FIGURE 23: Effect of Sc on $\theta'(0,\xi)$ ($\sigma = \gamma = 0.5$, $\mathrm{Pr} = 3$, $\lambda = 1$, $\hbar = -0.5$).

5. Conclusion

The paper extended the application of a bivariate spectral homotopy analysis method to solutions of systems of nonlinear partial differential equations. The test problem used was that of a three-dimensional unsteady laminar viscous flow with heat and mass transfer due to a stretching sheet in a rotating fluid. The transformed equations gave rise to a coupled nonlinear system of partial differential equations. Error analysis was performed in terms of $\hbar$ curves of the residual error of the solutions. Computations were also carried out to analyze the effect of the different flow parameters on the local skin-friction coefficient, the Nusselt number, and the Sherwood number. The buoyancy forces were found to be

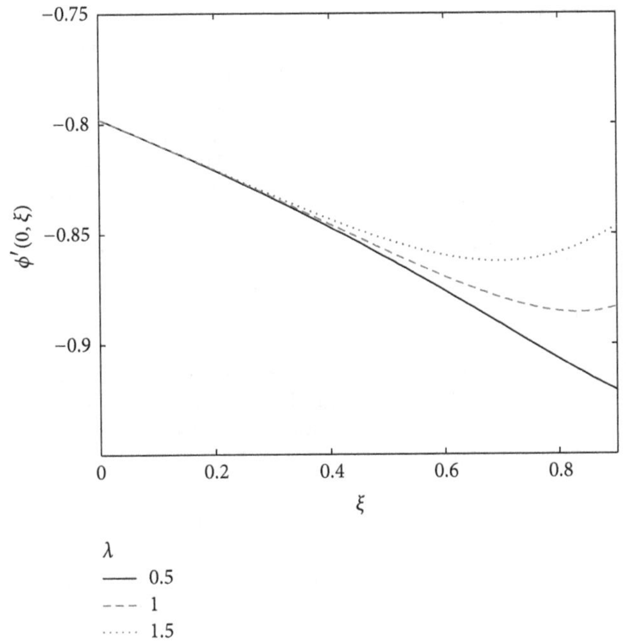

FIGURE 24: Effect of λ on $\phi'(0,\xi)$ ($\sigma = \gamma = 0.5$, $\mathrm{Pr} = 3$, $\mathrm{Sc} = 2$, $\hbar = -0.5$).

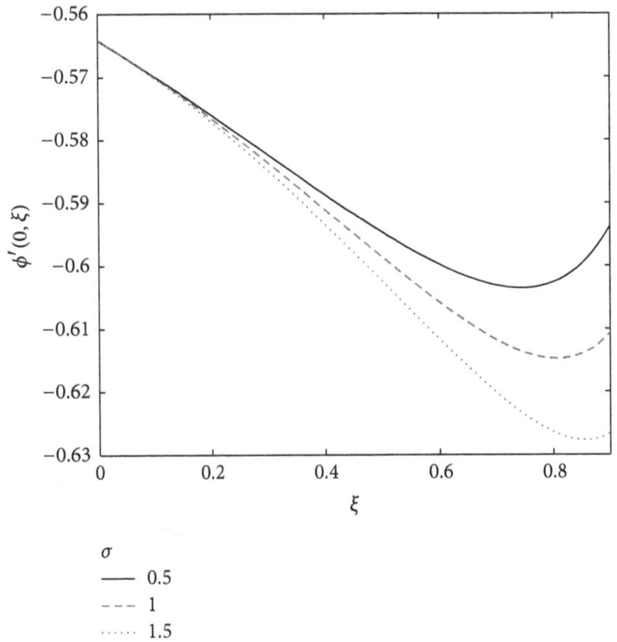

FIGURE 26: Effect of σ on $\phi'(0,\xi)$ ($\lambda = 1$, $\gamma = 0.5$, $\mathrm{Pr} = 3$, $\mathrm{Sc} = 2$, $\hbar = -0.5$).

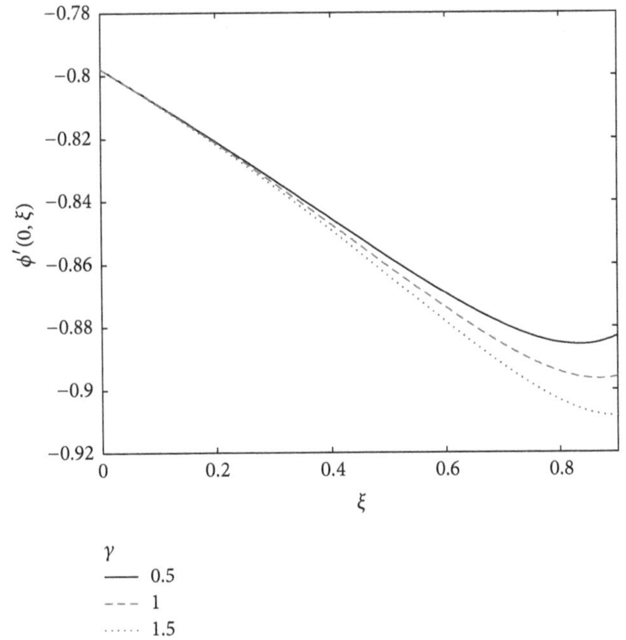

FIGURE 25: Effect of γ on $\phi'(0,\xi)$ ($\sigma = 0.5$, $\lambda = 1$, $\mathrm{Pr} = 3$, $\mathrm{Sc} = 2$, $\hbar = -0.5$).

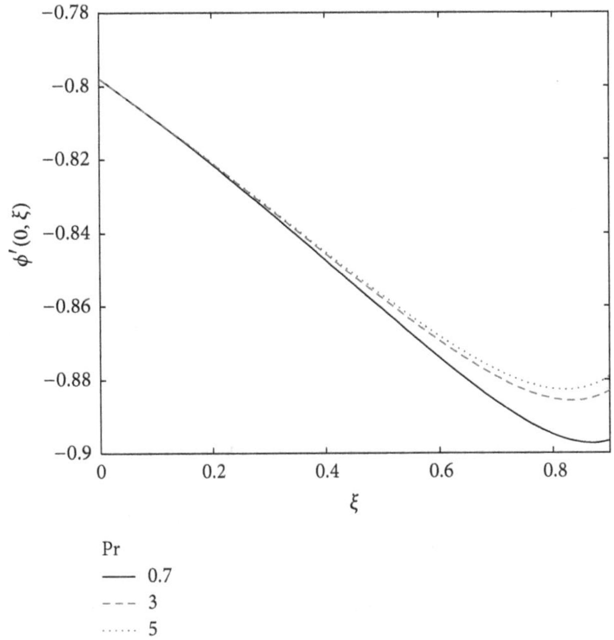

FIGURE 27: Effect of Pr on $\phi'(0,\xi)$ ($\sigma = \gamma = 0.5$, $\lambda = 1$, $\mathrm{Sc} = 2$, $\hbar = -0.5$).

supporting the velocity of the flow in both x- and y-directions in the process decreasing the local skin-friction coefficients in both directions. The buoyancy forces were also reported to increase both heat and mass transfer rates of the flow. Analysis of the other flow parameters proved to be consistent with similar findings reported in the literature. The study proved to be a success, and the bivariate spectral homotopy analysis method can be used to find solutions of coupled nonlinear systems of partial differential equations that are valid for the whole space and time domain.

Conflicts of Interest

The authors declare that there are no conflicts of interest regarding the publication of this paper.

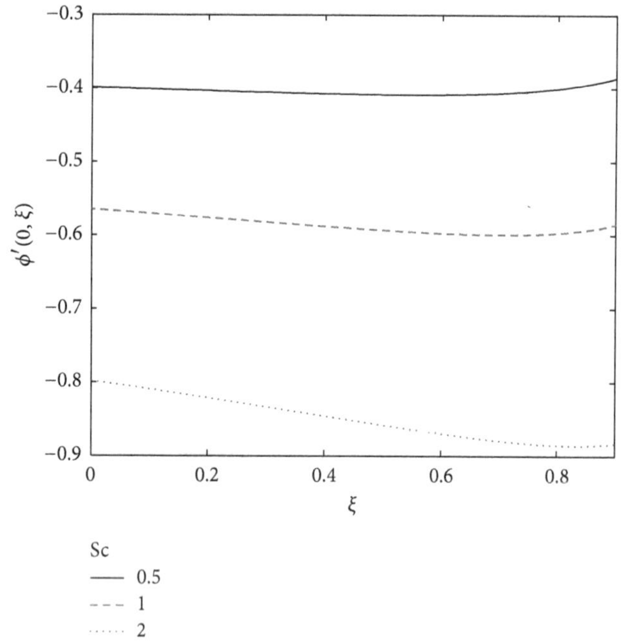

FIGURE 28: Effect of Sc on $\phi'(0,\xi)$ ($\sigma = \gamma = 0.5$, $\mathrm{Pr} = 3$, $\lambda = 1$, $\hbar = -0.5$).

References

[1] A. Ishak, R. Nazar, and I. Pop, "Unsteady mixed convection boundary layer flow due to a stretching vertical surface," *The Arabian Journal for Science and Engineering*, vol. 31, no. 2, pp. 165–182, 2006.

[2] L. J. Crane, "Flow past a stretching plate," *Journal of Applied Mathematics and Physics (ZAMP)*, vol. 21, no. 4, pp. 645–647, 1970.

[3] C. Y. Wang, "Stretching a surface in a roating fluid," *Journal of Applied Mathematics and Physics (ZAMP)*, vol. 39, pp. 177–185, 1988.

[4] Z. Abbas, T. Javed, M. Sajid, and N. Ali, "Unsteady MHD flow and heat transfer on a stretching sheet in a rotating fluid," *Journal of the Taiwan Institute of Chemical Engineers*, vol. 41, no. 6, pp. 644–650, 2010.

[5] V. Rajeswari and G. Nath, "Unsteady flow over a stretching surface in a rotating fluid," *International Journal of Engineering Science*, vol. 30, no. 6, pp. 747–756, 1992.

[6] R. Nazar, N. Amin, and I. Pop, "Unsteady boundary layer flow due to a stretching surface in a rotating fluid," *Mechanics Research Communications*, vol. 31, no. 1, pp. 121–128, 2004.

[7] Y. Tan and S.-J. Liao, "Series solution of three-dimensional unsteady laminar viscous flow due to a stretching surface in a rotating fluid," *Journal of Applied Mechanics*, vol. 74, no. 5, pp. 1011–1018, 2007.

[8] F. G. Awad, S. Motsa, and M. Khumalo, "Heat and mass transfer in unsteady rotating fluid flow with binary chemical reaction and activation energy," *PLoS ONE*, vol. 9, no. 9, Article ID e107622, 2014.

[9] K. Govardhan, B. Balaswamy, and N. Kishan, "Unsteady boundary layer flow due to a stretching porous surface in a rotating fluid," *Engineering Mechanics*, vol. 21, no. 4, pp. 269–277, 2014.

[10] M. E. H. Hafidzuddin and R. Nazar, "Numerical solutions of MHD rotating flow and heat transfer over a permeable shrinking sheet," *ScienceAsia*, vol. 40, pp. 58–62, 2014.

[11] S. S. Motsa, "On an interpolation based spectral homotopy analysis method for PDE based unsteady boundary layer flows," *Abstract and Applied Analysis*, Article ID 848170, 7 pages, 2014.

[12] I. Williams III and T. B. Rhyne, "Boundary layer development on a wedge impulsively set into motion," *SIAM Journal on Applied Mathematics*, vol. 38, no. 2, pp. 215–224, 1980.

[13] S. Liao, "Series solutions of unsteady boundary-layer flows over a stretching flat plate," *Studies in Applied Mathematics*, vol. 117, no. 3, pp. 239–263, 2006.

[14] S. Liao, "An analytic solution of unsteady boundary-layer flows caused by an impulsively stretching plate," *Communications in Nonlinear Science and Numerical Simulation*, vol. 11, no. 3, pp. 326–339, 2006.

[15] C. Canuto, M. Y. Hussaini, A. Quarteroni, and T. A. Zang, *Spectral Methods in Fluid Dynamics*, Springer, New York, NY, USA, 1988.

[16] L. N. Trefethen, *Spectral Methods in MATLAB*, vol. 10, Society for Industrial and Applied Mathematics (SIAM), Philadelphia, Pa, USA, 2000.

[17] S. Liao, *Beyond Perturbation: Introduction to Homotopy Analysis Method*, Chapman & Hall/CRC Press, Boca Raton, Fla, USA, 2003.

[18] S.-J. Liao, *Homotopy Analysis Method in Nonlinear Differential Equations*, Springer, Berlin, Germany, 2012.

6

Relation between Quaternion Fourier Transform and Quaternion Wigner-Ville Distribution Associated with Linear Canonical Transform

Mawardi Bahri and Muh. Saleh Arif Fatimah

Department of Mathematics, Hasanuddin University, Makassar 90245, Indonesia

Correspondence should be addressed to Mawardi Bahri; mawardibahri@gmail.com

Academic Editor: Guowei Yang

The quaternion Wigner-Ville distribution associated with linear canonical transform (QWVD-LCT) is a nontrivial generalization of the quaternion Wigner-Ville distribution to the linear canonical transform (LCT) domain. In the present paper, we establish a fundamental relationship between the QWVD-LCT and the quaternion Fourier transform (QFT). Based on this fact, we provide alternative proof of the well-known properties of the QWVD-LCT such as inversion formula and Moyal formula. We also discuss in detail the relationship among the QWVD-LCT and other generalized transforms. Finally, based on the basic relation between the quaternion ambiguity function associated with the linear canonical transform (QAF-LCT) and the QFT, we present some important properties of the QAF-LCT.

1. Introduction

The quaternion Fourier transform (QFT) is a nontrivial generalization of the real and complex classical Fourier transforms (FT) using quaternion algebra. Many useful properties of the QFT were obtained such as shift, modulation, convolution, correlation, differentiation, energy conservation, and uncertainty principle. It was first introduced in [1] for the analysis of 2D linear time-invariant partial differential systems and then applied in color image processing [2, 3]. It is a natural question to extend the QFT to the linear canonical transform (LCT) domains and then it is the so-called quaternionic linear transform (QLCT). This extension is constructed by substituting the kernel of the QFT with the kernel of the LCT. A number of useful properties of the QLCT have been investigated including shift, orthogonality relation, reconstruction formula, and Heisenberg uncertainty principle (see, for example, [4–6] and the references given therein).

In [7], the author studied that the fractional Fourier transform (FrFT) can be reduced to the classical Fourier transform. Based on this fact, some properties of the FrFT can be derived very easily from those of the classical Fourier transform by simple change variable. Recently, in [8], the authors developed this idea to derive an uncertainty principle associated with the quaternion linear canonical transform (QLCT) by using the fundamental relationship between the QLCT and the QFT [8]. In [9], the authors proposed the quaternion Wigner-Ville distribution associated with linear canonical transform (QWVD-LCT) and obtain its fundamental properties. In the present paper, we first establish the basic relationship between the QWVD-LCT and the QFT. We then show that some fundamental properties of the QWVD-LCT such as inversion formula and Moyal formula can be obtained by combining this relation and the properties of the QFT. We investigate that the QWVD-LCT can be reduced to the quaternion windowed Fourier transform and the continuous quaternion Fourier transform. We finally establish the relationship between the QAF-LCT and the QFT which enables us to derive some useful properties of the QAF-LCT.

2. Preliminaries

In the preliminaries we remind the reader of some facts of quaternions, the quaternion Fourier transform, and the quaternion linear canonical transform.

2.1. Basic Facts about Quaternions. The quaternion algebra over $\mathbb{R}$, denoted by $\mathbb{H}$, is an associative noncommutative four-dimensional algebra:

$$\mathbb{H} = \{q = q_0 + \mathbf{i}q_1 + \mathbf{j}q_2 + \mathbf{k}q_3;\ q_0, q_1, q_2, q_3 \in \mathbb{R}\}, \quad (1)$$

which obeys the following multiplication rules:

$$\begin{aligned} \mathbf{ij} &= -\mathbf{ji} = \mathbf{k}, \\ \mathbf{jk} &= -\mathbf{kj} = \mathbf{i}, \\ \mathbf{ki} &= -\mathbf{ik} = \mathbf{j}, \\ \mathbf{i}^2 &= \mathbf{j}^2 = \mathbf{k}^2 = \mathbf{ijk} = -1. \end{aligned} \quad (2)$$

For a quaternion $q = q_0 + \mathbf{i}q_1 + \mathbf{j}q_2 + \mathbf{k}q_3 \in \mathbb{H}$, q_0 is called the *scalar* part of q denoted by $\mathrm{Sc}(q)$ and $\mathbf{i}q_1 + \mathbf{j}q_2 + \mathbf{k}q_3$ is called the *vector* (or *pure*) part of q. The vector part of q is conventionally denoted by $\mathbf{q}$. Let $p, q \in \mathbb{H}$ and $\mathbf{p}, \mathbf{q}$ be their vector parts, respectively. Equation (2) yields the quaternionic multiplication qp as

$$qp = q_0p_0 - \mathbf{q}\cdot\mathbf{p} + q_0\mathbf{p} + p_0\mathbf{q} + \mathbf{q}\times\mathbf{p}, \quad (3)$$

where $\mathbf{q}\cdot\mathbf{p} = (q_1p_1 + q_2p_2 + q_3p_3)$ and $\mathbf{q}\times\mathbf{p} = \mathbf{i}(q_2p_3 - q_3p_2) + \mathbf{j}(q_3p_1 - q_1p_3) + \mathbf{k}(q_1p_2 - q_2p_1)$.

The quaternion conjugate of q, given by

$$\overline{q} = q_0 - \mathbf{i}q_1 - \mathbf{j}q_2 - \mathbf{k}q_3, \quad q_0, q_1, q_2, q_3 \in \mathbb{R}, \quad (4)$$

is an anti-involution; that is,

$$\overline{qp} = \overline{p}\,\overline{q}. \quad (5)$$

From (4), we obtain the norm or modulus of $q \in \mathbb{H}$ defined as

$$|q| = \sqrt{q\overline{q}} = \sqrt{q_0^2 + q_1^2 + q_2^2 + q_3^2}. \quad (6)$$

It is not difficult to see that

$$|qp| = |q|\,|p|, \quad \forall p, q \in \mathbb{H}. \quad (7)$$

Furthermore, it is easily seen that

$$|pqr| = |rqp|, \quad \forall p, q, q \in \mathbb{H}. \quad (8)$$

Using conjugate (4) and the modulus of q, we can define the inverse of $q \in \mathbb{H} \setminus \{0\}$ as

$$q^{-1} = \frac{\overline{q}}{|q|^2}, \quad (9)$$

which shows that $\mathbb{H}$ is a normed division algebra.

Now we observe that

$$\mathrm{Sc}(p\overline{q}) = \frac{1}{2}(p\overline{q} + q\overline{p}) = q_0p_0 + \mathbf{q}\cdot\mathbf{p}. \quad (10)$$

This leads to the cyclic multiplication; that is,

$$\mathrm{Sc}(pqr) = \mathrm{Sc}(rqp) = \mathrm{Sc}(qpr), \quad \forall p, q, r \in \mathbb{H}. \quad (11)$$

We define an inner product for quaternion-valued functions $f, g : \mathbb{R}^2 \to \mathbb{H}$ as follows:

$$(f, g) = \int_{\mathbb{R}^2} f(\mathbf{x})\overline{g(\mathbf{x})}d\mathbf{x}, \quad d\mathbf{x} = dx_1dx_2 \quad (12)$$

with symmetric real scalar part

$$\langle f, g\rangle = \frac{1}{2}[(f, g) + (g, f)] = \mathrm{Sc}\int_{\mathbb{R}^2} f(\mathbf{x})\overline{g}(\mathbf{x})\,d\mathbf{x}. \quad (13)$$

In particular, for $f = g$, we obtain the $L^2(\mathbb{R}^2; \mathbb{H})$-norm

$$\|f\| = \sqrt{\langle f, f\rangle} = \left(\int_{\mathbb{R}^2} |f(\mathbf{x})|^2\,d\mathbf{x}\right)^{1/2}. \quad (14)$$

This gives the $L^p(\mathbb{R}^2; \mathbb{H})$-norm

$$\|f\|_p = \left(\int_{\mathbb{R}^2} |f(\mathbf{x})|^p\,d\mathbf{x}\right)^{1/p}, \quad 1 \le p < \infty. \quad (15)$$

2.2. Quaternion Linear Canonical Transform and Its Basic Properties. In this section, we briefly discuss the definition of the two-sided QFT and the two-sided quaternion linear canonical transform (QLCT) (for simplicity of notation, we write the QFT and QLCT instead of the two-sided QFT and the two-sided QLCT, resp., in the next section). We further collect some basic properties of the QLCT, which will be very useful later on.

Definition 1. The QFT of $f \in L^1(\mathbb{R}^2; \mathbb{H})$ is the transform $\mathcal{F}_q\{f\} : \mathbb{R}^2 \to \mathbb{H}$ given by the integral

$$\mathcal{F}_q\{f\}(\boldsymbol{\omega}) = \int_{\mathbb{R}^2} e^{-\mathbf{i}\omega_1x_1} f(\mathbf{x})\,e^{-\mathbf{j}\omega_2x_2}d\mathbf{x}, \quad (16)$$

where $\mathbf{x} = x_1\mathbf{e}_1 + x_2\mathbf{e}_2$, $\boldsymbol{\omega} = \omega_1\mathbf{e}_1 + \omega_2\mathbf{e}_2$ and the quaternion exponential product $e^{-\mathbf{i}\omega_1x_1}e^{-\mathbf{j}\omega_2x_2}$ is the quaternion Fourier kernel. Here $\mathcal{F}_q$ is called the quaternion Fourier transform operator.

Definition 2. If $f \in L^1(\mathbb{R}^2; \mathbb{H})$ and $\mathcal{F}_q\{f\} \in L^1(\mathbb{R}^2; \mathbb{H})$, then the inverse transform of the QFT is given by

$$\begin{aligned} f(\mathbf{x}) &= \mathcal{F}_q^{-1}\left[\mathcal{F}_q\{f\}\right](\mathbf{x}) \\ &= \frac{1}{(2\pi)^2}\int_{\mathbb{R}^2} e^{\mathbf{i}\omega_1x_1}\mathcal{F}_q\{f\}(\boldsymbol{\omega})\,e^{\mathbf{j}\omega_2x_2}d\boldsymbol{\omega}, \end{aligned} \quad (17)$$

where $\mathcal{F}_q^{-1}$ is called the inverse QFT operator.

A useful property of the QFT is stated in the following lemma, which is needed to derive Moyal formula of the quaternion Wigner-Ville distribution associated with the linear canonical transform (QWVD-LCT).

Lemma 3 (QFT Parseval). *Let $f, g \in L^1(\mathbb{R}^2; \mathbb{H}) \cap L^2(\mathbb{R}^2; \mathbb{H})$. The relation between f, g and their QFT is given by*

$$\langle f, g\rangle = \left\langle \mathcal{F}_q\{f\}, \mathcal{F}_q\{g\}\right\rangle. \quad (18)$$

In particular, with $f = g$, we get the QFT version of the Plancherel formula; that is,

$$\|f\|^2 = \|\mathscr{F}_q\{f\}\|^2. \tag{19}$$

Definition 4 (QLCT definition). Suppose that $A_1 = (a_1, b_1, c_1, d_1)$ and $A_2 = (a_2, b_2, c_2, d_2)$ are real matrix parameters satisfying $\det(A_1) = \det(A_2) = 1$. The QLCT of a quaternion signal $f \in L^1(\mathbb{R}^2; \mathbb{H})$ is defined by

$$L^{\mathbb{H}}_{A_1,A_2}\{f\}(\boldsymbol{\omega}) = \int_{\mathbb{R}^2} K_{A_1}(x_1, \omega_1) f(\mathbf{x}) K_{A_2}(x_2, \omega_2)\, d\mathbf{x}, \tag{20}$$

where the kernel functions of the QLCT above are given by

$$K_{A_1}(x_1,\omega_1) = \begin{cases} \dfrac{1}{\sqrt{2\pi b_1}} e^{(\mathbf{i}/2)((a_1/b_1)x_1^2 - (2/b_1)x_1\omega_1 + (d_1/b_1)\omega_1^2 - \pi/2)}, & \text{for } b_1 \neq 0 \\ \sqrt{d_1} e^{\mathbf{i}(c_1 d_1/2)\omega_1^2}, & \text{for } b_1 = 0, \end{cases}$$

$$K_{A_2}(x_2,\omega_2) = \begin{cases} \dfrac{1}{\sqrt{2\pi b_2}} e^{(\mathbf{j}/2)((a_2/b_2)x_2^2 - (2/b_2)x_2\omega_2 + (d_2/b_2)\omega_2^2 - \pi/2)}, & \text{for } b_2 \neq 0 \\ \sqrt{d_2} e^{\mathbf{j}(c_2 d_2/2)\omega_2^2}, & \text{for } b_2 = 0. \end{cases} \tag{21}$$

From the definition of the QLCT, we can see easily that, when $b_1 b_2 = 0$ and $b_1 = b_2 = 0$, the QLCT of a signal is essentially a quaternion chirp multiplication. Therefore, in this work, we always assume that $b_1 b_2 \neq 0$.

Lemma 5. *The QLCT of a signal f can be seen as the QFT of the signal f in the following form:*

$$L^{\mathbb{H}}_{A_1,A_2}\{f\}(\boldsymbol{\omega}) = \frac{e^{-\mathbf{i}(\pi/4)}}{\sqrt{2\pi b_1}} \cdot e^{(\mathbf{i}d_1/2b_1)\omega_1^2} \mathscr{F}_q\left\{e^{\mathbf{i}(a_1/2b_1)x_1^2} f(\mathbf{x}) e^{\mathbf{j}(a_2/2b_2)x_2^2}\right\} \cdot \left(\frac{\omega_1}{b_1}, \frac{\omega_2}{b_2}\right) \frac{e^{-\mathbf{j}(\pi/4)}}{\sqrt{2\pi b_2}} e^{(\mathbf{j}d_2/2b_2)\omega_2^2}. \tag{22}$$

It is worth noting that if $A_1 = A_2 = (a_i, b_i, c_i, d_i) = (0, 1, -1, 0)$ for $i = 1, 2$, (22) will reduce to the QFT definition; that is,

$$\begin{aligned} L^{\mathbb{H}}_{A_1,A_2}\{f\}(\boldsymbol{\omega}) &= \int_{\mathbb{R}^2} \frac{e^{-\mathbf{i}(\pi/4)}}{\sqrt{2\pi}} e^{-\mathbf{i}\omega_1 x_1} f(\mathbf{x}) e^{-\mathbf{j}\omega_2 x_2} \frac{e^{-\mathbf{j}(\pi/4)}}{\sqrt{2\pi}} d\mathbf{x} \\ &= \frac{e^{-\mathbf{i}(\pi/4)}}{\sqrt{2\pi}} \mathscr{F}_q\{f\}(\boldsymbol{\omega}) \frac{e^{-\mathbf{j}(\pi/4)}}{\sqrt{2\pi}}. \end{aligned} \tag{23}$$

Theorem 6 (QLCT Parseval). *Two quaternion functions $f, h \in L^1(\mathbb{R}^2; \mathbb{H}) \cap L^2(\mathbb{R}^2; \mathbb{H})$ are related to their QLCT via the Parseval formula, given as*

$$\langle f, h\rangle = \left\langle L^{\mathbb{H}}_{A_1,A_2}\{f\}, L^{\mathbb{H}}_{A_1,A_2}\{h\}\right\rangle. \tag{24}$$

When $f = h$, we get

$$\|f\|^2 = \left\|L^{\mathbb{H}}_{A_1,A_2}\{f\}\right\|^2. \tag{25}$$

Proof. For a detailed proof of the above theorem, we refer the reader to [8]. □

3. Quaternion Wigner-Ville Distribution and Quaternion Ambiguity Function in Linear Canonical Transform Domains

Let us introduce the 2D quaternion Wigner-Ville distribution (QWVD) and quaternion ambiguity function QAF [10]. According to the QWVD and QAF definitions and the QLCT definition, we obtain a definition of the quaternion Wigner-Ville distribution associated with the linear canonical transform (QWVD-LCT) and the quaternion ambiguity function associated with the linear canonical transform (QAF-LCT) (see [9]). We establish the fundamental relationship between the QWVD-LCT and QFT. Applying this relation and the properties of the QFT, we in detail derive the inverse transform formula and Moyal's formula for the QWVD-LCT and the QAF-LCT, where the proof of the properties is quite different from one proposed in [9]. We also study the relationship among the QWVD-LCT, the quaternion windowed Fourier transform, and the continuous quaternion Fourier transform.

3.1. Main Properties of QWVD-LCT and Relation among QAF-LCT, QWFT, and CQWT

Definition 7. The cross quaternion Wigner-Ville distribution of two-dimensional functions (or signals) $f, g \in L^2(\mathbb{R}^2; \mathbb{H})$ is given by

$$\mathscr{W}_{f,g}(\mathbf{x}, \boldsymbol{\omega}) = \int_{\mathbb{R}^2} e^{-\mathbf{i}\omega_1\tau_1} f\left(\mathbf{x} + \frac{\boldsymbol{\tau}}{2}\right) \overline{g}\left(\mathbf{x} - \frac{\boldsymbol{\tau}}{2}\right) e^{-\mathbf{j}\omega_2\tau_2} d\boldsymbol{\tau}, \tag{26}$$

provided that the integral exists.

It should be remembered that the kernel of the cross QWVD in (26) does not commute with quaternion functions f and g so that several properties of the classical Wigner-Ville distribution (WVD) are not valid in the cross QWVD [10].

Definition 8. The cross quaternion ambiguity function of two-dimensional functions (or signals) $f, g \in L^2(\mathbb{R}^2; \mathbb{H})$ is given by

$$\mathscr{A}_{f,g}(\mathbf{x}, \boldsymbol{\omega}) = \int_{\mathbb{R}^2} e^{-\mathbf{i}\omega_1\tau_1} f\left(\boldsymbol{\tau} + \frac{\mathbf{x}}{2}\right) \overline{g}\left(\boldsymbol{\tau} - \frac{\mathbf{x}}{2}\right) e^{-\mathbf{j}\omega_2\tau_2} d\boldsymbol{\tau}, \tag{27}$$

provided that the integral exists.

Definition 9. Suppose that $A_1 = (a_1, b_1, c_1, d_1)$ and $A_2 = (a_2, b_2, c_2, d_2)$ are real matrix parameters satisfying $\det(A_1) = \det(A_2) = 1$. The cross QWVD-LCT and cross QAF-LCT of a quaternion signal $f, g \in L^1(\mathbb{R}^2; \mathbb{H})$ are defined by

$$\mathscr{W}_{f,g}^{A_1,A_2}(\mathbf{x}, \boldsymbol{\omega}) = \int_{\mathbb{R}^2} K_{A_1}(\tau_1, \omega_1) f\left(\mathbf{x} + \frac{\boldsymbol{\tau}}{2}\right) \cdot \overline{g}\left(\mathbf{x} - \frac{\boldsymbol{\tau}}{2}\right) K_{A_2}(\tau_2, \omega_2)\, d\boldsymbol{\tau} \tag{28}$$

$$\mathscr{A}_{f,g}^{A_1,A_2}(\mathbf{x}, \boldsymbol{\omega}) = \int_{\mathbb{R}^2} K_{A_1}(\tau_1, \omega_1) f\left(\boldsymbol{\tau} + \frac{\mathbf{x}}{2}\right) \cdot \overline{g}\left(\boldsymbol{\tau} - \frac{\mathbf{x}}{2}\right) K_{A_2}(\tau_2, \omega_2)\, d\boldsymbol{\tau}. \tag{29}$$

Here the kernel functions of the above transforms are given by

$$K_{A_1}(\tau_1, \omega_1) = \begin{cases} \frac{1}{\sqrt{2\pi b_1}} e^{(\mathbf{i}/2)((a_1/b_1)\tau_1^2 - (2/b_1)\tau_1\omega_1 + (d_1/b_1)\omega_1^2 - \pi/2)}, & \text{for } b_1 \neq 0 \\ \sqrt{d_1} e^{\mathbf{i}(c_1 d_1/2)\omega_1^2}, & \text{for } b_1 = 0, \end{cases}$$
$$K_{A_2}(\tau_2, \omega_2) = \begin{cases} \frac{1}{\sqrt{2\pi b_2}} e^{(\mathbf{j}/2)((a_2/b_2)\tau_2^2 - (2/b_2)\tau_2\omega_2 + (d_2/b_2)\omega_2^2 - \pi/2)}, & \text{for } b_2 \neq 0 \\ \sqrt{d_2} e^{\mathbf{j}(c_2 d_2/2)\omega_2^2}, & \text{for } b_2 = 0. \end{cases} \tag{30}$$

It can be directly seen that if we write $h_{f,g}(\mathbf{x}, \boldsymbol{\tau}) = f(\mathbf{x} + \boldsymbol{\tau}/2)\overline{g}(\mathbf{x} - \boldsymbol{\tau}/2)$, we immediately obtain

$$\begin{aligned} \mathscr{W}_{f,g}^{A_1,A_2}(\mathbf{x}, \boldsymbol{\omega}) &= \int_{\mathbb{R}^2} \frac{1}{\sqrt{2\pi b_1}} e^{(\mathbf{i}/2)((a_1/b_1)\tau_1^2 - (2/b_1)\tau_1\omega_1 + (d_1/b_1)\omega_1^2 - \pi/2)} f\left(\mathbf{x} + \frac{\boldsymbol{\tau}}{2}\right) \overline{g}\left(\mathbf{x} - \frac{\boldsymbol{\tau}}{2}\right) \\ &\quad \cdot \frac{1}{\sqrt{2\pi b_2}} e^{(\mathbf{j}/2)((a_2/b_2)\tau_2^2 - (2/b_2)\tau_2\omega_2 + (d_2/b_2)\omega_2^2 - \pi/2)} d\boldsymbol{\tau} \\ &= \int_{\mathbb{R}^2} \frac{1}{\sqrt{2\pi b_1}} e^{(\mathbf{i}/2)((a_1/b_1)\tau_1^2 - (2/b_1)\tau_1\omega_1 + (d_1/b_1)\omega_1^2 - \pi/2)} h_{f,g}(\mathbf{x}, \boldsymbol{\tau}) \frac{1}{\sqrt{2\pi b_2}} e^{(\mathbf{j}/2)((a_2/b_2)\tau_2^2 - (2/b_2)\tau_2\omega_2 + (d_2/b_2)\omega_2^2 - \pi/2)} d\boldsymbol{\tau} \\ &= L_{A_1,A_2}^{\mathbb{H}}\{h_{f,g}\}(\boldsymbol{\omega}). \end{aligned} \tag{31}$$

This tells us that the cross QWVD-LCT is in fact the QLCT of the function $h_{f,g}(\mathbf{x}, \boldsymbol{\tau})$ with respect to $\boldsymbol{\tau}$. This fact is very important in proving Moyal's formula for the cross QWVD-LCT. Similarly, we also get

$$\mathscr{A}_{f,g}^{A_1,A_2}(\mathbf{x}, \boldsymbol{\omega}) = L_{A_1,A_2}^{\mathbb{H}}\{R_{f,g}\}(\boldsymbol{\omega}), \tag{32}$$

where $R_{f,g}$ is given by

$$R_{f,g}(\mathbf{x}, \boldsymbol{\tau}) = f\left(\boldsymbol{\tau} + \frac{\mathbf{x}}{2}\right) \overline{g}\left(\boldsymbol{\tau} - \frac{\mathbf{x}}{2}\right). \tag{33}$$

The following result presents an inequality related to the cross QWVD-LCT.

Lemma 10. *Suppose that $f \in L^p(\mathbb{R}^2; \mathbb{H})$, $g \in L^q(\mathbb{R}^2; \mathbb{H})$ with $1/p + 1/q = 1$. Then we have*

$$\left|\mathscr{W}_{f,g}^{A_1,A_2}(\mathbf{x}, \boldsymbol{\omega})\right| \leq \frac{1}{2\pi\sqrt{b_1 b_2}} \|f\|_p \|g\|_q. \tag{34}$$

Proof. We straightforwardly obtain from (28) that

$$\begin{aligned} \left|\mathscr{W}_{f,g}^{A_1,A_2}(\mathbf{x}, \boldsymbol{\omega})\right| &= \left|\int_{\mathbb{R}^2} \frac{1}{\sqrt{2\pi b_1}} e^{(\mathbf{i}/2)((a_1/b_1)\tau_1^2 - (2/b_1)\tau_1\omega_1 + (d_1/b_1)\omega_1^2 - \pi/2)} f\left(\mathbf{x} + \frac{\boldsymbol{\tau}}{2}\right) \overline{g}\left(\mathbf{x} - \frac{\boldsymbol{\tau}}{2}\right) \times \frac{1}{\sqrt{2\pi b_2}} \right. \\ &\quad \left. \cdot e^{(\mathbf{j}/2)((a_2/b_2)\tau_2^2 - (2/b_2)\tau_2\omega_2 + (d_2/b_2)\omega_2^2 - \pi/2)} d\boldsymbol{\tau}\right| \leq \left(\int_{\mathbb{R}^2} \left|\frac{1}{\sqrt{2\pi b_1}} \right.\right. \\ &\quad \left.\left. \cdot e^{(\mathbf{i}/2)((a_1/b_1)\tau_1^2 - (2/b_1)\tau_1\omega_1 + (d_1/b_1)\omega_1^2 - \pi/2)} f\left(\mathbf{x} + \frac{\boldsymbol{\tau}}{2}\right)\right|^p d\boldsymbol{\tau}\right)^{1/p} \\ &\quad \times \left(\int_{\mathbb{R}^2} \left|\overline{g}\left(\mathbf{x} - \frac{\boldsymbol{\tau}}{2}\right) \frac{1}{\sqrt{2\pi b_2}} \right.\right. \\ &\quad \left.\left. \cdot e^{(\mathbf{j}/2)((a_2/b_2)\tau_2^2 - (2/b_2)\tau_2\omega_2 + (d_2/b_2)\omega_2^2 - \pi/2)}\right|^q d\boldsymbol{\tau}\right)^{1/q}. \end{aligned} \tag{35}$$

Letting $\mathbf{x} + \boldsymbol{\tau}/2 = \mathbf{y}$ and $\mathbf{x} - \boldsymbol{\tau}/2 = \mathbf{z}$, we immediately obtain

$$\begin{aligned} \left|\mathscr{W}_{f,g}^{A_1,A_2}(\mathbf{x}, \boldsymbol{\omega})\right| &= \frac{1}{2\pi\sqrt{b_1 b_2}} \left(\int_{\mathbb{R}^2} \left|f\left(\mathbf{x} + \frac{\boldsymbol{\tau}}{2}\right)\right|^p d\boldsymbol{\tau}\right)^{1/p} \cdot \left(\int_{\mathbb{R}^2} \left|\overline{g}\left(\mathbf{x} - \frac{\boldsymbol{\tau}}{2}\right)\right|^q d\boldsymbol{\tau}\right)^{1/q} \\ &= \frac{1}{2\pi\sqrt{b_1 b_2}} \left(\int_{\mathbb{R}^2} |f(\mathbf{y})|^p d\mathbf{y}\right)^{1/p} \cdot \left(\int_{\mathbb{R}^2} |\overline{g}(\mathbf{z})|^q d\mathbf{z}\right)^{1/q}. \end{aligned} \tag{36}$$

Hence, the result follows. □

Observe first that, for $p = q = 2$, (34) will reduce to

$$\left|\mathscr{W}_{f,g}^{A_1,A_2}(\mathbf{x}, \boldsymbol{\omega})\right| \leq \frac{1}{2\pi\sqrt{b_1 b_2}} \|f\| \|g\|, \tag{37}$$

which shows that $\mathscr{W}_{f,g}^{A_1,A_2}(\mathbf{x}, \boldsymbol{\omega})$ is bounded on $L^2(\mathbb{R}^2; \mathbb{H})$.

Lemma 11. *The cross QWVD-LCT of a signal f with matrix parameters $A_1 = (a_1, b_1, c_1, d_1)$ and $A_2 = (a_2, b_2, c_2, d_2)$*

can be reduced as the QFT of the signal f in the following form:

$$\mathscr{W}_{f,g}^{A_1,A_2}(\mathbf{x},\boldsymbol{\omega}) = \frac{e^{-\mathbf{i}(\pi/4)}}{\sqrt{2\pi b_1}} \cdot e^{(\mathbf{i}d_1/2b_1)\omega_1^2} \mathscr{F}_q\left\{e^{\mathbf{i}(a_1/2b_1)\tau_1^2} f\left(\mathbf{x}+\frac{\boldsymbol{\tau}}{2}\right)\overline{g}\left(\mathbf{x}-\frac{\boldsymbol{\tau}}{2}\right) \cdot e^{\mathbf{j}(a_2/2b_2)\tau_2^2}\right\}\left(\frac{\omega_1}{b_1},\frac{\omega_2}{b_2}\right)\frac{e^{-\mathbf{j}(\pi/4)}}{\sqrt{2\pi b_2}} e^{(\mathbf{j}d_2/2b_2)\omega_2^2}. \tag{38}$$

Proof. By calculation directly, we get

$$\begin{aligned}\mathscr{W}_{f,g}^{A_1,A_2}(\mathbf{x},\boldsymbol{\omega}) &= \frac{1}{\sqrt{2\pi b_1}} \cdot \int_{\mathbb{R}^2} e^{\mathbf{i}(1/2)((a_1/b_1)\tau_1^2-(2/b_1)\tau_1\omega_1+(d_1/b_1)\omega_1^2-\pi/2)} f\left(\mathbf{x}+\frac{\boldsymbol{\tau}}{2}\right)\overline{g}\left(\mathbf{x}-\frac{\boldsymbol{\tau}}{2}\right) \\ &\quad\cdot \frac{1}{\sqrt{2\pi b_2}} e^{\mathbf{j}(1/2)((a_2/b_2)\tau_2^2-(2/b_2)\tau_2\omega_2+(d_2/b_2)\omega_2^2-\pi/2)} d\boldsymbol{\tau} \\ &= \frac{e^{-\mathbf{i}(\pi/4)}}{\sqrt{2\pi b_1}} \cdot e^{\mathbf{i}(d_1/2b_1)\omega_1^2} \int_{\mathbb{R}^2} e^{-\mathbf{i}\tau_1(\omega_1/b_1)} \left(e^{\mathbf{i}(a_1/2b_1)\tau_1^2} f\left(\mathbf{x}+\frac{\boldsymbol{\tau}}{2}\right) \cdot \overline{g}\left(\mathbf{x}-\frac{\boldsymbol{\tau}}{2}\right) e^{\mathbf{j}(a_2/2b_2)\tau_2^2}\right) e^{-\mathbf{j}\tau_2(\omega_2/b_2)} d\boldsymbol{\tau} \\ &\quad\cdot \frac{e^{-\mathbf{j}(\pi/4)}}{\sqrt{2\pi b_2}} e^{\mathbf{j}(d_2/2b_2)\omega_2^2} = \frac{e^{-\mathbf{i}(\pi/4)}}{\sqrt{2\pi b_1}} \cdot e^{(\mathbf{i}d_1/2b_1)\omega_1^2} \mathscr{F}_q\left\{e^{\mathbf{i}(a_1/2b_1)\tau_1^2} f\left(\mathbf{x}+\frac{\boldsymbol{\tau}}{2}\right)\overline{g}\left(\mathbf{x}-\frac{\boldsymbol{\tau}}{2}\right) \cdot e^{\mathbf{j}(a_2/2b_2)\tau_2^2}\right\}\left(\frac{\omega_1}{b_1},\frac{\omega_2}{b_2}\right)\frac{e^{-\mathbf{j}(\pi/4)}}{\sqrt{2\pi b_2}} e^{(\mathbf{j}d_2/2b_2)\omega_2^2},\end{aligned} \tag{39}$$

where the last line follows directly from (28). □

For abbreviation, we use the notation

$$\widetilde{h_{f,g}}(\mathbf{x},\boldsymbol{\tau}) = \frac{e^{-\mathbf{i}(\pi/4)}}{\sqrt{2\pi b_1}} e^{\mathbf{i}(a_1/2b_1)\tau_1^2} f\left(\mathbf{x}+\frac{\boldsymbol{\tau}}{2}\right)\overline{g}\left(\mathbf{x}-\frac{\boldsymbol{\tau}}{2}\right) \cdot e^{\mathbf{j}(a_2/2b_2)\tau_2^2} \frac{e^{-\mathbf{j}(\pi/4)}}{\sqrt{2\pi b_2}}. \tag{40}$$

Therefore, we can write (38) in the form

$$\mathscr{F}_q\left\{\widetilde{h_{f,g}}(\mathbf{x},\boldsymbol{\tau})\right\}\left(\frac{\omega_1}{b_1},\frac{\omega_2}{b_2}\right) = e^{-(\mathbf{i}d_1/2b_1)\omega_1^2} \mathscr{W}_{f,g}^{A_1,A_2}(\mathbf{x},\boldsymbol{\omega}) e^{-(\mathbf{j}d_2/2b_2)\omega_2^2}. \tag{41}$$

In the below theorem by combining the properties of the QFT and the fundamental relation between the QFT and the QWVD-LCT, we provide a new proof of reconstruction formula for the QWVD-LCT.

Theorem 12 (reconstruction formula for QWVD-LCT). *The inverse transform of the cross the QWVD-LCT of the signal $f \in L^2(\mathbb{R}^2;\mathbb{H})$ is given by*

$$f(x) = \frac{1}{\overline{g}(\mathbf{0})} \int_{\mathbb{R}^2} K_{A_1^{-1}}(x_1,\omega_1) \mathscr{W}_{f,g}^{A_1,A_2}\left(\frac{\mathbf{x}}{2},\boldsymbol{\omega}\right) \cdot K_{A_2^{-1}}(x_2,\omega_2)\, d\boldsymbol{\omega} \tag{42}$$

provided that $\overline{g}(\mathbf{0}) \neq \mathbf{0}$.

Proof. From the inverse transform of the QFT (17), it follows that

$$\widetilde{h_{f,g}}(\mathbf{x},\boldsymbol{\tau}) = \frac{1}{(2\pi)^2} \int_{\mathbb{R}^2} e^{\mathbf{i}(\omega_1\tau_1/b_1)} \mathscr{F}_q\left\{\widetilde{h_{f,g}}(\mathbf{x},\boldsymbol{\tau})\right\} \cdot \left(\frac{\omega_1}{b_1},\frac{\omega_2}{b_2}\right) e^{\mathbf{j}(\omega_2\tau_2/b_2)} d\frac{\boldsymbol{\omega}}{\mathbf{b}}. \tag{43}$$

In view of (40) and (41), we easily get

$$\begin{aligned}&\frac{e^{-\mathbf{i}(\pi/4)}}{\sqrt{2\pi b_1}} e^{\mathbf{i}(a_1/2b_1)\tau_1^2} f\left(\mathbf{x}+\frac{\boldsymbol{\tau}}{2}\right)\overline{g}\left(\mathbf{x}-\frac{\boldsymbol{\tau}}{2}\right) e^{\mathbf{j}(a_2/2b_2)\tau_2^2} \cdot \frac{e^{-\mathbf{j}(\pi/4)}}{\sqrt{2\pi b_2}} = \frac{1}{(2\pi)^2} \int_{\mathbb{R}^2} e^{\mathbf{i}(\omega_1\tau_1/b_1)} \mathscr{F}_q\left\{\widetilde{h_{f,g}}(\mathbf{x},\boldsymbol{\tau})\right\} \\ &\quad\cdot \left(\frac{\omega_1}{b_1},\frac{\omega_2}{b_2}\right) e^{\mathbf{j}(\omega_2\tau_2/b_2)} d\frac{\boldsymbol{\omega}}{\mathbf{b}} = \frac{1}{(2\pi)^2} \cdot \int_{\mathbb{R}^2} e^{\mathbf{i}(\omega_1\tau_1/b_1)} e^{-(\mathbf{i}d_1/2b_1)\omega_1^2} \mathscr{W}_{f,g}^{A_1,A_2}(\mathbf{x},\boldsymbol{\omega}) \cdot e^{-(\mathbf{j}d_2/2b_2)\omega_2^2} e^{\mathbf{j}(\omega_2\tau_2/b_2)} d\frac{\boldsymbol{\omega}}{\mathbf{b}}.\end{aligned} \tag{44}$$

Subsequently,

$$\begin{aligned}f\left(\mathbf{x}+\frac{\boldsymbol{\tau}}{2}\right)\overline{g}\left(\mathbf{x}-\frac{\boldsymbol{\tau}}{2}\right) &= \int_{\mathbb{R}^2} \frac{e^{\mathbf{i}(\pi/4)}}{\sqrt{2\pi b_1}} e^{-\mathbf{i}(a_1/2b_1)\tau_1^2} e^{\mathbf{i}(\omega_1\tau_1/b_1)} e^{-(\mathbf{i}d_1/2b_1)\omega_1^2} \mathscr{W}_{f,g}^{A_1,A_2}(\mathbf{x},\boldsymbol{\omega}) \cdot \frac{e^{\mathbf{j}(\pi/4)}}{\sqrt{2\pi b_2}} e^{-(\mathbf{j}d_2/2b_2)\omega_2^2} e^{\mathbf{j}(\omega_2\tau_2/b_2)} e^{-\mathbf{j}(a_2/2b_2)\tau_2^2} d\boldsymbol{\omega} \\ &= \int_{\mathbb{R}^2} \frac{1}{\sqrt{2\pi b_1}} e^{-(\mathbf{i}/2)((a_1/b_1)\tau_1^2-(2/b_1)\tau_1\omega_1+(d_1/b_1)\omega_1^2-\pi/2)} \mathscr{W}_{f,g}^{A_1,A_2}(\mathbf{x},\boldsymbol{\omega}) \frac{1}{\sqrt{2\pi b_2}} e^{-(\mathbf{j}/2)((a_2/b_2)\tau_2^2-(2/b_2)\tau_2\omega_2+(d_2/b_2)\omega_2^2-\pi/2)} d\boldsymbol{\omega} \\ &= \int_{\mathbb{R}^2} K_{A_1^{-1}}(\tau_1,\omega_1) \mathscr{W}_{f,g}^{A_1,A_2}(\mathbf{x},\boldsymbol{\omega}) K_{A_2^{-1}}(\tau_2,\omega_2)\, d\boldsymbol{\omega}.\end{aligned} \tag{45}$$

Letting $\boldsymbol{\tau}/2 = \mathbf{x}$, the above expression will lead to

$$f(2\mathbf{x})\overline{g}(\mathbf{0}) = \int_{\mathbb{R}^2} K_{A_1^{-1}}(2x_1,\omega_1)\mathscr{W}_{f,g}^{A_1,A_2}(\mathbf{x},\boldsymbol{\omega}) \cdot K_{A_2^{-1}}(2x_2,\omega_2)\,d\boldsymbol{\omega}, \tag{46}$$

and the final result can be obtained by letting $2\mathbf{x} = \mathbf{s}$; that is,

$$f(\mathbf{s}) = \frac{1}{\overline{g}(\mathbf{0})}\int_{\mathbb{R}^2} K_{A_1^{-1}}(s_1,\omega_1)\mathscr{W}_{f,g}^{A_1,A_2}\left(\frac{\mathbf{s}}{2},\boldsymbol{\omega}\right) \cdot K_{A_2^{-1}}(s_2,\omega_2)\,d\boldsymbol{\omega}, \tag{47}$$

which completes the proof. □

We now observe that from (45), we straightforwardly obtain the time marginal property of the QWVD-LCT; that is,

$$f\left(\mathbf{x}+\frac{\boldsymbol{\tau}}{2}\right)\overline{g}\left(\mathbf{x}-\frac{\boldsymbol{\tau}}{2}\right) = \int_{\mathbb{R}^2}\frac{1}{\sqrt{2\pi b_1}}e^{(\mathbf{i}/2)((a_1/b_1)\tau_1^2-(2/b_1)\tau_1\omega_1+(d_1/b_1)\omega_1^2-\pi/2)}\mathscr{W}_{f,g}^{A_1,A_2}(\mathbf{x},\boldsymbol{\omega})\frac{1}{\sqrt{2\pi b_2}}e^{(\mathbf{j}/2)((a_2/b_2)\tau_2^2-(2/b_2)\tau_2\omega_2+(d_2/b_2)\omega_2^2-\pi/2)}\,d\boldsymbol{\omega}. \tag{48}$$

If we set $\boldsymbol{\tau} = 0$, the above identity will reduce to

$$f(\mathbf{x})\overline{g}(\mathbf{x}) = \int_{\mathbb{R}^2}\frac{e^{\mathbf{i}(\pi/4)}}{\sqrt{2\pi b_1}}e^{-\mathbf{i}(d_1/2b_1)\omega_1^2}\mathscr{W}_{f,g}^{A_1,A_2}(\mathbf{x},\boldsymbol{\omega}) \cdot \frac{e^{\mathbf{j}(\pi/4)}}{\sqrt{2\pi b_2}}e^{-\mathbf{j}(d_2/2b_2)\omega_2^2}\,d\boldsymbol{\omega}, \tag{49}$$

and if we set $\boldsymbol{\tau} = 0$ and $f = g$, (48) becomes

$$|f(\mathbf{x})|^2 = \int_{\mathbb{R}^2}\frac{e^{\mathbf{i}(\pi/4)}}{\sqrt{2\pi b_1}}e^{-\mathbf{i}(d_1/2b_1)\omega_1^2}\mathscr{W}_{f}^{A_1,A_2}(\mathbf{x},\boldsymbol{\omega}) \cdot \frac{e^{\mathbf{j}(\pi/4)}}{\sqrt{2\pi b_2}}e^{-\mathbf{j}(d_2/2b_2)\omega_2^2}\,d\boldsymbol{\omega}. \tag{50}$$

Integrating (49) and (50) with respect to the $\mathbf{x}$-variable gives

$$\begin{aligned}\int_{\mathbb{R}^2} f(\mathbf{x})\overline{g}(\mathbf{x})\,d\mathbf{x} &= \int_{\mathbb{R}^2}\int_{\mathbb{R}^2}\frac{e^{\mathbf{i}(\pi/4)}}{\sqrt{2\pi b_1}}e^{-\mathbf{i}(d_1/2b_1)\omega_1^2}\mathscr{W}_{f,g}^{A_1,A_2}(\mathbf{x},\boldsymbol{\omega}) \cdot \frac{e^{\mathbf{j}(\pi/4)}}{\sqrt{2\pi b_2}}e^{-\mathbf{j}(d_2/2b_2)\omega_2^2}\,d\boldsymbol{\omega}\,d\mathbf{x} \\ \int_{\mathbb{R}^2}|f(\mathbf{x})|^2\,d\mathbf{x} &= \int_{\mathbb{R}^2}\int_{\mathbb{R}^2}\frac{e^{\mathbf{i}(\pi/4)}}{\sqrt{2\pi b_1}}e^{-\mathbf{i}(d_1/2b_1)\omega_1^2}\mathscr{W}_{f}^{A_1,A_2}(\mathbf{x},\boldsymbol{\omega}) \cdot \frac{e^{\mathbf{j}(\pi/4)}}{\sqrt{2\pi b_2}}e^{-\mathbf{j}(d_2/2b_2)\omega_2^2}\,d\boldsymbol{\omega}\,d\mathbf{x}.\end{aligned} \tag{51}$$

Applying Moyal's formula for the QFT and the relation between the QFT and the QWVD-LCT, we establish general Moyal's formula of the QWVD-LCT (compared to [9]).

Theorem 13 (Moyal's formula for the QWVD-LCT). *Let $f_1, f_2, g_1, g_2 \in L^2(\mathbb{R}^2;\mathbb{H})$ be quaternion-valued signals. Then the following equation holds:*

$$\mathrm{Sc}\int_{\mathbb{R}^2}\int_{\mathbb{R}^2}\mathscr{W}_{f_1,g_1}^{A_1,A_2}(\mathbf{x},\boldsymbol{\omega})\overline{\mathscr{W}_{f_2,g_2}^{A_1,A_2}(\mathbf{x},\boldsymbol{\omega})}\,d\boldsymbol{\omega}\,d\boldsymbol{\tau} = \left\langle f_1(\overline{g_1},\overline{g_2}), f_2\right\rangle. \tag{52}$$

Proof. Applying Parseval's formula of the QFT (18) to $\boldsymbol{\omega}$-integral into the left-hand side of (52) yields

$$\begin{aligned}\int_{\mathbb{R}^2}\left\langle \widetilde{h_{f_1,g_1}}, \widetilde{h_{f_2,g_2}}\right\rangle d\boldsymbol{\tau} &= \int_{\mathbb{R}^2}\left\langle \mathscr{F}_q\left\{\widetilde{h_{f_1,g_1}}\right\}, \mathscr{F}_q\left\{\widetilde{h_{f_2,g_2}}\right\}\right\rangle d\boldsymbol{\tau} = \frac{1}{|b_1b_2|} \\ &\cdot\int_{\mathbb{R}^2}\mathrm{Sc}\int_{\mathbb{R}^2}\mathscr{F}_q\left\{\widetilde{h_{f_1,g_1}}\right\}\left(\frac{\boldsymbol{\omega}}{\mathbf{b}}\right) \cdot \overline{\mathscr{F}_q\left\{\widetilde{h_{f_2,g_2}}\right\}\left(\frac{\boldsymbol{\omega}}{\mathbf{b}}\right)}\,d\boldsymbol{\omega}\,d\boldsymbol{\tau}.\end{aligned} \tag{53}$$

Using the cyclic multiplication symmetry (11) yields

$$\begin{aligned}\int_{\mathbb{R}^2}\left\langle \widetilde{h_{f_1,g_1}}, \widetilde{h_{f_2,g_2}}\right\rangle d\boldsymbol{\tau} &= \frac{1}{|b_1b_2|} \cdot \mathrm{Sc}\int_{\mathbb{R}^2}\int_{\mathbb{R}^2} e^{-\mathbf{i}(d_1/2b_1)\omega_1^2}\mathscr{W}_{f_1,g_1}^{A_1,A_2}(\mathbf{x},\boldsymbol{\omega}) \\ &\cdot e^{-\mathbf{j}(d_2/2b_2)\omega_2^2}\overline{e^{-\mathbf{i}(d_1/2b_1)\omega_1^2}\mathscr{W}_{f_2,g_2}^{A_1,A_2}(\mathbf{x},\boldsymbol{\omega})\,e^{-\mathbf{j}(d_2/2b_2)\omega_2^2}}\,d\boldsymbol{\omega}\,d\boldsymbol{\tau} \\ &= \frac{1}{|b_1b_2|}\mathrm{Sc}\int_{\mathbb{R}^2}\int_{\mathbb{R}^2} e^{-\mathbf{i}(d_1/2b_1)\omega_1^2}\mathscr{W}_{f_1,g_1}^{A_1,A_2}(\mathbf{x},\boldsymbol{\omega}) \\ &\cdot e^{-\mathbf{j}(d_2/2b_2)\omega_2^2}e^{\mathbf{j}(d_2/2b_2)\omega_2^2}\overline{\mathscr{W}_{f_2,g_2}^{A_1,A_2}(\mathbf{x},\boldsymbol{\omega})}e^{\mathbf{i}(d_1/2b_1)\omega_1^2}\,d\boldsymbol{\omega}\,d\boldsymbol{\tau} \\ &= \frac{1}{|b_1b_2|}\mathrm{Sc}\int_{\mathbb{R}^2}\int_{\mathbb{R}^2}\mathscr{W}_{f_1,g_1}^{A_1,A_2}(\mathbf{x},\boldsymbol{\omega})\overline{\mathscr{W}_{f_2,g_2}^{A_1,A_2}(\mathbf{x},\boldsymbol{\omega})}\,d\boldsymbol{\omega}\,d\boldsymbol{\tau}.\end{aligned} \tag{54}$$

The right-hand side of the above identity can be rewritten in the form

$$\begin{aligned}\int_{\mathbb{R}^2}\left\langle \widetilde{h_{f_1,g_1}},\widetilde{h_{f_2,g_2}}\right\rangle d\boldsymbol{\tau} &= \mathrm{Sc}\int_{\mathbb{R}^2}\int_{\mathbb{R}^2}\widetilde{h_{f_1,g_1}}(\mathbf{x},\boldsymbol{\tau})\overline{\widetilde{h_{f_2,g_2}}(\mathbf{x},\boldsymbol{\tau})}d\mathbf{x}\,d\boldsymbol{\tau}\\ &= \mathrm{Sc}\Bigg(\int_{\mathbb{R}^2}\int_{\mathbb{R}^2}\frac{e^{-\mathrm{i}(\pi/4)}}{\sqrt{2\pi b_1}}e^{\mathrm{i}(a_1/2b_1)\tau_1^2}f_1\left(\mathbf{x}+\frac{\boldsymbol{\tau}}{2}\right)\overline{g_1}\left(\mathbf{x}-\frac{\boldsymbol{\tau}}{2}\right)e^{\mathrm{j}(a_2/2b_2)\tau_2^2}\frac{e^{-\mathrm{j}(\pi/4)}}{\sqrt{2\pi b_2}}\\ &\quad\times\overline{\frac{e^{-\mathrm{i}(\pi/4)}}{\sqrt{2\pi b_1}}e^{\mathrm{i}(a_1/2b_1)\tau_1^2}f_2\left(\mathbf{x}+\frac{\boldsymbol{\tau}}{2}\right)\overline{g_2}\left(\mathbf{x}-\frac{\boldsymbol{\tau}}{2}\right)\frac{e^{-\mathrm{j}(\pi/4)}}{\sqrt{2\pi b_2}}e^{\mathrm{j}(a_2/2b_2)\tau_2^2}}d\mathbf{x}\,d\boldsymbol{\tau}\Bigg)\\ &= \frac{1}{|b_1b_2|}\mathrm{Sc}\int_{\mathbb{R}^2}\int_{\mathbb{R}^2}f_1\left(\mathbf{x}+\frac{\boldsymbol{\tau}}{2}\right)\overline{g_1}\left(\mathbf{x}-\frac{\boldsymbol{\tau}}{2}\right)\overline{f_2\left(\mathbf{x}+\frac{\boldsymbol{\tau}}{2}\right)\overline{g_2}\left(\mathbf{x}-\frac{\boldsymbol{\tau}}{2}\right)}d\mathbf{x}\,d\boldsymbol{\tau}.\end{aligned} \tag{55}$$

Making the change of variables $\mathbf{y} = \mathbf{x} + \boldsymbol{\tau}/2$ and $\mathbf{z} = \mathbf{x} - \boldsymbol{\tau}/2$ and applying Fubini's theorem we obtain

$$\begin{aligned}&\int_{\mathbb{R}^2}\left\langle \widetilde{h_{f_1,g_1}},\widetilde{h_{f_2,g_2}}\right\rangle d\boldsymbol{\tau}\\ &= \frac{1}{|b_1b_2|}\mathrm{Sc}\int_{\mathbb{R}^2}\int_{\mathbb{R}^2}f_1(\mathbf{y})\overline{g_1}(\mathbf{z})g_2(\mathbf{z})\overline{f_2}(\mathbf{y})\,d\mathbf{y}\,d\mathbf{z}\\ &= \frac{1}{|b_1b_2|}\mathrm{Sc}\int_{\mathbb{R}^2}f_1(\mathbf{y})\int_{\mathbb{R}^2}\overline{g_1}(\mathbf{z})g_2(\mathbf{z})\,d\mathbf{z}\overline{f_2}(\mathbf{y})\,d\mathbf{y}\\ &= \frac{1}{|b_1b_2|}\left\langle f_1\left(\overline{g_1},\overline{g_2}\right),f_2\right\rangle.\end{aligned} \tag{56}$$

By comparing the last line of (54) with the last line of (56) finishes the proof of the theorem. □

Based on the above theorem, we obtain the following consequences:

(i) If $g_1 = g_2$, then

$$\begin{aligned}&\mathrm{Sc}\int_{\mathbb{R}^2}\int_{\mathbb{R}^2}\mathscr{W}_{f_1,g_1}^{A_1,A_2}(\mathbf{x},\boldsymbol{\omega})\overline{\mathscr{W}_{f_2,g_1}^{A_1,A_2}(\mathbf{x},\boldsymbol{\omega})}d\boldsymbol{\omega}\,d\boldsymbol{\tau}\\ &= \|g_1\|^2\langle f_1,f_2\rangle.\end{aligned} \tag{57}$$

(ii) If $f_1 = f_2$, then

$$\begin{aligned}&\mathrm{Sc}\int_{\mathbb{R}^2}\int_{\mathbb{R}^2}\mathscr{W}_{f_1,g_1}^{A_1,A_2}(\mathbf{x},\boldsymbol{\omega})\overline{\mathscr{W}_{f_1,g_2}^{A_1,A_2}(\mathbf{x},\boldsymbol{\omega})}d\boldsymbol{\omega}\,d\boldsymbol{\tau}\\ &= \left\langle f_1\left(\overline{g_1},\overline{g_2}\right),f_1\right\rangle.\end{aligned} \tag{58}$$

(iii) If $f_1 = f_2$ and $g_1 = g_2$, then

$$\int_{\mathbb{R}^2}\int_{\mathbb{R}^2}\left|\mathscr{W}_{f_1,g_1}^{A_1,A_2}(\mathbf{x},\boldsymbol{\omega})\right|^2 d\boldsymbol{\omega}\,d\boldsymbol{\tau} = \|f_1\|^2\|g_1\|^2. \tag{59}$$

3.2. Relationship between QAF-LCT and QWFT. In this subsection, we introduce the quaternionic windowed Fourier transform (QWFT) and obtain the relationship between the QAF-LCT and QWFT. Following [11, 12], we define the QWFT as follows.

Definition 14 (QWFT). Let $\phi \in L^2(\mathbb{R}^2;\mathbb{H})$ be a fixed nonzero quaternion window function. The QWFT of $f \in L^2(\mathbb{R}^2;\mathbb{H})$ with respect to ϕ is defined to be the quaternion function on phase space $\mathbb{R}^2 \times \mathbb{R}^2$ given by

$$\mathcal{S}_f(\mathbf{x},\boldsymbol{\omega}) = \int_{\mathbb{R}^2}e^{-\mathrm{i}\omega_1\tau_1}f(\mathbf{x})\overline{g}(\mathbf{x}-\boldsymbol{\tau})e^{-\mathrm{j}\omega_2\tau_2}d\boldsymbol{\tau}. \tag{60}$$

Before presenting the main result of this subsection, let us introduce the following *carrier* definition and its properties (see [13, 14]).

Definition 15 (carrier). Given two quaternions p and q, we define the right C_r and left C_l carrier operators as

$$\begin{aligned}C_r(p)q &= qp,\\ qC_l(p) &= pq.\end{aligned} \tag{61}$$

Lemma 16. *Carriers (61) above satisfy the following properties with $p \in \mathbb{H}$:*

$$\begin{aligned}\overline{C_r(p)} &= C_l(\overline{p}),\\ \overline{C_l(p)} &= C_r(\overline{p}).\end{aligned} \tag{62}$$

We now describe the relationship between the QAF-LCT and the QWFT in the following theorem.

Theorem 17. *The QAF-LCT of a quaternion signal $f \in L^2(\mathbb{R}^2;\mathbb{H})$ can be expressed by the QWFT in the form*

$$\begin{aligned}\mathscr{A}_{f,g}^{A_1,A_2}\left(\frac{\mathbf{x}}{2},\frac{\boldsymbol{\omega}}{2}\right) &= \frac{4e^{-\mathrm{i}(\pi/4)}}{\sqrt{2\pi b_1}}\\ &\quad\cdot e^{\mathrm{i}((d_1/8b_1)\omega_1^2+(1/2b_1)\omega_1x_1-(3a_1/2b_1)x_1^2)}\mathcal{S}_f\Bigg(\mathbf{x},\frac{\omega_1+2a_1x_1}{b_1},\\ &\quad\frac{\omega_2+2a_2x_2}{b_2}\Bigg)\times\frac{4e^{-\mathrm{j}(\pi/4)}}{\sqrt{2\pi b_2}}\\ &\quad\cdot e^{\mathrm{j}((d_2/8b_2)\omega_2^2+(1/2b_2)\omega_2x_2-(3a_2/2b_2)x_2^2)}.\end{aligned} \tag{63}$$

Proof. According to the definition of QWVD-LCT (28), we obtain the following by making the change of variable:

$$\mathscr{A}_{f,g}^{A_1,A_2}\left(\frac{\mathbf{x}}{2},\frac{\boldsymbol{\omega}}{2}\right)=\frac{4}{\sqrt{2\pi b_1}}$$
$$\cdot\int_{\mathbb{R}^2}e^{\mathbf{i}((a_1/2b_1)\tau_1^2-(1/2b_1)\tau_1\omega_1+(d_1/8b_1)\omega_1^2-\pi/2)}f\left(\frac{\boldsymbol{\tau}}{2}+\frac{\mathbf{x}}{2}\right)\overline{g}\left(\frac{\boldsymbol{\tau}}{2}-\frac{\mathbf{x}}{2}\right)\frac{1}{\sqrt{2\pi b_2}}e^{\mathbf{j}((a_2/2b_2)\tau_2^2-(1/2b_2)\tau_2\omega_2+(d_2/8b_2)\omega_2^2-\pi/2)}d\boldsymbol{\tau}$$
$$=\frac{4e^{-\mathbf{i}(\pi/4)}}{\sqrt{2\pi b_1}}e^{(\mathbf{i}d_1/8b_1)\omega_1^2}\int_{\mathbb{R}^2}e^{-\mathbf{i}(\tau_1-x_1/2)(\omega_1/b_1)}e^{\mathbf{i}(2a_1/b_1)(\tau_1-x_1/2)^2}f(\boldsymbol{\tau})\overline{g}(\boldsymbol{\tau}-\mathbf{x})e^{\mathbf{j}(2a_2/b_2)(\tau_2-x_2/2)^2}e^{-\mathbf{j}(\tau_2-x_2/2)(\omega_2/b_2)}d\boldsymbol{\tau} \quad (64)$$
$$\times\frac{4e^{-\mathbf{j}(\pi/4)}}{\sqrt{2\pi b_2}}e^{(\mathbf{j}d_2/8b_2)\omega_2^2}=\frac{4e^{-\mathbf{i}(\pi/4)}}{\sqrt{2\pi b_1}}e^{\mathbf{i}((d_1/8b_1)\omega_1^2+(1/2b_1)\omega_1x_1-(3a_1/2b_1)x_1^2)}\int_{\mathbb{R}^2}e^{-\mathbf{i}((\omega_1+2a_1x_1)/b_1)\tau_1}e^{\mathbf{i}(2a_1/b_1)(x_1-\tau_1)^2}f(\mathbf{x})\overline{g}(\mathbf{x}-\boldsymbol{\tau})$$
$$\times e^{\mathbf{j}(2a_2/b_2)(x_2-\tau_2)^2}e^{-\mathbf{j}((\omega_2+2a_2x_2)/b_2)\tau_2}d\boldsymbol{\tau}\frac{4e^{-\mathbf{j}(\pi/4)}}{\sqrt{2\pi b_2}}e^{\mathbf{j}((d_2/8b_2)\omega_2^2+(1/2b_2)\omega_2x_2-(3a_2/2b_2)x_2^2)}.$$

With the help of (61) and (62), we rewrite the above identity as

$$\mathscr{A}_{f,g}^{A_1,A_2}\left(\frac{\mathbf{x}}{2},\frac{\boldsymbol{\omega}}{2}\right)=\frac{4e^{-\mathbf{i}(\pi/4)}}{\sqrt{2\pi b_1}}$$
$$\cdot e^{\mathbf{i}((d_1/8b_1)\omega_1^2+(1/2b_1)\omega_1x_1-(3a_1/2b_1)x_1^2)}\int_{\mathbb{R}^2}e^{-\mathbf{i}((\omega_1+2a_1x_1)/b_1)\tau_1}f(\mathbf{x})$$
$$\times\overline{e^{-\mathbf{j}(2a_2/b_2)(x_2-\tau_2)^2}g(\mathbf{x}-\boldsymbol{\tau})\operatorname{Cr}\left(e^{-\mathbf{i}(2a_1/b_1)(x_1-\tau_1)^2}\right)} \quad (65)$$
$$\times e^{-\mathbf{j}((\omega_2+2a_2x_2)/b_2)\tau_2}d\boldsymbol{\tau}\frac{4e^{-\mathbf{j}(\pi/4)}}{\sqrt{2\pi b_2}}$$
$$\cdot e^{\mathbf{j}((d_2/8b_2)\omega_2^2+(1/2b_2)\omega_2x_2-(3a_2/2b_2)x_2^2)}.$$

According to the definition of QWFT (60), we choose the quaternion window function

$$g(\mathbf{x})=e^{-\mathbf{j}(2a_2/b_2)x_2^2}g(\mathbf{x})\operatorname{Cr}\left(e^{-\mathbf{i}(2a_1/b_1)x_1^2}\right) \quad (66)$$

and obtain

$$\mathscr{A}_{f,g}^{A_1,A_2}\left(\frac{\mathbf{x}}{2},\frac{\boldsymbol{\omega}}{2}\right)=\frac{4e^{-\mathbf{i}(\pi/4)}}{\sqrt{2\pi b_1}}$$
$$\cdot e^{\mathbf{i}((d_1/8b_1)\omega_1^2+(1/2b_1)\omega_1x_1-(3a_1/2b_1)x_1^2)}\mathcal{S}_f\left(\mathbf{x},\frac{\omega_1+2a_1x_1}{b_1},\right. \quad (67)$$
$$\left.\frac{\omega_2+2a_2x_2}{b_2}\right)\times\frac{4e^{-\mathbf{j}(\pi/4)}}{\sqrt{2\pi b_2}}$$
$$\cdot e^{\mathbf{j}((d_2/8b_2)\omega_2^2+(1/2b_2)\omega_2x_2-(3a_2/2b_2)x_2^2)}.$$

The proof is complete. □

3.3. Relationship between QAF-LCT and CQWT. Before proving the relationship between the QAF-LCT and the continuous quaternion wavelet transform (CQWT), we first introduce the definition of the CQWT (see [15–17]).

Definition 18 (CQWT). The CQWT of a quaternion function $f\in L^2(\mathbb{R}^2;\mathbb{H})$ with respect to the quaternion mother wavelet ψ is defined by

$$T_\psi f(a,\mathbf{b})=\int_{\mathbb{R}^2}f(\mathbf{x})\overline{\frac{1}{a}\psi\left(\frac{\mathbf{x}-\mathbf{b}}{a}\right)}d\mathbf{x}. \quad (68)$$

Here the family of the quaternion wavelets $\psi_{a,\mathbf{b}}$ is defined by

$$\psi_{a,\mathbf{b}}(\mathbf{x})=\frac{1}{a}\psi\left(\frac{\mathbf{x}-\mathbf{b}}{a}\right),\quad a\in\mathbb{R}^+. \quad (69)$$

This definition will lead to the following result.

Theorem 19. *The QAF-LCT of $f\in L^2(\mathbb{R}^2;\mathbb{H})$ can be reduced to the CQWT*

$$\mathscr{A}_{f,g}^{A_1,A_2}(\mathbf{x},\boldsymbol{\omega})=\frac{e^{-\mathbf{i}(\pi/4)}}{\sqrt{2\pi b_1}}$$
$$\cdot e^{\mathbf{i}(d_1/2b_1)\omega_1^2-\omega_1x_1/2b_1+a_1x_1^2/8b_1}T_\psi f(1,\mathbf{x})\frac{e^{-\mathbf{j}(\pi/4)}}{\sqrt{2\pi b_2}} \quad (70)$$
$$\cdot e^{\mathbf{j}(d_2/2b_2)\omega_2^2-\omega_2x_2/2b_2+a_2x_2^2/8b_2}.$$

Proof. Applying the definition of the QAF-LCT (29) and following [18], we easily get

$$\mathscr{A}_{f,g}^{A_1,A_2}(\mathbf{x},\boldsymbol{\omega})$$
$$=\frac{e^{-\mathbf{i}(\pi/4)}}{\sqrt{2\pi b_1}}\int_{\mathbb{R}^2}e^{\mathbf{i}((a_1/2b_1)\tau_1^2-(1/b_1)\tau_1\omega_1+(d_1/2b_1)\omega_1^2-\pi/2)}f\left(\boldsymbol{\tau}+\frac{\mathbf{x}}{2}\right)\overline{g}\left(\boldsymbol{\tau}-\frac{\mathbf{x}}{2}\right)\frac{e^{-\mathbf{j}(\pi/4)}}{\sqrt{2\pi b_2}}e^{\mathbf{j}((a_2/2b_2)\tau_2^2-(1/b_2)\tau_2\omega_2+(d_2/2b_2)\omega_2^2-\pi/2)}d\boldsymbol{\tau}$$

$$= \frac{e^{-\mathbf{i}(\pi/4)}}{\sqrt{2\pi b_1}} \int_{\mathbb{R}^2} e^{\mathbf{i}((a_1/2b_1)\tau_1^2-(1/b_1)\tau_1\omega_1+(d_1/2b_1)\omega_1^2-\pi/2)} f\left(\boldsymbol{\tau}+\frac{\mathbf{x}}{2}\right)\overline{g}\left(\boldsymbol{\tau}+\frac{\mathbf{x}}{2}-\mathbf{x}\right)\frac{e^{-\mathbf{j}(\pi/4)}}{\sqrt{2\pi b_2}} e^{\mathbf{j}((a_2/2b_2)\tau_2^2-(1/b_2)\tau_2\omega_2+(d_2/2b_2)\omega_2^2-\pi/2)} d\boldsymbol{\tau}$$

$$= \frac{e^{-\mathbf{i}(\pi/4)}}{\sqrt{2\pi b_1}} \int_{\mathbb{R}^2} e^{\mathbf{i}(d_1/2b_1)\omega_1^2} e^{\mathbf{i}(1/b_1)(\tau_1'-x_1/2)\omega_1} e^{\mathbf{i}(a_1/2b_1)(\tau_1'-x_1/2)^2\omega_1} f\left(\boldsymbol{\tau}'\right)\overline{g}\left(\boldsymbol{\tau}'-\mathbf{x}\right)$$

$$\times \frac{e^{-\mathbf{j}(\pi/4)}}{\sqrt{2\pi b_2}} e^{\mathbf{j}(d_2/2b_2)\omega_2^2} e^{\mathbf{j}(1/b_2)(\tau_2'-x_2/2)\omega_2} e^{\mathbf{j}(a_2/2b_2)(\tau_2'-x_2/2)^2\omega_1} d\boldsymbol{\tau}'$$

$$= \frac{e^{-\mathbf{i}(\pi/4)}}{\sqrt{2\pi b_1}} e^{\mathbf{i}((d_1/2b_1)\omega_1^2-\omega_1x_1/2b_1+a_1x_1^2/8b_1)} \int_{\mathbb{R}^2} e^{\mathbf{i}(a_1/2b_1)(\tau_1'-x_1)^2-\mathbf{i}(\tau_1'-x_1)((2\omega_1-a_1x_1)/2b_1)} f\left(\boldsymbol{\tau}'\right)\overline{g}\left(\boldsymbol{\tau}'-\mathbf{x}\right)$$

$$\times e^{\mathbf{j}(a_2/2b_2)(\tau_2'-x_2)^2-\mathbf{j}(\tau_2'-x_2)((2\omega_2-a_2x_2)/2b_2)} d\boldsymbol{\tau}' \frac{e^{-\mathbf{j}(\pi/4)}}{\sqrt{2\pi b_2}} e^{\mathbf{j}((d_2/2b_2)\omega_2^2-\omega_2x_2/2b_2+a_2x_2^2/8b_2)}. \tag{71}$$

Again applying (61) and (62), we obtain

$$\mathscr{A}_{f,g}^{A_1,A_2}(\mathbf{x},\boldsymbol{\omega}) = \frac{e^{-\mathbf{i}(\pi/4)}}{\sqrt{2\pi b_1}} e^{\mathbf{i}((d_1/2b_1)\omega_1^2-\omega_1x_1/2b_1+a_1x_1^2/8b_1)} \int_{\mathbb{R}^2} f\left(\boldsymbol{\tau}'\right)$$

$$\times \overline{e^{-\mathbf{j}(a_2/2b_2)(\tau_2'-x_2)^2+\mathbf{j}(\tau_2'-x_2)((2\omega_2-a_2x_2)/2b_2)} g\left(\boldsymbol{\tau}'-\mathbf{x}\right) \operatorname{Cr}\left(e^{-\mathbf{i}(a_1/2b_1)(\tau_1'-x_1)^2+\mathbf{i}(\tau_1'-x_1)((2\omega_1-a_1x_1)/2b_1)}\right)}$$

$$\times d\boldsymbol{\tau}' \frac{e^{-\mathbf{j}(\pi/4)}}{\sqrt{2\pi b_2}} e^{\mathbf{j}((d_2/2b_2)\omega_2^2-\omega_2x_2/2b_2+a_2x_2^2/8b_2)} \tag{72}$$

$$= \frac{e^{-\mathbf{i}(\pi/4)}}{\sqrt{2\pi b_1}} e^{\mathbf{i}((d_1/2b_1)\omega_1^2-\omega_1x_1/2b_1+a_1x_1^2/8b_1)} \int_{\mathbb{R}^2} f\left(\boldsymbol{\tau}'\right)\overline{\psi}\left(\boldsymbol{\tau}'-\mathbf{x}\right) d\boldsymbol{\tau}' \frac{e^{-\mathbf{j}(\pi/4)}}{\sqrt{2\pi b_2}} e^{\mathbf{j}((d_2/2b_2)\omega_2^2-\omega_2x_2/2b_2+a_2x_2^2/8b_2)}$$

$$= \frac{e^{-\mathbf{i}(\pi/4)}}{\sqrt{2\pi b_1}} e^{\mathbf{i}((d_1/2b_1)\omega_1^2-\omega_1x_1/2b_1+a_1x_1^2/8b_1)} T_\psi f(1,\mathbf{x}) \frac{e^{-\mathbf{j}(\pi/4)}}{\sqrt{2\pi b_2}} e^{\mathbf{j}((d_2/2b_2)\omega_2^2-\omega_2x_2/2b_2+a_2x_2^2/8b_2)},$$

where the family of the quaternion wavelets $\psi_{1,\mathbf{b}}$ is given by

$$\psi_{1,\mathbf{b}}(\mathbf{x}) = \overline{e^{-\mathbf{j}(a_2/2b_2)(\tau_2-x_2)^2+\mathbf{j}(\tau_2-x_2)((2\omega_2-a_2x_2)/2b_2)} g(\boldsymbol{\tau}-\mathbf{x}) \operatorname{Cr}\left(e^{-\mathbf{i}(a_1/2b_1)(\tau_1-x_1)^2+\mathbf{i}(\tau_1-x_1)((2\omega_1-a_1x_1)/2b_1)}\right)}. \tag{73}$$

This is the desired result. □

4. Relationship between QAF-LCT and QFT

By a similar argument as in the proof of the Lemma 11, one may establish the following lemma, which describes the basic relationship between the QAF-LCT and the QFT.

Lemma 20. *The cross QAF-LCT of a signal f with matrix parameters $A_1 = (a_1, b_1, c_1, d_1)$ and $A_2 = (a_2, b_2, c_2, d_2)$ can be reduced as the QFT of the signal f in the following form:*

$$\mathscr{A}_{f,g}^{A_1,A_2}(\mathbf{x},\boldsymbol{\omega}) = \frac{e^{-\mathbf{i}(\pi/4)}}{\sqrt{2\pi b_1}}$$

$$\cdot e^{(\mathbf{i}d_1/2b_1)\omega_1^2} \mathscr{F}_q\left\{e^{\mathbf{i}(a_1/2b_1)\tau_1^2} f\left(\boldsymbol{\tau}+\frac{\mathbf{x}}{2}\right)\overline{g}\left(\boldsymbol{\tau}-\frac{\mathbf{x}}{2}\right)\right. \tag{74}$$

$$\left.\cdot e^{\mathbf{j}(a_2/2b_2)\tau_2^2}\right\}\left(\frac{\omega_1}{b_1},\frac{\omega_2}{b_2}\right)\frac{e^{-\mathbf{j}(\pi/4)}}{\sqrt{2\pi b_2}} e^{(\mathbf{j}d_2/2b_2)\omega_2^2}.$$

Applying (74) and inversion formula for the QFT (17), we easily obtain the following theorem.

Theorem 21 (reconstruction formula for QAF-LCT). *The inverse transform of the cross the QAF-LCT of the signal* $f \in L^2(\mathbb{R}^2;\mathbb{H})$ *is given by*

$$f(x) = \frac{1}{\overline{g}(\mathbf{0})} \int_{\mathbb{R}^2} K_{A_1^{-1}}(x_1,\omega_1)\, \mathscr{A}_{f,g}^{A_1,A_2}\left(\frac{\mathbf{x}}{2},\boldsymbol{\omega}\right) \cdot K_{A_2^{-1}}(x_2,\omega_2)\, d\boldsymbol{\omega}, \tag{75}$$

provided that $\overline{g}(\mathbf{0}) \neq \mathbf{0}$.

In the same kind of reasoning as Theorem 13, one can easily derive Moyal's formula of the QAF-LCT as follows.

Theorem 22 (Moyal's formula for QAF-LCT). *Let* $f_1, f_2, g_1, g_2 \in L^2(\mathbb{R}^2;\mathbb{H})$ *be quaternion-valued signals. Then the following equation holds:*

$$\begin{aligned} &\mathrm{Sc} \int_{\mathbb{R}^2} \int_{\mathbb{R}^2} \mathscr{A}_{f_1,g_1}^{A_1,A_2}(\mathbf{x},\boldsymbol{\omega})\, \overline{\mathscr{A}_{f_2,g_2}^{A_1,A_2}(\mathbf{x},\boldsymbol{\omega})} d\boldsymbol{\omega}\, d\boldsymbol{\tau} \\ &\quad = \left\langle f_1\left(\overline{g_1},\overline{g_2}\right), f_2\right\rangle. \end{aligned} \tag{76}$$

5. Conclusion

In this paper, we have presented the quaternion Wigner-Ville distribution associated with linear canonical transform (QWVD-LCT) and the quaternion ambiguity function associated with linear canonical transform (QAF-LCT). We provide alternative proof of well-known properties of the QWVD-LCT and QAF-LCT such as inversion formula and Moyal formula. The proof of the properties can be obtained using the properties of the QFT and relationship among the QWVD-LCT, QAF-LCT, and the QFT. We also studied that the QAF-LCT can be reduced to the QWFT and the CQWT.

Conflicts of Interest

The authors declare that there are no conflicts of interest regarding the publication of this paper.

Acknowledgments

This research is partially supported by WCU Percepatan Publikasi Internasional Tahun 2016 (no. 3834/UN4.21/PL.09/2016) from the Hasanuddin University, Indonesia.

References

[1] T. A. Ell, "Quaternion-Fourier transforms for analysis of two-dimensional linear time-invariant partial differential systems," in *Proceedings of the 32nd IEEE Conference on Decision and Control. Part 2 (of 4)*, pp. 1830–1841, December 1993.

[2] T. A. Ell and S. J. Sangwine, "Hypercomplex Fourier transforms of color images," *IEEE Transactions on Image Processing*, vol. 16, no. 1, pp. 22–35, 2007.

[3] T. Bülow, *Hypercomplex Spectral Signal Representations for the Processing and Analysis of Images [Ph.D. thesis]*, University of Kiel, Germany, 1999.

[4] K. I. Kou, J.-Y. Ou, and J. Morais, "On uncertainty principle for quaternionic linear canonical transform," *Abstract and Applied Analysis*, vol. 2013, Article ID 725952, 2013.

[5] K. I. Kou and J. Morais, "Asymptotic behaviour of the quaternion linear canonical transform and the Bochner-Minlos theorem," *Applied Mathematics and Computation*, vol. 247, pp. 675–688, 2014.

[6] Y. Yang and K. Ian Kou, "Uncertainty principles for hypercomplex signals in the linear canonical transform domains," *Signal Processing*, vol. 95, pp. 67–75, 2014.

[7] A. I. Zayed, "On the relationship between the Fourier and fractional Fourier transforms," *IEEE Signal Processing Letters*, vol. 3, no. 12, pp. 310-311, 1996.

[8] M. Bahri and R. Ashino, "A simplified proof of uncertainty principle for quaternion linear canonical transform," *Abstract and Applied Analysis*, Article ID 5874930, Art. ID 5874930, 11 pages, 2016.

[9] X.-L. Fan, K. I. Kou, and M.-S. Liu, "Quaternion Wigner–Ville distribution associated with the linear canonical transforms," *Signal Processing*, vol. 130, pp. 129–141, 2017.

[10] M. Bahri, "On two-dimensional quaternion Wigner-Ville distribution," *Journal of Applied Mathematics*, Article ID 139471, Art. ID 139471, 13 pages, 2014.

[11] L. Akila and R. Roopkumar, "Multidimensional quaternionic Gabor transforms," *Advances in Applied Clifford Algebras*, vol. 26, no. 3, pp. 985–1011, 2016.

[12] Y. Fu, U. Kähler, and P. Cerejeiras, "The Balian-low theorem for the windowed quaternionic Fourier transform," *Advances in Applied Clifford Algebras*, vol. 22, no. 4, pp. 1025–1040, 2012.

[13] S. Hartmann, "Some results on the lattice parameters of quaternionic Gabor frames," *Advances in Applied Clifford Algebras*, vol. 26, no. 1, pp. 137–149, 2016.

[14] S. Hartmann, "Relaxed quaternionic Gabor expansions at critical density," *Mathematical Methods in the Applied Sciences*, vol. 40, no. 5, pp. 1666–1678, 2017.

[15] M. Bahri, R. Ashino, and R. Vaillancourt, "Two-dimensional quaternion wavelet transform," *Applied Mathematics and Computation*, vol. 218, no. 1, pp. 10–21, 2011.

[16] M. Bahri and R. Ashino, "A variation on uncertainty principle and logarithmic uncertainty principle for continuous quaternion wavelet transforms," *Abstract and Applied Analysis*, Article ID 3795120, Art. ID 3795120, 11 pages, 2017.

[17] P. Fletcher and S. J. Sangwine, "The development of the quaternion wavelet transform," *Signal Processing*, vol. 136, pp. 2–15, 2017.

[18] R. Tao, Y. E. Song, Z. J. Wang, and Y. Wang, "Ambiguity function based on the linear canonical transform," *IET Signal Processing*, vol. 6, no. 6, pp. 568–576, 2012.

7

Linear Programming Problem with Interval Type 2 Fuzzy Coefficients and an Interpretation for Its Constraints

A. Srinivasan and G. Geetharamani

Department of Mathematics, Anna University, Bharathidasan Institute of Technology (BIT) Campus, Tiruchirappalli, Tamilnadu 620 024, India

Correspondence should be addressed to A. Srinivasan; yeacheenu4phd@gmail.com

Academic Editor: Frank Werner

Interval type 2 fuzzy numbers are a special kind of type 2 fuzzy numbers. These numbers can be described by triangular and trapezoidal shapes. In this paper, first, perfectly normal interval type 2 trapezoidal fuzzy numbers with their left-hand and right-hand spreads and their core have been introduced, which are normal and convex; then a new type of fuzzy arithmetic operations for perfectly normal interval type 2 trapezoidal fuzzy numbers has been proposed based on the extension principle of normal type 1 trapezoidal fuzzy numbers. Moreover, in this proposal, linear programming problems with resources and technology coefficients are perfectly normal interval type 2 fuzzy numbers. To solve this kind of fuzzy linear programming problems, a method based on the degree of satisfaction (or possibility degree) of the constraints has been introduced. In this method the fulfillment of the constraints can be measured with the help of ranking method of fuzzy numbers. Optimal solution is obtained at different degree of satisfaction by using Barnes algorithm with the help of MATLAB. Finally, the optimal solution procedure is illustrated with numerical example.

1. Introduction

Linear programming and its applications are used in many fields of operations research. It is concerned with the optimization of a linear function while satisfying a set of linear equality and/or inequality constraints or restriction. In real world situation, a linear programming model involves a lot of parameters whose values are assigned by experts. However, both experts and decision makers often do not precisely know the value of those parameters. Therefore, it is useful to consider the knowledge of experts about the parameters as fuzzy data. To make this possible Zadeh [1] introduced fuzzy set theory. The theory proposes a mathematical technique for dealing with imprecise concepts and problems that have many possible solutions. The concept of fuzzy mathematical programming on general level was first proposed by Tanaka et al. [2] in the framework of the fuzzy decision-making in fuzzy environment given by Bellman and Zadeh [3].

The concept of fuzzy liner programming was first formulated by Zimmermann [4]. After this motivation and inspiration several authors such as Deng et al. discussed the fact that the fuzzy bilevel linear programming with multiple followers model solved this complex problem by using the fuzzy structured element method [5]. A new interior point method has been presented to solve fuzzy number linear programming problems using linear ranking function by Zhong et al. [6]. A fully fuzzy linear programming problem with fuzzy equality constraints has been discussed and solved using a compromise programming approach by Cheng et al. [7]. An intuitionistic fuzzy chance constraints model (CCM) based on possibility and necessity measures has been developed and proposed a new method for solving an intuitionistic fuzzy CCM using chance operators by Chakraborty et al. [8]. All the above authors and many others have considered various types of fuzzy linear programming problems and proposed several approaches for solving these problems. Among them, Le and Gogne introduced a class of linear programming problems based on the possibility and necessity relation [9]. Figueroa-García and Hernández proposed an extension of the fuzzy linear programming method which was proposed by Zimmermann to an interval type 2 fuzzy linear programming problem with linear membership function [10]. In another work García presented a general model for linear programming where its technological coefficients are assumed as interval type 2 fuzzy sets and it is solved through an α-*cuts* approach [11]. Furthermore García [12] presented a general model to handle uncertainty to right-hand side parameters of a linear programming model by interval type 2 fuzzy sets and trapezoidal membership functions to optimize by using classical algorithms. Chen and Lee [13] proposed the definition of possibility degree of trapezoidal interval type 2 fuzzy numbers and some arithmetic operations. Hu et al. [14] proposed a new approach based on possibility degree to solve multicriteria decision-making problems in which the criteria value takes the form of interval type 2 fuzzy number.

Based on these literatures' survey, with our best knowledge, however, none of them introduced fuzzy linear programming model with both the right-hand side (resources) and the technological coefficients being perfectly normal interval type 2 fuzzy numbers based on possibility. In this paper, first, we have presented a method to do Perfectly normal Interval Type 2 Fuzzy Number (PnIT2FN) arithmetic operation using well-known arithmetical operations on type 1 fuzzy numbers. Further the relation of possibility for PnIT2FN as well as its property are discussed. And a fuzzy linear programming model with both the right-hand side (resources) and the technological coefficients being PnIT2FNs has been proposed. By using Zadeh's extension principle, a pair of upper and lower linear problems are developed to calculate the optimal solution of upper and lower bounds of linear problems measure at different possibility level α.

The rest of this paper is organized as follows. Firstly Section 2 deals with some preliminary definitions which are given. In Section 3, the definitions of interval type 2 fuzzy sets and trapezoidal interval type 2 fuzzy number have been recalled. In Section 4, the new representation for PnIT2TrFN, its properties, and some arithmetic operations of PnIT2TrFN based on type 1 fuzzy number are presented. The possibility degree of PnIT2TrFN is discussed in Section 5. The fuzzy linear programming problem with the right-hand side and the technological coefficients model is presented in Section 6; Section 7 has numerical illustrations of proposed fuzzy linear programming model with PnIT2TrFN; finally Section 8 has the conclusion of the work.

2. Preliminaries

Definition 1 (see [15]). Let X be a nonempty set. A fuzzy set $\widetilde{A}$ in X is characterized by its membership function $\mu_{\widetilde{A}} : X \to [0,1]$ and $\mu_{\widetilde{A}}(x)$ is interpreted as the degree of membership of element x in fuzzy set $\widetilde{A}$ for each $x \in X$. It is clear that $\widetilde{A}$ is completely determined by the set of tuples

$$\widetilde{A} = \{(x, \mu_{\widetilde{A}}(x)) \mid x \in X\}. \quad (1)$$

Definition 2 (see [15]). Let $\widetilde{A}$ be a fuzzy subset of X: the support of $\widetilde{A}$, denoted by $\mathrm{Supp}(\widetilde{A})$, is the crisp subset of X whose elements all have nonzero membership grades in $\widetilde{A}$:

$$\mathrm{Supp}(\widetilde{A}) = \{x \in X \mid \mu_{\widetilde{A}}(x) > 0\}. \quad (2)$$

Definition 3 (see [15]). A fuzzy subset $\widetilde{A}$ of a classical set X is called normal if there exists $x \in X$ such that $\mu_{\widetilde{A}}(x) = 1$. Otherwise $\widetilde{A}$ is subnormal.

Definition 4. An α-*level* set (or α-*cut*) of a fuzzy set $\widetilde{A}$ of X is a nonfuzzy set denoted by $\widetilde{A}_\alpha$ and defined by

$$\widetilde{A}_\alpha = \begin{cases} \{x \in X \mid \mu_{\widetilde{A}}(x) \geq \alpha\}, & \text{if } \alpha > 0, \\ \mathrm{cl}\left(\mathrm{Supp}\,\widetilde{A}\right), & \text{if } \alpha = 0, \end{cases} \quad (3)$$

where $\mathrm{cl}(\mathrm{Supp}\widetilde{A})$ denotes the closure of the support of $\widetilde{A}$.

Definition 5 (see [15]). A fuzzy set $\widetilde{A}$ of X is called convex if $\widetilde{A}_\alpha$ is a convex subset of X for all $\alpha \in [0,1]$.

Definition 6 (see [15, 16]). A fuzzy number $\widetilde{A}$ is a fuzzy set of the real lines with a normal (fuzzy), convex, and continuous membership function of bounded support. Alternatively, the fuzzy subset $\widetilde{A}$ of $\mathbb{R}$ is called a fuzzy number if the following conditions are satisfied:

(i) $\widetilde{A}$ is normal; that is, there exists $x \in \mathbb{R}$ such that $\mu_{\widetilde{A}}(x) = 1$;

(ii) the membership function $\mu_{\widetilde{A}}(x)$ is quasiconcave; that is, $\mu_{\widetilde{A}}(\lambda x_1 + (1-\lambda)x_2) \geq \min\{\mu_{\widetilde{A}}(x_1), \mu_{\widetilde{A}}(x_2)\}$ for all $\alpha \in [0,1]$;

(iii) the membership function $\mu_{\widetilde{A}}(x)$ is upper semicontinuous; that is, $\{x \in \mathbb{R} : \mu_{\widetilde{A}}(x) \geq \alpha\}$ is a closed subset of $\mathbb{R}$ for all $\alpha \in [0,1]$;

(iv) the 0-*level* set $\widetilde{A}_{\alpha=0}$ is compact (closed and bounded in $\mathbb{R}$).

We denote by $F(\mathbb{R})$ the set of all fuzzy numbers. If $\widetilde{A}$ is a fuzzy number, then from Zadeh [1] α-*level* set $\widetilde{A}_\alpha$ is a convex set from condition (ii). Combining this fact with condition (iii), the α-*level* set $\widetilde{A}_\alpha$ is a compact and convex set for all $\alpha \in [0,1]$ (since $\widetilde{A}_{\alpha=0}$ is bounded, it says that $\widetilde{A}_\alpha \subseteq \widetilde{A}_{\alpha=0}$ is also bounded for all $\alpha \in (0,1]$). Therefore, we can write $\widetilde{A}_\alpha = [a_\alpha^L, a_\alpha^U]$.

Definition 7 (see [9]). The trapezoidal fuzzy number is fully determined by quadruples $(a^L, a^U, \alpha, \beta)$ of crisp numbers such that $a^L \leq a^U$, $\alpha \geq 0$, and $\beta \geq 0$, whose membership function can be denoted by

$$\mu_{\widetilde{A}}(x) = \begin{cases} \dfrac{(x - a^L + \alpha)}{\alpha}, & a^L - \alpha \leq x \leq a^L, \\ 1, & a^L \leq x \leq a^U, \\ -\dfrac{(x - a^U - \beta)}{\beta}, & a^U \leq x \leq a^U + \beta, \\ 0, & \text{otherwise}. \end{cases} \quad (4)$$

When $a^L = a^U$, the trapezoidal fuzzy number becomes a triangular fuzzy number. If $\alpha = \beta$, the trapezoidal fuzzy number becomes a symmetrical trapezoidal fuzzy number. $[a^L, a^U]$ is the core of $\widetilde{A}$, and $\alpha \geq 0$, $\beta \geq 0$ are the left-hand and right-hand spreads.

It can easily be shown that

$$\widetilde{A}_\alpha = \left[\alpha\left(a^L - \left(a^L - \alpha\right)\right) + \left(a^L - \alpha\right), -\alpha\left(\left(a^U + \beta\right) - a^U\right) + \left(a^U + \beta\right)\right], \tag{5}$$

and the support of $\widetilde{A}$ is $(a^L - \alpha, a^U + \beta)$.

2.1. Arithmetic Operations. In this subsection addition, subtraction, and scalar multiplication operation of trapezoidal fuzzy numbers are reviewed [9, 15].

Let $\widetilde{A} = (a^L, a^U, \alpha, \beta)$ and $\widetilde{B} = (b^L, b^U, \theta, \gamma)$ be two trapezoidal fuzzy numbers; then

$$\begin{aligned}
\widetilde{A} &= \left(a^L, a^U, \alpha, \beta\right),\\
\widetilde{B} &= \left(b^L, b^U, \gamma, \theta\right),\\
\widetilde{A} + \widetilde{B} &= \left(a^L + b^L, a^U + b^U, \alpha + \gamma, \beta + \theta\right),\\
\widetilde{A} - \widetilde{B} &= \left(a^L - b^U, a^U - b^L, \alpha + \theta, \beta + \gamma\right),\\
\lambda\widetilde{A} &= \begin{cases} \left(\lambda a^L, \lambda a^U, \lambda\alpha, \lambda\beta\right), & \lambda \geq 0,\\ \left(\lambda a^U, \lambda a^L, |\lambda|\,\beta, |\lambda|\,\alpha\right), & \lambda < 0. \end{cases}
\end{aligned} \tag{6}$$

3. Interval Type 2 Fuzzy Sets

Interval type 2 fuzzy sets (IT2FSs) play a central role in fuzzy sets as models for words and in engineering applications of type 2 fuzzy sets. These fuzzy sets are characterized by their footprints of uncertainty, which in turn are characterized by their boundaries upper and lower membership functions.

Definition 8 (see [17]). Type 2 fuzzy set $\widetilde{A}$ in the universe of discourse X can be represented by type 2 membership function $\mu_{\widetilde{A}}(x, u)$ as follows:

$$\widetilde{A} = \left\{\left((x, u), \mu_{\widetilde{A}}(x, u)\right) \mid \forall x \in X,\ \forall u \in J_x \subseteq [0, 1], 0 \leq \mu_{\widetilde{A}}(x, u) \leq 1\right\}, \tag{7}$$

where $J_x \subseteq [0, 1]$ is the primary membership function at x, and

$$\int_{u \in J_x} \frac{\mu_{\widetilde{A}}(x, u)}{u} \tag{8}$$

indicates the second membership at x. For discrete situations, $\int$ is replaced by $\sum$.

Definition 9 (see [17, 18]). Let $\widetilde{A}$ be type 2 fuzzy set in the universe of discourse X represented by type 2 membership function $\mu_{\widetilde{A}}(x, u)$. If all $\mu_{\widetilde{A}}(x, u) = 1$, then $\widetilde{A}$ is called IT2FS. IT2FS can be regarded as a special case of type 2 fuzzy set, which is defined as

$$\widetilde{A} = \int_{x \in X} \int_{u \in J_x} \frac{1}{(x, u)} = \int_{x \in X} \frac{\left[\int_{u \in J_x} 1/u\right]}{x}, \tag{9}$$

where x is the primary variable, $J_x \subseteq [0, 1]$ is the primary membership of x, u is the secondary variable, and $\int_{u \in J_x} 1/u$ is the secondary membership function at x.

It is obvious that IT2FS $\widetilde{A}$ defined on X is completely determined by the primary membership which is called the footprint of uncertainty, and the footprint of uncertainty can be expressed as follows:

$$\mathrm{FOU}\left(\widetilde{A}\right) = \bigcup_{x \in X} J_x = \bigcup_{x \in X} \left\{(x, u) \mid u \in J_x \subseteq [0, 1]\right\}. \tag{10}$$

Definition 10 (see [14, 19]). Let $\widetilde{A}$ be IT2FS; uncertainty in the primary membership of type 2 fuzzy set consists of a bounded region called the footprint of uncertainty, which is the union of all primary membership. Footprint of uncertainty is characterized by upper membership function and lower membership function. Both of the membership functions are type 1 fuzzy sets. Upper membership function is denoted by $\overline{\mu}_{\widetilde{A}}$ and lower membership function is denoted by $\underline{\mu}_{\widetilde{A}}$, respectively.

Definition 11 (see [14]). An interval type 2 fuzzy number is called trapezoidal interval type 2 fuzzy number where the upper membership function and lower membership function are both trapezoidal fuzzy numbers; that is,

$$\widetilde{A} = \left(A^L, A^U\right) = \left(\left(a_1^L, a_2^L, a_3^L, a_4^L; H_1\left(A^L\right), H_2\left(A^L\right)\right), \left(a_1^U, a_2^U, a_3^U, a_4^U; H_1\left(A^U\right), H_2\left(A^U\right)\right)\right), \tag{11}$$

where $H_j(A^L)$ and $H_j(A^U)$, $(j = 1, 2)$ denote membership values of the corresponding elements a_{j+1}^L and a_{j+1}^U $(j = 1, 2)$, respectively.

Definition 12 (see [13]). The upper membership function and lower membership function of IT2FSs are type 1 membership function, respectively.

Definition 13 (see [20]). IT2FS, $\widetilde{A}$, is said to be perfectly normal if both its upper and lower membership functions are normal. It is denoted by PnIT2FS; that is,

$$\sup \overline{\mu}_{\widetilde{A}}(x) = \sup \underline{\mu}_{\widetilde{A}} = 1. \tag{12}$$

4. Perfectly Normal IT2TrFN

In this section, the concept of PnIT2TrFN is discussed and it is the extension work of Chiao [21] that proposed the concept of trapezoidal interval type 2 fuzzy set, in which the upper membership function and the lower membership function are represented by trapezoidal fuzzy number, which is adopted for PnIT2TrFN. PnIT2FN is a fuzzy subset of the real line, which is both "normal" and "convex." The operations PnIT2FSs are very complex according to the decomposition theorem [22], and the IT2FSs are usually taken in some simplified formations in applications. In subsection, the arithmetic operation on PnIT2TrFNs is formulated by proposing the extension principle.

Definition 14. Let $\widetilde{A} = [A^L, A^U]$ be PnIT2FS on X. Let $\widetilde{A}^L = (a_2^L, a_3^L, \alpha_L, \beta_L)$ and $\widetilde{A}^U = (a_2^U, a_3^U, \alpha_U, \beta_U)$ be the lower and upper trapezoidal fuzzy number, respectively, with respect to $\widetilde{A}$ defined on the universe of discourse X, where $a_2^L \le a_3^L$, $a_2^U \le a_3^U$, $\alpha_L, \alpha_U \ge 0$, and $\beta_L, \beta_U \ge 0$. $[a_2^L, a_3^L]$ is the core of $\widetilde{A}^L$, and $\alpha_L, \beta_L \ge 0$ are the left-hand and right-hand spreads and $[a_2^U, a_3^U]$ is the core of $\widetilde{A}^U$, and $\alpha_U, \beta_U \ge 0$ are the left-hand and right-hand spreads. The membership functions of x in $\widetilde{A}^L$ and $\widetilde{A}^U$ are expressed as follows:

$$\underline{\mu}_{\widetilde{A}}(x) = \begin{cases} \dfrac{(x - a_2^L + \alpha_L)}{\alpha_L}, & a_2^L - \alpha_L \le x \le a_2^L, \\ 1, & a_2^L \le x \le a_3^L, \\ -\dfrac{(x - a_3^L - \beta_L)}{\beta_L}, & a_3^L \le x \le a_3^L + \beta_L, \\ 0, & \text{otherwise,} \end{cases}$$
$$\overline{\mu}_{\widetilde{A}}(x) = \begin{cases} \dfrac{(x - a_2^U + \alpha_U)}{\alpha_U}, & a_2^U - \alpha_U \le x \le a_2^U, \\ 1, & a_2^U \le x \le a_3^U, \\ -\dfrac{(x - a_3^U - \beta_U)}{\beta_U}, & a_3^U \le x \le a_3^U + \beta_U, \\ 0, & \text{otherwise.} \end{cases} \quad (13)$$

$\underline{\mu}_{\widetilde{A}}(x)$ and $\overline{\mu}_{\widetilde{A}}(x)$ are lower and upper bounds, respectively, of $\widetilde{A}$ (see Figure 1). Then, $\widetilde{A}$ is a PnIT2TrFN on X and is represented by the following: $\widetilde{A} = [A^L, A^U] = [(a_2^L, a_3^L, \alpha_L, \beta_L), (a_2^U, a_3^U, \alpha_U, \beta_U)]$. Obviously, if $a_2^L = a_3^L$ and $a_2^U = a_3^U$ the PnIT2TrFN reduces to the perfectly normal interval type 2 triangular fuzzy number (PnIT2TFN). If $A^L = A^U$, then PnIT2TrFN $\widetilde{A}$ becomes type 1 trapezoidal fuzzy number [13, 23].

Definition 15 (primary α-*cut* of PnIT2FS). The primary α-*cut* of PnIT2FS is ${}^{\alpha}\widetilde{A} = \{(x, u) \mid J_x \ge \alpha,\ u \in [0,1]\}$ which is bounded by two regions:

$$\begin{aligned} {}^{\alpha}\underline{\mu}_{\widetilde{A}}(x) &= \left\{\left(x, \underline{\mu}_{\widetilde{A}}(x)\right) \mid \underline{\mu}_{\widetilde{A}}(x) \ge \alpha, \forall \alpha \in [0,1]\right\}, \\ {}^{\alpha}\overline{\mu}_{\widetilde{A}}(x) &= \left\{(x, \overline{\mu}_{\widetilde{A}}(x)) \mid \overline{\mu}_{\widetilde{A}}(x) \ge \alpha, \forall \alpha \in [0,1]\right\}. \end{aligned} \quad (14)$$

Definition 16 (crisp bounds of PnIT2FN). The crisp boundary of the primary α-*cut* of PnIT2FN $\widetilde{A} = (A^L, A^U)$ is closed interval ${}^{\alpha^{cb}}\widetilde{A}$ which will be obtained as follows: A^L and A^U are the lower and upper interval valued bounds of $\widetilde{A}$. Also

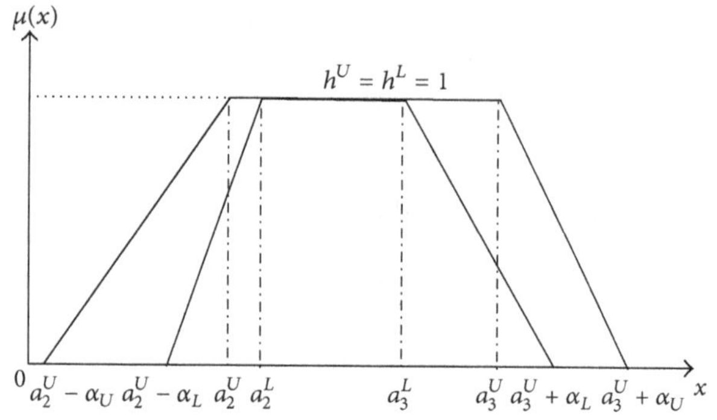

FIGURE 1: The lower trapezoidal membership function $\widetilde{A}^L$ and the upper trapezoidal membership function $\widetilde{A}^U$ of PnIT2FS $\widetilde{A}$.

the boundary of A_α^L and A_α^U can be defined as the boundary of the α-*cuts* of each interval type 1 fuzzy set:

$$\begin{aligned} A_\alpha^L &= \left[\inf_x {}^{\alpha}\underline{\mu}_{\widetilde{A}}(x), \sup_x {}^{\alpha}\underline{\mu}_{\widetilde{A}}(x)\right] = \left[a_l^L, a_u^L\right], \\ A_\alpha^U &= \left[\inf_x {}^{\alpha}\overline{\mu}_{\widetilde{A}}(x), \sup_x {}^{\alpha}\overline{\mu}_{\widetilde{A}}(x)\right] = \left[a_l^U, a_u^U\right], \\ {}^{\alpha^{cb}}\widetilde{A} &= \left[\inf_x \{{}^{\alpha}\mu_{\widetilde{A}}(x,u)\}, \sup_x \{{}^{\alpha}\mu_{\widetilde{A}}(x,u)\}\right] \\ &= \left[\left[a_l^U, a_l^L\right], \left[a_u^L, a_u^U\right]\right] \\ &= \left[\left[{}^{\alpha}\overline{a}^L, {}^{\alpha}\underline{a}^L\right], \left[{}^{\alpha}\underline{a}^R, {}^{\alpha}\overline{a}^R\right]\right] \end{aligned} \quad (15)$$

which is equivalent to, say,

$${}^{\alpha^{cb}}\mu_{\widetilde{A}} \in \left[\left[{}^{\alpha}\overline{a}^L, {}^{\alpha}\underline{a}^L\right], \left[{}^{\alpha}\underline{a}^R, {}^{\alpha}\overline{a}^R\right]\right]. \quad (16)$$

Evidently, for PnIT2TrFN from Figure 2

$${}^{\alpha}\overline{a}^L \le {}^{\alpha}\underline{a}^L \le {}^{\alpha}\underline{a}^R \le {}^{\alpha}\overline{a}^R. \quad (17)$$

Definition 17. A fuzzy number $\widetilde{A} = (A^L, A^U) = ((a_2^L, a_3^L, \alpha_L, \beta_L), (a_2^U, a_3^U, \alpha_U, \beta_U))$ is said to be nonnegative PnIT2TrFN if $\overline{\mu}_{\widetilde{A}}(x) = \underline{\mu}_{\widetilde{A}} = 0, \forall x > 0$.

Definition 18. A fuzzy number $\widetilde{A} = (A^L, A^U) = ((a_2^L, a_3^L, \alpha_L, \beta_L), (a_2^U, a_3^U, \alpha_U, \beta_U))$ is said to be nonpositive PnIT2TrFN if $\underline{\mu}_{\widetilde{A}}(x) = \overline{\mu}_{\widetilde{A}} = 0, \forall x < 0$.

Definition 19. A fuzzy number $\widetilde{A} = (A^L, A^U) = ((a_2^L, a_3^L, \alpha_L, \beta_L), (a_2^U, a_3^U, \alpha_U, \beta_U))$ is said to be zero PnIT2TrFN if $a_2^L = b_2^L = 0$, $a_3^L = b_3^L = 0$, $\alpha_L = \gamma_L = \beta_L = \theta_L = \alpha_U = \gamma_U = 0$, and $\beta_U = \theta_U = 0$.

Definition 20. PnIT2TrFNs $\widetilde{A} = ((a_2^L, a_3^L, \alpha_L, \beta_L), (a_2^U, a_3^U, \alpha_U, \beta_U))$ and $\widetilde{B} = ((b_2^L, b_3^L, \gamma_L, \theta_L), (b_2^U, b_3^U, \gamma_U, \theta_U))$ are said to be identically equal to $\widetilde{A} = \widetilde{B}$ if and only if $a_2^L = b_2^L$, $a_3^L = b_3^L$, $\alpha_L = \gamma_L$, $\beta_L = \theta_L$, $\alpha_U = \gamma_U$, and $\beta_U = \theta_U$.

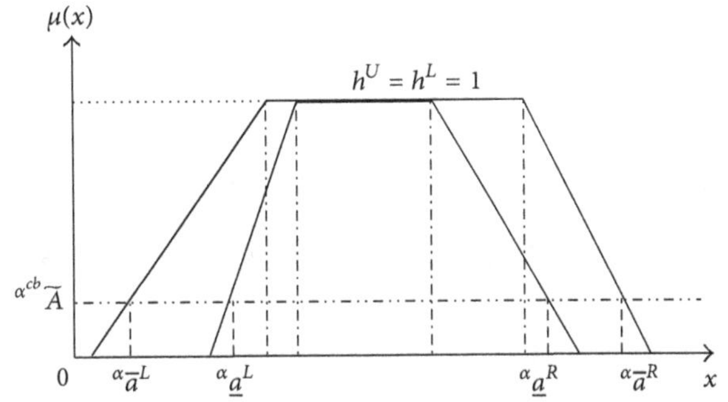

FIGURE 2: Crisp bounds of PnIT2TrFN.

4.1. Arithmetic Operations on PnIT2TrFN

Definition 21. If $\widetilde{A} = ((a_2^L, a_3^L, \alpha_L, \beta_L), (a_2^U, a_3^U, \alpha_U, \beta_U))$ and $\widetilde{B} = ((b_2^L, b_3^L, \gamma_L, \theta_L), (b_2^U, b_3^U, \gamma_U, \theta_U))$ are PnIT2TrFNs, then $\widetilde{C} = \widetilde{A} + \widetilde{B}$ is also PnIT2TrFN and defined by

$$\widetilde{C} = \left(\left(a_2^L + b_2^L, a_3^L + b_3^L, \alpha_L + \gamma_L, \beta_L + \theta_L\right), \left(a_2^U + b_2^U, a_3^U + b_3^U, \alpha_U + \gamma_U, \beta_U + \theta_U\right)\right). \quad (18)$$

Definition 22. If $\widetilde{A} = ((a_2^L, a_3^L, \alpha_L, \beta_L), (a_2^U, a_3^U, \alpha_U, \beta_U))$ and $\widetilde{B} = ((b_2^L, b_3^L, \gamma_L, \theta_L), (b_2^U, b_3^U, \gamma_U, \theta_U))$ are PnIT2TrFNs, then $\widetilde{C} = \widetilde{A} - \widetilde{B}$ is also PnIT2TrFN and defined by

$$\widetilde{C} = \left(\left[a_2^L - b_3^U, a_3^L - b_2^U, \alpha_L + \theta_U, \beta_L + \gamma_U\right], \left[a_2^U - b_3^L, a_3^U - b_2^L, \alpha_U + \theta_L, \beta_U + \gamma_L\right]\right). \quad (19)$$

Definition 23. Let $\lambda \in \mathbb{R}$. If $\widetilde{A} = ((a_2^L, a_3^L, \alpha_L, \beta_L), (a_2^U, a_3^U, \alpha_U, \beta_U))$ is PnIT2TrFN, then $\widetilde{C} = \lambda\widetilde{A}$ is also PnIT2TrFN and is given by

$$\widetilde{C} = \lambda\widetilde{A} = \begin{cases} \left(\left(\lambda a_2^L, \lambda a_3^L, \lambda\alpha_L, \lambda\beta_L\right), \left(\lambda a_2^U, \lambda a_3^U, \lambda\alpha_U, \lambda\beta_U\right)\right); & \text{if } \lambda \geq 0, \\ \left(\left(\lambda a_3^U, \lambda a_2^U, |\lambda|\,\beta_U, |\lambda|\,\alpha_U\right), \left(\lambda a_3^L, \lambda a_2^L, |\lambda|\,\beta_L, |\lambda|\,\alpha_L\right)\right); & \text{if } \lambda < 0. \end{cases} \quad (20)$$

5. The Possibility Degree of PnIT2TrFN

Comparison of fuzzy numbers is considered one of the most important topics in fuzzy logic theory. The early and most important work in the field of comparing fuzzy numbers has been presented by Dubois and Prade [24]. On the other hand, the dominance possibility indices, which have been introduced by Negi et al., were utilized in the field of fuzzy mathematical programming [25, 26]. The approach used in these fields was based on formulating a possibility function, whether in the case of trapezoidal fuzzy numbers or the case of triangular fuzzy numbers. In this paper, we are going to utilize the degree of possibility that the proposition stating that "A is less than or equal to B" is true which is proposed by Chen and Lee [13] for calculating the ranking values of perfectly normal trapezoidal interval type 2 fuzzy number. Here, the height of the upper membership function and lower membership function is considered as 1, so the modified proposition can be as follows.

Definition 24. Let $\widetilde{A} = ((a_2^L, a_3^L, \alpha_L, \beta_L), (a_2^U, a_3^U, \alpha_U, \beta_U))$ and $\widetilde{B} = ((b_2^L, b_3^L, \gamma_L, \theta_L), (b_2^U, b_3^U, \gamma_U, \theta_U))$ be two PnIT2TrFNs. Then the possibility degrees of lower and upper membership function are defined as follows:

$$\operatorname{Pos}\left(A^L \leq B^L\right) = \max\left(1 - \max\left(\frac{\max\left({}^{\alpha}\underline{a}^L - {}^{\alpha}\underline{b}^L, 0\right) + \max\left(\left(a_2^L - b_2^L\right), 0\right) + \max\left(\left(a_3^L - b_3^L\right), 0\right) + \max\left({}^{\alpha}\underline{a}^R - {}^{\alpha}\underline{b}^R, 0\right) + \left({}^{\alpha}\underline{a}^R - {}^{\alpha}\underline{b}^L\right)}{\left|{}^{\alpha}\underline{a}^L - {}^{\alpha}\underline{b}^L\right| + \left|a_2^L - b_2^L\right| + \left|a_3^L - b_3^L\right| + \left|{}^{\alpha}\underline{a}^R - {}^{\alpha}\underline{b}^R\right| + \left({}^{\alpha}\underline{b}^R - {}^{\alpha}\underline{b}^L\right) + \left({}^{\alpha}\underline{a}^R - {}^{\alpha}\underline{a}^L\right)}, 0\right), 0\right),$$

$$\operatorname{Pos}\left(A^U \leq B^U\right) = \max\left(1 - \max\left(\frac{\max\left({}^{\alpha}\overline{a}^L - {}^{\alpha}\overline{b}^L, 0\right) + \max\left(\left(a_2^U - b_2^U\right), 0\right) + \max\left(\left(a_3^U - b_3^U\right), 0\right) + \max\left({}^{\alpha}\overline{a}^R - {}^{\alpha}\overline{b}^R, 0\right) + \left({}^{\alpha}\overline{a}^R - {}^{\alpha}\overline{b}^L\right)}{\left|{}^{\alpha}\overline{a}^L - {}^{\alpha}\overline{b}^L\right| + \left|a_2^U - b_2^U\right| + \left|a_3^U - b_3^U\right| + \left|{}^{\alpha}\overline{a}^R - {}^{\alpha}\overline{b}^R\right| + \left({}^{\alpha}\overline{b}^R - {}^{\alpha}\overline{b}^L\right) + \left({}^{\alpha}\overline{a}^R - {}^{\alpha}\overline{a}^L\right)}, 0\right), 0\right). \tag{21}$$

Proposition 25. *Let* $\widetilde{A} = ((a_2^L, a_3^L, \alpha_L, \beta_L), (a_2^U, a_3^U, \alpha_U, \beta_U))$ *and* $\widetilde{B} = ((b_2^L, b_3^L, \gamma_L, \theta_L), (b_2^U, b_3^U, \gamma_U, \theta_U))$ *be two PnIT2FNs,* $\operatorname{Pos}(A^L \preceq B^L) \geq \alpha$*, if and only if* $\inf[A^L]_\alpha \leq \sup[B^L]_\alpha$*; that is,*

$$\operatorname{Pos}\left(A^L \preceq B^L\right) \geq \alpha \quad \text{iff } {}^{\alpha}\underline{a}^L \leq {}^{\alpha}\underline{b}^R. \tag{22}$$

Alternatively,

$$\operatorname{Pos}\left(A^L \preceq B^L\right) = \mu_{\operatorname{Pos}(A^L, B^L)} = \left(A^L \preceq^{\operatorname{Pos}} B^L\right). \tag{23}$$

An interpretation of the $\alpha-$ *relation associated with possibility when comparing fuzzy numbers* A^L *and* B^L *is as follows. For a given level of satisfaction* $\alpha \in [0,1]$*, a fuzzy number* A^L *is not better than* B^L *with respect to fuzzy relation* $\preceq^{\operatorname{Pos}}$*, if the smallest value of* A^L *with the degree of satisfaction (or possibility degree) being greater than or equal to* $\alpha-$ *is less than or equal to the largest value of* B^L *with the degree of satisfaction greater than or equal to* α*. Similar to the above*

$$\operatorname{Pos}\left(A^U \preceq B^U\right) \geq \alpha \quad \text{iff } {}^{\alpha}\overline{a}^L \leq {}^{\alpha}\overline{b}^R. \tag{24}$$

Proposition 26 (see [15, 23, 24]). *Let* $\widetilde{A} = ((a_2^L, a_3^L, \alpha_L, \beta_L), (a_2^U, a_3^U, \alpha_U, \beta_U))$ *and* $\widetilde{B} = ((b_2^L, b_3^L, \gamma_L, \theta_L), (b_2^U, b_3^U, \gamma_U, \theta_U))$ *be two PnIT2TrFNs; then*

$$\operatorname{Pos}\left(A^L \preceq B^L\right) = \begin{cases} 1, & b_3^L \geq a_2^L, \\ \dfrac{b_3^L - a_2^L + \theta_L + \alpha_L}{\theta_L + \alpha_L}, & 0 \leq a_2^L - b_3^L \leq \theta_L + \alpha_L, \\ 0, & a_2^L - b_3^L > \theta_L + \alpha_L; \end{cases} \tag{25}$$

see Figure 3:

$$\operatorname{Pos}\left(A^U \preceq B^U\right) = \begin{cases} 1, & b_3^U \geq a_2^U, \\ \dfrac{b_3^U - a_2^U + \theta_U + \alpha_u}{\theta_U + \alpha_U}, & 0 \leq a_2^U - b_3^U \leq \theta_U + \alpha_U, \\ 0, & a_2^U - b_3^U > \theta_U + \alpha_U; \end{cases} \tag{26}$$

see Figure 4.

Theorem 27. *Let* $\widetilde{A} = ((a_2^L, a_3^L, \alpha_L, \beta_L), (a_2^U, a_3^U, \alpha_U, \beta_U))$ *and* $\widetilde{B} = ((b_2^L, b_3^L, \gamma_L, \theta_L), (b_2^U, b_3^U, \gamma_U, \theta_U))$ *be two PnIT2TrFNs,* $p \in (0,1]$*,* $\operatorname{Pos}(\widetilde{A}^L \preceq \widetilde{B}^L) \geq p$*, if and only if* $b_3^L - a_2^L \geq (p-1)(\theta_L + \alpha_L)$*.*

Proof. If $p = 1$, then from $\operatorname{Pos}(\widetilde{A}^L \preceq \widetilde{B}^L) \geq 1$ one can get that $b_3^L \geq a_2^L$, and vice versa. If $0 < p < 1$, then $b_3^L < a_2^L$ and $b_3^L + \theta_L > a_2^L - \alpha_L$, $\operatorname{Pos}(\widetilde{A}^L \preceq \widetilde{B}^L) \geq p$, if and only if $(b_3^L - a_2^L + \theta_L + \alpha_L)/(\theta_L + \alpha_L) \geq p$; that is, $b_3^L - a_2^L \geq (p-1)(\theta_L + \alpha_L)$. □

Theorem 28. *Let* $\widetilde{A} = ((a_2^L, a_3^L, \alpha_L, \beta_L), (a_2^U, a_3^U, \alpha_U, \beta_U))$ *and* $\widetilde{B} = ((b_2^L, b_3^L, \gamma_L, \theta_L), (b_2^U, b_3^U, \gamma_U, \theta_U))$ *be two PnIT2TrFNs,* $p \in (0,1]$*,* $\operatorname{Pos}(\widetilde{A}^U \preceq \widetilde{B}^U) \geq p$*, if and only if* $b_3^U - a_2^U \geq (p-1)(\theta_U + \alpha_U)$*.*

Proof. If $p = 1$, then from $\operatorname{Pos}(\widetilde{A}^U \preceq \widetilde{B}^U) \geq 1$ one can get that $b_3^U \geq a_2^U$, and vice versa. If $0 < p < 1$, then $b_3^U < a_2^U$ and $b_3^U + \theta_U > a_2^U - \alpha_U$, $\operatorname{Pos}(\widetilde{A}^U \preceq \widetilde{B}^U) \geq p$, if and only if $(b_3^U - a_2^U + \theta_U + \alpha_U)/(\theta_U + \alpha_U) \geq p$; that is, $b_3^U - a_2^U \geq (p-1)(\theta_U + \alpha_U)$. □

6. Fuzzy Linear Programming

In this section, we propose a fuzzy linear programming model with the technological coefficients and right-hand side (resources) being PnIT2FN:

$$\begin{aligned} \operatorname*{Max/Min}_{x \in X} \quad & Z = cx \\ \text{S.t} \quad & Ax \preceq b, \; x \geq 0, \end{aligned} \tag{27}$$

where $A = (a_{ij})_{m \times n}$, $b = (b_1, b_2, \ldots, b_m)^T$, $c = (c_1, c_2, \ldots, c_n)$, $x = (x_1, x_2, \ldots, x_n)^T$, and a_{ij}, b_i are PnIT2FN and $x_j, c_j \in \mathbb{R}$ $(i = 1, 2, \ldots, m; j = 1, 2, \ldots, n)$. $\preceq$ is type 2 fuzzy order.

Definition 29. Consider a set of right-hand side (resources) parameters of a fuzzy linear programming problem defined as PnIT2FS $\widetilde{b}$ defined on the closed interval

$$\widetilde{b}_i \in \left[\left[{}^{\alpha}\overline{b}_i^L, {}^{\alpha}\underline{b}_i^L\right], \left[{}^{\alpha}\underline{b}_i^R, {}^{\alpha}\overline{b}_i^R\right]\right] \in \mathbb{R} \tag{28}$$

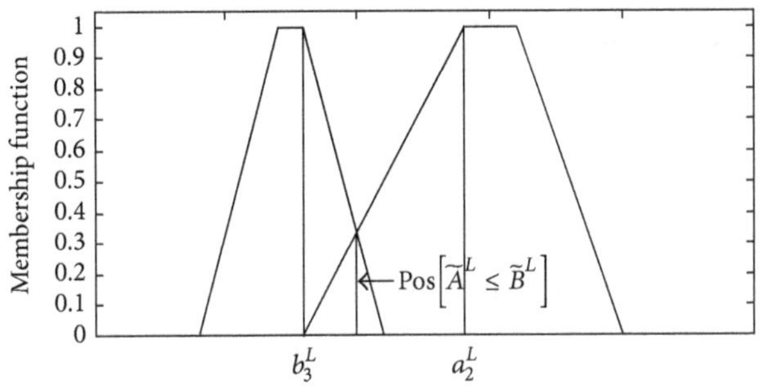

FIGURE 3: Pos[$\widetilde{A}^L \le \widetilde{B}^L$] < 1 because $\widetilde{A}^L > \widetilde{B}^L$.

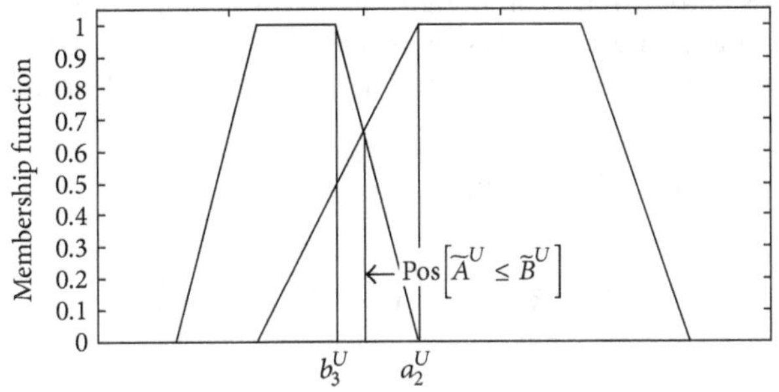

FIGURE 4: Pos[$\widetilde{A}^U \le \widetilde{B}^U$] < 1 because $\widetilde{A}^U > \widetilde{B}^U$.

and $i \in \mathbb{N}_n$. The membership function which represents the fuzzy space Supp($\tilde{b}_i$) is

$$\tilde{b}_i = \int_{b_i \in \mathbb{R}} \frac{\left[\int_{u \in J_{b_i}} 1/u\right]}{b_i}, \quad i \in \mathbb{N}_n, \; J_{b_i} \subseteq [0,1]. \tag{29}$$

Here, $\tilde{b}$ is bounded by both lower and upper primary membership function; namely,

$${}^{\alpha}\underline{\mu}_{\tilde{b}_i} = \left\{(b_i, u) \mid \underline{\mu}_{\tilde{b}_i} \ge \alpha\right\}, \tag{30}$$

with parameters ${}^{\alpha}\underline{b}_i^L$ and ${}^{\alpha}\underline{b}_i^R$, and

$${}^{\alpha}\overline{\mu}_{\tilde{b}_i} = \left\{(b_i, u) \mid \overline{\mu}_{\tilde{b}_i} \ge \alpha\right\} \tag{31}$$

with parameters ${}^{\alpha}\overline{b}_i^L$ and ${}^{\alpha}\overline{b}_i^R$.

Definition 30. Consider a technological coefficient of fuzzy linear programming problem defined as PnIT2FS $\tilde{a}_{ij}$ defined on the closed interval

$$\tilde{a}_{ij} \in \left[\inf_{a_{ij}} \left\{{}^{\alpha}\mu_{\tilde{a}_{ij}}(a_{ij}, u)\right\}, \sup_{a_{ij}} \left\{{}^{\alpha}\mu_{\tilde{a}_{ij}}(a_{ij}, u)\right\}\right] = \left[\left[{}^{\alpha}\overline{a}_{ij}^L, {}^{\alpha}\underline{a}_{ij}^L\right], \left[{}^{\alpha}\underline{a}_{ij}^R, {}^{\alpha}\overline{a}_{ij}^R\right]\right] \in \mathbb{R}, \quad i \in \mathbb{N}_n, \; j \in \mathbb{N}_m. \tag{32}$$

The membership function which represents the fuzzy space Supp($\tilde{a}_{ij}$) is

$$\tilde{a}_{ij} = \int_{\tilde{a}_{ij} \in \mathbb{R}} \frac{\left[\int_{u \in J_{\tilde{a}_{ij}}} 1/u\right]}{\tilde{a}_{ij}}, \quad i \in \mathbb{N}_n, \; j \in \mathbb{N}_m, \; J_{\tilde{a}_{ij}} \subseteq [0,1]. \tag{33}$$

Here, $\tilde{a}_{ij}$ is bounded by both lower and upper primary membership functions; namely

$${}^{\alpha}\underline{\mu}_{\tilde{a}_{ij}} = \left\{(a_{ij}, u) \mid \underline{\mu}_{\tilde{a}_{ij}} \ge \alpha\right\} \tag{34}$$

with parameters ${}^{\alpha}\underline{a}_{ij}^L$ and ${}^{\alpha}\underline{a}_{ij}^R$ and

$${}^{\alpha}\overline{\mu}_{\tilde{a}_{ij}} = \left\{(a_{ij}, u) \mid \overline{\mu}_{\tilde{a}_{ij}} \ge \alpha\right\} \tag{35}$$

with parameters ${}^{\alpha}\overline{a}_{ij}^L$ and ${}^{\alpha}\overline{a}_{ij}^R$.

Proposition 31. *Let $\tilde{a}_{ij} \in F(\mathbb{R})$, $x_j \ge 0$, $i \in \mathbb{N}_n$, $j \in \mathbb{N}_m$. Then the fuzzy set $\sum_{j=1}^n \tilde{a}_{ij} x_j$ defined by the extension principle is again a fuzzy number.*

Let $\tilde{P} = \preceq^{\text{Pos}}$ be a fuzzy relation [27] fuzzy extension of the usual binary relation $\le$ on $\mathbb{R}$. The fuzzy linear programming problem associated with a standard linear programming problem is denoted as

$$\begin{aligned} \text{Max/Min} \quad & Z = \sum_{j=1}^n c_j x_j \\ \text{S.t} \quad & \left(\sum_{j=1}^n \tilde{a}_{ij} x_j\right) \tilde{P} \tilde{b}_i, \quad i = 1, 2, \ldots, m \\ & x_j \ge 0, \quad j = 1, 2, \ldots, n. \end{aligned} \tag{36}$$

In (36) the value $\sum_{j=1}^n \tilde{a}_{ij} x_j \in F(\mathbb{R})$ is compared with a fuzzy number $\tilde{b}_i \in F(\mathbb{R})$ by sum fuzzy relation. The maximization of the objective function is denoted by Max $Z = \sum_{j=1}^n c_j x_j$.

Definition 32. Let $\mu_{\tilde{a}_{ij}} : \mathbb{R} \to [0,1]$ and $\mu_{\tilde{b}_i} : \mathbb{R} \to [0,1]$, $i \in \mathbb{N}_n$, $j \in \mathbb{N}_m$, be membership function of fuzzy numbers $\tilde{a}_{ij}$ and $\tilde{b}_i$, respectively. Let $\tilde{P}$ be a fuzzy extension of a binary relation $\le$ on $\mathbb{R}$. A fuzzy set $\widetilde{X}$, whose membership function $\mu_{\widetilde{X}}$ is defined as $x \in \mathbb{R}^n$ by

$$\mu_{\widetilde{X}} = \begin{cases} \min\left\{\mu_{\tilde{P}}\left(\tilde{a}_{11}x_1 + \cdots + \tilde{a}_{1n}x_n, \tilde{b}_1\right), \ldots, \mu_{\tilde{P}}\left(\tilde{a}_{m1}x_1 + \cdots + \tilde{a}_{mn}x_n, \tilde{b}_m\right)\right\}, & \text{if } x_j \ge 0, \; j = 1, 2, \ldots, n, \\ 0, & \text{otherwise,} \end{cases} \tag{37}$$

is called the fuzzy set of feasible regions of the fuzzy linear programming problem (36). For $\alpha \in (0,1]$, a vector $x \in [\widetilde{X}]_\alpha$ is called the *α-feasible* solution of the fuzzy linear programming problem.

Notice that the feasible region $\widetilde{X}$ of the fuzzy linear programming problem is a fuzzy set. On the other hand, *α-feasible* solution is a vector belonging to the *α-cut* of the feasible region $\widetilde{X}$. If all the coefficients $\widetilde{a}_{ij}$ and $\widetilde{b}_i$ are crisp fuzzy number, that is, they are equivalent to the corresponding crisp fuzzy number, then the fuzzy feasible region is equivalent to the set of all feasible solutions of the corresponding classical linear programming.

6.1. Type 2 Fuzzy Linear Programming Problem. Let us consider the following fuzzy linear programming problem with right-hand side (resources) and technology coefficients are PNIT2TrFNs:

$$\begin{aligned} \text{Max/Min} \quad & Z = \sum_{j=1}^{n} c_j x_j \\ \text{S.t} \quad & \sum_{j=1}^{n} \widetilde{a}_{ij} x_j \le \widetilde{b}_i, \quad i = 1, 2, \ldots, m \\ & x_j \ge 0, \quad j = 1, 2, \ldots, n, \end{aligned} \tag{38}$$

where $\widetilde{a}_{ij} = ((\widetilde{a}_{ij})^L, (\widetilde{a}_{ij})^U) = (((a_{ij})_2^L, (a_{ij})_3^L, (\alpha_{ij})_L, (\beta_{ij})_L), ((a_{ij})_2^U, (a_{ij})_3^U, (\alpha_{ij})_U, (\beta_{ij})_U))$, $\widetilde{b}_i = ((\widetilde{b}_i)^L, (\widetilde{b}_i)^U) = (((b_i)_2^L, (b_i)_3^L, (\gamma_i)_L, (\theta_i)_L), ((b_i)_2^U, (b_i)_3^U, (\gamma_i)_U, (\theta_i)_U))$ are PnIT2TrFNs, c_i are crisp coefficients of the objective, and x_i are the decision variable.

From Definitions 15, 21, 22, and 23 and (38),

$$\begin{aligned} \text{Max/Min} \quad & Z = \sum_{j=1}^{n} c_j x_j \\ \text{S.t} \quad & \sum_{j=1}^{n} \widetilde{a}_{ij}^L x_j \le \widetilde{b}_i^L, \quad i = 1, 2, \ldots, m \\ & x_j \ge 0, \quad j = 1, 2, \ldots, n, \\ \text{Max/Min} \quad & Z = \sum_{j=1}^{n} c_j x_j \\ \text{S.t} \quad & \sum_{j=1}^{n} \widetilde{a}_{ij}^U x_j \le \widetilde{b}_i^U, \quad i = 1, 2, \ldots, m \\ & x_j \ge 0, \quad j = 1, 2, \ldots, n. \end{aligned} \tag{39}$$

Based on the relation of possibility we can rewrite (39) as follows:

$$\begin{aligned} \text{Max/Min} \quad & Z = \sum_{j=1}^{n} c_j x_j \\ \text{S.t} \quad & \widetilde{a}_{i1}^L x_1 + \widetilde{a}_{i2}^L x_2 + \cdots + \widetilde{a}_{in}^L x_n \preceq^{\text{Pos}} \widetilde{b}_i^L, \\ & \qquad i = 1, 2, \ldots, m, \\ & x_j \ge 0, \quad j = 1, 2, \ldots, n, \\ \text{Max/Min} \quad & Z = \sum_{j=1}^{n} c_j x_j \\ \text{S.t} \quad & \widetilde{a}_{i1}^U x_1 + \widetilde{a}_{i2}^U x_2 + \cdots + \widetilde{a}_{in}^U x_n \preceq^{\text{Pos}} \widetilde{b}_i^U, \\ & \qquad i = 1, 2, \ldots, m, \\ & x_j \ge 0, \quad j = 1, 2, \ldots, n. \end{aligned} \tag{40}$$

With respect to constraints of linear programming possibility, the optimal solution is completely determined by degrees of possibility. So, we can obtain the optimal solution of linear programming possibility at *p-cut* levels by solving the two crisp linear programming problems using Barnes algorithm:

$$\begin{aligned} \text{Max/Min} \quad & Z = \sum_{j=1}^{n} c_j x_j \\ \text{S.t} \quad & \text{Pos}\left(\widetilde{a}_{i1}^L x_1 + \widetilde{a}_{i2}^L x_2 + \cdots + \widetilde{a}_{in}^L x_n \preceq \widetilde{b}_i^L\right) \\ & \ge p, \\ & \qquad i = 1, 2, \ldots, m, \\ & x_j \ge 0, \quad j = 1, 2, \ldots, n, \\ \text{Max/Min} \quad & Z = \sum_{j=1}^{n} c_j x_j \\ \text{S.t} \quad & \text{Pos}\left(\widetilde{a}_{i1}^U x_1 + \widetilde{a}_{i2}^U x_2 + \cdots + \widetilde{a}_{in}^U x_n \preceq \widetilde{b}_i^U\right) \\ & \ge p, \\ & \qquad i = 1, 2, \ldots, m, \\ & x_j \ge 0, \quad j = 1, 2, \ldots, n. \end{aligned} \tag{41}$$

From Theorems 27 and 28 we have

$$\begin{aligned} \underline{\left(LPP_p\right)} = \text{Max/Min} \quad & Z = \sum_{j=1}^{n} c_j x_j \\ \text{S.t} \quad & \underline{b}^U - \underline{A}^L x \ge (p-1)\left(\underline{\alpha} x + \underline{\theta}\right), \\ & x_j \ge 0, \ j = 1, 2, \ldots, n, \\ \overline{\left(LPP_p\right)} = \text{Max/Min} \quad & Z = \sum_{j=1}^{n} c_j x_j \\ \text{S.t} \quad & \overline{b}^U - \overline{A}^L x \ge (p-1)\left(\overline{\alpha} x + \overline{\theta}\right), \\ & x_j \ge 0, \ j = 1, 2, \ldots, n, \end{aligned} \tag{42}$$

where $\underline{A}^L = \{(a_{ij})_2^L\}$, $\overline{A}^L = \{(a_{ij})_2^U\}$, $\underline{b}^U = \{(b_i)_3^L\}$, $\overline{b}^U = \{(b_i)_3^U\}$, $\underline{\alpha} = \{(\alpha_{ij})_L\}$, $\overline{\alpha} = \{(\alpha_{ij})_U\}$, $\underline{\beta} = \{(\beta_{ij})_L\}$, $\overline{\beta} = \{(\beta_{ij})_U\}$, $\underline{\gamma} = \{(\gamma_i)_L\}$, $\overline{\gamma} = \{(\gamma_i)_U\}$, $\underline{\theta} = \{(\theta_i)_L\}$, $\overline{\theta} = \{(\theta_i)_U\}$, and $x = x_j$, $j = 1, 2, \ldots, n$, $i = 1, 2, \ldots, m$.

The feasible regions of $\underline{(LPP_p)}$ and $\overline{(LPP_p)}$ are denoted by

$$\underline{X}_p^P = \{x \in R^n \mid \underline{b}^U - \underline{A}^L x \geq (p-1)(\underline{\alpha}x + \underline{\theta})\},$$
$$\overline{X}_p^P = \{x \in R^n \mid \overline{b}^U - \overline{A}^L x \geq (p-1)(\overline{\alpha}x + \overline{\theta})\}. \tag{43}$$

Assume the optimal solutions of $\underline{(LPP_p)}$ and $\overline{(LPP_p)}$ are $\underline{Z}_p^P$ and $\overline{Z}_p^P$, respectively.

Theorem 33. *For linear programming $\underline{(LPP_p)}$ and programming $\overline{(LPP_p)}$, $0 \leq p_1 \leq p_2 \leq 1$, and then $\underline{X}_{p_2}^P \subseteq \underline{X}_{p_1}^P$, $\overline{X}_{p_2}^P \subseteq \overline{X}_{p_1}^P$ and $\underline{Z}_{p_1}^P \geq \underline{Z}_{p_2}^P$, $\overline{Z}_{p_1}^P \geq \overline{Z}_{p_2}^P$.*

Proof. If $x \in \underline{X}_{p_2}^P$, we have $\underline{b}^U - \underline{A}^L x \geq (p_2-1)(\underline{\alpha}x + \underline{\theta})$, from $\underline{\beta} > 0$, $\underline{\gamma} > 0$, and $x > 0$; then $\underline{b}^U - \underline{A}^L x \geq (p_2-1)(\underline{\alpha}x + \underline{\theta}) \geq (p_1-1)(\underline{\alpha}x+\underline{\theta})$, so $x \in \underline{X}_{p_1}^P$; that is, $\underline{X}_{p_2}^P \subseteq \underline{X}_{p_1}^P$ and $\underline{Z}_{p_1}^P \geq \underline{Z}_{p_2}^P$. In a similar way $\overline{X}_{p_2}^P \subseteq \overline{X}_{p_1}^P$ and $\overline{Z}_{p_1}^P \geq \overline{Z}_{p_2}^P$. □

Theorem 34. *For linear programming $\underline{(LPP_p)}$ and programming $\overline{(LPP_p)}$, $0 \leq p_1 \leq p_2 \leq 1$, and then $\underline{X}_{p_2}^P \subseteq \overline{X}_{p_1}^P$ and $\underline{Z}_{p_1}^P \geq \overline{Z}_{p_2}^P$.*

Proof. If $x \in \underline{X}_{p_2}^P$, we have $\underline{b}^U - \underline{A}^L x \geq (p_2-1)(\underline{\alpha}x + \underline{\theta})$, from $\underline{\beta} > 0$, $\underline{\gamma} > 0$, and $x > 0$; then $\underline{b}^U - \underline{A}^L x \geq (p_2-1)(\underline{\alpha}x + \underline{\theta}) \geq (p_1-1)(\overline{\alpha}x + \overline{\theta})$, so $x \in \overline{X}_{p_1}^P$; that is, $\underline{X}_{p_2}^P \subseteq \overline{X}_{p_1}^P$ and $\overline{Z}_{p_1}^P \geq \underline{Z}_{p_2}^P$. □

From Theorems 33 and 34, we obtain that $\underline{Z}_p^P, \overline{Z}_p^P$ constitute the lower and upper bound of fuzzy objective value at *level p*. Then the objective value of Z_p^P at level p is the mean of $[\underline{Z}_p^P, \overline{Z}_p^P]$.

7. Numerical Illustration

An application of the proposed method is introduced with an example, where all its technological coefficients and right-hand side (resources) are defined as PnIT2FNs. The optimal solution will be obtained in terms of crisp for the following problem.

The company plans to manufacture two types of products. The selling prices for these products are as follows: P_1 costs Rs. 1,750 per unit and P_2 costs Rs. 2,000 per unit. Daily production volume of each type of these products is constrained by available man hours and available machine hours. The production specifications for the given problem situation are presented in as shown in Table 1:

$$\begin{aligned} \text{Max} \quad & Z = c_1x_1 + c_2x_2 \\ \text{S.t} \quad & \tilde{a}_{11}x_1 + \tilde{a}_{12}x_2 \preceq \tilde{b}_1 \\ & \tilde{a}_{21}x_1 + \tilde{a}_{22}x_2 \preceq \tilde{b}_2 \\ & x_j \geq 0, \quad j = 1, 2, \end{aligned} \tag{44}$$

Table 1

Resource	Resource requirement/unit P_1	P_2	Availability
Man hours (in minutes)	Around 4	Around 8	Around 32
Machine hours (in minutes)	Around 6	Around 4	Around 36

where $\tilde{a}_{11} = [(230, 250, 20, 10), (230, 250, 30, 20)]$, $\tilde{a}_{12} = [(460, 500, 40, 20), (460, 500, 60, 40)]$, $\tilde{a}_{21} = [(345, 375, 30, 15), (345, 375, 45, 30)]$, $\tilde{a}_{22} = [(230, 250, 20, 10), (230, 250, 30, 20)]$, $\tilde{b}_1 = [(1840, 2000, 160, 80), (1840, 2000, 240, 160)]$, $\tilde{b}_2 = [(2070, 2250, 180, 90), (2070, 2250, 270, 180)]$, $c_1 = 1750$, and $c_2 = 2000$. From (42) we have the following crisp optimal problem based on the level p:

$$\begin{aligned} \text{Max} \quad & \underline{Z}_p^P = 1750x_1 + 2000x_2 \\ \text{S.t} \quad & (210 + 20p)x_1 + (420 + 40p)x_2 \\ & \leq 2080 - 80p \\ & (315 + 30p)x_1 + (210 + 20p)x_2 \\ & \leq 2340 - 90p \\ & x_1, x_2 \geq 0 \end{aligned} \tag{45}$$

$$\begin{aligned} \text{Max} \quad & \overline{Z}_p^P = 1750x_1 + 2000x_2 \\ \text{S.t} \quad & (200 + 30p)x_1 + (400 + 60p)x_2 \\ & \leq 2160 - 160p \\ & (300 + 45p)x_1 + (200 + 30p)x_2 \\ & \leq 2430 - 180p \\ & x_1, x_2 \geq 0. \end{aligned} \tag{46}$$

For different cut level p, we can get different optimal solution and denote by $\underline{Z}_p^P, \overline{Z}_p^P$ the optimal solution (see Figure 5 and Table 2) [28] of the crisp programming (45) and programming (46), respectively.

8. Conclusion

In this paper a method to do perfectly normal interval type 2 fuzzy arithmetic operation using extension principle is presented and defined the possibility degrees of upper and lower membership function to compare perfectly normal interval type 2 fuzzy numbers. These functions are defined based on the strength of upper and lower membership function of perfectly normal interval type 2 fuzzy numbers. Meanwhile, some properties and theorems were proved. Then, the lower and upper fuzzy satisfaction based on possibility degrees of

TABLE 2: Optimal solution at different possibility levels.

p-cut	0.1	0.2	0.3	0.4	0.5	0.6	0.7	0.8	0.9	1
$\underline{Z}_p^P$	14354.9528	14165.8878	13980.324	13798.1651	13619.3181	13443.6937	13271.2053	13101.7699	12935.3069	12771.7393
$\overline{Z}_p^P$	15512.3153	15172.3301	14842.1051	14521.2264	14209.3022	13905.9633	13610.8597	13323.6607	13044.0529	12771.7393
Z_p^P	14933.6341	14669.1090	14411.2146	14159.6958	13914.3102	13674.8285	13441.0325	13212.7153	12989.6799	12771.7393

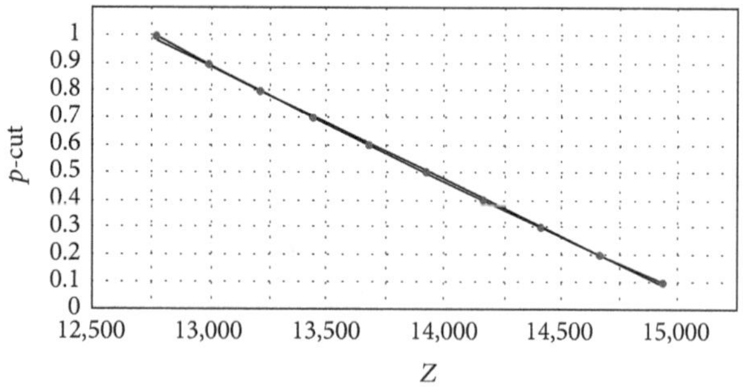

FIGURE 5: Optimal solution at different possibility levels.

membership function, respectively, are defined. These fuzzy satisfactions create two nonstrict order relations on the set of overlapping intervals. Also, linear programming problem was introduced with resources and technology coefficients of perfectly normal interval type 2 fuzzy numbers. Then, the possibility degrees of membership function were applied in order to interpret inequality constraints with interval coefficients. According to the definition of possibility degrees of membership function and their properties, the inequality constraints with interval coefficients were reduced in their satisfactory crisp equivalent forms. Finally, the decision maker can get the crisp optimal solution of the problem for every grade $\alpha \in [0, 1]$ which was obtained by using Barnes algorithm and then mean value is selected according to the decision maker optimistic attitude.

Conflict of Interests

The authors declare that there is no conflict of interests regarding the publication of this paper.

References

[1] L. A. Zadeh, "Fuzzy sets," *Information and Computation*, vol. 8, pp. 338–353, 1965.

[2] H. Tanaka, T. Okuda, and K. Asai, "On fuzzy-mathematical programming," *Journal of Cybernetics*, vol. 3, no. 4, pp. 37–46, 1973.

[3] R. E. Bellman and L. A. Zadeh, "Decision-making in a fuzzy environment," *Management Science*, vol. 17, pp. B141–B164, 1970/71.

[4] H.-J. Zimmermann, "Fuzzy programming and linear programming with several objective functions," *Fuzzy Sets and Systems*, vol. 1, no. 1, pp. 45–55, 1978.

[5] S. Deng, L. Zhou, and X. Wang, "Solving the fuzzy bilevel linear programming with multiple followers through structured element method," *Mathematical Problems in Engineering*, vol. 2014, Article ID 418594, 6 pages, 2014.

[6] Y.-H. Zhong, Y.-L. Jia, D. Chen, and Y. Yang, "Interior point method for solving fuzzy number linear programming problems using linear ranking function," *Journal of Applied Mathematics*, vol. 2013, Article ID 795098, 9 pages, 2013.

[7] H. Cheng, W. Huang, and J. Cai, "Solving a fully fuzzy linear programming problem through compromise programming," *Journal of Applied Mathematics*, vol. 2013, Article ID 726296, 10 pages, 2013.

[8] D. Chakraborty, D. K. Jana, and T. K. Royl, "A new approach to solve intuitionistic fuzzy optimization problem using possibility, necessity, and credibility measures," *International Journal of Engineering Mathematics*, vol. 2014, Article ID 593185, 12 pages, 2014.

[9] H. Le and Z. Gogne, "Fuzzy linear programming with possibility and necessity relation," in *Fuzzy Information and Engineering 2010*, vol. 78 of *Advances in Intelligent and Soft Computing*, pp. 305–311, Springer, Berlin, Germany, 2010.

[10] J. C. Figueroa-García and G. Hernández, "A method for solving linear programming models with interval type-2 fuzzy constraints," *Pesquisa Operacional*, vol. 34, no. 1, pp. 73–89, 2014.

[11] J. C. F. García, "A general model for linear programming with interval type-2 fuzzy technological coefficients," in *Proceedings of the IEEE Annual Meeting of the North American Fuzzy Information Processing Society (NAFIPS '12)*, pp. 1–4, Berkeley, Calif, USA, August 2012.

[12] J. C. F. García, "Linear programming with interval type-2 fuzzy right hand side parameters," in *Proceedings of the Annual Meeting of the North American Fuzzy Information Processing Society (NAFIPS '08)*, pp. 1–6, IEEE, New York, NY, USA, May 2008.

[13] S.-M. Chen and L.-W. Lee, "Fuzzy multiple attributes group decision-making based on the ranking values and the arithmetic operations of interval type-2 fuzzy sets," *Expert Systems with Applications*, vol. 37, no. 1, pp. 824–833, 2010.

[14] J. Hu, Y. Zhang, X. Chen, and Y. Liu, "Multi-criteria decision making method based on possibility degree of interval type-2 fuzzy number," *Knowledge-Based Systems*, vol. 43, pp. 21–29, 2013.

[15] R. Fuller, *Fuzzy Reasoning and Fuzzy Optimization*, Turku Centre for Computer Science, 1998.

[16] H.-C. Wu, "Duality theory in fuzzy linear programming problems with fuzzy coefficients," *Fuzzy Optimization and Decision Making*, vol. 2, no. 1, pp. 61–73, 2003.

[17] J. Qin and X. Liu, "Frank aggregation operators for triangular interval type-2 fuzzy set and its application in multiple attribute group decision making," *Journal of Applied Mathematics*, vol. 2014, Article ID 923213, 24 pages, 2014.

[18] Z. M. Zhang and S. H. Zhang, "A novel approach to multi attribute group decision making based on trapezoidal interval type-2 fuzzy soft sets," *Applied Mathematical Modelling*, vol. 37, no. 7, pp. 4948–4971, 2013.

[19] Y. Gong, "Fuzzy multi-attribute group decision making method based on interval type-2 fuzzy sets and applications to global supplier selection," *International Journal of Fuzzy Systems*, vol. 15, no. 4, pp. 392–400, 2013.

[20] D. S. Dinagar and A. Anbalagan, "A new type-2 fuzzy number arithmetic using extension principle," in *Proceedings of the 1st International Conference on Advances in Engineering, Science and Management (ICAESM '12)*, pp. 113–118, March 2012.

[21] K.-P. Chiao, "Multiple criteria group decision making with triangular interval type-2 fuzzy sets," in *Proceedings of the IEEE International Conference on Fuzzy Systems (FUZZ '11)*, pp. 2575–2582, IEEE, Taipei, Taiwan, June 2011.

[22] J. M. Mendel, R. I. John, and F. Liu, "Interval type-2 fuzzy logic systems made simple," *IEEE Transactions on Fuzzy Systems*, vol. 14, no. 6, pp. 808–821, 2006.

[23] X. Liu, "Measuring the satisfaction of constraints in fuzzy linear programming," *Fuzzy Sets and Systems*, vol. 122, no. 2, pp. 263–275, 2001.

[24] D. Dubois and H. Prade, "Ranking fuzzy numbers in the setting of possibility theory," *Information Sciences*, vol. 30, no. 3, pp. 183–224, 1983.

[25] D. S. Negi and E. S. Lee, "Possibility programming by the comparison of fuzzy numbers," *Computers & Mathematics with Applications*, vol. 25, no. 9, pp. 43–50, 1993.

[26] M. G. Iskander, "Comparison of fuzzy numbers using possibility programming: comments and new concepts," *Computers & Mathematics with Applications*, vol. 43, no. 6-7, pp. 833–840, 2002.

[27] J. Ramik, "Duality in fuzzy linear programming with possibility and necessity relations," *Fuzzy Sets and Systems*, vol. 157, no. 10, pp. 1283–1302, 2006.

[28] G. R. Lindfield and J. E. T. Penny, *Numerical methods Using Matlab*, Academic Press, 3rd edition, 2013.

8

Boltzmann's Six-Moment One-Dimensional Nonlinear System Equations with the Maxwell-Auzhan Boundary Conditions

A. Sakabekov[1] and Y. Auzhani[2]

[1]*Research Laboratory "Applied Modeling Oil and Gas Fields", Kazakh-British Technical University, 59 Tole Bi Street, Almaty 050000, Kazakhstan*
[2]*Al-Farabi Kazakh National University, 71 Al-Farabi Avenue, Almaty 050040, Kazakhstan*

Correspondence should be addressed to A. Sakabekov; a.sakabekov@kbtu.kz

Academic Editor: Peter G. L. Leach

We prove existence and uniqueness of the solution of the problem with initial and Maxwell-Auzhan boundary conditions for nonstationary nonlinear one-dimensional Boltzmann's six-moment system equations in space of functions continuous in time and summable in square by a spatial variable. In order to obtain a priori estimation of the initial and boundary value problem for nonstationary nonlinear one-dimensional Boltzmann's six-moment system equations we get the integral equality and then use the spherical representation of vector. Then we obtain the initial value problem for Riccati equation. We have managed to obtain a particular solution of this equation in an explicit form.

1. Introduction

In case of one-atom gas any macroscopic system during process of its evolution to an equilibrium state passes 3 stages: initial transition period, described in terms of a full function of system distribution; the kinetic period, by means of a one-partial distribution function; and the hydrodynamic period, by means of five first moments of the distribution function. In kinetic regime the behavior of a rarefied gas in space of time and velocity is described by Boltzmann's equation. It is known from gas dynamics that in most of the encountered problems there is no need to use detailed microscopic gas description with help of the distribution function. Therefore it is natural to look for a less detailed description using macroscopic hydrodynamic variables (density, hydrodynamic velocity, temperature, etc.). As these variables are defined in terms of moments of the distribution function, we are faced with the problem of analyzing the various moments of Boltzmann's equation.

Note that Boltzmann's moment equations are intermediate between Boltzmann (kinetic theory) and hydrodynamic levels of description of state of the rarefied gas and form class of nonlinear partial differential equations. Existence of such class of equations was noticed by Grad [1, 2] in 1949. He obtained the moment system by expanding the particle distribution function in Hermite polynomials near the local Maxwell distribution. Grad used Cartesian coordinates of velocities and Grad's moment system contained as coefficients such unknown hydrodynamic characteristics like density, temperature, average speed, and so forth. Formulation of boundary conditions for Grad's system is almost impossible, as the characteristic equations for various approximations of Grad's hyperbolic system contain unknown parameters like density, temperature, and average speed. However, 13- and 20-moment Grad equations are widely used in solving many problems of the kinetic theory of gases and plasma.

Boltzmann's equation is equivalent to an infinite system of differential equations relative to the moments of the particle distribution function in the complete system of eigenfunctions of linearized operator. As a rule we limit study by finite moment system of equations because solving infinite system of equations does not seem to be possible.

Finite system of moment equations for a specific task with a certain degree of accuracy replaces Boltzmann's equation. It is necessary, also roughly, to replace the boundary conditions for the particle distribution function by a number of macroscopic conditions for the moments; that is, there arises the problem of boundary conditions for a finite system of equations that approximate the microscopic boundary conditions for Boltzmann's equation. The question of boundary conditions for a finite system of moment equations can be divided into two parts: how many conditions must be imposed and how they should be prepared. From microscopic boundary conditions for Boltzmann's equation there can be obtained an infinite set of boundary conditions for each type of decomposition. However, the number of boundary conditions is determined not by the number of moment equations; that is, it is impossible, for example, to take as many boundary conditions as equations, although the number of moment equations affects the number of boundary conditions. In addition, the boundary conditions must be consistent with the moment equations and the resulting problem must be correct. Boundary condition problems arise in the following tasks: (1) moment boundary conditions in rarefied gas slip-flow problems; (2) definition of boundary conditions on the surfaces of streamlined rarefied gas; (3) prediction aerospace aerodynamic characteristics of aircraft at very high speeds and at high altitudes and so forth.

In work [3] we have obtained the moment system which differs from Grad's system of equations. We used spherical velocity coordinates and decomposed the distribution function into a series of eigenfunctions of the linearized collision operator [4, 5], which is the product of Sonine polynomials and spherical functions. The resulting system of equations, which correspond to the partial sum of series and which we called Boltzmann's moment system of equations, is a nonlinear hyperbolic system in relation to the moments of the particles distribution function.

The structure of Boltzmann's moment system of equations corresponds to the structure of Boltzmann's equation; namely, the differential part of the resulting system is linear in relation to the moments of the distribution function and nonlinearity is included as moments of collision integral [6].

The linearity of differential part of Boltzmann's moment system of equations simplifies the task of formulation of the boundary conditions. In work [3] a homogeneous boundary condition for particles distribution function was approximated and proved the correctness of initial and boundary value problem for nonlinear nonstationary Boltzmann's moment system of equations in three-dimensional space. In work [7] the initial and boundary value problem for one-dimensional nonstationary Boltzmann's equation with boundary conditions of Maxwell was approximated by a corresponding problem for Boltzmann's moment system of equations. The boundary conditions for Boltzmann's moment system of equations were called Maxwell-Auzhan conditions.

In work [8] a systematic nonperturbative derivation of a hierarchy of closed systems of moment equations corresponding to any classical theory has been presented. This paper is a fundamental work where closed systems of moment equations describe a transition regime. Moreover, hydrodynamical model is used to describe charge transport in a generic compound semiconductor. Compound semiconductors have found wide use in the microelectronic industry. The evolution equation for macroscopic variables is obtained by taking moments of the transport equation [9–13].

The study of various problems for Boltzmann's moment system of equations is an important and actual task in the theory of a rarefied gas and other applications of the moment system equations. The correctness of initial and boundary problems for Boltzmann's moment system of equations with Maxwell-Auzhan boundary conditions is being studied for the first time.

2. Existence and Uniqueness of the Solutions of Initial and Boundary Value Problem for Six-Moment One-Dimensional Boltzmann's System of Equations with Maxwell-Auzhan Boundary Conditions

In this section we prove the existence and uniqueness of solutions of the initial and boundary value problem for six-moment one-dimensional Boltzmann's system of equations with Maxwell-Auzhan boundary conditions in space of functions, continuous in time and summable in square by spatial variable. Note that the theorem of the existence of a global solution in time of the initial and boundary value problem for 3-dimensional nonlinear Boltzmann's equation with boundary conditions of Maxwell is proved in work [14].

We write in an expanded form system of one-dimensional Boltzmann's moment equations in the kth approximation, which corresponds to decomposition of the particle distribution function by eigenfunctions of the linearized collision operator

$$\frac{\partial f_{nl}}{\partial t} + \frac{1}{\alpha}\frac{\partial}{\partial x}\left[l \left(\sqrt{\frac{2(n+l+1/2)}{(2l-1)(2l+1)}} f_{n,l-1} - \sqrt{\frac{2(n+1)}{(2l-1)(2l+1)}} f_{n+1,l-1} \right) + (l+1) \cdot \left(\sqrt{\frac{2(n+l+1/2)}{(2l+1)(2l+3)}} f_{n,l+1} - \sqrt{\frac{2n}{(2l+1)(2l+3)}} f_{n-1,l+1} \right) \right] = I_{nl}, \tag{1}$$

$$2n + l = 0, 1, \ldots, k,$$

where the moments I_{nl} of a nonlinear collision operator are expressed through coefficients of Talmi and Klebsh-Gordon [15, 16] and have the form

$$I_{nl} = \sum \langle N_3 L_3 n_3 l_3 : l \mid nl00 : l \rangle \cdot \langle N_3 L_3 n_3 l_3 : l \mid n_1 l_1 n_2 l_2 : l \rangle \left(\frac{l_1 0 l_2 0}{l0} \right) \left(\sigma_{l_3} - \sigma_0 \right) \cdot f_{n_1 l_1} f_{n_2 l_2}, \tag{2}$$

where $\langle N_3 L_3 n_3 l_3 : l \mid n_1 l_1 n_2 l_2 : l \rangle$ are generalized coefficients of Talmi, $(l_1 0 l_2 0 / l0)$ are Klebsh-Gordon coefficients, $\alpha = 1/\sqrt{R\theta}$ is the constant, R is Boltzmann's constant, and θ is the ideal gas temperature.

For generalized coefficients of Talmi there exists a table [15] for each value of quantum number $\xi = 2n + l$ from 0 to 6. Moreover, there is a program on IBM for calculation of generalized coefficients of Talmi.

If in (1) $2n+l$ takes values from 0 to 3, we get Boltzmann's moment system equations in the third approximation. We write it in an expanded form:

$$\begin{gathered}
\frac{\partial f_{00}}{\partial t}+\frac{1}{\alpha}\frac{\partial f_{01}}{\partial x}=0,\\
\frac{\partial f_{02}}{\partial t}+\frac{1}{\alpha}\frac{\partial}{\partial x}\left(\frac{2}{\sqrt{3}}f_{01}+\frac{3}{\sqrt{5}}f_{03}-\frac{2\sqrt{2}}{\sqrt{15}}f_{11}\right)=J_{02},\\
\frac{\partial f_{10}}{\partial t}+\frac{1}{\alpha}\frac{\partial}{\partial x}\left(-\sqrt{\frac{2}{3}}f_{01}+\sqrt{\frac{5}{3}}f_{11}\right)=0,\\
\frac{\partial f_{01}}{\partial t}+\frac{1}{\alpha}\frac{\partial}{\partial x}\left(f_{00}+\frac{2}{\sqrt{3}}f_{02}-\sqrt{\frac{2}{3}}f_{10}\right)=0,\\
\frac{\partial f_{03}}{\partial t}+\frac{1}{\alpha}\frac{\partial}{\partial x}\frac{3}{\sqrt{5}}f_{02}=J_{03},\\
\frac{\partial f_{11}}{\partial t}+\frac{1}{\alpha}\frac{\partial}{\partial x}\left(-\frac{2\sqrt{2}}{\sqrt{15}}f_{02}+\sqrt{\frac{5}{3}}f_{10}\right)=J_{11},\\
x\in(-a,a),\ t>0,
\end{gathered}\tag{3}$$

where $f_{00} = f_{00}(t,x)$ and $f_{01} = f_{01}(t,x),\ldots,f_{11} = f_{11}(t,x)$ are the moments of the particle distribution function; $J_{02} = (\sigma_2 - \sigma_0)(f_{00}f_{02} - f_{01}^2/\sqrt{3})/2$, $J_{03} = (1/4)(\sigma_3 + 3\sigma_1 - 4\sigma_0)f_{00}f_{03} + (1/4\sqrt{5})(2\sigma_1 + \sigma_0 - 3\sigma_3)f_{01}f_{02}$, and $J_{11} = (\sigma_1 - \sigma_0)(f_{00}f_{11} + (1/2)\sqrt{5/3}f_{10}f_{01} - (\sqrt{2}/\sqrt{15})f_{01}f_{02})$ are moments of the collision integral, where σ_0, σ_1, σ_2, and σ_3 are the Fourier coefficients of the cross section expansion by the Legendre polynomials. The first, third, and fourth equations of system (3) correspond to mass conservation law, momentum conservation law, and energy conservation law, respectively.

We study the correctness of initial and boundary value problem for six-moment one-dimensional Boltzmann's system equations

$$\begin{gathered}
\frac{\partial u}{\partial t}+A\frac{\partial w}{\partial x}=J_1(u,w)\\
\frac{\partial w}{\partial t}+A'\frac{\partial u}{\partial x}=J_2(u,w),\\
t\in(0,T],\ x\in(-a,a),
\end{gathered}\tag{4}$$

$$\begin{gathered}
u|_{t=0}=u_0(x),\\
w|_{t=0}=w_0(x),\\
x\in[-a,a],
\end{gathered}\tag{5}$$

$$\begin{aligned}
(Aw^- + Bu^-)|_{x=-a} &= \frac{1}{\beta}(Aw^+ - Bu^+)|_{x=-a}\\
&\quad-\frac{(1-\beta)\eta}{\alpha\beta\sqrt{\pi}}F,\quad t\in[0,T],
\end{aligned}\tag{6}$$

$$\begin{aligned}
(Aw^- - Bu^-)|_{x=a} &= \frac{1}{\beta}(Aw^+ + Bu^+)|_{x=a}\\
&\quad+\frac{(1-\beta)\eta}{\alpha\beta\sqrt{\pi}}F,\quad t\in[0,T],
\end{aligned}\tag{7}$$

where

$$A=\frac{1}{\alpha}\begin{pmatrix}1&0&0\\ \frac{2}{\sqrt{3}}&\frac{3}{\sqrt{5}}&-\frac{2\sqrt{2}}{\sqrt{15}}\\ -\sqrt{\frac{2}{3}}&0&\sqrt{\frac{5}{3}}\end{pmatrix},$$

$$B=\frac{1}{\alpha\sqrt{\pi}}\begin{pmatrix}\sqrt{2}&\sqrt{\frac{2}{3}}&-\frac{1}{\sqrt{3}}\\ \sqrt{\frac{2}{3}}&2\sqrt{2}&-1\\ -\frac{1}{\sqrt{3}}&-1&3\sqrt{2}\end{pmatrix}\tag{8}$$

$$\begin{gathered}
J_1(u,w)=(0,J_{02},0)',\\
J_2(u,w)=(0,J_{03},J_{11})',\\
u=(f_{00},f_{02},f_{10})',\ w=(f_{01},f_{03},f_{11})',\\
F=\left(\frac{1}{4\sqrt{2}},\frac{1}{8\sqrt{6}},\frac{1}{8\sqrt{3}}\right)'.
\end{gathered}$$

A' is the transpose matrix; B is the positive definite matrix; $u_0(x) = (f_{00}^0(x), f_{02}^0(x), f_{10}^0(x))'$ and $w_0(x) = (f_{01}^0(x), f_{03}^0(x), f_{11}^0(x))'$ are given initial vector-functions; w^+, u^+ are vector moments of falling to the boundary particle distribution function; w^-, u^- are vector moments of reflecting from the boundary particle distribution function. Equation (4) is vector matrix form of system equations (3).

It is possible to check through direct calculations that

$$\det A_1=\det\begin{pmatrix}0&A\\A'&0\end{pmatrix}\neq 0,\tag{9}$$

and matrix A_1 has three positive and the same number of negative nonzero eigenvalues. From (6)-(7) it follows that the number of boundary conditions on the left and right ends of interval $(-a,a)$ is equal to the number of positive and negative eigenvalues of the matrix A_1.

Thus system (4) is a symmetric hyperbolic nonlinear partial differential equations system. Let us show that J_{02}

is a sign-nondefined square form. It is easy to check that $f_{00}f_{02} - (f_{01}^2/\sqrt{3}) = (CU, U)$, where $U = (u, w)'$ and

$$C = \begin{pmatrix} 0 & \frac{1}{2} & 0 & 0 & 0 & 0 \\ \frac{1}{2} & 0 & 0 & 0 & 0 & 0 \\ 0 & 0 & 0 & 0 & 0 & 0 \\ 0 & 0 & 0 & -\frac{1}{\sqrt{3}} & 0 & 0 \\ 0 & 0 & 0 & 0 & 0 & 0 \\ 0 & 0 & 0 & 0 & 0 & 0 \end{pmatrix}. \quad (10)$$

Eigenvalues of matrix C are $-1/2$, $-1/\sqrt{3}$, 0, 0, 0, and 1/2. Therefore J_{02} is a sign-nondefined square form. Similarly, we can show that J_{03} and J_{11} are also sign-nondefined square forms.

For problem (4)–(7) the following theorem takes place.

Theorem 1. *If $U_0 = (u_0(x), w_0(x)) \in L^2[-a, a]$, then problem (4)-(7) has a unique solution in domain $[-a, a] \times [0, T]$ belonging to the space $C([0, T]; L^2[-a, a])$; moreover*

$$\left\| \|U\|_{L^2[-a,a]} - r_1 \right\|_{C[0,T]} \le C_1 \left(\|U_0\|_{L^2[-a,a]} - r_1(0) \right), \quad (11)$$

where C_1 is a constant independent from U and $T \sim O(\|U_0\|_{L^2[-a,a]} - r_1(0))^{-1})$, $r_1(t)$ is the particular solution of Riccati equation (15) in an explicit form.

Proof. Let $U_0 \in L^2[-a, a]$. Let us prove estimation (11). We multiply the first equation in system (4) by u and the second equation by w and perform integration from $-a$ to a:

$$\frac{1}{2}\frac{d}{dt}\int_{-a}^{a} [(u, u) + (w, w)]\, dx + \int_{-a}^{a} \left[\left(A\frac{\partial w}{\partial x}, u \right) + \left(A'\frac{\partial u}{\partial x}, w \right) \right] dx = \int_{-a}^{a} [(J_1, u) + (J_2, w)]\, dx. \quad (12)$$

After integration by parts we get

$$\frac{1}{2}\frac{d}{dt}\int_{-a}^{a} [(u, u) + (w, w)]\, dx + (u^-, Aw^-)|_{x=a} - (u^-, Aw^-)|_{x=-a} = \int_{-a}^{a} [(J_1, u) + (J_2, w)]\, dx. \quad (13)$$

Taking into account boundary conditions (6)-(7) we rewrite equality (13) in the following form:

$$\begin{aligned}\frac{1}{2}\frac{d}{dt}\int_{-a}^{a} [(u, u) + (w, w)]\, dx + (Bu^-, u^-)|_{x=a} + (Bu^-, u^-)|_{x=-a} - \frac{1}{\beta}\left((Aw^+ - Bu^+), u^-\right)|_{x=-a} \\ + \frac{1}{\beta}\left((Aw^+ + Bu^+), u^-\right)|_{x=a} + (F_1, u^-)|_{x=a} + (F_1, u^-)|_{x=-a} \\ = \int_{-a}^{a} [(J_1(u, w), u) + (J_2(u, w), w)]\, dx,\end{aligned} \quad (14)$$

where $F_1 = ((1 - \beta)\eta/\alpha\beta\sqrt{\pi})F$.

Let us use the spherical representation [17] of vector $U(t, x) = r(t)\omega(t, x)$, where $\omega(t, x) = (\omega_1(t, x), \omega_2(t, x))'$, $r(t) = \|U(t, \cdot)\|_{L^2[-a,a]}$, and $\|\omega\|_{L^2[-a,a]} = 1$. Furthermore assume that $\omega(t, x) \in C^2[0, T]$, where the value of T we define below.

Substituting the values $u = r(t)\omega_1(t, x)$ and $w = r(t)\omega_2(t, x)$ into (14) we have that

$$\frac{dr}{dt} + rP(t) = r^2Q(t) - f(t), \quad (15)$$

where

$$\begin{aligned} P(t) &= (B\omega_1^-, \omega_1^-)|_{x=a} + (B\omega_1^-, \omega_1^-)|_{x=-a} \\ &+ \frac{1}{\beta}\left[(A\omega_2^+, \omega_1^-)|_{x=a} + (B\omega_1^+, \omega_1^-)|_{x=a} + (B\omega_1^+, \omega_1^-)|_{x=-a} - (A\omega_2^+, \omega_1^-)|_{x=-a} \right], \\ Q(t) &= \int_{-a}^{a} [(J_1(\omega_1, \omega_2), \omega_1) + (J_2(\omega_1, \omega_2), \omega_2)]\, dx, \\ f(t) &= (F_1, \omega_1^-)|_{x=a} + (F_1, \omega_1^-)|_{x=-a}. \end{aligned} \quad (16)$$

Let us study (15) with an initial condition

$$r(0) = \|U_0\| = \|U_0\|_{L^2[-a,a]}. \quad (17)$$

Let $r_1(t)$ be a particular solution of Riccati equation (15). Then the general solution of (15) has the form

$$r(t) = r_1(t) + \exp\left(\int_0^t (2Q(\tau) r_1(\tau) - P(\tau))\, d\tau \right) \left[C - \int_0^t Q(\tau) \cdot \exp\left(\int_0^\tau (2Q(\xi) r_1(\xi) - P(\xi))\, d\xi \right) d\tau \right]^{-1}. \quad (18)$$

Hence, taking into consideration initial condition (17) we obtain

$$r(t) = r_1(t) + \exp\left(\int_0^t (2Q(\tau)r_1(\tau) - P(\tau))\,d\tau\right)\left[\frac{1}{\|U_0\| - r_1(0)} - \int_0^t Q(\tau)\exp\left(\int_0^\tau (2Q(\xi)r_1(\xi) - P(\xi))\,d\xi\right)d\tau\right]^{-1}. \quad (19)$$

If $R(t) \equiv \int_0^t Q(\tau)\exp(\int_0^\tau (2Q(\xi)r_1(\xi) - P(\xi))d\xi)d\tau \leq 0\ \forall t$, then the second term of $r(t)$ is bounded $\forall t \in [0, +\infty)$ under the condition that difference $\|U_0\| - r_1(0)$ is positive. Further we suppose a difference $\|U_0\| - r_1(0)$ is a positive number. Let $R(t) > 0$. We denote by T_1 the moment of time at which

$$\frac{1}{\|U_0\| - r_1(0)} - \int_0^{T_1} Q(\tau)\exp\left(\int_0^\tau (2Q(\xi)r_1(\xi) - P(\xi))\,d\xi\right)d\tau = 0. \quad (20)$$

Then $r(t) - r_1(t)$ is bounded $\forall t \in [0, T]$, where $T < T_1$ and $T_1 \sim O(1/(\|U_0\| - r_1(0)))$. Analyzing structure of $r_1(t)$ (see (31)) from equality (20) we obtain $Q(t) > 0$ for any $t \in [0, T]$.

Now we found one particular solution of (15). The nonlinear Riccati equation can always be reduced to a second-order linear ordinary differential equation [18]

$$y'' + f_1(t)\,y' + f_2(t)\,y = 0, \quad (21)$$

where

$$y(t) = \exp\left(-\int_0^t Q(\tau)\,r(\tau)\,d\tau\right),$$
$$f_1(t) = \frac{Q(t)\,P(t) - Q'(t)}{Q(t)}, \quad (22)$$
$$f_2(t) = Q(t)\,f(t).$$

Solution of this equation will lead to a solution of $r(t) = -y'/yQ(t)$ of the original Riccati equation. By substitution

$$y(t) = z(t)\exp\left(-\frac{1}{2}\int_0^t f_1(\tau)\,d\tau\right) \quad (23)$$

we write (21) in the following form [18]:

$$z''(t) + a(t)\,z(t) = 0, \quad (24)$$

where $a(t) = f_2(t) - (1/2)f_1'(t) - (1/4)f_1^2(t)$.

If instead of $f_1(t)$ and $f_2(t)$ we substitute values from (22), then we get

$$a(t) = Qf(t) - \frac{2\left(Q^2P' - Q''Q + Q'^2\right) + \left(QP - Q'\right)^2}{4Q^2}. \quad (25)$$

In work [19] it was proven that function

$$z(t) = 1 + \sum_{k=1}^{\infty} b_k(t) \quad (26)$$

is the solution of (24), where

$$b_1(t) = -\int_0^t \int_0^\xi a(\tau)\,d\tau\,d\xi,$$
$$b_k(t) = -\int_0^t \int_0^\xi a(\tau)\,b_{k-1}(\tau)\,d\tau\,d\xi \quad (27)$$

and the series on the right hand side of (26) uniformly converge.

Then

$$y = \left(1 + \sum_{k=1}^{\infty} b_k(t)\right)\exp\left(-\frac{1}{2}\int_0^t f_1(\tau)\,d\tau\right),$$
$$r_1(t) = -\frac{y'}{yQ} = -\frac{z'(t) - (1/2)\,z(t)\,f_1(t)}{z(t)\,Q(t)} \quad (28)$$

is a particular solution of Riccati equation (15).

Since $z'(0) = 0$, value of

$$r_1(0) = -\frac{z'(0) - (1/2)\,z(0)\,f_1(0)}{z(0)\,Q(0)} = \frac{f_1(0)}{2Q(0)}. \quad (29)$$

Now instead of $f_1(0)$ we substitute its value $P(0) - Q'(0)/Q(0)$ and get $r_1(0) = (P(0)Q(0) - Q'(0))/2Q^2(0)$.

The value of $Q(t)$ with $t = 0$ is

$$Q(0) = \int_{-a}^{a} \left((J_1(\omega_1, \omega_2), \omega_1) + (J_2(\omega_1, \omega_2), \omega_2)\right)\Big|_{t=0} dx > 0, \quad (30)$$

since $Q(t) > 0$ for any $t \in [0, T]$.

An expansion form of the function

$$r_1(t) = \frac{1}{Q(t)}\left[\frac{1}{z(t)}\int_0^t a(\tau)\,z(\tau)\,d\tau + \frac{Q(t)\,P(t) - Q'(t)}{2Q(t)}\right]. \quad (31)$$

It is not difficult to prove using immediate substitution that function $r_1(t)$ satisfies Riccati equation (15).

From equality (19) we obtain

$$\left|r(t) - r_1(t)\right| \leq C\left(\|U_0\| - r_1(0)\right). \quad (32)$$

Hence

$$\left|\|U(t,\cdot)\|_{L^2[-a,a]} - r_1(t)\right| \leq C\left(\|U_0\| - r_1(0)\right),$$
$$\left\|\|U\|_{L^2[-a,a]} - r_1\right\|_{C[0,T]} \leq C\left(\|U_0\| - r_1(0)\right). \quad (33)$$

We take as interval of the solution of problem (15)–(17) segment $[0; 1/(\|U_0\| - r_1(0))]$, since the integrand $Q(\tau)\exp(\int_0^\tau (2Q(\xi)r_1(\xi) - P(\xi))d\xi)$ is bounded. Hence $\forall t \in [0, T]$ (where $T \approx T_1$) a priori estimation (11) takes place.

Now we prove the existence of solution for (4)–(7) using Galerkin method. Let $\{v_l(x)\}_{l=1}^{\infty}$ be a basis in space $L^2[-a, a]$, where the dimension of vector $v_l(x)$ is equal to the dimension of vector U. For each m we define an approximate solution U_m of (4)–(7) as follows:

$$U_m = \sum_{j=1}^{m} c_{jm}(t)\, v_j(x), \tag{34}$$

$$\begin{aligned}&\int_{-a}^{a} \left(\left(\frac{\partial U_m}{\partial t} + A_1 \frac{\partial U_m}{\partial x} \right), v_i(x) \right) dx \\ &\quad = \int_{-a}^{a} (J(U_m), v_i(x))\, dx, \quad i = \overline{1, m},\ t \in (0, T],\end{aligned} \tag{35}$$

$$U_m|_{t=0} = U_{0m}(x), \quad x \in R, \tag{36}$$

$$\begin{aligned}&(Aw_m^- \mp Bu_m^-)|_{x=\mp a} \\ &\quad = \frac{1}{\beta}(Aw_m^+ \pm Bu_m^+)|_{x=\pm a} \pm \frac{(1-\beta)\eta}{\alpha\beta\sqrt{\pi}} F,\end{aligned} \tag{37}$$

where U_{0m} is the orthogonal projection in L^2 of the function U_0 on the subspace, spanned by $v_1, \ldots, v_m$,

$$J(U_m) = (J_1(u_m, w_m), J_2(u_m, w_m))'. \tag{38}$$

We represent $v_j(x)$ in form $v_j(x) = (v_j^{(1)}, v_j^{(2)})'$, where

$$\begin{aligned} v_j^{(1)} &= (v_{j1}, v_{j2}, v_{j3})', \\ v_j^{(2)} &= (v_{j4}, v_{j5}, v_{j6})'. \end{aligned} \tag{39}$$

The coefficients $c_{jm}(t)$ are determined from the equations

$$\begin{aligned}&\sum_{j=1}^{m} \left\{ \frac{dc_{jm}}{dt} \int_{-a}^{a} (v_j, v_i)\, dx + c_{jm} \left[\left(Bv_j^{-(1)}, v_i^{-(1)}\right)\Big|_{x=a} \right.\right. \\ &\quad + \left(Bv_j^{-(1)}, v_i^{-(1)}\right)\Big|_{x=-a} + \left(Bv_i^{-(1)}, v_j^{-(1)}\right)\Big|_{x=a} \\ &\quad + \left(Bv_i^{-(1)}, v_j^{-(1)}\right)\Big|_{x=-a} \\ &\quad + \frac{1}{\beta}\left(\left(Av_i^{+(2)} + Bv_i^{+(1)}\right), v_j^{-(1)}\right)\Big|_{x=a} \\ &\quad - \frac{1}{\beta}\left(\left(Av_i^{+(2)} - Bv_i^{+(1)}\right), v_j^{-(1)}\right)\Big|_{x=-a} \\ &\quad + \frac{1}{\beta}\left(\left(Av_j^{+(2)} + Bv_j^{+(1)}\right), v_i^{-(1)}\right)\Big|_{x=a} \\ &\quad - \frac{1}{\beta}\left(Av_j^{+(2)} - Bv_j^{+(1)}, v_i^{-(1)}\right)\Big|_{x=-a} + \left(F_1, v_i^{-(1)}\right)\Big|_{x=a} \\ &\quad + \left(F_1, v_i^{-(1)}\right)\Big|_{x=-a} + \left(F_1, v_j^{-(1)}\right)\Big|_{x=a} \\ &\quad \left. + \left(F_1, v_j^{-(1)}\right)\Big|_{x=-a} \right] \\ &\quad \left. - \int_{-a}^{a} \left(\left(A \frac{\partial v_i^{(2)}}{\partial x}, v_j^{(1)} \right) + \left(A' \frac{\partial v_i^{(1)}}{\partial x}, v_j^{(2)} \right) \right) dx \right\} \\ &\quad = \int_{-a}^{a} \left(J\left(\sum_{j=1}^{m} c_{jm} v_j \right), v_i \right) dx, \\ &\qquad i = \overline{1, m},\ t \in (0, T]\end{aligned} \tag{40}$$

$$c_{im}(0) = d_{im}, \quad i = \overline{1, m}, \tag{41}$$

where d_{im} is the ith component of U_{0m}.

We multiply (35) by $c_{im}(t)$ and sum over i from 1 to m:

$$\begin{aligned}&\int_{-a}^{a} \left(\left(\frac{\partial U_m}{\partial t} + A_1 \frac{\partial U_m}{\partial x} \right), U_m \right) dx \\ &\quad = \int_{-a}^{a} (J(U_m), U_m)\, dx.\end{aligned} \tag{42}$$

With help of the above shown arguments we now prove that $r_m(t)$ is bounded in some time interval $[0, T_m]$, where $U_m(t, x) = r_m(t)\omega_m(t, x)$, $T_m \approx O((\|U_{0m}\| - r_{m1}(0))^{-1})$, $T_m \geq T\ \forall m$, and

$$\begin{aligned}&\left\| \|U_m\|_{L^2[-a,a]} - r_{m1} \right\|_{C[0,T]} \\ &\quad \leq C_2 \left(\|U_0\|_{L^2[-a,a]} - r_1(0) \right),\end{aligned} \tag{43}$$

where C_2 is constant and independent from m and $r_{m1}(t)$ is a particular solution of Riccati equation about $r_m(t)$.

Problem (40)-(41) represents a Cauchy problem for the ordinary system of differential equations. Existence of the solution of problem (40)-(41) follows from theory of ordinary system of differential equations (Picard's existence and uniqueness theorem) [20]. Hence the existence of the solution of problem (34)–(37) follows.

Thus, according to estimation (43) the sequence $\{U_m - r_{m1}\}$ of approximate solutions of problem (5)–(7) is uniformly bounded in function space $C([0, T]; L^2[-a, a])$. Moreover, the homogeneous system of equations $\tau E + (1/\alpha)A\xi$ with respect to τ, ξ has only a trivial solution. Then it follows from results in [21] that $U_m - r_{m1} \to U - r_1$ is week in $C([0, T]; L^2[-a, a])$ and $J(U_{m,}) \to J(U)$ is week in $C([0, T]; L^2[-a, a])$ as $m \to \infty$. Further it can be shown with standard methods that limit element is a weak solution of problem (5)–(7).

The uniqueness of solution of problem (5)–(7) is proved by contradiction. Let problem (5)–(7) have two different solutions u_1, w_1 and u_2, w_2. We denote them again by $u = u_1 - u_2$

and $w = w_1 - w_2$. Then with respect to new values of u and w we obtain the following problem:

$$\frac{\partial u}{\partial t} + A\frac{\partial w}{\partial x} = J_1(u_1, w_1) - J_1(u_2, w_2)$$
$$\frac{\partial w}{\partial t} + A'\frac{\partial u}{\partial x} = J_2(u_1, w_1) - J_2(u_2, w_2), \qquad (44)$$
$$x \in (-a, a)$$

$$u|_{t=0} = 0,$$
$$w|_{t=0} = 0, \qquad (45)$$

$$(Aw^- + Bu^-)|_{x=-a} = \frac{1}{\beta}(Aw^+ - Bu^+)|_{x=-a},$$
$$(Aw^- - Bu^-)|_{x=a} = \frac{1}{\beta}(Aw^+ + Bu^+)|_{x=a}. \qquad (46)$$

We prove that solution of problem (44)–(46) is trivial. Hence the uniqueness of the solution of problem (5)–(7) follows.

Using the method which was mentioned before we get (see equality (14))

$$\frac{1}{2}\frac{d}{dt}\int_{-a}^{a}[(u,u) + (w,w)]\,dx + (Bu^-, u^-)|_{x=a} + (Bu^-, u^-)|_{x=-a} - \frac{1}{\beta}((Aw^+ - Bu^+), u^-)|_{x=-a} + \frac{1}{\beta}((Aw^+ + Bu^+), u^-)|_{x=a} = \int_{-a}^{a}[((J_1(u_1, w_1) - J_1(u_2, w_2)), u) + ((J_2(u_1, w_1) - J_2(u_2, w_2)), w)]\,dx. \qquad (47)$$

Let us transform the integrand on the right side of the equality as

$$((J_1(u_1, w_1) - J_1(u_2, w_2)), u) = \frac{\sigma_2 - \sigma_0}{2}\left[\left(f_{00,1}f_{02,1} - \frac{f_{01,1}^2}{\sqrt{3}}\right) - \left(f_{00,2}f_{02,2} - \frac{f_{01,2}^2}{\sqrt{3}}\right)\right](f_{02,1} - f_{02,2}) = \frac{\sigma_2 - \sigma_0}{2}\left[(f_{00,1}f_{02,1} - f_{00,2}f_{02,2}) - \frac{1}{\sqrt{3}}(f_{01,1}^2 - f_{01,2}^2)\right](f_{02,1} - f_{02,2}) = \frac{\sigma_2 - \sigma_0}{2}\left[(f_{00,1} - f_{00,2})f_{02,1} + f_{00,2}(f_{02,1} - f_{02,2}) - \frac{1}{\sqrt{3}}(f_{01,1} - f_{01,2})(f_{01,1} + f_{01,2})\right](f_{02,1} - f_{02,2});$$

$$((J_2(u_1, w_1) - J_2(u_2, w_2)), u) = ((J_{03}(u_1, w_1) - J_{03}(u_2, w_2)), (f_{03,1} - f_{03,2})) + ((J_{11}(u_1, w_1) - J_{11}(u_2, w_2)), (f_{11,1} - f_{11,2})) = \left(\frac{1}{4}(\sigma_3 + 3\sigma_1 - 4\sigma_0)[(f_{00,1} - f_{00,2})f_{03,1} + f_{00,2}(f_{03,1} - f_{03,2})] + \frac{1}{4\sqrt{5}}(2\sigma_1 + \sigma_0 - 3\sigma_3)[(f_{01,1} - f_{01,2})f_{02,1} + f_{01,2}(f_{02,1} - f_{02,2})]\right)(f_{03,1} - f_{03,2}) + (\sigma_1 - \sigma_0)\cdot\left((f_{00,1} - f_{00,2})f_{11,1} + f_{00,2}(f_{11,1} - f_{11,2}) + 0.5\sqrt{\frac{5}{2}}[(f_{10,1} - f_{10,2})f_{01,1} + f_{10,2}(f_{01,1} - f_{01,2})] - \sqrt{\frac{2}{15}}[(f_{01,1} - f_{01,2})f_{02,1} + f_{01,2}(f_{02,1} - f_{02,2})]\right)(f_{11,1} - f_{11,2}). \qquad (48)$$

Once again we use spherical representation of $u = r(t)w_1(t, x)$ and $w = r(t)w_2(t, x)$, where $r(t) = \|U(t,\cdot)\|_{L^2[-a,a]}$. Then concerning $r(t)$ we obtain new initial value problem

$$\frac{dr}{dt} + rP(t) = r^2Q_1(t), \qquad (49)$$
$$r(0) = 0, \qquad (50)$$

where $P(t)$ has the same value as in (15) and

$$Q_1(t) = \frac{\sigma_2 - \sigma_0}{2}\int_{-a}^{a}\left[(\omega_{00,1} - \omega_{00,2})\omega_{02,1} + \omega_{00,2}(\omega_{02,1} - \omega_{02,2}) - \frac{1}{\sqrt{3}}(\omega_{01,1} - \omega_{01,2})(\omega_{01,1} + \omega_{01,2})\right](\omega_{02,1} - \omega_{02,2})\,dx + \frac{1}{4}(\sigma_3 + 3\sigma_1 - 4\sigma_0)\int_{-a}^{a}[(\omega_{00,1} - \omega_{00,2})\omega_{03,1} + +\omega_{00,2}(\omega_{03,1} - \omega_{03,2})](\omega_{03,1} - \omega_{03,2})\,dx + \frac{1}{4\sqrt{5}}(2\sigma_1 + \sigma_0 - 3\sigma_3)\cdot\int_{-a}^{a}[(\omega_{01,1} - \omega_{01,2})\omega_{02,1} + \omega_{01,2}(\omega_{02,1} - \omega_{02,2})]\cdot(\omega_{03,1} - \omega_{03,2})\,dx + (\sigma_1 - \sigma_0)\int_{-a}^{a}\left[(\omega_{00,1} - \omega_{00,2})\omega_{11,1} + \omega_{00,2}(\omega_{11,1} - \omega_{11,2}) + \frac{1}{2}\right.$$

$$\cdot \sqrt{\frac{5}{2}} \left[\left(\omega_{10,1} - \omega_{10,2} \right) \omega_{01,1} + \omega_{10,2} \left(\omega_{01,1} - \omega_{01,2} \right) \right] - \sqrt{\frac{2}{15}} \left[\left(\omega_{01,1} - \omega_{01,2} \right) \omega_{02,1} + \omega_{01,2} \left(\omega_{02,1} - \omega_{02,2} \right) \right] \cdot \left(\omega_{11,1} - \omega_{11,2} \right) \Bigg] dx. \quad (51)$$

The general solution of (49) is

$$r(t) = \exp\left(-\int_0^t P(\tau)\, d\tau\right) \cdot \left[C - \int_0^t Q_1(\tau) \exp\left(-\int_0^\tau P(\xi)\, d\xi\right) d\tau\right]^{-1}. \quad (52)$$

Solution of (49), which is satisfying homogeneous condition (50), is trivial; that is, $r(t) = 0$. Hence, $\|U\|_{C([0,T];L^2[-a,a])} = 0$ and $u_1 = u_2$ and $w_1 = w_2$.

The theorem is proved. □

Competing Interests

The authors declare that they have no competing interests.

References

[1] H. Grad, "On the kinetic theory of rarefied gases," *Communications on Pure and Applied Mathematics*, vol. 2, no. 4, pp. 331–407, 1949.

[2] H. Grad, "Principle of the kinetic theory of gases," in *Thermodynamics of Gases*, vol. 12 of *Handbuch der Physik*, pp. 205–294, Springer, Berlin, Germany, 1958.

[3] A. Sakabekov, *Initial-Boundary Value Problems for the Boltzmann's Moment System Equations Gylym, Almaty*, 2002.

[4] C. Cercignani, *Theory and Application of the Boltzmann Equation*, Instituto di Matematica, Milano, Italy, 1975.

[5] M. N. Kogan, *Dynamic of Rarefied Gas*, Nauka, Moscow, Russia, 1967.

[6] K. Kumar, "Polynomial expansions in kinetic theory of gases," *Annals of Physics*, vol. 37, no. 1, pp. 113–141, 1966.

[7] A. Sakabekov and Y. Auzhani, "Boundary conditions for the onedimensional nonlinear nonstationary Boltzmann's moment system equations," *Journal of Mathematical Physics*, vol. 55, Article ID 123507, 2014.

[8] C. D. Levermore, "Moment closure hierarchies for kinetic theories," *Journal of Statistical Physics*, vol. 83, no. 5-6, pp. 1021–1065, 1996.

[9] G. Mascali and V. Romano, "A hydrodynamical model for holes in silicon semiconductors: the case of non-parabolic warped bands," *Mathematical and Computer Modelling*, vol. 53, no. 1-2, pp. 213–229, 2011.

[10] G. Mascali and V. Romano, "A non parabolic hydrodynamical subband model for semiconductors based on the maximum entropy principle," *Mathematical and Computer Modelling*, vol. 55, no. 3-4, pp. 1003–1020, 2012.

[11] V. D. Camiola and V. Romano, "2DEG-3DEG charge transport model for MOSFET based on the maximum entropy principle," *SIAM Journal on Applied Mathematics*, vol. 73, no. 4, pp. 1439–1459, 2013.

[12] G. Alì, G. Mascali, V. Romano, and R. C. Torcasio, "A hydrodynamical model for covalent semiconductors with a generalized energy dispersion relation," *European Journal of Applied Mathematics*, vol. 25, no. 2, pp. 255–276, 2014.

[13] V. D. Camiola and V. Romano, "Hydrodynamical model for charge transport in graphene," *Journal of Statistical Physics*, vol. 157, no. 6, pp. 1114–1137, 2014.

[14] S. Mischler, "Kinetic equations with Maxwell boundary conditions," *Annales Scientifiques de l'Ecole Normale Superieure*, vol. 43, no. 5, pp. 719–760, 2010.

[15] V. G. Neudachin and U. F. Smirnov, *Nucleon Association of Easy Kernel*, Nauka, Moscow, Russia, 1969.

[16] M. Moshinsky, *The Harmonic Oscillator in Modern Physics: from Atoms to Quarks*, 1960.

[17] S. I. Pokhozhaev, "On an approach to nonlinear equation," *Doklady Akademii Nauk SSSR*, vol. 247, pp. 1327–1331, 1979.

[18] E. Kamke, *Differentialgleichungen Lösungsmethoden und Lösungen*, I. Gewohuliche Differentialgleichungen, B.G.Teubner, Leipzig, Germany, 1977.

[19] A. Tungatarov and D. K. Akhmed-Zaki, "Cauchy problem for one class of ordinary differential equations," *International Journal of Mathematical Analysis*, vol. 6, no. 14, pp. 695–699, 2012.

[20] M. Tenenbaum and H. Pollard, *Ordinary Differential Equations*, Harper and Row, New York, NY, USA, 1963.

[21] L. Tartar, "Compensated compactness and applications to partial differential equations," in *Proceedings of the Non-Linear Analysis and Mechanics, Heriot-Watt Symposium*, R. J. Knops, Ed., vol. 4 of *Research Notes in Math*, pp. 136–212, 1979.

9

An Analysis of a Semelparous Population Model with Density-Dependent Fecundity and Density-Dependent Survival Probabilities

Arild Wikan

School of Business and Economics, The Arctic University of Norway, Campus Harstad, Havnegata 5, 9480 Harstad, Norway

Correspondence should be addressed to Arild Wikan; arild.wikan@uit.no

Academic Editor: Urmila Diwekar

A discrete age-structured semelparous Leslie matrix model where density dependence is included both in the fecundity and in the survival probabilities is analysed. Depending on strength of density dependence, we show in the precocious semelparous case that the nonstationary dynamics may indeed be rich, ranging from SYC (a dynamical state where the whole population is in one age class only) dynamics to cycles of low period where all age classes are populated. Quasiperiodic and chaotic dynamics have also been identified. Moreover, outside parameter regions where SYC dynamics dominates, we prove that the transfer from stability to instability goes through a supercritical Neimark–Sacker bifurcation, and it is further shown that when the population switches from possessing a precocious to a delayed semelparous life history both stability properties and the possibility of periodic dynamics become weaker.

1. Introduction

Within the framework of nonlinear discrete age-structured population models, the dynamical properties and behaviour of a great variety of species may be explored. Such species may possess an iteroparous life history which means that individuals in several age classes of the population are fertile, or they may possess a semelparous life history which is characterized by the property that only individuals of the last age class are fertile.

Regarding iteroparity, most population models (both age-structured and stage-structured) focus on cases where non-linearities (density dependence) are built into the fecundity elements and not into the survivals. Particularly in fishery models, this has often been motivated by the assumption that most density effects are present only in the first year of life. Examples of theoretical studies which deal with nonstationary and chaotic behaviour as well as behaviour linked to concrete species may be found in [1–7]. In case of ergodic properties we refer to [8–10]; see also [11]. Another strategy is to assume constant fecundity terms and non-linear year-to-year survival probabilities, compare [12–16]. Depending on functional form of the nonlinear terms, a crude conclusion found in several of the papers referred to above is that, in case of small population densities, the models possess a stable nontrivial equilibrium where all age classes are populated. At higher population densities there may be nonstationary, periodic, and chaotic dynamics of stunning complexity.

Now, turning to the semelparous case with nonlinear fecundity element and constant survival probabilities a rather peculiar phenomenon has been detected, compare [4]. Here, in contrast to the iteroparous case, the nontrivial equilibrium tends to be unstable in large parameter regions, also in case of low population densities. Instead one finds, as time $t \to \infty$, that a cyclic state is attained where the whole population is in just one single age class at each time step. Such behaviour, which is called synchronization or SYC (single year class) dynamics, has been detected among insects; see [17]. Hoppensteadt and Keller [18] presented a model for the 17-year cicada (magicicada) which included both predation and intraspecific competition and in [19] cicada dynamics is further explored. Regarding biennials and possible SYC dynamics we refer to [20]. In [21] SYC phenomena and related MYC (multiple year class) dynamics are considered while Kon [22] discusses in a general context conditions for SYC dynamics to occur in matrix models, also compare [23]. However, if we take the opposite approach, constant fecundity term, and nonlinear year-to-year survivals probabilities, SYC dynamics appear to be a rare event; see [16]. Instead, the nontrivial equilibrium is stable whenever the population size is sufficiently small and the nonstationary dynamics has a strong resemblance of 4-cycles, either exact or approximate, the chaotic regime included.

In contrast to most of the papers quoted above, the purpose of this paper is to study the combined effect of nonlinear recruitment and nonlinear survival probabilities in semelparous population models. Several general results about possible SYC dynamics and stability properties of the nontrivial equilibrium in such problems may be obtained in [23] (two age classes) and [24] (three age classes). In case of the more general setting where an arbitrary number of age classes is considered we refer to [25, 26]. Regarding our study, we shall restrict the analysis to the case where both the fecundity and the survival elements depend on the total population and, moreover, the functional form of these elements is of Ricker type. As we prove, under such restrictions, it is possible to obtain explicit thresholds for secondary bifurcations of flip and Neimark–Sacker type and we may also prove (in some cases) that bifurcations involved are of supercritical nature. We may also investigate how the dynamics reported earlier will change as more density-dependent terms are included. For example, given SYC cycles, will the cycles persist if the strength of density dependence in the survival terms is included? If not, what kind of qualitative dynamics is it then possible to obtain? Assuming cycles of low period where all age classes are populated, does the inclusion of density dependence in the fecundity terms act in a stabilizing or destabilizing fashion? Such questions are difficult to address in the general models presented in [24–26].

The paper is organized in the following way. In Section 2 we present the model, compute equilibria, and derive the nth order eigenvalue equation which we need in order to perform stability and bifurcation analysis. This is followed (Sections 3 and 4) by a rigorous analysis of possible dynamic outcomes in two and three age class models, respectively. Finally, in Section 5, we unify and discuss results when the number of age classes exceeds three.

2. The Model, Fixed Points, and Stability

First we establish the model. At time t we split the population x_t into n distinct nonoverlapping age classes $\overline{x}_t = (x_1, \ldots, x_n)^T$ where the total population is given by $x = x_1 + \cdots + x_n$. Next, we introduce the transition matrix

$$A = \begin{pmatrix} 0 & 0 & \cdots & 0 & f \\ p_1 & 0 & & & 0 \\ 0 & p_2 & & & 0 \\ & & \ddots & & \\ 0 & & 0 & p_{n-1} & 0 \end{pmatrix}, \qquad (1)$$

where f is the average fecundity of a member of the last age class at time t. p_i denotes the survival probabilities, the (year-to-year) survival from age class $i-1$ to i. In contrast to most papers the assumption here is that both the fecundity and the survivals are nonlinear terms. Thus, we write f as $f = F\exp(-\alpha x)$ where the constant $F > 1$ and the survivals p_i as $p_i = P\exp(-\beta_i x)$ where the constant P satisfies $0 < P \le 1$. The parameters $\alpha \ge 0$, $\beta_i \ge 0$ may be regarded as parameters that measure the strength of density dependence. The relation between $\overline{x}$ at two consecutive time steps is then expressed as

$$\overline{x}_{t+1} = A\overline{x}_t \qquad (2a)$$

which may also be formulated as a nonlinear map of the form

$$f : \Re^n \longrightarrow \Re^n$$
$$\overline{x} \longrightarrow A\overline{x}. \qquad (2b)$$

Besides the trivial fixed point $\overline{x} = (0, \ldots, 0)^T$ maps (2a) and (2b) also possess a unique nontrivial point $\overline{x}^* = (x_1^*, \ldots, x_n^*)^T$. The latter may be expressed as

$$(x_1^*, \ldots, x_i^*, \ldots, x_n^*) = \left(\frac{1}{D}x^*, \ldots, \frac{P^{i-1}a^{\beta_1+\cdots+\beta_{i-1}}}{D} \cdot x^*, \ldots, \frac{P^{n-1}a^{\beta_1+\cdots+\beta_{n-1}}}{D}x^*\right). \qquad (3)$$

The quantities a and D are defined as

$$a = R_0^{-(1/(\alpha+\beta_1+\cdots+\beta_{n-1}))},$$
$$D = 1 + Pa^{\beta_1} + P^2a^{\beta_1+\beta_2} + \cdots + P^{n-1}a^{\beta_1+\cdots+\beta_{n-1}}, \qquad (4)$$

where $R_0 = FP^{n-1}$ and $R_0 > 1$ is assumed throughout the paper in order to have a feasible equilibrium. The total equilibrium population x^* is given as

$$x^* = \frac{1}{\alpha + \sum_{i=1}^{n-1}\beta_i}\ln R_0. \qquad (5)$$

In order to investigate stability we linearize about the fixed point. This gives birth to the nth order eigenvalue equation

$$\lambda^n + a_1\lambda^{n-1} + a_2\lambda^{n-2} + \cdots + a_{n-1}\lambda + a_n = 0, \qquad (6)$$

where the coefficients $a_1, \ldots, a_n$ satisfy

$$\begin{pmatrix} a_1 \\ a_2 \\ a_3 \\ \vdots \\ a_{n-1} \\ a_n \end{pmatrix} = \begin{pmatrix} \alpha & \beta_1 & \beta_2 & & \beta_{n-2} & \beta_{n-1} \\ \beta_{n-1} & \alpha & \beta_1 & & \beta_{n-3} & \beta_{n-2} \\ \beta_{n-2} & \beta_{n-1} & \alpha & & \beta_{n-4} & \beta_{n-3} \\ \vdots & & & \ddots & & \vdots \\ \beta_2 & \beta_3 & \beta_4 & & \alpha & \beta_1 \\ \beta_1 & \beta_2 & \beta_3 & & \beta_{n-1} & \alpha \end{pmatrix} \begin{pmatrix} x_1^* \\ x_2^* \\ x_3^* \\ \vdots \\ x_{n-1}^* \\ x_n^* \end{pmatrix} + \begin{pmatrix} 0 \\ 0 \\ 0 \\ \vdots \\ 0 \\ -1 \end{pmatrix}. \quad (7)$$

$\overline{x}^*$ is a locally stable hyperbolic fixed point as long as all eigenvalues λ of (6) are located inside the unit circle in the complex plane.

There are three ways in which $\overline{x}^*$ may fail to be stable. It may lose its hyperbolicity when λ crosses the unit circle through 1 which in the general case leads to a saddle node bifurcation, alternatively through -1 which gives birth to a flip (period doubling) bifurcation, or it may fail to be hyperbolic as a pair of complex-valued eigenvalues cross the unit circle. Then a Neimark–Sacker bifurcation occurs. The Jury criteria, see the book by Murray [27], provide conditions for all eigenvalues to satisfy $|\lambda| < 1$.

3. Two Age Classes

Let $n = 2$ in maps (2a) and (2b). Then we have

$$(x_1, x_2) \longrightarrow \left(Fe^{-\alpha x} x_2, Pe^{-\beta_1 x} x_1\right). \quad (8)$$

$R_0 = FP$, $R_0 > 1$, and $x^* = (\alpha + \beta_1)^{-1} \ln R_0$. Moreover, the fixed point becomes

$$(x_1^*, x_2^*) = \left(\frac{1}{1 + Pa^{\beta_1}} x^*, \frac{Pa^{\beta_1}}{1 + Pa^{\beta_1}} x^*\right) \quad (9)$$

and the eigenvalue equation may be cast in the form

$$\lambda^2 + a_1 \lambda + a_2 = 0, \quad (10)$$

where $a_1 = \alpha x_1^* + \beta_1 x_2^*$ and $a_2 = \beta_1 x_1^* + \alpha x_2^* - 1$.

Fixed point (9) is stable whenever the Jury criteria $1 + a_1 + a_2 > 0$, $1 - a_1 + a_2 > 0$, and $1 - |a_2| > 0$ hold, that is, as long as

$$(\alpha + \beta_1) x^* > 0, \quad (11a)$$

$$(\beta_1 - \alpha)\left(1 - Pa^{\beta_1}\right) x_1^* > 0, \quad (11b)$$

$$2 - (\beta_1 x_1^* + \alpha x_2^*) > 0, \quad (11c)$$

respectively.

There are two cases to consider: (A) the case $\alpha > \beta_1$, which means that the strength of density dependence in the fecundity is stronger than the strength of density dependence in the survival, and (B) the case $\beta_1 > \alpha$.

Considering (A), it is clear from (11b) that there does not exist any stable fixed point. Moreover, since (11b) fails as an eigenvalue crosses the unit circle through -1 it is natural to search for a 2-cycle which should be stable provided x^* is small. Evidently, such a 2-cycle must be obtained from

$$\begin{aligned} x_{1,t+2} &= R_0 e^{-\beta_1 (x_{1,t} + x_{2,t})} e^{-\alpha (x_{1,t+1} + x_{2,t+1})} x_{1,t} \\ x_{2,t+2} &= R_0 e^{-\alpha (x_{1,t} + x_{2,t})} e^{-\beta_1 (x_{1,t+1} + x_{2,t+1})} x_{2,t} \end{aligned} \quad (12)$$

and here there are two possibilities:

(1) $x_t = x_{t+1}$ which leads to the trivial 2-cycle where the unstable fixed point (x_1^*, x_2^*) is the only point the cycle.

(2) The points are on the form $(A, 0)$ or $(0, B)$. In this case it follows from (12) that A and B must satisfy the equations

$$A - \frac{1}{P} e^{-\beta_1 A} B = 0 \quad (13a)$$

$$\beta_1 A - \alpha B + \ln R_0 = 0 \quad (13b)$$

and by finding B from (13b) and substitute back into (13a) we arrive at

$$\alpha P e^{\beta_1 A} = \beta_1 + \frac{\ln R_0}{A}. \quad (13c)$$

Geometrically, it is now easy to see that the graph of the left hand side of (13c) and that of the right side have a unique intersection point lying in the first (positive) quadrant. In the special case $\beta_1 = 0$ we obtain $A = (\alpha P)^{-1} \ln R_0$ and $B = \alpha^{-1} \ln R_0$. Hence, there exists a 2-cycle on the form (SYC form)

$$\begin{aligned} &\left(\frac{1}{\alpha P} \ln R_0, 0\right), \\ &\left(0, \frac{1}{\alpha} \ln R_0\right) \end{aligned} \quad (14)$$

and as shown in [4] this cycle is stable in case of F small (P fixed). In Figure 1(a) we show an orbit starting at $(x_{10}, x_{20}) \neq (0, 0)$ which settles on the 2-cycle (14). If we continue to increase F, we find that (14) goes unstable and cycles of

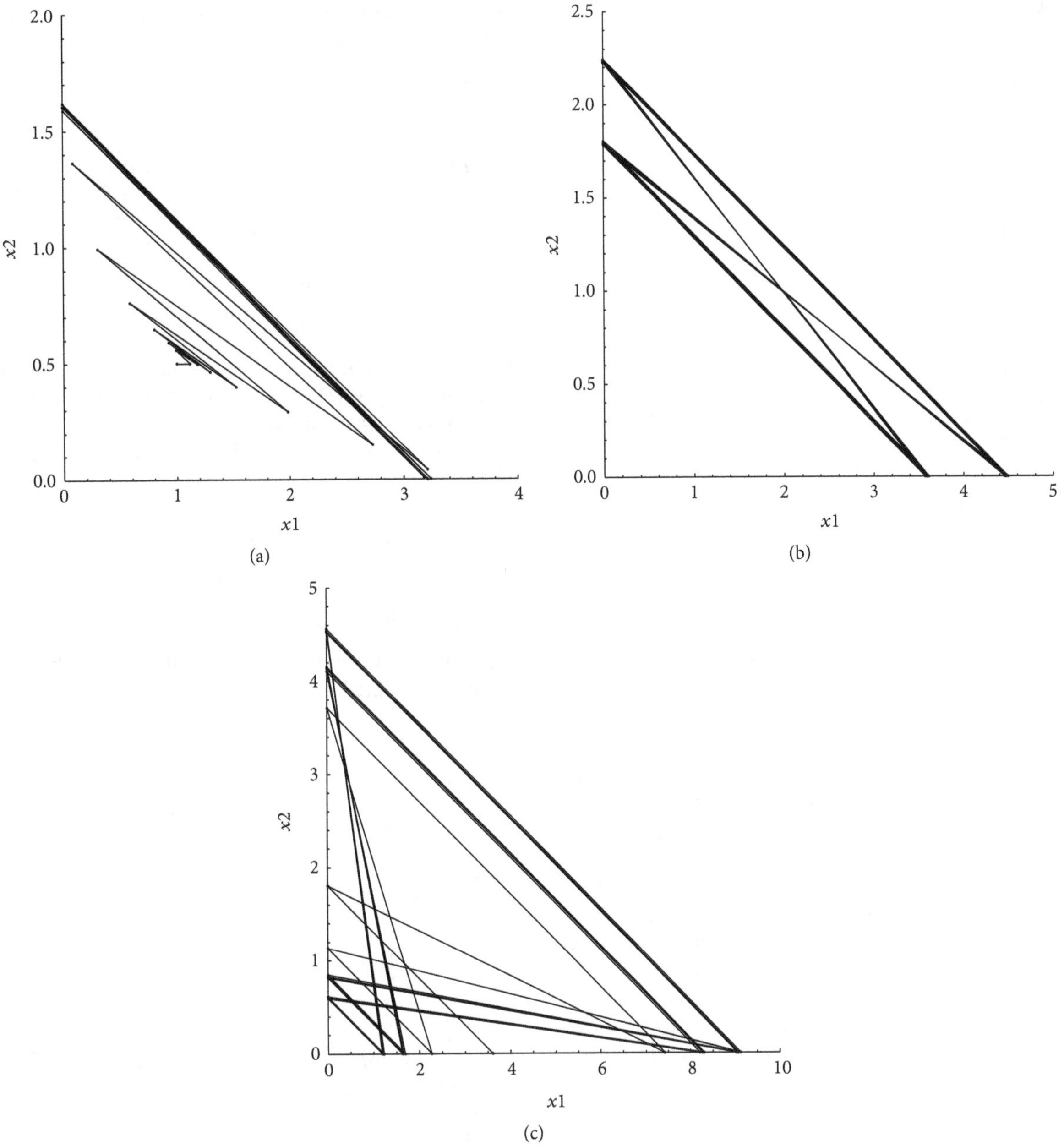

FIGURE 1: (a) An orbit approaching the SYC 2-cycle $(P^{-1}\ln R_0.0)$, $(0, \ln R_0)$. Parameter values $(\alpha, \beta) = (1, 0)$, $(P, F) = (0.5, 10)$. (b) A 4-cycle on SYC form. $(\alpha, \beta) = (1, 0)$, $(P, F) = (0.5, 15)$. (c) Chaotic SYC dynamics. $(\alpha, \beta) = (1, 0)$, $(P, F) = (0.5, 25)$.

period 2^k. $k = 2, 3, \ldots$, are established through successive flip bifurcations. These cycles, which are all on SYC form, are stable in smaller and smaller regions as F is increased. Eventually, the dynamics becomes chaotic but we emphasize that it is on SYC form also in the chaotic regime. These scenarios are demonstrated in Figures 1(b) and 1(c). Actually, we have not accounted for what happens when $R_0 = 1$ (or $x^* = 0$). Here, compare (10), the eigenvalues are 1 and -1, respectively, and both the positive equilibrium and the SYC 2-cycle bifurcate forward. For proofs and details we refer to [4, 23–26].

When $0 < \beta_1 < \alpha$ we observe much of the same SYC dynamics as in the $\beta_1 = 0$ case. However, we may in a sense argue that an increase of β_1 acts in a stabilizing fashion. Indeed, if $\beta_1 = 0$ and $F = 25$, map (8) generates chaotic dynamics. On the other hand, if $\beta_1 = 0.8$ and $F = 25$ ($P = 0.5$ in both cases) the outcome is a stable period 4-cycle on SYC form as shown in Figure 2.

Since the boundary of the positive cone is always invariant for semelparous Leslie matrix models of any dimension, see [26], initial conditions of the form $(v, 0)$ or $(0, w)$ in the 2-dimensional case will always produce SYC dynamics. However, if $\beta_1 \to \alpha$ and $(x_{10}, x_{2,0}) \neq (v, 0)$ or $(0, w)$ the dynamics occurs in the vicinity (mostly as a stable 2-cycle, not on SYC form) of the unstable fixed point

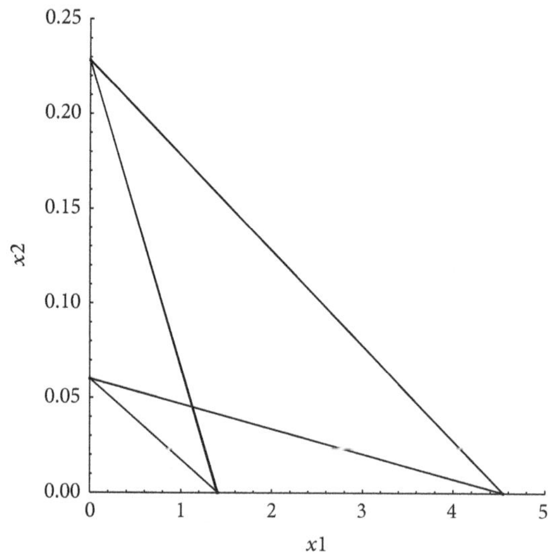

FIGURE 2: 4-periodic SYC dynamics. $(\alpha, \beta) = (1, 0.8)$ and $(P, F) = (0.5, 25)$.

$$(x_1^*, x_2^*) = \left(\frac{\sqrt{F}}{\sqrt{F} + \sqrt{P}} x^*, \frac{\sqrt{P}}{\sqrt{F} + \sqrt{P}} x^* \right), \tag{15}$$

where $x^* = (2\alpha)^{-1} \ln R_0$. Hence, $0 < \beta_1 < \alpha$ does not necessarily imply SYC dynamics although it is the most likely outcome. A further discussion is postponed to Section 5.

Next, consider (B) (the case $0 < \alpha < \beta_1$). Our first observation is that the Jury criteria (11a) and (11b) will never be violated. Moreover, in case of sufficiently small equilibrium populations x_1^*, x_2^* the left hand side of (11c) will be positive as well. Consequently, there exists a region in parameter space where (9) is a hyperbolic stable fixed point. However, if we increase the value of x^* (by increasing F) such that (11c) turns into an equality, (x_1^*, x_2^*) undergoes a Neimark–Sacker bifurcation, loses its hyperbolicity, and becomes unstable at the threshold

$$\ln R_0 = (\alpha + \beta_1) \frac{2\left(1 + Pa^{\beta_1}\right)}{\beta_1 + \alpha P a^{\beta_1}} \tag{16}$$

while the corresponding modulus 1 solutions of the eigenvalue equation (10) may be cast in the form

$$\lambda = -\frac{\alpha + \beta_1 P a^{\beta_1}}{\beta_1 + \alpha P a^{\beta_1}} \pm \frac{\sqrt{\left(\beta_1^2 - \alpha^2\right)\left(1 - P^2 a^{2\beta_1}\right)}}{\beta_1 + \alpha P a^{\beta_1}} i. \tag{17}$$

As is well known, bifurcations may be of both supercritical and subcritical nature. If a fixed point shall undergo a supercritical bifurcation it means that an eigenvalue (pair of eigenvalues) λ must cross the unit circle outwards at instability and in the Neimark–Sacker case that an attracting quasiperiodic orbit restricted to an invariant curve is created beyond instability threshold. Now, considering our bifurcation, we first express map (8), using the abbreviation $b = \sqrt{(\beta_1^2 - \alpha^2)(1 - P^2 a^{2\beta_1})}$, as

$$\begin{pmatrix} u \\ v \end{pmatrix} \longrightarrow \begin{pmatrix} -\dfrac{\alpha + \beta_1 P a^{\beta_1}}{\beta_1 + \alpha P a^{\beta_1}} & -\dfrac{b}{\beta_1 + \alpha P a^{\beta_1}} \\ \dfrac{b}{\beta_1 + \alpha P a^{\beta_1}} & -\dfrac{\alpha + \beta_1 P a^{\beta_1}}{\beta_1 + \alpha P a^{\beta_1}} \end{pmatrix} \begin{pmatrix} u \\ v \end{pmatrix} + \begin{pmatrix} f(u, v) \\ g(u, v) \end{pmatrix}, \tag{18}$$

where $f(u, v)$ and $g(u, v)$ contain second- and third-order terms of u and v. (Details of how (18) is derived and explicit formulas of $f(u, v)$ and $g(u, v)$ may be found in Appendix.)

Following Wan [28], the bifurcation will be supercritical if

$$\gamma = -\operatorname{Re}\left[\frac{(1 - 2\lambda)\bar{\lambda}^2}{1 - \lambda} \xi_{11} \xi_{20}\right] - \frac{1}{2}\left|\xi_{11}\right|^2 - \left|\xi_{02}\right|^2 + \operatorname{Re}\left(\bar{\lambda} \xi_{21}\right) \tag{19}$$

is negative and that $d|\lambda|/dF > 0$ at bifurcation. The quantities in (19) are defined as

$$\begin{aligned}
\xi_{11} &= \frac{1}{4}\left[\left(f_{uu} + f_{vv}\right) + i\left(g_{uu} + g_{vv}\right)\right] \\
\xi_{20} &= \frac{1}{8}\left[\left(f_{uu} - f_{vv} + 2g_{uv}\right) + i\left(g_{uu} - g_{vv} - 2f_{uv}\right)\right] \\
\xi_{02} &= \frac{1}{8}\left[\left(f_{uu} - f_{vv} - 2g_{uv}\right) + i\left(g_{uu} - g_{vv} + 2f_{uv}\right)\right] \\
\xi_{21} &= \frac{1}{16}\left[\left(f_{uuu} + f_{uvv} + g_{uuv} + g_{vvv}\right)\right. \\
&\quad \left. + i\left(g_{uuu} + g_{uvv} - f_{uuv} - f_{vvv}\right)\right].
\end{aligned} \tag{20}$$

Due to the complexity of $f(u, v)$ and $g(u, v)$ it is out of reach to compute the sign of γ in the most general case. However, it works in the important special case $\alpha = 0$. Indeed, (18) simplifies to

$$\begin{pmatrix} u \\ v \end{pmatrix} \longrightarrow \begin{pmatrix} -\dfrac{1}{F} & -\dfrac{b}{F} \\ \dfrac{b}{F} & -\dfrac{1}{F} \end{pmatrix} \begin{pmatrix} u \\ v \end{pmatrix} + \begin{pmatrix} f(u, v) \\ g(u, v) \end{pmatrix}, \tag{21}$$

where $b = \beta_1 \sqrt{F^2 - 1}$ and

$$\begin{aligned}
f(u, v) &= \beta_1 b u v - \frac{1}{2} F b^2 u v^2 + \frac{1}{b} \frac{F^2 b^2}{\beta_1} v^3, \\
g(u, v) &= \frac{\beta_1}{bF} f(u, v).
\end{aligned} \tag{22}$$

Moreover, ξ_{11} becomes zero so γ may be found as

$$\gamma = -\left|\xi_{02}\right|^2 + \operatorname{Re}\left(\bar{\lambda} \xi_{21}\right) = -\frac{\beta_1^2}{16}\left[\beta_1^2 + \left(F^2 - 1\right)\right] \tag{23}$$

which is clearly negative.

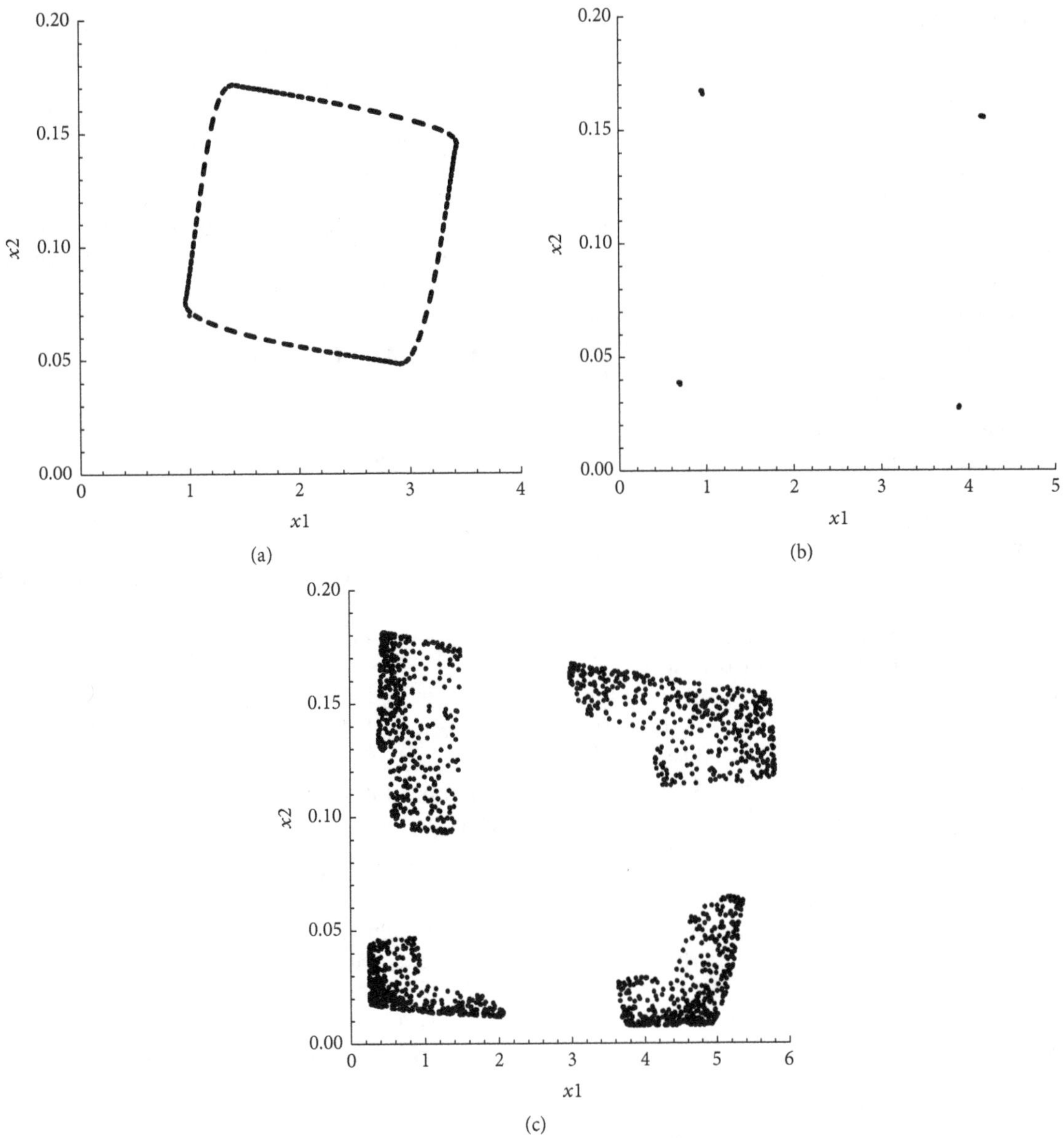

FIGURE 3: (a) An invariant curve. $(\alpha, \beta) = (0, 1)$ and $(P, F) = (0.5, 20)$. (b) 4-periodic dynamics generated by map (8). $(\alpha, \beta) = (0, 1)$ and $(P, F) = (0.5, 25)$. (c) Chaotic dynamics generated by (8). $(\alpha, \beta) = (0, 1)$ and $(P, F) = (0.5, 32)$.

Finally, at threshold $x_2^* = 2/F$, compare (11c) or (16)

$$\frac{d|\lambda|}{dF} = \frac{F+2}{2F(F+1)} > 0. \tag{24}$$

Hence, we have proved that the Neimark–Sacker bifurcation is of supercritical nature.

Still, assuming $\alpha = 0$, let us now focus on the dynamics on the invariant curve. Whenever $F > F_T$ and $|F - F_T|$ is small where F_T is the value of F at threshold (for a given value of P, F_T is found from $P = (F_T)^{-1}\exp(2(1 + F_T)/F_T))$, we find a quasiperiodic orbit which fills the invariant curve with points; see Figure 3(a). In case of larger values of F we observe a stable 4-cycle, comapre Figure 3(b), and as we continue to increase F stable orbits of period $4 \cdot 2^k$, $k = 1, 2, \ldots$, are the outcomes. (Note, however, that the points in these orbits are clustered in such a way that one from an observational point of view may argue that we have almost 4-periodic dynamics in these cases as well.) The smaller P is, the larger the interval $(F_{\min}, F_{\max})$ becomes where the 4-periodic structure occurs. Eventually, in case of even larger fecundity values, the dynamics becomes chaotic, but even in the chaotic regime a certain kind of 4-periodicity is preserved in the sense that the chaotic attractor is divided into 4 disjoint subsets that are visited once every fourth iteration; see Figure 3(c).

The reason why we have all this 4-periodicity may be understood along the following line. Once the invariant curve is established we may regard map (8) ($\alpha = 0$) restricted to the curve as topological equivalent to a circle map. Associated with a circle map there is a rotation number σ, and whenever σ is rational we have an orbit of finite period. If σ is an irrational number the orbit is quasiperiodic.

Now, at bifurcation threshold the modulus 1 eigenvalues are

$$\lambda = -\frac{1}{F} \pm \frac{1}{F}\sqrt{F^2 - 1}i \qquad (25)$$

and since F is large at threshold, λ is located close to the imaginary axis in the left half plane. Following Guckenheimer and Holmes [29], $\arg\lambda$ gives asymptotic information about the rotation number; thus $\sigma \approx 1/4$. This accounts for the 4-periodicity observed. Finally, note that when an orbit becomes exact periodic as a result of changing parameter F, the implicit function theorem guarantees that there exists an interval around that specific parameter value where the periodicity is preserved as well.

Next, consider the case $0 < \alpha < \beta$. The location of the modulus 1 eigenvalues is now given by (17) and they are not so close to the imaginary axis anymore. For "small" values of α we find the invariant curve, but the 4-periodic dynamics reported above ($\alpha = 0$) is absent. Instead, as a result of increasing F, the invariant curve becomes kinked and not topological equivalent to a circle. We may also describe the dynamics by use of the Lyapunov exponent L of the orbit generated by (8) [30, 31]. In Figure 4(a) we display the values of L in the F range $10 < F < 50$ when $(\alpha, \beta) = (0.1, 1.0)$ and $P = 0.5$. Whenever $10 < F < 20.341$, $L < 0$. In this interval the fixed point is stable. When $20.341 \leq F < 29.3$ the dynamics is quasiperiodic and restricted to invariant attracting curves and the corresponding Lyapunov exponent is $L = 0$. $L < 0$ in the parameter window $30.5 < F < 32.7$. Periodic dynamics of period 11 is the outcome here. Finally, when F exceeds 32.7 and also in a tiny interval just below 30.5 we find that $L > 0$ which means that the dynamics is chaotic. These findings are also visualized in Figure 4(b) where the dynamics is shown for selected values of F. From bottom to top we recognize stable fixed points, invariant curves, kinked curves, 11-periodic dynamics, and chaos.

In case of intermediate values of α, $(\alpha, \beta) = (0.5, 10)$ we observe a significant change of dynamics. Here, there exists a critical F value F_C, $F_C < F_T$, where the third iterate of map (8) undergoes a saddle node bifurcation which results in two large amplitude 3-cycles, one stable and one unstable. Thus, in the interval $F_C < F < F_T$ we find coexistence between two stable attractors, the stable fixed point (9) and the stable 3-cycle. Consequently, the ultimate fate of an orbit depends on the initial condition but since the trapping region of the 3-cycle appears to be (much) larger, 3-periodic dynamics is the most likely outcome. Beyond F_T there is an interval $F_T < F < F_K$ where the invariant curve established at F_T and the stable 3-cycle coexists. At $F = F_K$ the invariant curve disappears as it is hit by the unstable 3-cycle, and if $F > F_K$, $|F - F_K|$ small, the 3-cycle is the only attractor. Similar phenomena have also been found in iteroparous population models, first by Guckenheimer et al. [1], later in [4]. At even higher fecundity values the stable 3-cycle undergoes a Neimark–Sacker bifurcation which leads to three invariant curves which are visited once every third iteration. This is followed by kinked curves, periodic dynamics, and chaos through further enlargement of F. Values of L as well as the dynamics reported above are shown in Figures 4(c) and 4(d).

If α is large but $\alpha < \beta$, we find that the dynamics qualitatively is quite similar to the case where α is small, compare Figures 4(e) and 4(f) where $(\alpha, \beta) = (0.8, 1.0)$ and $P = 0, 5$. However, there are several parameter values where the dynamics is periodic but most of these have almost no widths. The exception is the first window where the dynamics is 5-periodic. A final comment is that the larger α is, $(\alpha < \beta)$, the higher F is at bifurcation threshold (16), so one may argue that the strength of density dependence in the fecundity $(\alpha < \beta)$ acts in a stabilizing way.

4. Three Age Classes

In this section we study the map

$$\begin{aligned} f &: \mathfrak{R}^3 \longrightarrow \mathfrak{R}^3 \\ (x_1, x_2, x_3) &\longrightarrow \left(Fe^{-\alpha x}x_3, Pe^{-\beta_1 x}x_1, Pe^{-\beta_2 x}x_2\right). \end{aligned} \qquad (26)$$

The nontrivial fixed point (cf. (3)) may be expressed as

$$(x_1^*, x_2^*, x_3^*) = \left(\frac{1}{D}x^*, \frac{Pa^{\beta_1}}{D}x^*, \frac{P^2a^{\beta_1+\beta_2}}{D}x^*\right), \qquad (27)$$

where $a = R_0^{-(1/(\alpha+\beta_1+\beta_2))}$, $D = 1 + Pa^{\beta_1} + P^2a^{\beta_1+\beta_2}$, $R_0 = FP^2$, $R_0 > 1$, and $x^* = (\alpha + \beta_1 + \beta_2)^{-1}\ln R_0$.

From (6) it follows that the coefficient in the eigenvalue equation

$$\lambda^3 + a_1\lambda^2 + a_2\lambda + a_3 = 0 \qquad (28)$$

may be written as $a_1 = \alpha x_1^* + \beta_1 x_2^* + \beta_2 x_3^*$, $a_2 = \alpha x_2^* + \beta_2 x_1^* + \beta_1 x_3^*$, and $a_3 = \alpha x_3^* + \beta_1 x_1^* + \beta_2 x_2^* - 1$.

Equilibrium (27) is locally asymptotic stable as long as the Jury criteria $1+a_1+a_2+a_3 > 0$, $1-a_1+a_2-a_3 > 0$, $1-|a_3| > 0$, and $|1-a_3^2| - |a_2 - a_2a_1| > 0$ hold. Hence, the stability criteria are

$$(\alpha + \beta_1 + \beta_2)x^* > 0 \qquad (29a)$$

$$\begin{aligned} 2 &- (\alpha + \beta_1 - \beta_2)x_1^* - (\beta_1 + \beta_2 - \alpha)x_2^* \\ &- (\alpha - \beta_1 + \beta_2)x_3^* > 0 \end{aligned} \qquad (29b)$$

$$2 - (\beta_1 x_1^* + \beta_2 x_2^* + \alpha x_3^*) > 0 \qquad (29c)$$

$$\begin{aligned} &(2\beta_1 - \alpha - \beta_2)x_1^* + (2\beta_2 - \alpha - \beta_1)x_2^* \\ &+ (2\alpha - \beta_1 - \beta_2)x_3^* + (\alpha\beta_1 - \beta_1^2)x_1^{*2} \\ &+ (\beta_1\beta_2 - \beta_2^2)x_2^{*2} + (\alpha\beta_2 - \alpha^2)x_3^{*2} \\ &+ (\beta_1^2 + \alpha\beta_2 - 2\beta_1\beta_2)x_1^*x_2^* \\ &+ (\alpha^2 + \beta_1\beta_2 - 2\alpha\beta_1)x_1^*x_3^* \\ &+ (\alpha\beta_1 + \beta_2^2 - 2\alpha\beta_2)x_2^*x_3^* > 0. \end{aligned} \qquad (29d)$$

A final observation from (28) is that when $x^* = 0$ all eigenvalues $\lambda = \exp[(2k\pi/3)i]$, $k = 0, 1, 2$, are located on

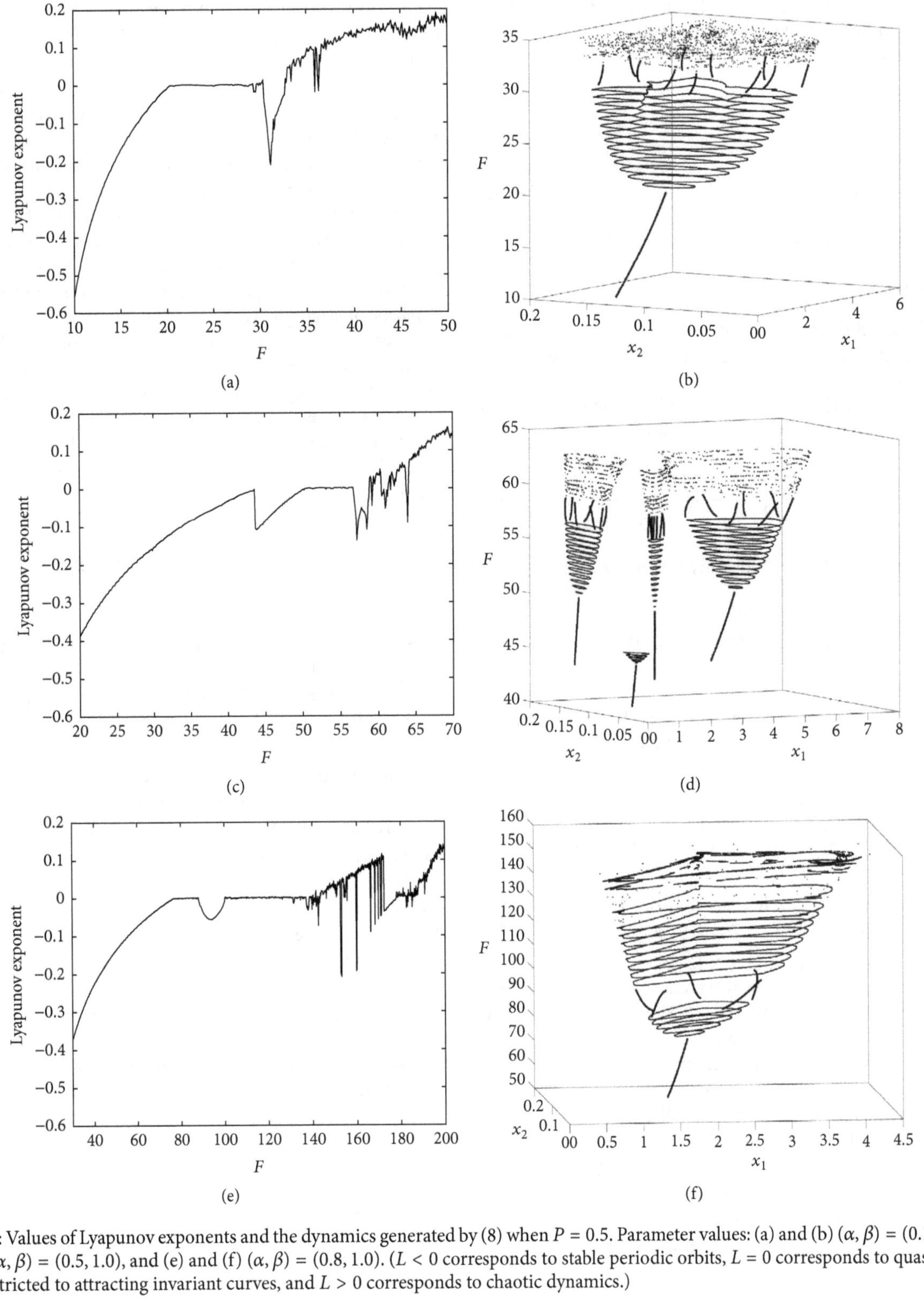

FIGURE 4: Values of Lyapunov exponents and the dynamics generated by (8) when $P = 0.5$. Parameter values: (a) and (b) $(\alpha, \beta) = (0.1, 1.0)$, (c) and (d) $(\alpha, \beta) = (0.5, 1.0)$, and (e) and (f) $(\alpha, \beta) = (0.8, 1.0)$. ($L < 0$ corresponds to stable periodic orbits, $L = 0$ corresponds to quasiperiodic orbits restricted to attracting invariant curves, and $L > 0$ corresponds to chaotic dynamics.)

the boundary of the unit circle. Thus, if all eigenvalues move inside the unit circle when x^* is increased, then (27) is a hyperbolic stable equilibrium in case of x^* small. If at least one eigenvalue moves out, (27) is unstable. The Jury criteria (29a), (29b), (29c), and (29d) will help us to decide. Whenever (29c) and (29d) fail, a pair of complex valued eigenvalues will have modulus larger (or equal to) than 1 and consequently have a location outside (or on) the unit circle. If (29a) and (29b) fail, an eigenvalue crosses the unit circle through 1 or −1, respectively.

Moreover, due to the complexity of the Jury criteria we limit the discussion in the rest of this section to $\beta_1 = \beta_2 = \beta$. Then the Jury criteria may be expressed as

$$(\alpha + 2\beta)\, x^* > 0 \tag{30a}$$

$$2 - \alpha\left(x_1^* - x_2^* + x_3^*\right) - 2\beta x_2^* > 0 \tag{30b}$$

$$2 - \alpha x_3^* - \beta\left(x_1^* + x_2^*\right) > 0 \tag{30c}$$

$$(\beta - \alpha)\left\{\left(x_1^* - x_3^*\right)\left[1 - \left(\alpha x_3^* + \beta\left(x_1^* + x_2^*\right)\right)\right] + x_2^* - x_3^*\right\} > 0. \tag{30d}$$

Criteria (30a), (30b), and (30c) obviously hold in case of small equilibrium populations x^*. Regarding (30d) the same is true only as long as $\beta > \alpha$.

(A) First, consider $\beta < \alpha$. Then (30d) fails whenever x^* is small but it holds for those x^* which satisfy

$$x^* > \frac{\left(1 + Pa^{\beta} - 2P^2a^{2\beta}\right) D}{\left(1 - P^2a^{2\beta}\right)\left[\alpha P^2a^{2\beta} + \beta\left(1 + Pa^{\beta}\right)\right]} = u. \tag{31a}$$

On the other hand, (30b) holds only as long as

$$x^* < \frac{2D}{\alpha\left(1 - Pa^{\beta} + P^2a^{2\beta}\right) + 2\beta Pa^{\beta}} = v. \tag{31b}$$

Therefore, if x^* satisfies the inequality $u < x^* < v$, then (x_1^*, x_2^*, x_3^*) is stable. Moreover, if this shall be the case, the difference $u - v$ must be strictly negative which is equivalent to say that

$$\frac{\alpha - 2\beta + \sqrt{(2\beta - \alpha)(10\beta - 13\alpha)}}{2(3\alpha - 2\beta)} < Pa^{\beta} < 1. \tag{32}$$

The most transparent case to discuss is $\beta = 0$. Then $u = (1 + P - 2P^2)(1 + P + P^2)(P^2 - P^4)^{-1}$, $v = 2(1 + P + P^2)(1 - P + P^2)^{-1}$, and from (32) we find $t_1 = (1 + \sqrt{13})/6 < P < 1$ in order to ensure that $u - v < 0$. Now, if we choose a P value larger than t_1, say $P = 0.9$, we find $u = 4.93$, $v = 5.96$ and finally if $170.92 < F < 476.64$ we obtain $u < x^* < v$. Hence, there exists a parameter region where the fixed point is stable.

Turning to the dynamics, SYC dynamics is the only outcome in the region where (30d) fails. In case of small F values there exists a stable 3-cycle where the points in the cycle are $(\alpha^{-1}P^{-2}x^*, 0, 0)$, $(0, \alpha^{-1}P^{-1}x^*, 0)$, and $(0, 0, \alpha^{-1}x^*)$. For higher fecundity values there are cycles of period $3 \cdot 2^k$ as well as chaotic dynamics. In the intermediate region where all criteria (30a)–(30d) are valid, we find coexistence between the fixed point and chaotic SYC dynamics. Whether an orbit shall converge towards (x_1^*, x_2^*, x_3^*) or settle on a chaotic attractor on SYC form depends on the initial condition. If we continue to increase F we find that (30b) fails and as a result the fixed point will undergo a (supercritical) flip bifurcation. This gives birth to a tiny parameter region where SYC dynamics coexists with a stable 2-cycle not on SYC form. Through further enlargement of F, SYC dynamics is the only outcome.

In [4] it was proved that there also exists a 3-cycle where two age classes are populated at each time. Within our framework the points in such a cycle are found to be $(\alpha^{-1}P^{-1}(1 - P^2)z^*, \alpha^{-1}P(1 - P)z^*, 0)$, $(0, \alpha^{-1}(1 - P^2)z^*, \alpha^{-1}P^2(1 - P)z^*)$, and $(\alpha^{-1}(1 - P)z^*, 0, \alpha^{-1}P(1 - P^2)z^*)$ where $z^* = (1 - P^2)^{-1}x^*$. In [21] it is proved that the cycle is unstable and further, in a more general context, it is shown in [24] that such 2-class cycles may be embedded in a heteroclinic cycle which may be attracting. Indeed, depending of the kind of competition within and between age classes, the analysis in [24] accounts for possible loops and cycles. For example, if the competition is asymmetric the bifurcating invariant loop consists of a single year class 3-cycle with a synchronous two-year class orbit that heteroclinically connect the phases of this 3-cycle. Thus, all orbits on such a heteroclinic cycle approach the single year class 3-cycle. This mechanism was also found in Bulmer's original work on periodical cicadas. For further reading, compare [17, 24]. A final point is that when $P = 1$, the eigenvalue equation may be cast in the form

$$\lambda^3 + \frac{1}{3}\ln R_0 \lambda^2 + \frac{1}{3}\ln R_0 \lambda + \frac{1}{3}\ln R_0 - 1 = 0, \tag{33}$$

where the dominant roots are $\exp[\pm(2\pi/3)i]$. This means that independent of population size there will always be 3-cyclic dynamics in the sense that the total population at every point in the cycle equals x^* but the structure of the points is on SYC form.

Turning to the case $0 < \beta < \alpha$, it is still possible to find combinations of P and F such that the fixed point is stable but the parameter interval where this occurs appears to be (very) small. Indeed, assuming $\beta = \alpha/100$, the choice $Pa^{\beta} = 0.8$ satisfies (32) and if $P = 0.845$ it is possible to find F such that $Pa^{\beta} = 0.8$ as well as $u < x^* < v$. However, if we change to $P = 0.84$ or 0.85 it does not work. Therefore, our conclusion is that SYC dynamics dominates almost completely in the region $0 < \beta < \alpha$.

(B) Next, consider $\beta > \alpha$. In this case all criteria (30a)–(30d) hold whenever x^* is sufficiently small. Thus, there exists a parameter region where (27) is stable. Instability is introduced through a Neimark–Sacker bifurcation when $x^* = u$ (cf. (31a)) or equivalently when

$$P = \sqrt{\frac{1}{F}\exp\left[(\alpha + 2\beta)\frac{\left(1 + Pa^{\beta} - 2P^2a^{2\beta}\right)\left(1 + Pa^{\beta} + P^2a^{2\beta}\right)}{\left(1 - P^2a^{2\beta}\right)\left(\alpha P^2a^{2\beta} + \beta\left(1 + Pa^{\beta}\right)\right)}\right]}. \tag{34}$$

Regarding the eigenvalue equation (28), coefficients a_1 and a_2 are always positive and at threshold (34), $a_3 = Pa^{\beta}(1 + Pa^{\beta})^{-1}$ which is positive too. Consequently, there are no changes of signs between the coefficients in (28) which implies that there are two complex modulus 1 solutions and one real negative solution which necessarily must be $\lambda = -Pa^{\beta}(1 - Pa^{\beta})^{-1}$ and clearly $-1 < \lambda < 0$. These findings allow us to express (28) as

$$\left[\lambda^2 + \frac{\alpha\left(1 + Pa^{\beta} - P^2a^{2\beta}\right) + 2\beta P^2a^{2\beta}}{\alpha P^2a^{2\beta} + \beta\left(1 + Pa^{\beta}\right)}\lambda + 1\right] \cdot \left(\lambda + \frac{Pa^{\beta}}{1 + Pa^{\beta}}\right) = 0 \tag{35}$$

and the location of eigenvalue at threshold is given by the complex roots.

Now, scrutinizing the special case $\alpha = 0$, $\beta > 0$, we find that instability threshold (34) becomes

$$P = \sqrt{\frac{1}{F}\exp\left[\frac{2\left(F + \sqrt{F} + 1\right)\left(F + \sqrt{F} - 2\right)}{\left(F + \sqrt{F}\right)(F - 1)}\right]} \tag{36}$$

and moreover that the complex modulus 1 solutions of (35) may be expressed as

$$\lambda = -\frac{\sqrt{F} - 1}{\sqrt{F}(F - 1)} \pm \sqrt{1 - \frac{\left(\sqrt{F} - 1\right)^2}{F(F - 1)^2}}\, i. \tag{37}$$

From (36) it follows that F is large at bifurcation threshold; thus the eigenvalue (37) is located in the left half plane close to the imaginary axis (in fact even closer than in the corresponding 2-age class model); therefore $\arg\lambda \approx \pi/2$ here too. In order to visualize the dynamics we have studied the case $P = 0.7$ in somewhat more detail. In Figure 5(a), where $F = 24$, we show a quasiperiodic orbit which is restricted to a invariant curve. When $F = 30$, see Figure 5(b), an exact 4-period orbit is established through the frequency locking mechanism. In case of higher F values, the fourth iterate of (24) undergoes a Neimark–Sacker bifurcation which results in four invariant curves as displayed in Figure 5(c) ($F = 39$). Through further enlargement of F the dynamics becomes chaotic, see Figure 5(d) where $F = 43$.

Next, assume $\alpha > 0$. Whenever α is small, $\alpha \ll \beta$, we observe qualitatively the same kind of dynamics as just reported, that is, four periodicities either exact or approximate beyond instability threshold. Through further increase of α the tendency towards 4-periodic dynamics becomes less pronounced and gradually disappears completely. Moreover, we experience that the fixed point may be stable for much higher fecundity values than when α is small. The nonstationary dynamics is restricted to invariant curves, not topologically equivalent to circles, compare Figure 5(e), or chaotic. The curves appear to be weakly attracting.

5. Discussion

In the previous sections we have analysed different versions of two and three-dimensional maps. We shall now consider the more general situation where there are n age classes and since the parameter space is huge we will limit the discussion to the case where we have the same "strength" of survival between any two age classes. Hence, we let $\beta = \beta_1 = \cdots = \beta_{n-1}$ which means that the eigenvalue equation (6) may be cast in the form

$$\begin{aligned}\lambda^n &+ \left[\alpha x_1^* + \beta\left(\sum_{i=2}^{n} x_i^*\right)\right]\lambda^{n-1} \\ &+ \left[\alpha x_2^* + \beta\left(\sum_{\substack{i=1\\ i\neq 2}}^{n} x_i^*\right)\right]\lambda^{n-2} \\ &+ \left[\alpha x_3^* + \beta\left(\sum_{\substack{i=1\\ i\neq 3}}^{n} x_i^*\right)\right]\lambda^{n-3} + \cdots \\ &+ \left[\alpha x_{n-1}^* + \beta\left(\sum_{\substack{i=1\\ i\neq n-1}}^{n} x_i^*\right)\right]\lambda \\ &+ \left[\alpha x_n^* + \beta\left(\sum_{i=1}^{n-1} x_i^*\right)\right] - 1 = 0.\end{aligned} \tag{38}$$

First, let us comment on the parameter region where $\alpha \geq \beta$. Whenever x^* is small the fixed point is always unstable. This is proved in a general setting in [26] which also applies to the case under consideration in this paper. Next, suppose that n is even and $\lambda = -1$. Then left hand side of (38) may be expressed as $(x_1^* - x_2^*)(\beta - \alpha) + (x_3^* - x_4^*)(\beta - \alpha) + \cdots + (x_{n-1}^* - x_n^*)(\beta - \alpha)$ which is clearly ≤ 0. On the other hand, when $\lambda \to -\infty$, then left hand side of (38) $\to +\infty$; thus there must be a root $\tilde{\lambda} < -1$ of the equation which actually proves that the fixed point $(x_1^*, \ldots, x_n^*)$ will always be unstable. When n is odd the argument presented above does not hold, compare our analysis of the three age class model. However, it is possible to show that the fixed point is unstable given that x^* is sufficiently large. To this end, assume that n is odd and $\lambda = -1$. Then left hand side of (38) is positive as long as

$$x^* > \frac{2\sum_{j=0}^{n-1}\left(Pa^{\beta}\right)^j}{\alpha\sum_{j=0}^{n-1}(-1)^j\left(Pa^{\beta}\right)^j} \tag{39}$$

and we observe that when $\lambda \to -\infty$, $\lambda^n \to -\infty$ too, so evidently there must exist an eigenvalue $\tilde{\lambda} < -1$ where left hand side of (38) is zero. Hence, there is no stable fixed point in this case.

As the number of age classes increases there may also be cycles where more than one age class is populated at each time. Such cycles, referred to as multiple year class, MYC, cycles have already been identified in the $n = 3$ case ($\beta = 0$). In [21] where no density dependence in the survivals is assumed, MYC dynamics is further discussed and a major finding is that stability properties of such cycles on the whole are similar to the stability properties of $(x_1^*, \ldots, x_n^*)$ (i.e., unstable). Consequently, assuming $\alpha > \beta$, we conclude that the nontrivial fixed point (3) will always be unstable except for small parameter windows when n is odd. As demonstrated, such windows may be hard to find when $0 < \beta < \alpha$. Regarding the dynamics, the general case is SYC dynamics.

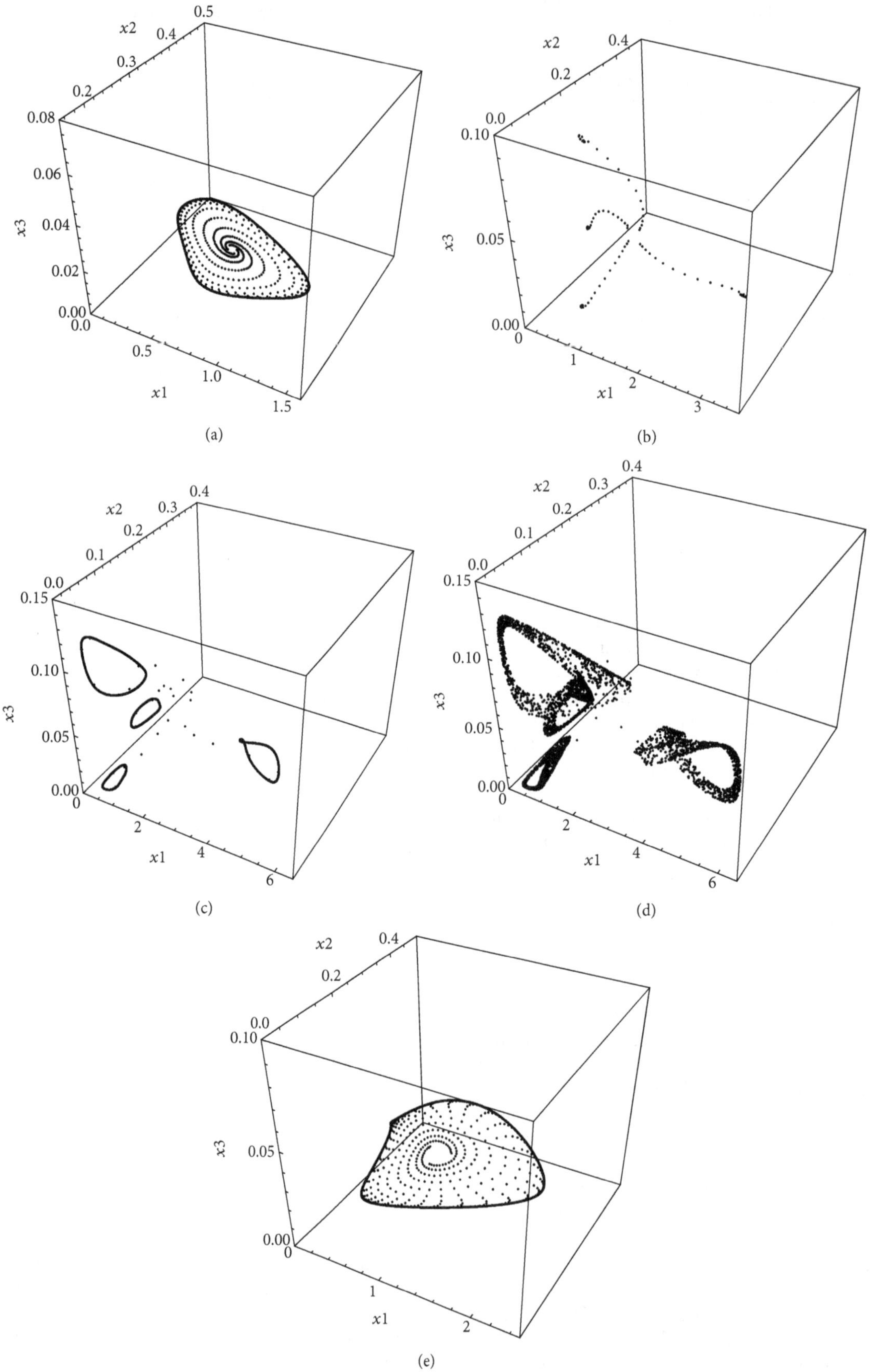

FIGURE 5: (a) A quasiperiodic orbit which settles on an invariant curve. $(\alpha, \beta) = (0, 1)$ and $(P, F) = (0.7, 24)$. (b) Convergence towards a 4-cycle. $(\alpha, \beta) = (0, 1)$ and $(P, F) = (0.7, 30)$. (c) 4 invariant curves which are visited once every fourth iteration. $(\alpha, \beta) = (0, 1)$ and $(P, F) = (0.7, 39)$. (d) Chaotic dynamics. $(\alpha, \beta) = (0, 1)$ and $(P, F) = (0.7, 43)$. (e) Dynamics in the case $(\alpha, \beta) = (0.9, 1)$ and $(P, F) = (0.7, 95)$.

When x^* is small, there is an attracting n cycle on SYC form, as x^* is increased we observe cycles of period $2^k n$, $k = 1, 2, \ldots$, as well as chaotic dynamics. In parameter windows where (3) is stable there is coexistence between the fixed point and chaotic dynamics on SYC form. Considering the simplest case $(\alpha, \beta) = (1, 0)$ the n cycle may be written as

$$\begin{gathered}\left(P^{-(n-1)}x^*, 0, \ldots, 0\right), \ldots, \left(0, \ldots, 0, P^{-(i-1)}x^*, 0, \ldots, 0\right),\\ \ldots, \left(0, \ldots, x^*\right).\end{gathered} \tag{40}$$

An increase of β ($\beta < \alpha$) acts in a stabilizing fashion in the sense that the higher β is, the higher F value is necessary in order to generate chaotic dynamics.

From a biological point of view it is not obvious how one should interpret SYC dynamics. As far as we know, Bulmer [17] is the first who has noticed SYC dynamics in theoretical insect models. He explains its presence by saying that competition between age classes is more severe than competition within age classes. Bulmer's argument is further strengthened by the findings in [24], see Theorem 9, where it is shown mathematically that strong competition within age classes in general will lead to an equilibrium where all age classes are populated (i.e., an equilibrium of form (3) in our model) while strong competition between age classes destabilizes and promotes oscillations with missing age classes. Another argument is presented in [21]. However, it should be emphasized that the conclusion obtained there is found from a simple model.

In the remaining part of the paper we shall deal exclusively with the case $\beta > \alpha$. Denoting left hand side of (38) for $P(\lambda)$ it is immediately clear from our previous analysis that when n is even the Jury criterion $(-1)^n P(-1) > 0$ will always be satisfied. Hence, there is no flip bifurcation in this case. When n is odd the value of x^* at threshold $(-1)^n P(-1) = 0$ is found to be

$$x^* = \frac{2D}{2\beta P a^{\beta} \sum_{j=0}^{(n-3)/2} \left(P a^{\beta}\right)^{2j} + \alpha \sum_{j=0}^{n-1} (-1)^j \left(P a^{\beta}\right)^j} \tag{41}$$

and if we substitute back into (38) the constant term becomes

$$a_n = \frac{2\beta \sum_{j=0}^{(n-3)/2} \left(P a^{\beta}\right)^{2j} - \alpha \sum_{j=0}^{b-1} (-1)^j \left(P a^{\beta}\right)^j}{K}, \tag{42}$$

where K is the denumerator of (41). Moreover, since $a_n > 1$ we may rule out the flip here too. Consequently, the only possible transfer from stability to instability goes through a (supercritical) Neimark–Sacker bifurcation for any $n \geq 2$.

There are two more questions which we find natural to discuss. (A) Will the 4-periodic dynamics observed when $n = 2$ and $n = 3$ also persist in case of larger values of n and (B) what happens to the size of the stable parameter region when the number of age classes increases?

Regarding (A), assume $n = 4$ and $(\alpha, \beta) = (0, 1)$ (the values of α and β with the most pronounced 4-periodic dynamics when $n = 2$ and $n = 3$). From the Jury criteria (and a lengthy calculation!) it now follows that the size of x^* at instability threshold must satisfy

$$\begin{aligned}&\left(D - P^3\right)\left[\left(D - P^3\right)P + (D - 1)\right]x^{*2}\\ &\quad + D\left[1 + P^2 + 4P^4 - 2D(1 + 2P)\right]x^* + 4PD^2\\ &= 0.\end{aligned} \tag{43}$$

Thus, for given values of P we may compute x^* from (43), substitute back into (38) and solve in order to find the location of the modulus 1 eigenvalues λ at threshold. For example, if $P = 0.6$ we find $\lambda_{1,2} = 0.111928 \pm 0.993716i$ ($\lambda_3 = 0.203731$, $\lambda_4 = 0.915659$) while $P = 0.2$ gives $\lambda_{1,2} = 0.110914 \pm 0.99383i$ ($\lambda_3 = 0.605291$, $\lambda_4 = 0.916205$). Now, the crucial observation from calculations as above is that $\lambda_{1,2}$ (as opposed to the $n = 2$ and $n = 3$ cases) are located in the right half of the complex plane and not particularly close to the imaginary axis. Hence, $|\arg \lambda_{1,2} - \pi/2|$ is not small, so the tendency towards 4-periodic dynamics is much weaker than in $n = 2$, $n = 3$ cases, a result which has also been verified through several simulations. Therefore, our analysis clearly suggests that periodic dynamics of low period, in particular 4-periodic dynamics, is limited to species with a few number of age classes. If only the first age class is not fertile and density effects occur in the survival terms exclusively, we may find 4-periodic dynamics also for larger values of n; see [15].

Finally consider (B) and let $x^*_{(2)}(P)$, $x^*_{(3)}(P)$, and $x^*_{(4)}(P)$ be the size of x^* at bifurcation threshold in the $n = 2, 3$ and 4 cases, respectively. From (16), (31a), (31b), and (43) it now follows that $x^*_{(i)}(P)$, $i = 2, 3, 4$, are increasing functions of P, $0 < P < 1$, and moreover, that $2 < x^*_{(2)}(P) < 4$, $1 < x^*_{(3)}(P) < 9/4$, and $0 < x^*_{(4)}(P) < 4/3$. These findings are presented in Figure 6. Hence, for a given value of P, $x^*_{(4)}(P) < x^*_{(3)}(P) < x^*_{(2)}(P)$ which means that the value of x^* at threshold becomes smaller as the number of age classes is increased; that is, an increase of n acts as a destabilizing effect. This argument is further increased if we continue to increase n. Indeed, when $n = 5$ we have verified numerically that the fixed point is unstable in case of P sufficiently small. Only quasiperiodic orbits have been identified. In [32] where a discrete stage-structured population model was analysed, the authors concluded that species

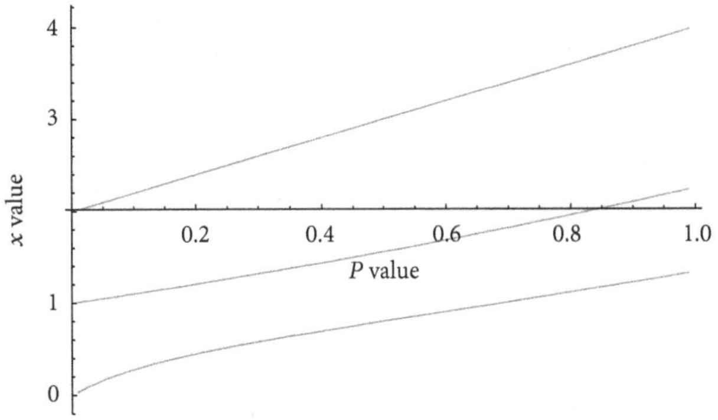

FIGURE 6: The curves $x^*_{(i)}$ at threshold. From bottom to top: $x^*_{(4)}$, $x^*_{(3)}$, and $x^*_{(2)}$.

who possess delayed semelparous life histories tend to be more stable than species who possess precocious semelparous life histories. Results from the analysis of age-structured models in [16] show a tendency in the opposite direction. The findings in this paper support the latter, but we feel that more work has to be done before one may conclude this issue.

Appendix

Considering map (8), $\beta > \alpha$, we find that the Jacobian at bifurcation threshold (11c) or (16) may be written as

$$J = \begin{pmatrix} -\dfrac{2\alpha}{\beta_1 + \alpha Pa^{\beta_1}} & \dfrac{\beta - \alpha Pa^{\beta_1}}{Pa^{\beta_1}\left(\beta_1 + \alpha Pa^{\beta_1}\right)} \\ Pa^{\beta_1}\dfrac{\alpha Pa^{\beta_1} - \beta_1}{\beta_1 + \alpha Pa^{\beta_1}} & -\dfrac{2\beta_1 Pa^{\beta_1}}{\beta_1 + \alpha Pa^{\beta_1}} \end{pmatrix} \tag{A.1}$$

and as shown in the main text

$$\lambda = -\frac{\alpha + \beta_1 Pa^{\beta_1}}{\beta_1 + \alpha Pa^{\beta_1}} \pm \frac{b}{\beta_1 + \alpha Pa^{\beta_1}} i, \tag{A.2}$$

where $b = \sqrt{(\beta_1^2 - \alpha^2)(1 - P^2 a^{2\beta_1})}$. The corresponding eigenvector becomes

$$\{w_1, w_2\} = \left\{\frac{\alpha - \beta_1 Pa^{\beta_1}}{Pa^{\beta_1}\left(\beta_1 - \alpha Pa^{\beta_1}\right)} - \frac{b}{Pa^{\beta_1}\left(\beta_1 - \alpha Pa^{\beta_1}\right)} i, 1 + Oi\right\}. \tag{A.3}$$

Further, define the matrix (cf. (A.3))

$$T = \begin{pmatrix} \dfrac{\alpha - \beta_1 Pa^{\beta_1}}{Pa^{\beta_1}\left(\beta - \alpha Pa^{\beta_1}\right)} & \dfrac{b}{Pa^{\beta_1}\left(\beta_1 - \alpha Pa^{\beta_1}\right)} \\ 1 & 0 \end{pmatrix}. \tag{A.4}$$

Then, after expanding the components of (8) up to third order, translating the bifurcation to the origin through the change of coordinates $(\widetilde{x}_1, \widetilde{x}_2) = (x_1 - x_1^*, x_2 - x_2^*)$, and finally applying the transformations

$$\begin{pmatrix} \widetilde{x}_1 \\ \widetilde{x}_2 \end{pmatrix} = T \begin{pmatrix} u \\ v \end{pmatrix}$$
$$\begin{pmatrix} u \\ v \end{pmatrix} = T^{-1} \begin{pmatrix} \widetilde{x}_1 \\ \widetilde{x}_2 \end{pmatrix} \tag{A.5}$$

map (8) may be cast into standard form at the bifurcation as (18) where

$$\begin{aligned} f(u, v) = {} & \frac{\alpha\beta_1 b^2}{\left(\beta_1 + \alpha Pa^{\beta_1}\right)\left(\beta_1 - \alpha Pa^{\beta_1}\right)^2} u^2 \\ & + \frac{\beta_1 b\left[\beta_1^2 - \alpha^2 - \alpha^2\left(1 - P^2 a^{2\beta_1}\right)\right]}{\left(\beta_1 + \alpha Pa^{\beta_1}\right)\left(\beta_1 - \alpha Pa^{\beta_1}\right)^2} uv \\ & - \frac{\alpha\beta_1 b^2}{\left(\beta_1 + \alpha Pa^{\beta_1}\right)\left(\beta_1 - \alpha Pa^{\beta_1}\right)^2} v^2 + \frac{Pa^{\beta_1}}{\beta_1 + \alpha Pa^{\beta_1}} \\ & \times \left\{\left[\frac{1}{6}\beta_1^2\left(3\alpha Pa^{\beta_1} + \beta_1\right)k^3 + \alpha\beta_1^2 Pa^{\beta}k^2\right.\right. \\ & \left. + \frac{1}{2}\beta_1^2\left(\alpha Pa^{\beta_1} - \beta_1\right)k - \frac{\beta_1^3}{3}\right]u^2 \\ & + \left[\frac{1}{6}\beta_1^2\left(3\alpha Pa^{\beta_1} + \beta_1\right)3k^2 l + 2\beta_1^2\alpha Pa^{\beta_1}kl\right. \\ & \left. + \frac{1}{2}\beta_1^2\left(\alpha Pa^{\beta_1} - \beta_1\right)l\right]u^2 v \\ & + \left[\frac{1}{6}\beta_1^2\left(3\alpha Pa^{\beta_1} + \beta_1\right)3kl^2 + \beta_1^2\alpha Pa^{\beta_1}l^2\right]uv^2 \\ & \left. + \frac{1}{6}\beta_1^2\left(3\alpha Pa^{\beta_1} + \beta_1\right)l^3 v^3\right\} \end{aligned} \tag{A.6}$$

$$\begin{aligned} g(u, v) = {} & -\frac{\alpha - \beta_1 Pa^{\beta_1}}{b} f(u.v) \\ & - \frac{\alpha^2 b}{Pa^{\beta_1}\left(\beta_1 + \alpha Pa^{\beta_1}\right)\left(\beta_1 - \alpha Pa^{\beta_1}\right)^2} u^2 \\ & + \frac{\alpha\left[\alpha^2 - \beta_1^2 + \alpha^2\left(1 - P^2 a^{2\beta_1}\right)\right]}{Pa^{\beta_1}\left(\beta_1 + \alpha Pa^{\beta}\right)\left(\beta_1 \alpha Pa^{\beta_1}\right)} uv \\ & + \frac{\alpha^2 b}{Pa^{\beta_1}\left(\beta_1 + \alpha Pa^{\beta_1}\right)\left(\beta_1 \alpha Pa^{\beta_1}\right)} v^2 \\ & + \frac{\beta_1 - \alpha Pa^{\beta_1}}{b\left(\beta_1 + \alpha Pa^{\beta_1}\right)} \times \left\{\left[-\frac{1}{3}\alpha^3 Pa^{\beta_1}k^3\right.\right. \\ & \left. + \frac{1}{2}\alpha^2\left(\beta_1 - \alpha Pa^{\beta_1}\right)k^2 + \alpha^2\beta_1 k\right]u^3 + \left[\alpha^3 Pa^{\beta_1}k^2 l\right. \\ & \left. + \alpha^2\left(\beta_1 - \alpha Pa^{\beta_1}\right)kl + \alpha^2\beta_1 l\right]u^2 v + \left[-\alpha^3 Pa^{\beta}kl^2\right. \\ & \left.\left. + \frac{1}{2}\alpha^2\left(\beta_1 - \alpha Pa^{\beta_1}\right)l^2\right]uv^2 - \frac{1}{3}\alpha^3 Pa^{\beta_1}l^3 v^3\right\}. \end{aligned} \tag{A.7}$$

k and l are defined as

$$k = \frac{\alpha - \beta_1 Pa^{\beta_1}}{Pa^{\beta_1}\left(\beta_1 - \alpha Pa^{\beta_1}\right)}$$
$$l = \frac{b}{Pa^{\beta_1}\left(\beta_1 - \alpha Pa^{\beta_1}\right)}, \tag{A.8}$$

respectively.

Conflicts of Interest

The author declares that there are no conflicts of interest regarding the publication of this paper.

Acknowledgments

The author thanks Ørjan Kristensen for computer assistance. Financial support from the publication fund of UiT The Arctic University of Norway is also gratefully acknowledged.

References

[1] J. Guckenheimer, G. Oster, and A. Ipaktchi, "The dynamics of density dependent population models," *Journal of Mathematical Biology*, vol. 4, no. 2, pp. 101–147, 1977.

[2] S. A. Levin and C. P. Goodyear, "Analysis of an age-structured fishery model," *Journal of Mathematical Biology*, vol. 9, no. 3, pp. 245–274, 1980.

[3] J. A. Silva and T. G. Hallam, "Effects of delay, truncations and density dependence in reproduction schedules on stability of nonlinear Leslie matrix models," *Journal of Mathematical Biology*, vol. 31, no. 4, pp. 367–395, 1993.

[4] A. Wikan and E. Mjølhus, "Overcompensatory recruitment and generation delay in discrete age-structured population models," *Journal of Mathematical Biology*, vol. 35, no. 2, pp. 195–239, 1996.

[5] A. Wikan and A. Eide, "An analysis of a nonlinear stage-structured cannibalism model with application to the northeast arctic cod stock," *Bulletin of Mathematical Biology*, vol. 66, no. 6, pp. 1685–1704, 2004.

[6] I. Ugarcovici and H. Weiss, "Chaotic dynamics of a nonlinear density dependent population model," *Nonlinearity*, vol. 17, no. 5, pp. 1689–1711, 2004.

[7] W. Govaerts and R. K. Ghaziani, "Numerical bifurcation analysis of a nonlinear stage structured cannibalism population model," *Journal of Difference Equations and Applications*, vol. 12, no. 10, pp. 1069–1085, 2006.

[8] J. M. Cushing, "Nonlinear matrix models and population dynamics," *Natural Resource Modeling*, vol. 2, no. 4, pp. 539–580, 1988.

[9] J. M. Cushing, "A strong ergodic theorem for some nonlinear matrix models for the dynamics of structured populations," *Natural Resource Modeling*, vol. 3, no. 3, pp. 331–357, 1989.

[10] K. M. Crowe, "A nonlinear ergodic theorem for discrete systems," *Journal of Mathematical Biology*, vol. 32, no. 3, pp. 179–191, 1994.

[11] R. Kon, Y. Saito, and Y. Takeuchi, "Permanence of single-species stage-structured models," *Journal of Mathematical Biology*, vol. 48, no. 5, pp. 515–528, 2004.

[12] D. L. Deangelis, L. J. Svoboda, S. W. Christensen, and D. S. Vaughan, "Stability and return times of Leslie matrices with density-dependent survival: applications to fish populations," *Ecological Modelling*, vol. 8, no. C, pp. 149–163, 1980.

[13] R. A. Desharnais and L. Liu, "Stable demographic limit cycles in laboratory populations of tribolium castaneum," *Journal of Animal Ecology*, vol. 56, no. 3, pp. 885–906, 1987.

[14] T. V. Burkey and N. C. Stenseth, "Population dynamics of territorial species in seasonal and patchy environments," *Oikos*, vol. 69, no. 1, pp. 47–53, 1994.

[15] A. Wikan, "Four-periodicity in Leslie matrix models with density dependent survival probabilities," *Theoretical Population Biology*, vol. 53, no. 2, pp. 85–97, 1998.

[16] A. Wikan, "Age or stage structure? a comparison of dynamic outcomes from discrete age- and stage-structured population models," *Bulletin of Mathematical Biology*, vol. 74, no. 6, pp. 1354–1378, 2012.

[17] M. G. Bulmer, "Periodical insects," *The American Naturalist*, vol. 111, no. 982, pp. 1099–1117, 1977.

[18] F. C. Hoppensteadt and J. B. Keller, "Synchronization of periodical cicada emergences," *Science*, vol. 194, no. 4262, pp. 335–337, 1976.

[19] H. Behncke, "Periodical cicadas," *Journal of Mathematical Biology*, vol. 40, no. 5, pp. 413–431, 2000.

[20] N. V. Davydova, O. Diekmann, and S. A. van Gils, "Year class coexistence or competitive exclusion for strict biennials?" *Journal of Mathematical Biology*, vol. 46, no. 2, pp. 95–131, 2003.

[21] E. Mjølhus, A. Wikan, and T. Solberg, "On synchronization in semelparous populations," *Journal of Mathematical Biology*, vol. 50, no. 1, pp. 1–21, 2005.

[22] R. Kon, "Nonexistence of synchronous orbits and class coexistence in matrix population models," *SIAM Journal on Applied Mathematics*, vol. 66, no. 2, pp. 616–626, 2005.

[23] J. M. Cushing, "Nonlinear semelparous Leslie models," *Mathematical Biosciences and Engineering*, vol. 3, no. 1, pp. 17–36, 2006.

[24] J. M. Cushing, "Three stage semelparous Leslie models," *Journal of Mathematical Biology*, vol. 59, no. 1, pp. 75–104, 2009.

[25] J. M. Cushing and S. M. Henson, "Stable bifurcations in semelparous Leslie models," *Journal of Biological Dynamics*, vol. 6, no. suppl. 2, pp. 80–102, 2012.

[26] J. M. Cushing, "On the fundamental bifurcation theorem for semelparous Leslie models," in *Dynamics, Games And Science*, J. P. Bourguignon, R. Jeltsch, A. Pinto, and M. Viana, Eds., vol. 1 of *CIM Ser. Math. Sci.*, pp. 215–251, Springer, Cham, Berlin, Germany, 2015.

[27] J. D. Murray, *Mathematical Biology: Spatial Models And Biomedical Applications*, vol. 18, Springer, Berlin, Germany, 3rd edition, 2003.

[28] Y. H. Wan, "Computation of the stability condition for the Hopf bifurcation of diffeomorphisms on R_2," *SIAM Journal on Applied Mathematics*, vol. 34, no. 1, pp. 167–175, 1978.

[29] J. Guckenheimer and P. Holmes, *Nonlinear Oscillations, Dynamical Systems, and Bifurcation of Vector Fields*, Springer, NY, USA, 2002.

[30] J. Jost, *Dynamical Systems*, Universitext, Springer-Verlag, Berlin, Germany, 2005.

[31] H. N. Agiza, E. M. ELabbasy, H. EL-Metwally, and A. A. Elsadany, "Chaotic dynamics of a discrete prey-predator model with holling type II," *Nonlinear Analysis: Real World Applications*, vol. 10, no. 1, pp. 116–129, 2009.

[32] M. G. Neubert and H. Caswell, "Density-dependent vital rates and their population dynamic consequences," *Journal of Mathematical Biology*, vol. 41, no. 2, pp. 103–121, 2000.

10

Shape Preserving Interpolation Using C^2 Rational Cubic Spline

Samsul Ariffin Abdul Karim[1] and Kong Voon Pang[2]

[1] *Fundamental and Applied Sciences Department, Universiti Teknologi PETRONAS, Bandar Seri Iskandar, 32610 Seri Iskandar, Perak Darul Ridzuan, Malaysia*
[2] *School of Mathematical Sciences, Universiti Sains Malaysia (USM), 11800 Minden, Penang, Malaysia*

Correspondence should be addressed to Samsul Ariffin Abdul Karim; samsul_ariffin@petronas.com.my

Academic Editor: Francisco J. Marcellán

This paper discusses the construction of new C^2 rational cubic spline interpolant with cubic numerator and quadratic denominator. The idea has been extended to shape preserving interpolation for positive data using the constructed rational cubic spline interpolation. The rational cubic spline has three parameters α_i, β_i, and γ_i. The sufficient conditions for the positivity are derived on one parameter γ_i while the other two parameters α_i and β_i are free parameters that can be used to change the final shape of the resulting interpolating curves. This will enable the user to produce many varieties of the positive interpolating curves. Cubic spline interpolation with C^2 continuity is not able to preserve the shape of the positive data. Notably our scheme is easy to use and does not require knots insertion and C^2 continuity can be achieved by solving tridiagonal systems of linear equations for the unknown first derivatives $d_i, i = 1, \ldots, n-1$. Comparisons with existing schemes also have been done in detail. From all presented numerical results the new C^2 rational cubic spline gives very smooth interpolating curves compared to some established rational cubic schemes. An error analysis when the function to be interpolated is $f(t) \in C^3[t_0, t_n]$ is also investigated in detail.

1. Introduction

Spline interpolation has been used extensively in many research disciplines such as in car design and airplane fuselage. Univariate and bivariate spline can be used to approximate or interpolate the given finite data sets. Even though the cubic spline has second-order parametric continuity, C^2, it has some weakness such that the interpolating curves may give few unwanted behavior of the original data due to the existing wiggles along some interval. This uncharacteristic behavior may destroy the data. If the given data is positive cubic spline may give some negative values along the whole interval where the interpolating curves will lie below x-axis. For some application any negativity is unacceptable. For example, the wind speed, solar energy, and rainfall received are always having positive values and any negativity values need to be avoided as it may destroy any important information that may exist in the original data. Similarly if the data is monotone, then the resulting interpolating curves also must be monotone too. Furthermore if the given data is convex, the rational cubic spline interpolation should be able to maintain the shape of the original data. Thus shape preserving interpolation is important in computer graphics and computer aided geometric design (CAGD).

Due to the fact that cubic spline is not able to produce completely the positive, monotone, and convex interpolating curves on entire given interval, many researchers have proposed several methods and idea to preserve the positivity, monotonicity, and convexity of the data. Fritsch and Carlson [1] and Dougherty et al. [2] have discussed the monotonicity, positivity, and convexity preserving by using cubic spline interpolation by modifying the first derivative values in which the shape violation is found. Butt and Brodlie [3] and Brodlie and Butt [4] have used cubic spline interpolation to preserve the positivity and convexity of the finite data by inserting extra knots in the interval in which the positivity and/or convexity is not preserved by the cubic spline. Their methods did not give any extra freedom to the user in controlling the final shape of the interpolating curves. In order to change the final shape of the interpolating curves, the user needs to change the given data. Another thing is that their methods require the modification of the first derivative parameters. Sarfraz [5], Sarfraz et al. [6, 7], and Abbas [8] studied the use of rational cubic interpolant for preserving the positive data. Meanwhile M. Z. Hussain and M. Hussain [9] studied positivity preserving for curves and surfaces by utilizing rational cubic spline with quadratic denominator. In works by Hussain et al. [10] and Sarfraz et al. [11] the rational cubic spline with quadratic denominator has been used for positivity, monotonicity, and convexity preserving with C^2 continuity. Hussain et al. [10] have only one free parameter meanwhile Sarfraz et al. [11] have no free parameter. Abbas [8] and Abbas et al. [12] have discussed the positivity by using new C^2 rational cubic spline with two free parameters. Another paper concerning C^2 rational spline can be found by Delbourgo [13], Gregory [14], and Delbourgo and Gregory [15]. Karim and Kong [16–19] have proposed new C^1 rational cubic spline (cubic/quadratic) with three parameters where two of them are free parameters. The rational cubic spline has been successfully applied to the local control of the interpolating functions, positivity, monotonicity, and convexity preserving as well as the derivative control including an error analysis when the function to be interpolated is $f(t) \in C^3[t_0, t_n]$. Motivated by the works of Tian et al. [20], Abbas et al. [12], Hussain et al. [10], and Sarfraz et al. [11], in this paper the authors will proposed new C^2 rational cubic spline for positivity, monotonicity, and convexity preserving and data constrained modeling. Under some circumstances our C^2 rational cubic spline will give new C^2 rational cubic spline based on rational cubic spline defined by Tian et al. [20]. Numerical comparison between C^2 rational cubic spline and the works of Hussain et al. [10], Abbas [8], Abbas et al. [12], and Sarfraz et al. [11] also has been made comprehensively. From all presented numerical results shape preserving interpolation by using the new C^2 rational cubic spline gives comparable results with existing rational cubic spline schemes. The main scientific contribution of this paper is summarized as follows:

(i) In this paper C^2 rational cubic spline (cubic/quadratic) with three parameters has been used for positivity, monotonicity, and convexity preserving and constrained data modeling while in works by Karim and Kong [17–19], M. Z. Hussain and M. Hussain [9], and Sarfraz [5] the degree of smoothness attained is C^1.

(ii) Hussain et al. [10] and Sarfraz et al. [11] discussed the positivity by using C^2 rational cubic spline (cubic/quadratic) with two parameters with one or no free parameter while our rational cubic spline has two free parameters. Even though Abbas et al. [12] also proposed C^2 rational cubic spline (cubic/quadratic) with two parameters, their rational spline is different form our rational cubic spline. Furthermore it was noticed that schemes of Abbas et al. [12] may not be able to produce completely positive interpolating curves with C^2 continuity.

(iii) When $\gamma_i = 0$, we may obtain the new C^2 rational cubic spline with two parameters, an extension to the original C^1 rational cubic spline of Tian et al. [20]. Thus our C^2 rational cubic spline gives a larger class of rational cubic spline which also includes the rational cubic spline of Tian et al. [20].

(iv) Our rational scheme is local while in Lamberti and Manni [21] their scheme is global. Furthermore our rational scheme works well for both equally and unequally spaced data while the rational spline interpolant by Duan et al. [22] and Bao et al. [23] only works for equally spaced data.

(v) Numerical comparisons between the C^2 rational cubic spline and the existing schemes such as Hussain et al. [10], Sarfraz et al. [11], and Abbas et al. [12] for positivity preserving and C^1 rational spline of Karim and Kong [17–19] also have been done comprehensively.

(vi) Our method also does not require any knots insertion. Meanwhile the cubic spline interpolation by Butt and Brodlie [3], Brodlie and Butt [4], and Fiorot and Tabka [24] requires knots insertion in the interval where the interpolating curves produce the negative values (lies below x-axis) for positive data, nonmonotone interpolating curves (for monotone data), and nonconvex interpolating curves for convex data.

(vii) This paper utilized the rational cubic spline meanwhile in Dube and Tiwari [25], Pan and Wang [26], and Ibraheem et al. [27] the rational trigonometric spline is used in place of standard rational cubic spline. Thus no trigonometric functions are involved. Therefore the method is not computationally expensive.

The remainder of the paper is organized as follows. Section 2 introduces the new C^2 rational cubic spline with three parameters, with some discussion on the methods to estimate the first derivatives values as well as shape controls of the rational cubic spline interpolation. Meanwhile Section 3 discusses the positivity preserving by using C^2 rational cubic spline together with numerical demonstrations as well as comparison with some existing schemes including error analysis. Section 4 is devoted for research discussion. Finally a summary and conclusions are given in Section 5.

2. C^2 Rational Cubic Spline Interpolant

This section will introduce C^2 rational cubic spline interpolant with three parameters. Originally this rational cubic spline has been initiated by Karim and Kong [18]. The main difference is that in this paper the rational cubic spline has C^2 continuity while in work by Karim and Kong [18] it has C^1 continuity. We begin with the definition of C^2 rational cubic spline interpolant, given set of data points $\{(x_i, f_i),\ i = 0, 1, \ldots, n\}$ such that $x_0 < x_1 < \cdots < x_n$. Let $h_i = x_{i+1} - x_i$, $\Delta_i = (f_{i+1} - f_i)/h_i$ and $\theta = (x - x_i)/h_i$, where $0 \le \theta \le 1$. For

$x \in [x_i, x_{i+1}]$, $i = 0, 1, 2, \ldots, n-1$, the rational cubic spline interpolant with three parameters is defined as follows:

$$s(x) \equiv s_i(x) = \frac{A_{i0}(1-\theta)^3 + A_{i1}\theta(1-\theta)^2 + A_{i2}\theta^2(1-\theta) + A_{i3}\theta^3}{(1-\theta)^2\alpha_i + \theta(1-\theta)(2\alpha_i\beta_i + \gamma_i) + \theta^2\beta_i}. \quad (1)$$

The following conditions will assure that the rational cubic spline interpolant in (1) has C^2 continuity:

$$\begin{aligned} s(x_i) &= f_i, \\ s(x_{i+1}) &= f_{i+1}, \\ s^{(1)}(x_i) &= d_i, \\ s^{(1)}(x_{i+1}) &= d_{i+1}, \\ s^{(2)}(x_{i+}) &= s^{(2)}(x_{i-}), \end{aligned} \quad (2)$$

where $s^{(1)}(x_i)$ and $s^{(2)}(x_i)$ denote the first- and second-order derivative with respect to x, respectively. Meanwhile the notations $s^{(2)}(x_{i+})$ and $s^{(2)}(x_{i-})$ correspond to the right and left second derivatives values. Furthermore d_i denotes the derivative value which is given at the knot x_i, $i = 0, 1, 2, \ldots, n$.

By using (2), the required C^2 rational cubic spline interpolant with three parameters defined by (1) has the unknowns A_{ij}, $j = 0, 1, 2, 3$, and is given as follows:

$$\begin{aligned} A_{i0} &= \alpha_i f_i, \\ A_{i1} &= (2\alpha_i\beta_i + \alpha_i + \gamma_i) f_i + \alpha_i h_i d_i, \\ A_{i2} &= (2\alpha_i\beta_i + \beta_i + \gamma_i) f_{i+1} - \beta_i h_i d_{i+1}, \\ A_{i3} &= \beta_i f_{i+1}. \end{aligned} \quad (3)$$

The parameters $\alpha_i, \beta_i > 0$, $\gamma_i \geq 0$ are used to control the final shape of the interpolating curves. From work by Karim and Kong [19], the second-order derivative $s^{(2)}(x)$ is given as

$$s^{(2)}(x) = \frac{\sum_{j=0}^{3} C_{ij}(1-\theta)^{3-j}\theta^j}{h_i[Q_i(\theta)]^3}, \quad (4)$$

where

$$\begin{aligned} C_{i0} &= 2\alpha_i^2(\gamma_i(\Delta_i - d_i) - \beta_i(d_{i+1} - \Delta_i) \\ &\quad - 2\alpha_i\beta_i(d_i - \Delta_i)), \\ C_{i1} &= 6\alpha_i^2\beta_i(\Delta_i - d_i), \\ C_{i2} &= 6\alpha_i\beta_i^2(d_{i+1} - \Delta_i), \\ C_{i3} &= 2\beta_i^2[\gamma_i(d_{i+1} - \Delta_i) \\ &\quad - \alpha_i(\Delta_i - d_i - 2\beta_i(d_{i+1} - \Delta_i))]. \end{aligned} \quad (5)$$

Now C^2 continuity, $s^{(2)}(x_{i+}) = s^{(2)}(x_{i-})$, $i = 1, 2, \ldots, n-1$, will give

$$\frac{2[\gamma_{i-1}(d_i - \Delta_{i-1}) - \alpha_{i-1}(\Delta_{i-1} - d_{i-1} - 2\beta_{i-1}(d_i - \Delta_{i-1}))]}{h_{i-1}\beta_{i-1}} = \frac{2(\gamma_i(\Delta_i - d_i) - \beta_i(d_{i+1} - \Delta_i) - 2\alpha_i\beta_i(d_i - \Delta_i))}{h_i\alpha_i}. \quad (6)$$

Now (6) provides the following system of linear equations that can be used to compute the first derivative parameters, d_i, $i = 1, 2, \ldots, n-1$, such that

$$a_i d_{i-1} + b_i d_i + c_i d_{i+1} = e_i, \quad i = 1, 2, \ldots, n-1, \quad (7)$$

with

$$\begin{aligned} a_i &= h_i\alpha_{i-1}\alpha_i, \\ b_i &= h_i\alpha_i(\gamma_{i-1} + 2\alpha_{i-1}\beta_{i-1}) + h_{i-1}\beta_{i-1}(\gamma_i + 2\alpha_i\beta_i), \\ c_i &= h_{i-1}\beta_{i-1}\beta_i, \\ e_i &= h_i\alpha_i(\gamma_{i-1} + \alpha_{i-1} + 2\alpha_{i-1}\beta_{i-1})\Delta_{i-1} \\ &\quad + h_{i-1}\beta_{i-1}(\gamma_i + \beta_i + 2\alpha_i\beta_i)\Delta_i. \end{aligned} \quad (8)$$

The system of linear equations given by (7) is strictly tridiagonal and has a unique solution for the unknown derivative parameters d_i, $i = 1, 2, \ldots, n-1$, for all $\alpha_i, \beta_i > 0$, $\gamma_i \geq 0$. The system in (7) gives $n-1$ linear equations for $n+1$ unknown derivative values. Thus two more equations are required in order to obtain the unique solution in (7). The following is common choices for the end points condition, that is, d_0 and d_n:

$$\begin{aligned} s^{(1)}(x_0) &= d_0, \\ s^{(1)}(x_n) &= d_n. \end{aligned} \quad (9)$$

By using (3), C^2 rational cubic spline interpolant in (1) can be reformulated and is given as follows:

$$s(x) \equiv s_i(x) = \frac{P_i(\theta)}{Q_i(\theta)}, \quad (10)$$

where

$$\begin{aligned} P_i(\theta) &= \alpha_i f_i(1-\theta)^3 \\ &\quad + ((2\alpha_i\beta_i + \alpha_i + \gamma_i) f_i + \alpha_i h_i d_i)\theta(1-\theta)^2 \\ &\quad + ((2\alpha_i\beta_i + \beta_i + \gamma_i) f_{i+1} - \beta_i h_i d_{i+1})\theta^2(1-\theta) \\ &\quad + \beta_i f_{i+1}\theta^3, \\ Q_i(\theta) &= (1-\theta)^2\alpha_i + \theta(1-\theta)(2\alpha_i\beta_i + \gamma_i) + \theta^2\beta_i. \end{aligned} \quad (11)$$

The data-dependent sufficient conditions on parameters γ_i will be developed in order to preserve the positivity, data constrained, monotonicity, and convexity on the entire interval

$[x_i, x_{i+1}]$, $i = 0, 1, 2, \ldots, n-1$. The remaining parameters α_i and β_i can be used to refine the resulting interpolating curves. Thus the rational cubic spline provides greater flexibility to the user in controlling the final shape of the interpolating curves.

Theorem 1 (C^2 rational cubic spline interpolant). *The rational cubic spline with three parameters defined by (1) is $C^2 \in [x_0, x_n]$ if there exist positive parameters $\alpha_i, \beta_i > 0$, $\gamma_i \geq 0$, and d_i, $i = 1, 2, \ldots, n-1$, that satisfy (7).*

The choice of the end point derivatives d_0 and d_n depended on the original data which are chosen as follows.

Choice 1 (Geometric Mean Method (GMM)). Consider

$$d_0 = \begin{cases} 0, & \Delta_0 = 0 \text{ or } \Delta_{2,0} = 0 \\ \Delta_0^{(1+h_0/h_1)} \Delta_{2,1}^{(-h_0/h_1)} & \text{otherwise,} \end{cases}$$

$$d_n = \begin{cases} 0 & \Delta_{n-1} = 0 \text{ or } \Delta_{n,n-2} = 0 \\ \Delta_{n-1}^{(1+h_{n-1}/h_{n-2})} \Delta_{n,n-2}^{(-h_{n-1}/h_{n-2})} & \text{otherwise.} \end{cases} \quad (12)$$

Choice 2 (Arithmetic Mean Method (AMM)). Consider

$$d_0 = \Delta_0 + (\Delta_0 - \Delta_1)\left(\frac{h_0}{h_0 + h_1}\right),$$
$$d_n = \Delta_{n-1} + (\Delta_{n-1} - \Delta_{n-2})\left(\frac{h_{n-1}}{h_{n-1} + h_{n-2}}\right). \quad (13)$$

Delbourgo and Gregory [29] give more details about the method that can be used to estimate the first derivative value. In this paper the AMM will be used to estimate the end point derivatives d_0 and d_n, respectively.

Some observation and shape control analysis of the new C^2 rational cubic spline interpolant defined by (10) are given as follows:

(1) When $\alpha_i > 0$, $\beta_i > 0$, and $\gamma_i = 0$, the rational interpolant in (10) reduces to the rational spline of the form cubic/quadratic by Tian et al. [20] and we may obtain C^2 rational cubic spline with two parameters, an extension to C^1 rational cubic spline originally proposed by Tian et al. [20]. Thus by rewriting C^2 condition in (7), we can obtain C^2 rational cubic of Tian et al. [20].

(2) When $\alpha_i = \beta_i = 1$; $\gamma_i = 0$, the rational cubic interpolant in (1) is just a standard cubic Hermite spline with C^1 continuity that may not be able to completely preserve the positivity of the data [18]:

$$s(x) = (1-\theta)^2 (1-2\theta) f_i + \theta^2 (3-2\theta) f_{i+1} + \theta (1-\theta)^2 d_i - \theta^2 (1-\theta) d_{i+1}, \quad (14)$$

for $i = 0, 1, \ldots, n-1$.

TABLE 1: A data from work by Sarfraz et al. [6].

i	0	1	1	3	4
x_i	0	2	3	9	11
f_i	0.5	1.5	7	9	13
h_i	2	1	6	2	
Δ_i	0.5	5.5	1/3	2	
d_i (C^1)	−2.833	3.833	4.7619	1.5833	2.4167
d_i (C^2)	−2.833	4.3223	4.9828	1.2146	2.4167

(3) Furthermore the rational interpolant in (1) can be written as [18]

$$s(x) = (1-\theta) f_i + \theta f_{i+1} + \frac{h_i \theta (1-\theta) [\alpha_i (d_i - \Delta_i)(1-\theta) + \beta_i (\Delta_i - d_{i+1}) \theta]}{Q_i(\theta)}. \quad (15)$$

(4) Obviously when $\alpha_i \to 0$, $\beta_i \to 0$, or $\gamma_i \to \infty$, the rational interpolant in (10) converges to following straight line:

$$\lim_{\alpha_i, \beta_i \to 0, \gamma_i \to \infty} s(x) = (1-\theta) f_i + \theta f_{i+1}. \quad (16)$$

Either the decrease of the parameters α_i and β_i or the increase of γ_i will reduce the rational cubic spline to a linear interpolant. Figure 1 shows this example. We test shape control analysis by using the data from work by Sarfraz et al. [6] given in Table 1.

To show the difference between C^2 rational cubic spline with three parameters and C^1 rational cubic spline of Karim and Kong [17–19], we choose $\alpha_i = \beta_i = 1$; $\gamma_i = 2$ for both cases. The main difference is that to generate C^1 rational interpolating curves the first derivative parameter d_i, $i = 0, 1, 2, \ldots, n$, is calculated by using Arithmetic Mean Method (AMM); meanwhile to generate C^2 rational cubic spline with three parameters, the first derivative parameter d_i, $i = 1, 2, \ldots, n-1$, is calculated by solving (7) with suitable choices of the end point derivatives d_0 and d_n.

Remark 2. If $\Delta_i - d_i = 0$ or $d_{i+1} - \Delta_i = 0$, then $d_i = d_{i+1} = \Delta_i$. In this case the rational interpolant in (10) will be linear in the corresponding interval or region; that is,

$$s(x) = (1-\theta) f_i + \theta f_{i+1}. \quad (17)$$

Figure 2 shows the shape control using the rational cubic spline for the given data in Table 1.

Figure 2(f) shows the combination of Figures 2(a), 2(d), and 2(e). Meanwhile Figure 2(g) shows the combination of Figures 2(a), 2(b), and 2(c), respectively.

Clearly the curves approach to the straight line if $\alpha_i \to 0$, $\beta_i \to 0$, or $\gamma_i \to \infty$. Furthermore decreases in the value of β_i will pull the curves upward and vice versa. This shape control will be useful for shape preserving interpolation as well as local control of the interpolating curves.

Meanwhile from Figure 1(d), it can be seen clearly that C^2 rational cubic spline interpolation (shown as red color) is

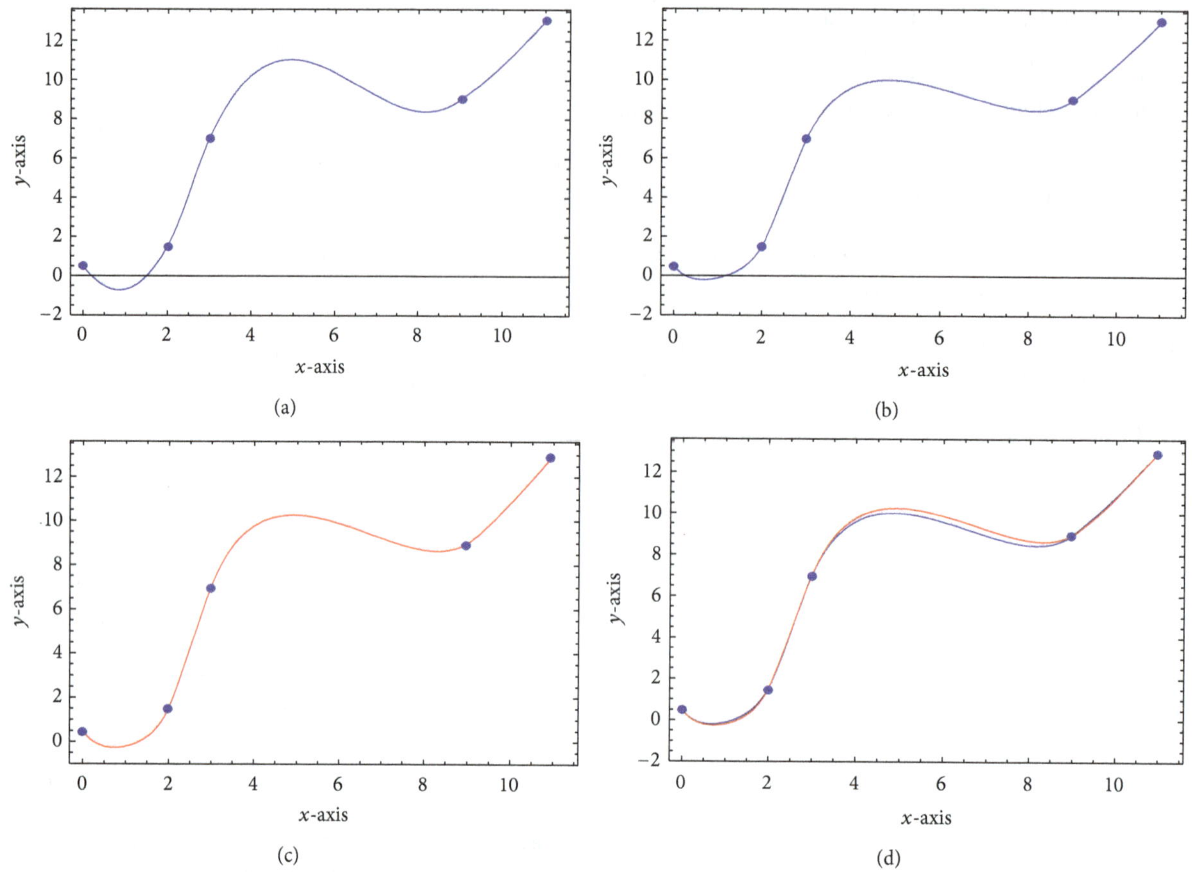

FIGURE 1: Interpolating curve for data in Table 1. (a) Default C^1 cubic spline with $\alpha_i = \beta_i = 1$ and $\gamma_i = 0$. (b) C^1 rational cubic spline with $\alpha_i = \beta_i = 1$ and $\gamma_i = 2$. (c) C^2 rational cubic spline by using derivative calculated from tridiagonal equation (7) with $\alpha_i = \beta_i = 1$ and $\gamma_i = 2$. (d) The graphs of (b) C^1 rational cubic spline (blue) and (c) C^2 rational cubic spline (red).

smoother than C^1 rational cubic spline interpolation (shown as black color). Thus the new C^2 rational cubic spline provides good alternative to the existing C^1 and C^2 rational cubic spline.

Remark 3. Since the constructed C^2 rational cubic spline has three parameters, then how do we choose the parameter values? To answer this question, the choices of the parameters totally depend on the data that are under considerations by the main user. The main difference between our C^2 rational cubic spline and C^2 cubic spline interpolation is that, in order to change the final shape of the interpolating curves, our schemes are only required to change the parameters values without the need to change the data points itself. But there are no free parameters in the descriptions of C^2 cubic spline interpolation.

3. Positivity Preserving Using C^2 Rational Cubic Spline Interpolation

In this section, the positivity preserving by using the proposed C^2 rational cubic spline interpolation defined by (10) will be discussed in detail. We follow the same idea of Karim and Kong [18] and Abbas et al. [12] such that simple data-dependent conditions for positivity are derived on one parameter γ_i while the remaining parameters α_i and β_i are free to be utilized. The main objective is that, in order to preserve the positivity of the positive data, the rational cubic spline interpolant must be positive on the entire given interval. The simple way to achieve it is by finding the automated choice of the shape parameter γ_i. We begin by giving the definition of strictly positive data.

Given the strictly positive set of data (x_i, f_i), $i = 0, 1, \ldots, n$, $x_0 < x_1 < \cdots < x_n$, such that

$$f_i > 0, \quad i = 0, 1, \ldots, n. \tag{18}$$

Now from (10), the rational cubic spline will preserve the positivity of the data if and only if $P_i(\theta) > 0$ and $Q_i(\theta) > 0$. Since for all $\alpha_i, \beta_i > 0$ and $\gamma_i \geq 0$, the denominator $Q_i(\theta) > 0$, $i = 0, 1, \ldots, n-1$. Thus $s(x) > 0$ if and only if $P_i(\theta) > 0$, $i = 0, 1, \ldots, n-1$. The cubic polynomial $P_i(\theta)$, $i = 0, 1, \ldots, n-1$ can be written as follows [18]:

$$P_i(\theta) = B_i\theta^3 + C_i\theta^2 + D_i\theta + E_i, \tag{19}$$

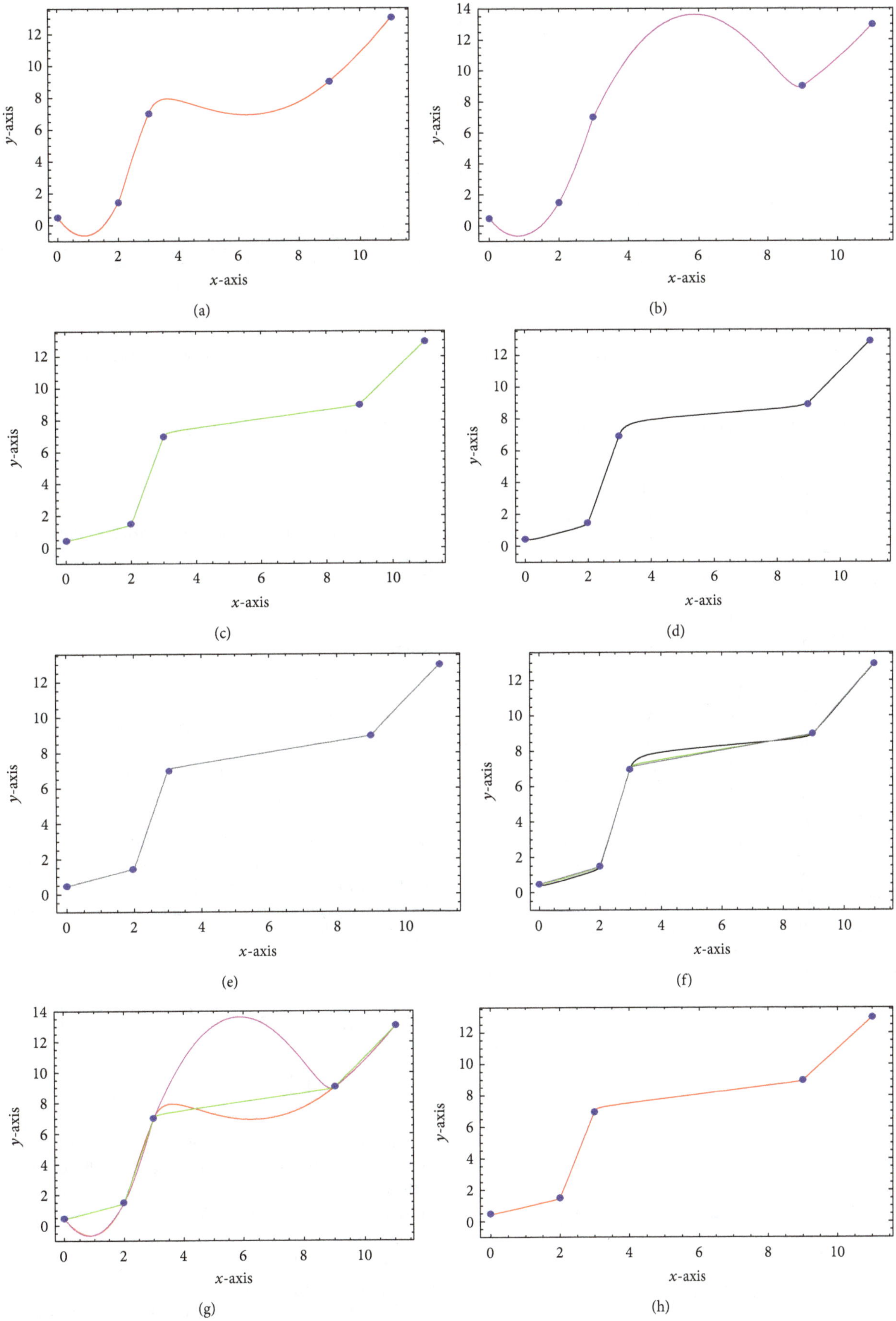

Figure 2: Continued.

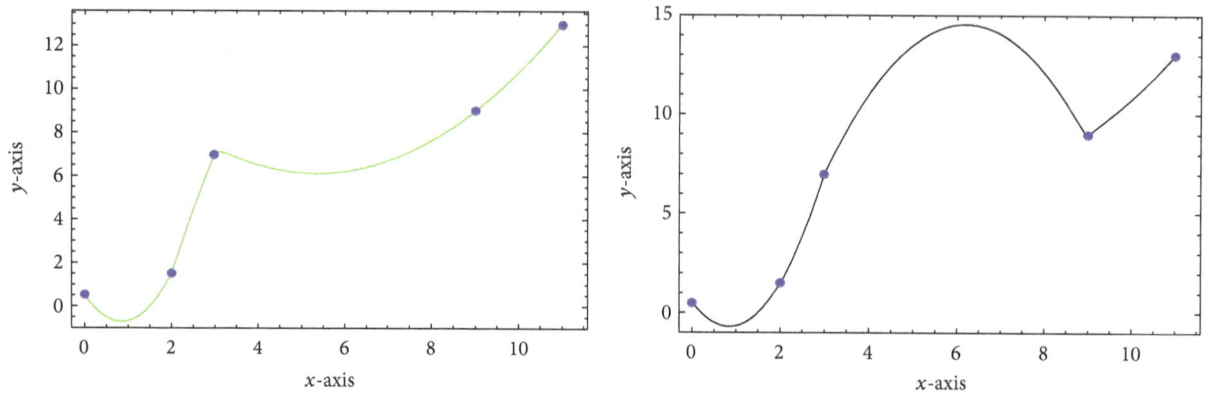

(i) (j)

FIGURE 2: Shape control of rational cubic interpolating curves with various values of shape parameters list in Table 2.

TABLE 2: Shape parameters value for Figure 2.

Figure 2	i	0	1	2	3
	α_i	0.1	0.1	0.1	0.1
Figure 2(a)	β_i	1	1	1	1
	γ_i	1	1	1	1
	α_i	1	1	1	1
Figure 2(b)	β_i	0.1	0.1	0.1	0.1
	γ_i	1	1	1	1
	α_i	1	1	1	1
Figure 2(c)	β_i	1	1	1	1
	γ_i	100	100	100	100
	α_i	5	5	5	5
Figure 2(d)	β_i	5	5	5	5
	γ_i	100	100	100	100
	α_i	5	5	5	5
Figure 2(e)	β_i	5	5	5	5
	γ_i	1000	1000	1000	1000
	α_i	0.01	0.01	0.01	0.01
Figure 2(h)	β_i	0.01	0.01	0.01	0.01
	γ_i	1	1	1	1
	α_i	0.01	0.01	0.01	0.01
Figure 2(i)	β_i	1	1	1	1
	γ_i	1	1	1	1
	α_i	1	1	1	1
Figure 2(j)	β_i	0.01	0.01	0.01	0.01
	γ_i	1	1	1	1

where

$$\begin{aligned}B_i &= \alpha_i h_i d_i + \beta_i h_i d_{i+1} + 2\alpha_i\beta_i f_i + \gamma_i f_i - 2\alpha_i\beta_i f_{i+1} \\ &\quad - \gamma_i f_{i+1}, \\ C_i &= -2\alpha_i h_i d_i - \beta_i h_i d_{i+1} + \alpha_i f_i - 4\alpha_i\beta_i f_i - 2\gamma_i f_i \\ &\quad + \beta_i f_{i+1} + 2\alpha_i\beta_i f_{i+1} + \gamma_i f_{i+1}, \\ D_i &= \alpha_i h_i d_i - 2\alpha_i f_i + 2\alpha_i\beta_i f_i + \gamma_i f_i, \\ E_i &= \alpha_i f_i. \end{aligned} \tag{20}$$

From Schmidt and Hess [30], with variable substitution $\theta = s/(1+s)$, $s \geq 0$, $P_i(\theta)$, $i = 0, 1, \ldots, n-1$, can be rewritten as follows:

$$P_i(s) = As^3 + Bs^2 + Cs + D, \tag{21}$$

where

$$\begin{aligned} A &= \beta_i f_{i+1}, \\ B &= \left(2\alpha_i\beta_i + \beta_i + \gamma_i\right) f_{i+1} - \beta_i h_i d_{i+1}, \\ C &= \left(2\alpha_i\beta_i + \alpha_i + \gamma_i\right) f_i - \alpha_i h_i d_i, \\ D &= \alpha_i f_i. \end{aligned} \tag{22}$$

Theorem 4. *For strictly positive data defined in (18), C^2 rational cubic interpolant defined over the interval $[x_0, x_n]$ is positive if in each subinterval $[x_i, x_{i+1}]$, $i = 0, 1, \ldots, n-1$, the involving parameters α_i, β_i, and γ_i satisfy the following sufficient conditions:*

$$\begin{aligned} &\alpha_i, \beta_i > 0, \\ &\gamma_i > \operatorname{Max}\left\{0, -\alpha_i\left[\frac{h_i d_i + (2\beta_i + 1) f_i}{f_i}\right], \right. \\ &\left. \beta_i\left[\frac{h_i d_{i+1} - (2\alpha_i + 1) f_{i+1}}{f_{i+1}}\right]\right\}. \end{aligned} \tag{23}$$

Remark 5. The sufficient condition for positivity preserving by using C^2 rational cubic interpolant is different from the sufficient condition for positivity preserving by using C^1 rational cubic interpolant of Karim and Kong [18]. To achieve C^2 continuity, the derivative parameters d_i, $i = 1, 2, \ldots, n-1$, must be calculated from C^2 condition given in (7); meanwhile to achieve C^1 continuity, the derivative parameters d_i, $i = 0, 1, 2, \ldots, n$, are estimated by using standard approximation methods such as Arithmetic Mean Method (AMM).

TABLE 3: A positive data from work by Hussain et al. [10].

i	0	1	2	3
x_i	0.0	1.0	1.70	1.80
f_i	0.25	1.0	11.10	25

TABLE 4: Numerical results.

i	0	1	2	3
$d_i\ (C^1)$	−7.296	8.796	123.429	154.570
Δ_i	0.75	14.429	139	
α_i	0.5	0.5	0.5	
β_i	0.5	0.5	0.5	
γ_i	13.84	3.14	0.25	
$d_i\ (C^2)$	−7.296	2.108	82.5421	154.570

Remark 6. The sufficient condition in (23) can be rewritten as

$$\alpha_i, \beta_i > 0,$$
$$\gamma_i = \lambda_i + \text{Max}\left\{0, -\alpha_i\left[\frac{h_i d_i + (2\beta_i + 1) f_i}{f_i}\right], \beta_i\left[\frac{h_i d_{i+1} - (2\alpha_i + 1) f_{i+1}}{f_{i+1}}\right]\right\}, \quad \lambda_i > 0. \tag{24}$$

The following is an algorithm that can be used to generate C^2 positivity-preserving curves.

Algorithm 7 (C^2 positivity preserving). Consider the following:

(1) Input the data points $\{x_i, f_i\}_{i=0}^{n}$, d_0, and d_n.

(2) For $i = 0, 1, \ldots, n-1$, calculate the parameter γ_i using (24) with suitable choices of $\alpha_i > 0$, $\beta_i > 0$, and $\lambda_i > 0$.

(3) For $i = 1, 2, \ldots, n-1$, calculate the value of derivative, d_i, by solving (7) by using *LU* decomposition and so forth.

(4) For $i = 0, 1, \ldots, n-1$, construct the piecewise positive C^2 rational interpolating curves defined by (10).

3.1. Numerical Demonstrations. In this section several numerical results for positivity preserving by using C^2 rational interpolating curves will be shown including comparison with existing rational cubic spline schemes. Two sets of positive data taken from works by Hussain et al. [10] and Sarfraz et al. [28] were used.

Figures 3 and 4 show the positivity preserving by using C^2 rational cubic spline for data in Tables 3 and 5, respectively. Figures 3(a) and 4(a) show the default cubic Hermite spline polynomial for data in Tables 3 and 5, respectively. Numerical results of Figure 3(d) were obtained by using the parameters from the data in Table 4. Meanwhile numerical results of

TABLE 5: A Positive data from work by Sarfraz et al. [28].

i	0	1	2	3	4	5	6
x_i	2	3	7	8	9	13	14
f_i	10	2	3	7	2	3	10

TABLE 6: Numerical results.

i	0	1	2	3	4	5	6
$d_i\ (C^1)$	−9.65	−6.35	3.25	0	−3.95	5.65	8.35
Δ_i	−8	2.25	4.0	−0.5	0.25	7	
α_i	2.5	2.5	2.5	2.5	2.5	2.5	
β_i	2.5	2.5	2.5	2.5	2.5	2.5	
γ_i	0.1	16.85	0.1	0.1	4.85	0.1	
$d_i\ (C^2)$	−9.65	−4.86	3.34	−0.48	−4.057	5.25	8.35

Figure 4(e) were generated by using the data in Table 6. It can be seen clearly that the positivity preserving by using our C^2 rational cubic spline gives more smooth results compared with the works of Hussain et al. [10] and Abbas et al. [12]. The graphical results in Figure 3(e) were very smooth and visually pleasing. Meanwhile for the positive data given in Table 5, our C^2 rational cubic spline gives comparable results with the works of Abbas et al. [12], Hussain et al. [10], and Sarfraz et al. [11]. The final resulting positive curves by using the proposed C^2 rational cubic spline interpolant are slightly different between the works of Abbas et al. [12], Hussain et al. [10], and Sarfraz et al. [11]. Finally Figure 5 shows the examples of positive interpolating by using Fritsch and Carlson [1] cubic spline schemes that are well documented in Matlab as PCHIP. It can be seen clearly that shape preserving by using PCHIP does not give smooth results and is not visually pleasing enough. Some of the interpolating curves tend to overshot on some interval that the interpolating curves are tight when compared with our work in this paper. For instance, in Figure 4(b), the interpolating curves are very tight and not visually pleasing.

Error Analysis. In this section, the error analysis for the function to be interpolated is $f(t) \in C^3[t_0, t_n]$ using our C^2 rational cubic spline which will be discussed in detail. Note that the constructed rational cubic spline with three parameters is a local interpolant and without loss of generality, we may just consider the error on the subinterval $I_i = [t_i, t_{i+1}]$. By using Peano Kernel Theorem [31] the error of interpolation in each subinterval $I_i = [t_i, t_{i+1}]$ is defined as

$$R[f] = f(t) - P_i(t) = \frac{1}{2}\int_{t_i}^{t_{i+1}} f^{(3)} R_t\left[(t-\tau)_+^2\right] d\tau, \tag{25}$$

where

$$R_t\left[(t-\tau)_+^2\right] = \begin{cases} r(\tau, t), & t_i < \tau < t, \\ s(\tau, t), & t < \tau < t_{i+1} \end{cases} \tag{26}$$

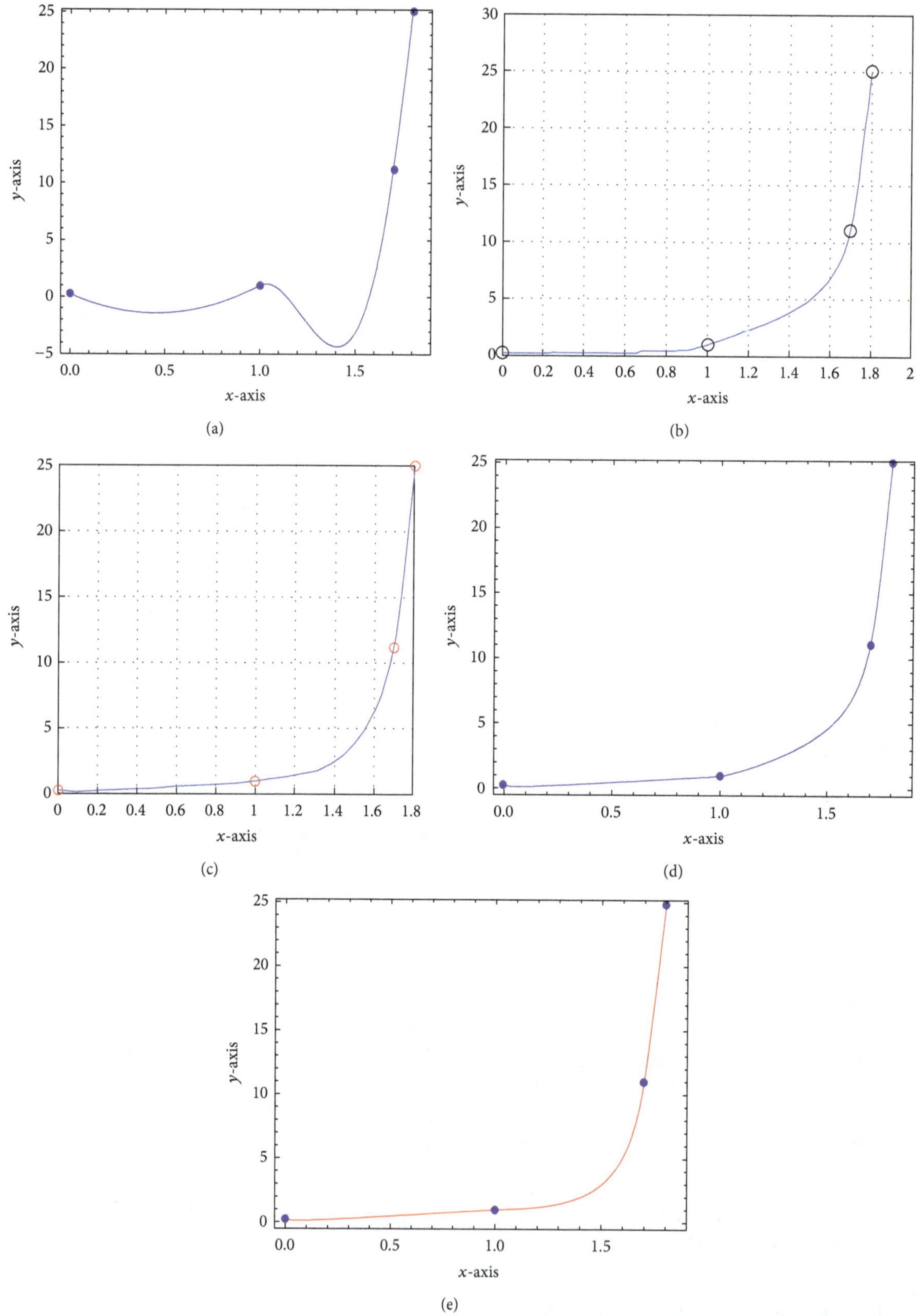

FIGURE 3: Comparison of (a) cubic Hermite spline curve, (b) Hussain et al. [10], (c) Abbas et al. [12], and positivity-preserving rational cubic spline with values of shape parameters: (d) $\alpha_i = \beta_i = 0.5$; (e) $\alpha_i = \beta_i = 2$.

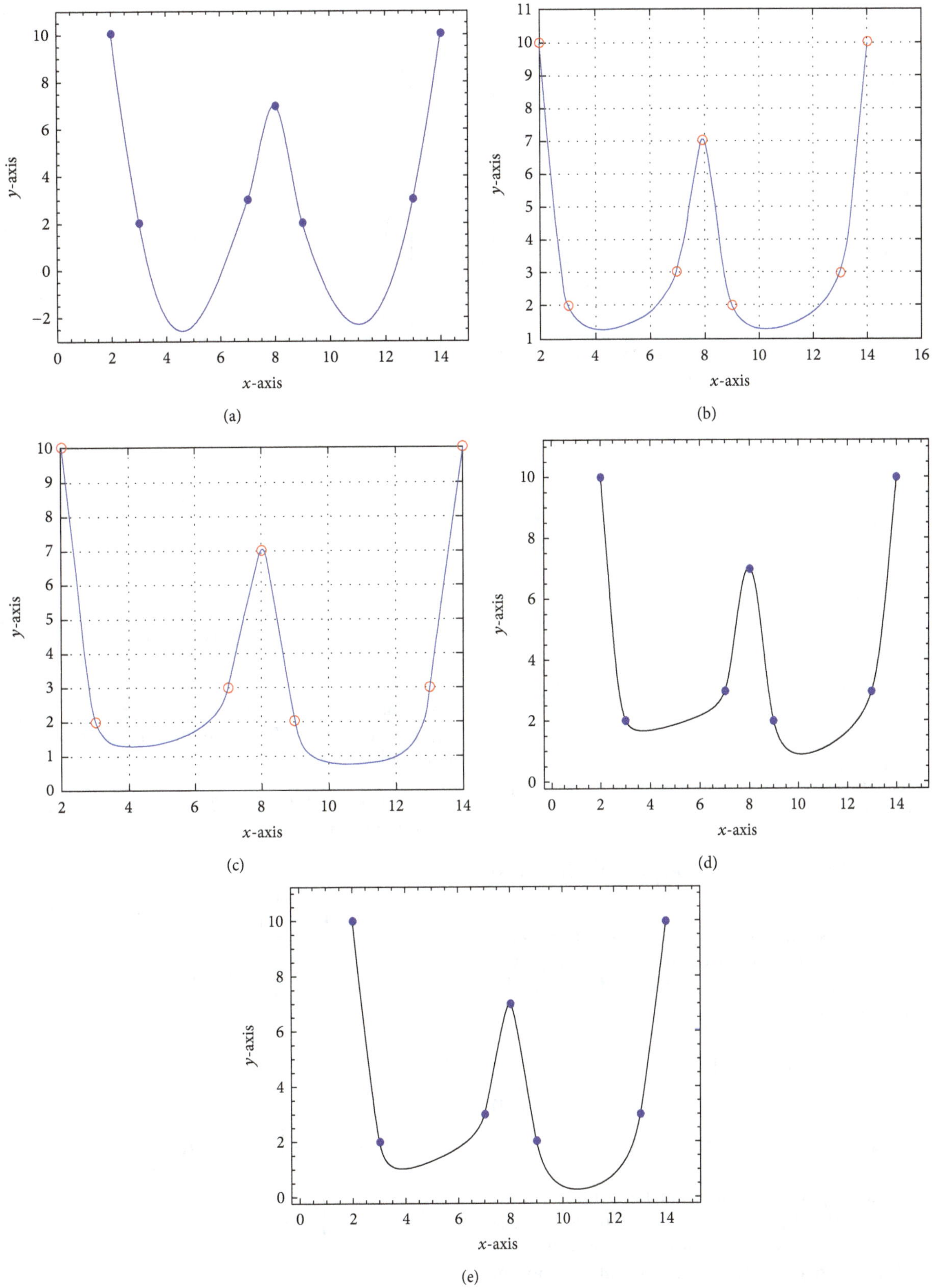

FIGURE 4: Comparison of (a) cubic Hermite spline curve, (b) Sarfraz et al. [11], (c) Abbas et al. [12], and positivity-preserving rational cubic spline with values of shape parameters: (d) $\alpha_i = \beta_i = 0.5$; (e) $\alpha_i = \beta_i = 2.5$.

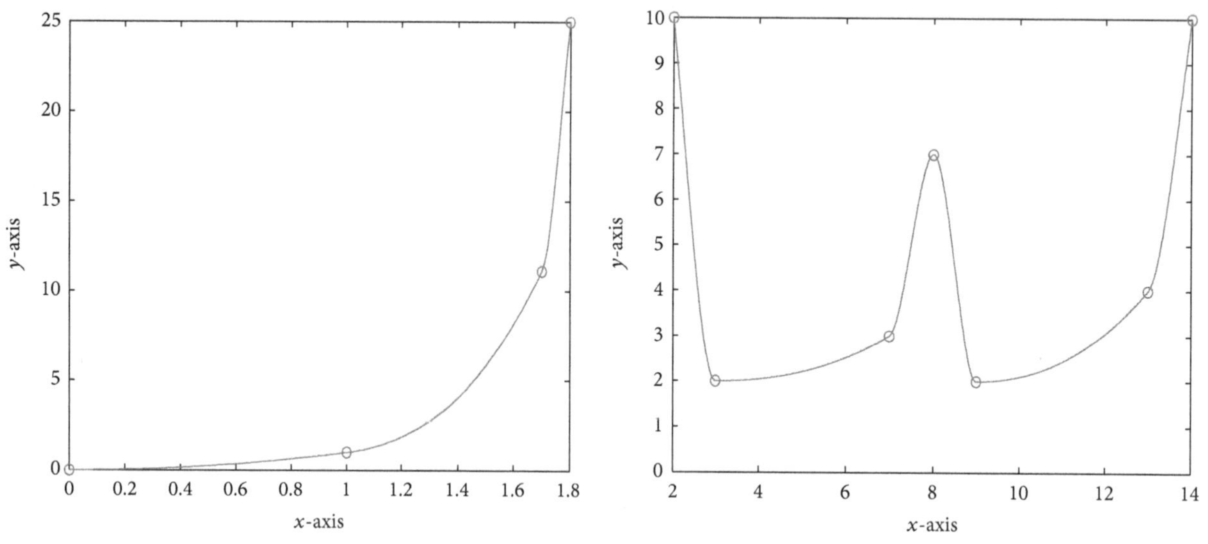

(a) (b)

FIGURE 5

with

$$r(\tau,t) = (t-\tau)^2 + s(\tau,t),$$
$$s(\tau,t) = -\frac{\left[\left\{(2\alpha_i\beta_i + \beta_i + \gamma_i)\zeta^2 - 2h_i\beta_i\zeta\right\}\theta^2(1-\theta) + \beta_i\zeta^2\theta^3\right]}{Q_i(\theta)} \tag{27}$$

with $\zeta = (t_{i+1} - \tau)$.

The absolute error in each subinterval $I_i = [x_i, x_{i+1}]$ is given as follows:

$$|f(t) - P_i(t)| \le \frac{1}{2}\left\|f^{(3)}(\tau)\right\| \int_{t_i}^{t_{i+1}} R_t\left[(t-\tau)_+^2\right] d\tau. \tag{28}$$

In order to enable us to derive the error analysis of C^2 rational cubic spline interpolation, we need to study the properties of the kernel functions $r(\tau,t)$ and $s(\tau,t)$ and evaluate the following definite integrals:

$$\int_{t_i}^{t_{i+1}} R_t\left[(t-\tau)_+^2\right] d\tau = \int_{t_i}^{t} |r(\tau,x)|\, d\tau + \int_{t}^{t_{i+1}} |s(\tau,x)|\, d\tau. \tag{29}$$

To simplify the integrals in (29) we begin by finding the roots of $r(t,t) = 0$, $r(\tau,t) = 0$, and $s(\tau,t) = 0$, respectively. It is easy to see that the roots of $r(t,t) = 0$ in $[0,1]$ are $\theta = 0, 1$ and $\theta^* = 1 - \beta_i/\rho_i$, where $\rho_i = 2\alpha_i\beta_i + \gamma_i$. Meanwhile the roots of $r(\tau,t) = 0$ are

$$\tau_k = x - \frac{\theta h_i\left(\theta\rho_i + (-1)^{k+1} H\right)}{\alpha_i + \theta\rho_i}, \quad k = 1, 2, \tag{30}$$

where

$$H = \sqrt{(\rho_i - \beta_i)(\alpha_i + \rho_i\theta) - \alpha_i\rho_i\theta}. \tag{31}$$

The roots of $s(\tau,t) = 0$ are $\tau_3 = t_{i+1}$, $\tau_4 = t_{i+1} - 2\beta_i h_i(1-\theta)/(\beta_i + (1-\theta)\rho_i)$. Thus the following three cases can be obtained.

Case 1. For $0 \le \beta_i/\rho_i \le 1$, $0 < \theta < \theta^*$, then (28) takes the following form:

$$|f(t) - P_i(t)| \le \frac{1}{2}\left\|f^{(3)}(\tau)\right\| \int_{t_i}^{t_{i+1}} R_t\left[(t-\tau)_+^2\right] d\tau = \frac{1}{2}\left\|f^{(3)}(\tau)\right\| G_i(\tau) \tag{32}$$

with

$$G_i(\tau) = \left\{\int_{t_i}^{t} -r(\tau,t)\, d\tau + \int_{t}^{\tau_4} -s(\tau,t)\, d\tau + \int_{\tau_4}^{t_{i+1}} s(\tau,t)\, d\tau\right\}. \tag{33}$$

Hence

$$|f(t) - P_i(t)| \le \left\|f^{(3)}(\tau)\right\| \zeta_1(\alpha_i, \beta_i, \gamma_i, \theta), \tag{34}$$

where

$$\zeta_1(\alpha_i,\beta_i,\gamma_i,\theta) = -\frac{h_i^3\theta^3}{3} + \frac{\beta_i h_i^3\theta^3}{3Q_i(\theta)} - \frac{2(\rho_i+\beta_i)\left(8h_i^3\theta^2(1-\theta)^4\right)}{3(\rho_i(1-\theta)+\beta_i)^3 Q_i(\theta)} + \frac{8\beta_i^2\left(h_i^3\theta^2(1-\theta)^2\right)}{3(\rho_i(1-\theta)+\beta_i)^2 Q_i(\theta)} - \frac{16\beta_i h_i^3\theta^3(1-\theta)^3}{3(\rho_i(1-\theta)+\beta_i)^3 Q_i(\theta)} + \frac{h_i^3(1-\theta)\theta^2}{3Q_i(\theta)}[\rho_i - 2\beta_i]. \quad (35)$$

Case 2. For $0 \le \beta_i/\rho_i \le 1$, $\theta^* < \theta < 1$, then (28) takes the following form:

$$|f(t) - P_i(t)| \le \frac{1}{2}\left\|f^{(3)}(\tau)\right\| \int_{t_i}^{t_{i+1}} R_t\left[(t-\tau)_+^2\right] d\tau = \frac{1}{2}\left\|f^{(3)}(\tau)\right\| G_2(\tau) \quad (36)$$

with

$$G_2(\tau) = \left\{\int_{t_i}^{\tau_2} -r(\tau,t)\,d\tau + \int_{\tau_2}^{t} r(\tau,t)\,d\tau + \int_{t}^{t_{i+1}} s(\tau,t)\,d\tau\right\}. \quad (37)$$

Hence

$$|f(t) - P_i(t)| \le \left\|f^{(3)}(\tau)\right\| \zeta_2(\alpha_i,\beta_i,\gamma_i,\theta), \quad (38)$$

where

$$\zeta_2(\alpha_i,\beta_i,\gamma_i,\theta) = \frac{2h_i^3\theta^3(\theta\rho_i - H)^3}{3} - \frac{2\beta_i h_i^3\theta^3[G_3(\theta)]^3}{3Q_i(\theta)} - \frac{2(\rho_i+\beta_i)h_i^3(1-\theta)\theta^2[G_3(\theta)]^3}{3Q_i(\theta)} + \frac{2\beta_i h_i^3(1-\theta)\theta^2[G_3(\theta)]^3}{Q_i(\theta)} \quad (39)$$

with $G_3(\theta) = (1-\theta) + \theta(\theta\rho_i - H)/(\alpha_i + \rho_i\theta)$.

Case 3. For $\beta_i/\rho_i > 1$, $0 < \theta < 1$, then (28) takes the following form:

$$|f(t) - P_i(t)| \le \frac{1}{2}\left\|f^{(3)}(\tau)\right\| \int_{t_i}^{t_{i+1}} R_t\left[(t-\tau)_+^2\right] d\tau = \frac{1}{2}\left\|f^{(3)}(\tau)\right\| \left\{\int_{t_i}^{t} r(\tau,t)\,d\tau + \int_{t}^{t_{i+1}} s(\tau,t)\,d\tau\right\} = \left\|f^{(3)}(\tau)\right\| \zeta_3(\alpha_i,\beta_i,\gamma_i,\theta), \quad (40)$$

where

$$\zeta_3(\alpha_i,\beta_i,\gamma_i,\theta) = \frac{h_i^3\theta^3}{3} - \frac{(\rho_i+\beta_i)h_i^3(1-\theta)\theta^2}{3Q_i(\theta)} + \frac{\beta_i h_i^3(1-\theta)\theta^2}{Q_i(\theta)} - \frac{\beta_i h_i^3\theta^3}{3Q_i(\theta)}. \quad (41)$$

Theorem 8. *The error for the interpolating rational cubic spline interpolant defined by (1) in each subinterval $I_i = [t_i, t_{i+1}]$, when $f(t) \in C^3[t_0, t_n]$, is given by*

$$|f(t) - P_i(t)| \le \frac{1}{2}\left\|f^{(3)}(\tau)\right\| \int_{t_i}^{t_{i+1}} R_t\left[(t-\tau)_+^2\right] d\tau = \frac{1}{2}\left\|f^{(3)}(\tau)\right\| \left\{\int_{t_i}^{t} r(\tau,t)\,d\tau + \int_{t}^{t_{i+1}} s(\tau,t)\,d\tau\right\} = \left\|f^{(3)}(\tau)\right\| c_i, \quad (42)$$

with

$$c_i = \max \zeta(\alpha_i,\beta_i,\gamma_i,\theta),$$
$$\zeta(\alpha_i,\beta_i,\gamma_i,\theta) = \begin{cases} \zeta_1(\alpha_i,\beta_i,\gamma_i,\theta), & 0 \le \theta \le \theta^* \\ \zeta_2(\alpha_i,\beta_i,\gamma_i,\theta), & \theta^* \le \theta \le 1 \\ \zeta_3(\alpha_i,\beta_i,\gamma_i,\theta), & 0 \le \theta \le 1. \end{cases} \quad (43)$$

Remark 9. When $\alpha_i = 1$, $\beta_i = 1$, and $\gamma_i = 0$, the interpolation function defined by (1) is reduced to the standard cubic Hermite interpolation with

$$\zeta_1(\alpha_i,\beta_i,\gamma_i,\theta) = \frac{4\theta^2(1-\theta)^3}{\left(3(3-2\theta)^2\right)}, \quad 0 \le \theta \le \frac{1}{2},$$
$$\zeta_2(\alpha_i,\beta_i,\gamma_i,\theta) = \frac{4\theta^3(1-\theta)^2}{\left(3(1+2\theta)^2\right)}, \quad \frac{1}{2} \le \theta \le 1, \quad (44)$$
$$\zeta_3(\alpha_i,\beta_i,\gamma_i,\theta) = 0, \quad 0 \le \theta \le 1,$$

and $c_i = 1/96$. This is standard error for cubic Hermite polynomial spline.

4. Discussions

From all numerical results presented in Section 3.1, it can be seen clearly that the proposed C^2 rational cubic spline works very well and it is comparable with existing schemes such as Hussain et al. [10], Sarfraz et al. [11], and Abbas et al. [12] for positivity preserving. Furthermore similar to the construction of rational cubic of Abbas et al. [8] our C^2 rational cubic spline interpolation also has three parameters where two are free parameters. But based on our numerical experiments, it was noticed that schemes of Abbas et al. [12] may not be able to produce completely C^2 positive interpolating curves. Meanwhile the sufficient condition for positivity, monotonicity, and convexity preserving and data constrained are derived on the remaining parameter. One of the advantages by using our C^2 rational cubic spline is that when $\gamma_i = 0$ we may obtain the new variant of C^2 rational cubic spline of Hussain and Ali [32], M. Z. Hussain and M. Hussain [9], and Tian et al. [20] for positivity- and convexity-preserving interpolation. Thus our C^2 rational cubic spline has a larger spline class compared to the works by Abbas et al. [12]. Furthermore the error analysis when the function to be interpolated is $f(t) \in C^3[t_0, t_n]$ also has been derived in detail.

5. Conclusions

In this paper the new C^2 rational cubic spline with three parameters has been introduced. It is an extension to the work of Karim and Pang [16]. To achieve C^2 continuity at the join knots x_i, $i = 1, 2, \ldots, n-1$, the first derivative value, d_i, is calculated by solving systems of linear equation (tridiagonal) that is strictly positive and the solution is unique. Shape control of the new C^2 rational cubic interpolation with numerical examples also was presented. Finally C^2 rational cubic spline has been used for positivity preserving including a comparison with existing schemes. From the numerical results clearly our C^2 rational cubic spline gives comparable results with existing schemes. Finally work in parametric shape preserving is underway by the authors.

Competing Interests

The authors declare that there are no competing interests regarding the publication of this paper.

Acknowledgments

Samsul Ariffin Abdul Karim would like to acknowledge Universiti Teknologi PETRONAS (UTP) for the financial support received in the form of a research grant: Short Term Internal Research Funding (STIRF) no. 0153AA-D91 including the Mathematica Software.

References

[1] F. N. Fritsch and R. E. Carlson, "Monotone piecewise cubic interpolation," *SIAM Journal on Numerical Analysis*, vol. 17, no. 2, pp. 238–246, 1980.

[2] R. L. Dougherty, A. S. Edelman, and J. M. Hyman, "Nonnegativity-, monotonicity-, or convexity-preserving cubic and quintic Hermite interpolation," *Mathematics of Computation*, vol. 52, no. 186, pp. 471–494, 1989.

[3] S. Butt and K. W. Brodlie, "Preserving positivity using piecewise cubic interpolation," *Computers and Graphics*, vol. 17, no. 1, pp. 55–64, 1993.

[4] K. W. Brodlie and S. Butt, "Preserving convexity using piecewise cubic interpolation," *Computers and Graphics*, vol. 15, no. 1, pp. 15–23, 1991.

[5] M. Sarfraz, "Visualization of positive and convex data by a rational cubic spline interpolation," *Information Sciences*, vol. 146, no. 1–4, pp. 239–254, 2002.

[6] M. Sarfraz, S. Butt, and M. Z. Hussain, "Visualization of shaped data by a rational cubic spline interpolation," *Computers & Graphics*, vol. 25, no. 5, pp. 833–845, 2001.

[7] M. Sarfraz, M. Z. Hussain, and A. Nisar, "Positive data modeling using spline function," *Applied Mathematics and Computation*, vol. 216, no. 7, pp. 2036–2049, 2010.

[8] M. Abbas, *Shape preserving data visualization for curves and surfaces using rational cubic functions [Ph.D. thesis]*, School of Mathematical Sciences, Universiti Sains Malaysia, Penang, Malaysia, 2012.

[9] M. Z. Hussain and M. Hussain, "Visualization of data subject to positive constraint," *Journal of Information and Computing Sciences*, vol. 1, no. 3, pp. 149–160, 2006.

[10] M. Z. Hussain, M. Sarfraz, and T. S. Shaikh, "Shape preserving rational cubic spline for positive and convex data," *Egyptian Informatics Journal*, vol. 12, no. 3, pp. 231–236, 2011.

[11] M. Sarfraz, M. Z. Hussain, T. S. Shaikh, and R. Iqbal, "Data visualization using shape preserving C2 rational spline," in *Proceeding of 5th International Conference on Information Visualisation*, pp. 528–533, London, UK, July 2011.

[12] M. Abbas, A. A. Majid, M. N. H. Awang, and J. M. Ali, "Positivity-preserving C2 rational cubic spline interpolation," *ScienceAsia*, vol. 39, no. 2, pp. 208–213, 2013.

[13] R. Delbourgo, "Accurate C^2 rational interpolants in tension," *SIAM Journal on Numerical Analysis*, vol. 30, no. 2, pp. 595–607, 1993.

[14] J. A. Gregory, "Shape preserving spline interpolation," *Computer-Aided Design*, vol. 18, no. 1, pp. 53–57, 1986.

[15] R. Delbourgo and J. A. Gregory, "C^2 rational quadratic spline interpolation to monotonic data," *IMA Journal of Numerical Analysis*, vol. 3, no. 2, pp. 141–152, 1983.

[16] S. A. A. Karim and K. V. Pang, "Local control of the curves using rational cubic spline," *Journal of Applied Mathematics*, vol. 2014, Article ID 872637, 12 pages, 2014.

[17] S. A. A. Karim and K. V. Pang, "Monotonicity-preserving using rational cubic spline interpolation," *Research Journal of Applied Sciences*, vol. 9, no. 4, pp. 214–223, 2014.

[18] S. A. A. Karim and V. P. Kong, "Shape preserving interpolation using rational cubic spline," *Research Journal of Applied Sciences, Engineering and Technology*, vol. 8, no. 2, pp. 167–168, 2014.

[19] S. A. A. Karim and V. P. Kong, "Convexity-preserving using rational cubic spline interpolation," *Research Journal of Applied Sciences, Engineering and Technology*, vol. 8, no. 3, pp. 312–320, 2014.

[20] M. Tian, Y. Zhang, J. Zhu, and Q. Duan, "Convexity-preserving piecewise rational cubic interpolation," *Journal of Information and Computational Science*, vol. 2, no. 4, pp. 799–803, 2005.

[21] P. Lamberti and C. Manni, "Shape-preserving C^2 functional interpolation via parametric cubics," *Numerical Algorithms*, vol. 28, no. 1–4, pp. 229–254, 2001.

[22] Q. Duan, L. Wang, and E. H. Twizell, "A new C^2 rational interpolation based on function values and constrained control of the interpolant curves," *Applied Mathematics and Computation*, vol. 161, no. 1, pp. 311–322, 2005.

[23] F. Bao, Q. Sun, and Q. Duan, "Point control of the interpolating curve with a rational cubic spline," *Journal of Visual Communication and Image Representation*, vol. 20, no. 4, pp. 275–280, 2009.

[24] J.-C. Fiorot and J. Tabka, "Shape-preserving C^2 cubic polynomial interpolating splines," *Mathematics of Computation*, vol. 57, no. 195, pp. 291–298, 1991.

[25] M. Dube and P. Tiwari, "Convexity preserving C^2 rational quadratic trigonometric spline," in *Proceedings of the International Conference of Numerical Analysis and Applied Mathematics (ICNAAM '12)*, vol. 1479, pp. 995–998, September 2012.

[26] Y.-J. Pan and G.-J. Wang, "Convexity-preserving interpolation of trigonometric polynomial curves with a shape parameter," *Journal of Zhejiang University SCIENCE A*, vol. 8, no. 8, pp. 1199–1209, 2007.

[27] F. Ibraheem, M. Hussain, M. Z. Hussain, and A. A. Bhatti, "Positive data visualization using trigonometric function," *Journal of Applied Mathematics*, vol. 2012, Article ID 247120, 19 pages, 2012.

[28] M. Sarfraz, M. Z. Hussain, and F. S. Chaudary, "Shape preserving cubic spline for data visualization," *Computer Graphics and CAD/CAM*, vol. 1, pp. 185–193, 2005.

[29] R. Delbourgo and J. A. Gregory, "The determination of derivative parameters for a monotonic rational quadratic interpolant," *IMA Journal of Numerical Analysis*, vol. 5, no. 4, pp. 397–406, 1985.

[30] J. W. Schmidt and W. Hess, "Positivity of cubic polynomials on intervals and positive spline interpolation," *BIT Numerical Mathematics*, vol. 28, no. 2, pp. 340–352, 1988.

[31] M. H. Schultz, *Spline Analysis*, Prentice-Hall, Englewood-Cliffs, NJ, USA, 1973.

[32] M. Z. Hussain and J. M. Ali, "Positivity preserving piecewise rational cubic interpolation," *Matematika*, vol. 22, no. 2, pp. 147–153, 2006.

11

Numerical Solution of Second-Order Fredholm Integrodifferential Equations with Boundary Conditions by Quadrature-Difference Method

Chriscella Jalius and Zanariah Abdul Majid

Institute for Mathematical Research, Universiti Putra Malaysia (UPM), 43400 Serdang, Selangor, Malaysia

Correspondence should be addressed to Zanariah Abdul Majid; zana_majid99@yahoo.com

Academic Editor: Mehmet Sezer

In this research, the quadrature-difference method with Gauss Elimination (GE) method is applied for solving the second-order of linear Fredholm integrodifferential equations (LFIDEs). In order to derive an approximation equation, the combinations of Composite Simpson's 1/3 rule and second-order finite-difference method are used to discretize the second-order of LFIDEs. This approximation equation will be used to generate a system of linear algebraic equations and will be solved by using Gauss Elimination. In addition, the formulation and the implementation of the quadrature-difference method are explained in detail. Finally, some numerical experiments were carried out to examine the accuracy of the proposed method.

1. Introduction

Integrodifferential equation is an equation where the unknown function appears under the sign of integration and it contains the derivatives of the unknown function. In this study, we will focus on second-order linear Fredholm integrodifferential equations (LFIDEs) and they can be defined as follows:

$$y''(x) = q(x)\,y'(x) + p(x)\,y(x) + g(x) + \lambda \int_a^b k(x,t)\,y(t)\,dt \tag{1}$$

with Dirichlet boundary conditions

$$y(a) = A_1, \quad y(b) = B_1, \tag{2}$$

where the functions $q(x)$, $p(x)$, and $g(x)$ and kernel $k(x,t)$ are known, λ is a real parameter, and a, b are constant, while $y(x)$ is an unknown function to be determined.

In this study, we have discovered four types of problems in second-order LFIDEs such as the following.

Type 1. When $p(x) = 0$ and $q(x) = 0$

$$y''(x) = g(x) + \lambda \int_a^b k(x,t)\,y(t)\,dt. \tag{3}$$

Type 2. When $p(x) \neq 0$ and $q(x) = 0$

$$y''(x) = p(x)\,y(x) + g(x) + \lambda \int_a^b k(x,t)\,y(t)\,dt. \tag{4}$$

Type 3. When $p(x) = 0$ and $q(x) \neq 0$

$$y''(x) = q(x)\,y'(x) + g(x) + \lambda \int_a^b k(x,t)\,y(t)\,dt. \tag{5}$$

Type 4. When $p(x) \neq 0$ and $q(x) \neq 0$

$$y''(x) = q(x)\,y'(x) + p(x)\,y(x) + g(x) + \lambda \int_a^b k(x,t)\,y(t)\,dt. \tag{6}$$

Nowadays, there are many researchers who study in the field of integrodifferential equations since it has emerged in many scientific and physical engineering applications.

Numerical approaches were practiced in engineering field in order to obtain an approximation solution for problem (1). Therefore, many authors have studied the numerical solutions of Fredholm integrodifferential equations. Several numerical methods were used such as Adomian's Decomposition method [1], Generalised minimal residual method (GMRES) [2], CAS Wavelet method [3], compact finite difference [4], Lagrange interpolation method [5], Spline Collocation [6], differential transformation method [7], variational iteration method [8], Tau method [9], Legendre collocation matrix method [10], and quadrature-difference method [11, 12]. LFIDEs are usually difficult to solve analytically. Therefore, in this study numerical approaches are practiced in order to obtain an approximation solution for problem (1).

In this paper, we will focus on Types 1 and 2 only. Therefore, we will introduce a new combination of quadrature-difference method, namely, Composite Simpson's 1/3 rule with second-order finite difference (CSFD2), and it will be implemented to discretize Types 1 and 2 to generate a system of linear equations. In addition, Gauss Elimination (GE) will be used for solving the linear algebraic equations produced by the discretization of the second-order LFIDEs.

2. Formulation of Approximation Equations

In this section, quadrature and second-order finite-difference methods are applied to discretize the second-order of LFIDEs to construct approximation equations.

2.1. Derivation of the Quadrature Method. In order to form the system of linear equations, quadrature method is applied to approximate the integral parts in second-order of LFIDEs. Therefore, Composite Simpson's 1/3 rule will be used in this study and it can be expressed as follows:

$$\int_a^b y(x)\,dx = \frac{h}{3}\left[y(a) + 4\sum_{j=1}^{n/2} y(x_{2j-1}) + 2\sum_{j=1}^{(n/2)-1} y(x_{2j}) + y(b)\right] - \frac{b-a}{180}h^4 y^{(4)}(\mu), \quad (7)$$

where $x_j = a + jh$ $(j = 0, 1, 2, \ldots, n)$ are the abscissas of the partition points of the integration in interval $[a, b]$. $((b-a)/180)h^4 y^{(4)}(\mu)$ is the error term of (7) where $a \le \mu \le b$. In the next section, the formulation of the approximation equations for problem LFIDEs will be further discussed.

The constant step size, h in (7), can be defined as

$$h = \frac{b-a}{n}, \quad (8)$$

where n is the number of subintervals in the interval $[a, b]$.

Therefore, if Composite Simpson's 1/3 rule with n subintervals is applied in (3) and (4), then the integration parts in second-order of LFIDEs is approximated by

$$\begin{aligned}\int_a^b k(x,t)\,y(t)\,dt \approx{}& \frac{h}{3}\left[k(x,t_0)\,y(t_0)\right.\\ &+ 4k(x,t_1)\,y(t_1) + 2k(x,t_2)\,y(t_2)\\ &+ 4k(x,t_3)\,y(t_3) + 2k(x,t_4)\,y(t_4) + \cdots\\ &+ 4k(x,t_{n-3})\,y(t_{n-3}) + 2k(x,t_{n-2})\,y(t_{n-2})\\ &\left.+ 4k(x,t_{n-1})\,y(t_{n-1}) + k(x,t_n)\,y(t_n)\right].\end{aligned} \quad (9)$$

2.2. Derivation of the Finite-Difference Method. In solving second-order LFIDEs, the differential part will be approximated by using second-order derivative of finite-difference method which is fourth-order accuracy. Therefore, finite-difference method can be derived as follows.

Forward difference is

$$\begin{aligned}y''(x_i) ={}& \frac{1}{12h^2}\left[35y(x_i) - 104y(x_{i+1}) + 114y(x_{i+2})\right.\\ &\left.- 56y(x_{i+3}) + 11y(x_{i+4})\right] + O(h^4)\end{aligned} \quad (10)$$

for $i = 1$.

Central difference is

$$\begin{aligned}y''(x_i) ={}& \frac{1}{12h^2}\left[-y(x_{i-2}) + 16y(x_{i-1}) - 30y(x_i)\right.\\ &\left.+ 16y(x_{i+1}) - y(x_{i+2})\right] + O(h^4)\end{aligned} \quad (11)$$

for $i = 2, \ldots, n-2$.

Backward difference is

$$\begin{aligned}y''(x_i) ={}& \frac{1}{12h^2}\left[35y(x_i) - 104y(x_{i-1}) + 114y(x_{i-2})\right.\\ &\left.- 56y(x_{i-3}) + 11y(x_{i-4})\right] + O(h^4)\end{aligned} \quad (12)$$

for $i = n-1$, where h is the constants step size as defined in (8). Therefore, the differential term in second-order LFIDEs will be approximated by using equations (10), (11), and (12).

3. Formulation of Composite Simpson's 1/3 and Second-Order Finite-Difference Method

The formulation of Composite Simpson's 1/3 rule and second-order finite-difference method (CSFD2) for second-order LFIDEs is elaborated in this section. In order to generate the system of linear algebraic equations, equation in (9), (10), (11), and (12) will be substituted into (4).

3.1. Formulation of CSFD2 for Types 1 and 2. A system of linear algebraic equations will be generated in this section. The difference between Types 1 and 2 is only in the function of $p(x)y(x)$ which is $p(x) \neq 0$ in Type 2, while in Type 1 $p(x) = 0$. Therefore, substitute (9) into (4) and we will get

$$\begin{aligned}y''(x_i) ={}& p(x_i)\,y(x_i) + g(x_i) + \frac{\lambda h}{3}\left[k(x_i,t_0)\,y(t_0)\right.\\ &+ 4k(x_i,t_1)\,y(t_1) + 2k(x_i,t_2)\,y(t_2)\\ &+ 4k(x_i,t_3)\,y(t_3) + 2k(x_i,t_4)\,y(t_4) + \cdots\\ &+ 4k(x_i,t_{n-3})\,y(t_{n-3}) + 2k(x_i,t_{n-2})\,y(t_{n-2})\\ &\left.+ 4k(x_i,t_{n-1})\,y(t_{n-1}) + k(x_i,t_n)\,y(t_n)\right].\end{aligned} \quad (13)$$

Accordingly, for $y''(x_i) = y''(t_i) = y_i''$ and $k(x_i, t_j) = k_{i,j}$, the relation in (13) becomes

$$
\begin{aligned}
y_i'' = p_i y_i + g_i + \frac{\lambda h}{3} [k_{i,0} y_0 + 4k_{i,1} y_1 + 2k_{i,2} y_2 \\
+ 4k_{i,3} y_3 + 2k_{i,4} y_4 + \cdots + 4k_{i,n-3} y_{n-3} + 2k_{i,n-2} y_{n-2} \\
+ 4k_{i,n-1} y_{n-1} + k_{i,n} y_n]
\end{aligned} \tag{14}
$$

where $i = 1, 2, \ldots, n-1$.

Next, we substitute (10), (11), and (12) into (14); then we will have

$$
\begin{aligned}
\frac{35y_i - 104y_{i+1} + 114y_{i+2} - 56y_{i+3} + 11y_{i+4}}{12h^2} = p_i y_i \\
+ g_i + \frac{\lambda h}{3} [k_{i,0} y_0 + 4k_{i,1} y_1 + 2k_{i,2} y_2 + 4k_{i,3} y_3 + \cdots \\
+ 4k_{i,n-3} y_{n-3} + 2k_{i,n-2} y_{n-2} + 4k_{i,n-1} y_{n-1} + k_{i,n} y_n]
\end{aligned} \tag{15}
$$

for $i = 1$.

$$
\begin{aligned}
\frac{-y_{i-2} + 16y_{i-1} - 30y_i + 16y_{i+1} - y_{i+2}}{12h^2} = p_i y_i + g_i \\
+ \frac{\lambda h}{3} [k_{i,0} y_0 + 4k_{i,1} y_1 + 2k_{i,2} y_2 + 4k_{i,3} y_3 + \cdots \\
+ 4k_{i,n-3} y_{n-3} + 2k_{i,n-2} y_{n-2} + 4k_{i,n-1} y_{n-1} + k_{i,n} y_n]
\end{aligned} \tag{16}
$$

for $i = 2, \ldots, n-2$.

$$
\begin{aligned}
\frac{35y_i - 104y_{i-1} + 114y_{i-2} - 56y_{i-3} + 11y_{i-4}}{12h^2} = p_i y_i \\
+ g_i + \frac{\lambda h}{3} [k_{i,0} y_0 + 4k_{i,1} y_1 + 2k_{i,2} y_2 + 4k_{i,3} y_3 + \cdots \\
+ 4k_{i,n-3} y_{n-3} + 2k_{i,n-2} y_{n-2} + 4k_{i,n-1} y_{n-1} + k_{i,n} y_n]
\end{aligned} \tag{17}
$$

for $i = n-1$.

Lastly, from (15), (16), and (17), the linear system equations for $y_1, y_2, \ldots, y_{n-2}, y_{n-1}$ are generated. Therefore, the generated system of linear algebraic equations can be written in matrix form of

$$
\mathbf{AY} = \mathbf{F}, \tag{18}
$$

where

$$
\mathbf{A} = \begin{bmatrix}
a_{1,1} + \alpha & b_{1,2} - 104 & a_{1,1} + 114 & b_{1,4} - 56 & \cdots & b_{1,n-2} & a_{1,n-1} \\
a_{2,1} + 16 & b_{2,2} - \beta & a_{2,3} + 16 & b_{2,4} - 1 & \cdots & b_{2,n-2} & a_{2,n-1} \\
a_{3,1} - 1 & b_{3,2} + 16 & a_{3,3} - \beta & b_{3,4} + 16 & \cdots & b_{3,n-2} & a_{3,n-1} \\
a_{4,1} & b_{4,2} - 1 & a_{4,3} + 16 & b_{4,4} - \beta & \cdots & b_{4,n-2} & a_{3,n-1} \\
\vdots & \vdots & \vdots & \vdots & \vdots & \vdots & \vdots \\
a_{n-4,1} & b_{n-4,2} & \cdots & b_{n-4,n-4} - \beta & a_{n-4,n-3} + 16 & b_{n-4,n-2} - 1 & a_{n-4,n-1} \\
a_{n-3,1} & b_{n-3,2} & \cdots & b_{n-3,n-4} + 16 & a_{n-3,n-3} - \beta & b_{n-3,n-2} + 16 & a_{n-3,n-1} - 1 \\
a_{n-2,1} & b_{n-2,2} & \cdots & b_{n-2,n-4} - 1 & a_{n-2,n-3} + 16 & b_{n-2,n-2} - \beta & a_{n-2,n-1} + 16 \\
a_{n-1,1} & b_{n-1,2} & \cdots & b_{n-1,n-4} - 56 & a_{n-1,n-3} + 114 & b_{n-1,n-2} - 104 & a_{n-1,n-1} + \alpha
\end{bmatrix}, \tag{19}
$$

where

$$
\begin{aligned}
a_{i,j} &= -16\lambda h^3 k_{i,j}, \\
b_{i,j} &= -8\lambda h^3 k_{i,j}, \\
\alpha &= 35 - 12h^2 p_i, \\
\beta &= 30 + 12h^2 p_i,
\end{aligned}
$$

$$
\mathbf{Y} = \begin{bmatrix} y_1 \\ y_2 \\ y_3 \\ y_4 \\ \vdots \\ y_{n-4} \\ y_{n-3} \\ y_{n-2} \\ y_{n-1} \end{bmatrix},
$$

$$
\mathbf{F} = \begin{bmatrix}
12h^2 f_1 + (4\lambda h^3 k_{1,0}) y_0 + (4\lambda h^3 k_{1,n}) y_n \\
12h^2 f_2 + (4\lambda h^3 k_{2,0} + 1) y_0 + (4\lambda h^3 k_{2,n}) y_n \\
12h^2 f_3 + (4\lambda h^3 k_{3,0}) y_0 + (4\lambda h^3 k_{3,n}) y_n \\
12h^2 f_4 + (4\lambda h^3 k_{4,0}) y_0 + (4\lambda h^3 k_{4,n}) y_n \\
\vdots \\
12h^2 f_{n-4} + (4\lambda h^3 k_{n-4,0}) y_0 + (4\lambda h^3 k_{n-4,n}) y_n \\
12h^2 f_{n-3} + (4\lambda h^3 k_{n-3,0}) y_0 + (4\lambda h^3 k_{n-3,n}) y_n \\
12h^2 f_{n-2} + (4\lambda h^3 k_{n-2,0}) y_0 + (4\lambda h^3 k_{n-2,n} + 1) y_n \\
12h^2 f_{n-1} + (4\lambda h^3 k_{n-1,0}) y_0 + (4\lambda h^3 k_{n-1,n}) y_n
\end{bmatrix}, \tag{20}
$$

where **A** is a coefficient matrix, **F** is a given function, and **Y** is an unknown function which is to be determined. Gauss Elimination method will be used in order to solve the system of linear algebraic equations.

4. Results and Discussion

The performance of the method described in the previous section was tested through numerical examples of second-order LFIDEs. In addition, the examples were performed by using algorithm in C language.

Example 1 (Type 1).

$$y''(x) = x - 2 + 60\int_0^1 (x-t)\, y(t)\, dt, \quad 0 \le x \le 1, \quad (21)$$

with the boundary condition

$$y(0) = 0, \qquad y(1) = 0 \quad (22)$$

and the exact solution

$$y(x) = x(x-1)^2. \quad (23)$$

Example 2 (Type 1).

$$y''(x) = e^x - x + \int_0^1 xty(t)\, dt, \quad 0 \le x \le 1, \quad (24)$$

with the boundary condition

$$y(0) = 0, \qquad y(1) = e^1 \quad (25)$$

and the exact solution

$$y(x) = e^x. \quad (26)$$

Example 3 (Type 2).

$$y''(x) = 9y(x) + \frac{e^{-15}-1}{3} + \int_0^5 y(t)\, dt, \quad 0 \le x \le 5, \quad (27)$$

with the boundary condition

$$y(0) = 0, \qquad y(5) = e^{-15} \quad (28)$$

and the exact solution

$$y(x) = e^{-3x}. \quad (29)$$

Notations used in Tables 1, 2, and 3 are as follows:

HSGS (CD-RT): half-sweep quadrature-difference schemes (central difference, Repeated Trapezoid), [12].

HSGS (CD-RS): half-sweep quadrature-difference schemes (central difference, Repeated Simpson), [12].

B-Spline: Cubic B-Spline collocation method, [6].

DiffTrans: differential transformation method, [7].

Table 1: Comparison between CSFD2 (GE), HSGS (CD-RT), and HSGS (CD-RS) for Example 1.

x	CSFD2 (GE)		HSGS (CD-RS)	HSGS (CD-RT)
	$N = 24$	$N = 48$	$N = 24$	$N = 24$
0.125	1.51×10^{-6}	9.45×10^{-8}		
0.250	2.26×10^{-6}	1.41×10^{-7}	$\mathbf{3.25 \times 10^{-6}}$	$\mathbf{2.89 \times 10^{-3}}$
0.375	$\mathbf{2.41 \times 10^{-6}}$	$\mathbf{1.51 \times 10^{-7}}$		
0.500	2.13×10^{-6}	1.33×10^{-7}	$N = 48$	$N = 48$
0.625	1.58×10^{-6}	9.87×10^{-8}		
0.750	9.31×10^{-7}	5.82×10^{-8}	$\mathbf{2.41 \times 10^{-7}}$	$\mathbf{7.91 \times 10^{-4}}$
0.875	3.49×10^{-7}	2.18×10^{-8}		

Table 2: Comparison between CSFD2 (GE), DiffTrans, and B-Spline method for Example 2.

x	CSFD2 (GE)		DiffTrans	B-Spline
	$N = 10$	$N = 20$	$N = 10$	$N = 10$
0.1	4.31×10^{-5}	1.21×10^{-6}	$0.00 \times 10^{+0}$	4.39×10^{-6}
0.2	3.07×10^{-5}	6.82×10^{-7}	4.63×10^{-4}	1.81×10^{-5}
0.3	1.43×10^{-5}	1.41×10^{-7}	1.43×10^{-3}	4.24×10^{-5}
0.4	2.37×10^{-6}	4.02×10^{-7}	2.94×10^{-3}	7.83×10^{-5}
0.5	1.91×10^{-5}	9.45×10^{-7}	5.05×10^{-3}	1.27×10^{-4}
0.6	3.59×10^{-5}	1.49×10^{-6}	7.81×10^{-3}	1.91×10^{-4}
0.7	5.28×10^{-5}	2.04×10^{-6}	1.13×10^{-2}	2.70×10^{-4}
0.8	6.93×10^{-5}	2.59×10^{-6}	1.55×10^{-2}	3.67×10^{-4}
0.9	$\mathbf{7.94 \times 10^{-5}}$	$\mathbf{3.12 \times 10^{-6}}$	$\mathbf{2.05 \times 10^{-2}}$	$\mathbf{4.85 \times 10^{-4}}$

Tau method: Tau method with piecewise approximate solution of degree 5, [9].

CSFD2 (GE): the proposed method, Composite Simpson's 1/3 with second-order finite-difference method (Gauss Elimination).

Examples 1, 2, and 3 are solved by using the proposed method which is CSFD2 (GE) and the numerical results are shown in Tables 1, 2, and 3. In Table 1, the same methods are used to compare the absolute errors using a different number of subintervals. The maximum absolute errors of the proposed method also compared with the existing methods such as HSGS (CD-RT) and HSGS (CD-RS). As shown in Table 1, when the number of subintervals increases, the absolute errors become smaller. The results for the proposed method are more accurate compared to the HSGS (CD-RT) while the proposed method is slightly better compared to HSGS (CD-RS).

In Table 2, the numerical results of CSFD2 (GE) for Example 2 are compared with the results from differential transformation method and Cubic B-Spline method. CSFD2 (GE) gives more accurate results than the other two existing methods. In Table 3, the maximum absolute errors for CSFD2 (GE) for Example 3 are slightly better compared to Tau method.

5. Conclusion

The new combinations of Composite Simpson's 1/3 rule and second-order finite-difference method with Gauss Elimination CSFD2 (GE) were implemented in this study in

TABLE 3: Comparison between CSFD2 (GE) and Tau method for Example 3.

x	CSFD2 (GE)		Tau	
	$N = 10$	$N = 20$	$N = 10$	$N = 20$
0.5	$\mathbf{1.28 \times 10^{-1}}$	$\mathbf{3.88 \times 10^{-2}}$		
1.0	3.47×10^{-2}	6.56×10^{-2}		
1.5	2.83×10^{-3}	9.48×10^{-4}		
2.0	5.30×10^{-3}	2.64×10^{-3}		
2.5	7.20×10^{-3}	3.03×10^{-3}	$\mathbf{8.31 \times 10^{-1}}$	4.71×10^{-2}
3.0	7.64×10^{-3}	3.11×10^{-3}		
3.5	7.73×10^{-3}	3.10×10^{-3}		
4.0	7.70×10^{-3}	3.01×10^{-3}		
4.5	7.08×10^{-3}	2.57×10^{-3}		

order to solve second-order LFIDEs for Types 1 and 2. In conclusion, this study shows that CSFD2 (GE) calculates better approximations to the exact solution of second-order LFIDEs. Therefore, we introduce CSFD2 (GE) as a simple and more accurate method. In future research, the concept of the proposed method can also be used for Types 3 and 4 of second-order LFIDEs and also for higher order of LFIDEs problems. Moreover, the accuracy and efficiency of the proposed method also can be improved.

Competing Interests

The authors declare that there is no conflict of interests regarding the publication of this paper.

Acknowledgments

The authors gratefully acknowledge the financial support given from Malaysian Ministry of Education for MyBrain15 (My Master) Scholarship and Graduate Research Fund (GRF) by Universiti Putra Malaysia.

References

[1] A. R. Vahidi, E. Babolian, G. Asadi Cordshooli, and Z. Azimzadeh, "Numerical solution of Fredholm integro-differential equation by Adomian's decomposition method," *International Journal of Mathematical Analysis*, vol. 3, no. 33–36, pp. 1769–1773, 2009.

[2] E. Aruchunan and J. Sulaiman, "Numerical solution of second-order linear fredholm integro-differential equation using generalized minimal residual method," *American Journal of Applied Sciences*, vol. 7, no. 6, pp. 780–783, 2010.

[3] H. Danfu and S. Xufeng, "Numerical solution of integro-differential equations by using CAS wavelet operational matrix of integration," *Applied Mathematics and Computation*, vol. 194, no. 2, pp. 460–466, 2007.

[4] J. Zhao and R. M. Corless, "Compact finite difference method for integro-differential equations," *Applied Mathematics and Computation*, vol. 177, no. 1, pp. 271–288, 2006.

[5] M. T. Rashed, "Lagrange interpolation to compute the numerical solutions of differential, integral and integro-differential equations," *Applied Mathematics and Computation*, vol. 151, no. 3, pp. 869–878, 2004.

[6] N. Ebrahimi and J. Rashidinia, "Spline collocation for Fredholm and Volterra integro-differential equations," *International Journal of Mathematical Modelling & Computations*, vol. 4, no. 3, pp. 289–298, 2014.

[7] P. Darania and A. Ebadian, "A method for the numerical solution of the integro-differential equations," *Applied Mathematics and Computation*, vol. 188, no. 1, pp. 657–668, 2007.

[8] R. Saadati, B. Raftari, H. Adibi, S. M. Vaezpour, and S. Shakeri, "A comparison between the Variational Iteration method and Trapezoidal rule for solving linear integro-differential equations," *World Applied Sciences Journal*, vol. 4, no. 3, pp. 321–325, 2008.

[9] S. M. Hosseini and S. Shahmorad, "Numerical piecewise approximate solution of Fredholm integro-differential equations by the Tau method," *Applied Mathematical Modelling*, vol. 29, no. 11, pp. 1005–1021, 2005.

[10] S. Yalcinbas, M. S. Sezer, and H. H. Sorkun, "Legendre polynomial solutions of high-order linear Fredholm integro-differential equations," *Applied Mathematical Model*, vol. 27, pp. 145–154, 2003.

[11] A. I. Fedotov, "Quadrature-difference methods for solving linear and nonlinear singular integro-differential equations," *Nonlinear Analysis*, vol. 71, no. 12, pp. e303–e308, 2009.

[12] E. Aruchunan and J. Sulaiman, "Half-sweep quadrature-difference schemes with iterative method in solving linear fredholm integro-differential equations," *Progress in Applied Mathematics*, vol. 5, no. 1, pp. 11–21, 2013.

12

Steady Flow of Couple-Stress Fluid in Constricted Tapered Artery: Effects of Transverse Magnetic Field, Moving Catheter, and Slip Velocity

Hamzah Bakhti[1] **and Lahcen Azrar**[1,2]

[1] *MMC Team, Department of Mathematics, Faculty of Sciences & Techniques, AEU, BP 416 Tangier, Morocco*
[2] *MMC Team, Department of Mechanical Engineering, Faculty of Engineering, KAU, Jeddah, Saudi Arabia*

Correspondence should be addressed to Hamzah Bakhti; bakhti_hamzah@outlook.com

Academic Editor: Taha Sochi

Steady flow of a couple-stress fluid in constricted tapered artery has been studied under the effects of transverse magnetic field, moving catheter, and slip velocity. With the help of Bessel's functions, analytic expressions for axial velocity, flow rate, impedance, and wall shear stress have been obtained. It is of interest to note that these solutions can be used for different types of fluid flow in tubes and not only the case of blood. The effects of various geometric parameters, the parameters arising out of the fluid considered and the magnetic field, are discussed by considering the slip velocity, the catheter velocity, and tapering angle. The study of the above model is very important as it has direct applications in the treatment of cardiovascular diseases.

1. Introduction

Catheters are semirigid, thin tubes made from medical grade materials serving a broad range of functions and can be inserted in the body to detect and identify diseases or perform a surgical procedure inside the heart, brain, arms, legs, or lungs [1]. But the problem is how we can know that the catheter is going the right way or where is it. Most hospitals use X-ray machine, mounted above the couch to check the catheter's progress. Low-dose X-rays are used to monitor the progress of the catheter tip. But if we want to navigate the catheter's passage in real time, some ten to twenty X-ray images are made every second. Even though the radiation dose involved is very low, no radiation at all is much healthier for both the patient and the medical staff who run the risk of being exposed to the radiation on daily basis. Researchers developed a magnetic navigation system for medical instruments such as catheters and guide wires. The catheters with magnetic tips are steered within the patient, without the need for an electrophysiologist to maneuver the catheter or guide wire placement manually. A magnetic sensor on the tip of the medical instrument measures an external magnetic field and reports exactly where the tip of the instrument is located [2].

Each year, heart disease is at the top of the list of the country's most serious health problems. Statistics show that cardiovascular disease in some countries, like America, is the first health problem and the leading cause of death (recent statistics released by the American Heart Association). Catheters are often used to treat stenosis (partial occlusion of the blood vessel), aneurysm (dilation of the blood vessel resulting in stretching of the vessel wall), and embolism (complete occlusion of a blood vessel by a blood clot or some other particle), Figure 1. The catheters may contain a balloon that can be used to stretch the blood vessel or a rinsing mechanism or laser device to remove embolism. They are also used to measure blood pressure in situ. Sometimes a catheter is used to deliberately create embolism, for example, to stop the flow of blood to a tumor in order to mortify it. An aneurysm can also be treated by means of embolisation [3].

The study of blood flow through different types of arteries is of considerable importance in many cardiovascular diseases; one of them is atherosclerosis. The blood flow through

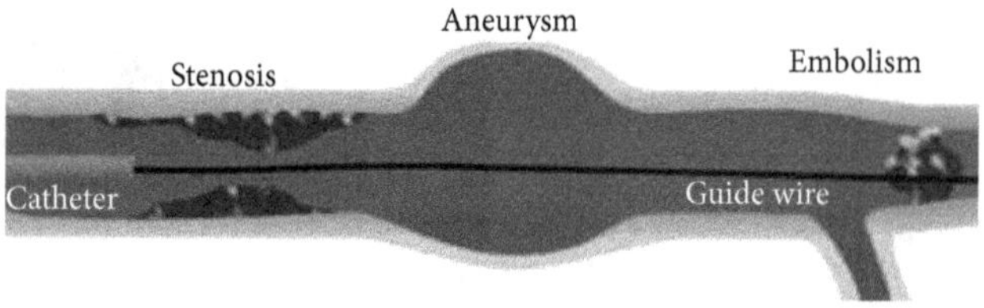

Figure 1: Catheters are used to treat stenosis (partial occlusion of the blood vessel), aneurysm (dilation of the blood vessel resulting in stretching of the vessel wall), and embolism (complete occlusion of a blood vessel by a blood clot or some other particle) (a reproduction from Delft Outlook 2004).

an artery has drawn the attention of researchers for a long time up to now, due to its great importance in medical sciences. Under normal conditions, blood flow in the human circulatory system depends upon the pumping action of the heart and this produces a pressure gradient throughout the arterial network.

Many researchers have studied blood flow in the artery by considering blood as either Newtonian or non-Newtonian fluids; since blood is a suspension of red cells in plasma, it behaves as a non-Newtonian fluid at low shear rate. In the stenosed condition, substantial reduction in the lumen of an artery results in size effects (ratio of haematocrit to vessel diameter), which influences flow characteristics significantly. To study the size effect in the fluid flow, Stokes [4], Eringen [5], and Cowin [6] have proposed continuum model. Micro-continuum structure of the fluid is also referred to as couple-stress fluid which is proposed by Stokes. The foreground of couple-stress fluid is to introduce size dependent effects as mentioned above that is not present in other classical viscous fluids. Because of its mathematical simplicity, it has been widely used by number of researchers; Chaturani and Upadhya [7] have studied the pulsatile flow of couple-stress fluid through circular tubes. The Poiseuille flow of couple-stress fluid has been examined by Chaturani and Rathod [8].

The application of Magnetohydrodynamics in physiological problems is of growing interest and it is very important from both theoretical and practical points of view. The application of Magnetohydrodynamics in physiological problems is of growing interest. The flow of blood can be controlled by applying appropriate quantity of magnetic field. Kollin [9] has coined the idea of electromagnetic field in the medical research for the first time in the year 1936. Korchevskii and Marochnik [10] have discussed the possibility of regulating the blood movement in human system by applying magnetic field. Rao and Deshikachar [11] have investigated the effect of transverse magnetic field in physiological type of flow, through a uniform circular pipe. Vardanyan [12] showed that the application of magnetic field reduces the speed of blood flow. A mathematical model for two-layer pulsatile flow of blood with microorganism in a uniform tube under low Reynolds number and magnetic effect has been studied by Rathod and Gayatri [13]. Thus, all these researchers have reported that the effect of magnetic field reduces the velocity of blood.

There are many treatments available for diagnosing and treating constricted vessels. Catheterization (thin, flexible tube) is one of them, in which balloon angioplasty is a specialized form of catheterization. These procedures are widely used in the medical field for treating the atherosclerosis. Insertion of the catheter in a tube creates an annular region between inner wall of the artery and outer wall of the catheter which influences the flow field such as pressure distribution and shear stress at the wall. In view of its immense importance, the effect of the catheter on physiological parameters was discussed by the researchers [14, 15].

The shapes of the stenosis in the above aforesaid studies have been considered to be radially symmetric or asymmetric. But while stenosis is maturing, it may grow up in series manner, overlapping with each other, and it would appear like x-shape. Riahi et al. [16] observed the pressure gradient force, flow velocity, impedance, and wall shear stress in the overlapping stenotic zone at the critical height and at the throats of such stenosis. Here they considered steady nature of flow. Srivastava and Mishra [17] explored the arterial blood flow through an overlapping stenosis by treating the blood as a Casson fluid. They figured impedance and shear stress for different stenosis heights. Chakravarty and Mandal [18] discussed the effects of the overlapping stenosis under low shear rate flow.

The presence of red cell slip at the vessel wall was recommended theoretically by Vand [19], experimentally by Bennett [20] and Nubar [21], Chaturani and Biswas [22], and so forth. They used slip velocity at the wall in their analysis. Lately, Ponalagusamy [23, 25] has developed mathematical models for blood flow through stenosed arterial segment, by taking a slip velocity condition at the constricted wall. Thus, it is very appropriate to consider slip velocity at the wall of the stenosed artery in blood flow modeling. The catheter is actually moving into the body, so it is recommended to consider a nonnull velocity at the catheter wall; in this paper, we consider that the catheter is moving in the z-axis with a fixed velocity.

In the present analysis a mathematical model for the steady blood flow through tapered stenosed artery under the influence of a moving catheter, slip velocity, and a magnetic field is presented by considering blood as a couple-stress fluid in a circular tube. It is assumed that the magnetic field along the radius of the pipe is present, no external electric field is imposed, and magnetic Reynolds number is very small. The motivation for studying this problem is to understand the blood flow in an artery under the effect of magnetic field alongside with the catheter inserted into the blood vessel also when the fatty plaques of cholesterol and artery clogging blood clots are formed in the lumen of the artery.

The main aim of this work is to study these phenomena, obtain analytic expressions for axial velocity and shear stress, and also study the effect of magnetic field (Hartmann number H) and couple-stress parameter (β) on the velocity and the effects of Hartmann number on the fluid velocity. Hence, the present mathematical model gives a simple form of velocity expression for the blood flow so that it will help not only people working in the field of physiological fluid dynamics but also the medical practitioners.

2. Problem Formulation

Let us consider a two-dimensional steady flow of blood through a rigid tapered stenosed tube by considering blood as an electrically conducting, incompressible, couple-stress fluid. The magnetic field is acting along the radius of the tube. The magnetic Reynolds number of the flow is assumed to be sufficiently small that the induced magnetic and electric fields can be neglected [24]. The catheter is assumed to be moving in the z-axis direction.

Mathematical model of blood flow through a tapered stenosed arterial segment in the presence of a moving catheter and a magnetic field is to be built to study the impact of various geometric, Hartman, and fluid parameters on physiological parameters. The geometry of the tapered stenosed artery is shown in Figures 2 and 3 and is expressed mathematically with inputs from [26] as

$$R(z)=\begin{cases}(R_0+\zeta z)\left(1-\dfrac{\epsilon n^{n/(n-1)}}{(n-1)L_0^n}\left(L_0^{n-1}(z-d)-(z-d)^n\right)\right), & d\le z\le d+L_0\\ R_0+\zeta z, & \text{otherwise,}\end{cases}\tag{1}$$

where R_0 is the radius of the annular region in case of nontapered artery in the nonconstricted domain, R_c is the radius of the catheter, L_0 is the stenosis length, d indicates the location of stenosis, ϵ is the maximum height of the stenosis into the lumen, and $\zeta=\tan\phi$ is the tapering parameter which represents the slope of the tapered vessel with ϕ being the tapering angle. $\phi<0$, $\phi>0$, and $\phi=0$ are for converging taper, diverging taper, and no-taper, respectively.

The current density $\mathbf{J}$ is expressed by

$$\mathbf{J}=\sigma(\mathbf{E}+\mathbf{V}\times\mathbf{B}),\tag{2}$$

where $\mathbf{E}$ is the electric field intensity, σ is the electrical conductivity, $\mathbf{B}$ is the magnetic flux intensity, and $\mathbf{V}$ is the velocity vector. In the momentum equation, the electromagnetic force $\mathbf{F}_m$ is included and is defined as

$$\mathbf{F}_m=\mathbf{J}\times\mathbf{B}=\sigma(\mathbf{E}+\mathbf{V}\times\mathbf{B})\times\mathbf{B}.\tag{3}$$

The conservation equations which govern the couple-stress fluid flow including a Lorentz force can be written in the following form:

$$\begin{aligned}&\nabla\cdot\mathbf{V}=0,\\ &\rho(\mathbf{V}\cdot\nabla)\mathbf{V}=-\nabla p+(\lambda+\mu)\nabla\nabla\cdot\mathbf{V}+\eta\Delta\nabla\nabla\cdot\mathbf{V}\\ &\qquad-\eta\Delta^2\mathbf{V}+\mu\Delta\mathbf{V}+\sigma(\mathbf{V}\times\mathbf{B})\times\mathbf{B}+\rho\mathbf{f}\\ &\qquad+\frac{1}{2}\nabla\times(\rho\mathbf{l}),\end{aligned}\tag{4}$$

where Δ is the Laplacian operator. For couple-stress fluid, shear stress tensor is not symmetric. The force stress tensor $\boldsymbol{\tau}$ and the couple-stress tensor $\mathbf{M}$ that arises in the theory of couple-stress fluids are given by

$$\begin{aligned}\boldsymbol{\tau}&=(-p+\lambda\nabla\mathbf{V})\mathbf{I}+\mu\left(\nabla\mathbf{V}+(\nabla\mathbf{V})^T\right)+\frac{1}{2}\mathbf{I}\\ &\quad\times(\nabla\mathbf{M}+\rho\mathbf{l}),\\ \mathbf{M}&=m\mathbf{I}+2\eta(\nabla(\nabla\times\mathbf{V}))+2\eta'(\nabla(\nabla\times\mathbf{V}))^T,\end{aligned}\tag{5}$$

where ρ is the density, $\mathbf{V}$ is velocity vector, p is the pressure, λ and μ are the viscosity coefficients, and η and η' are couple-stress coefficients. $\mathbf{f}$ is a body force and $\mathbf{l}$ is a body couple moment. Further, the materials constants μ, λ, η, and η' satisfy the following inequalities:

$$\begin{aligned}\mu&\ge 0,\\ 3\lambda+2\mu&\ge 0,\\ \eta&\ge 0,\\ \eta&\ge\eta'.\end{aligned}\tag{6}$$

As the flow is steady and incompressible, in the absence of body force and body couple moment (4) reduce to

$$\begin{aligned}&\nabla\cdot\mathbf{V}=0,\\ &\rho(\mathbf{V}\cdot\nabla)\mathbf{V}=-\nabla p+\mu\Delta\mathbf{V}-\eta\Delta^2\mathbf{V}+\sigma(\mathbf{V}\times\mathbf{B})\times\mathbf{B}.\end{aligned}\tag{7}$$

The flow is considered to take place under the influence of externally applied magnetic field in axial direction z. Under these assumptions, the governing equations may be written in the cylindrical coordinates system as follows.

Equation of continuity is

$$\frac{\partial u}{\partial r}+\frac{u}{r}+\frac{\partial w}{\partial z}=0.\tag{8}$$

Equation of radial momentum is

$$\begin{aligned}\rho\left(u\frac{\partial u}{\partial r}+w\frac{\partial u}{\partial z}\right)&=-\frac{\partial p}{\partial r}+\mu\left(\frac{\partial^2u}{\partial r^2}+\frac{1}{r}\frac{\partial u}{\partial r}-\frac{u}{r^2}\right.\\ &\quad\left.+\frac{\partial^2u}{\partial z^2}\right)-\eta\left(\frac{\partial^4u}{\partial r^4}+2\frac{\partial^4u}{\partial r^2\partial z^2}+\frac{\partial^4u}{\partial z^4}+\frac{2}{r}\frac{\partial^3u}{\partial r^3}\right.\\ &\quad\left.+\frac{2}{r}\frac{\partial^3u}{\partial r\partial z^2}-\frac{3}{r^2}\frac{\partial^2u}{\partial r^2}-\frac{2}{r^2}\frac{\partial^2u}{\partial z^2}+\frac{3}{r^3}\frac{\partial u}{\partial r}-\frac{3}{r^4}u\right).\end{aligned}\tag{9}$$

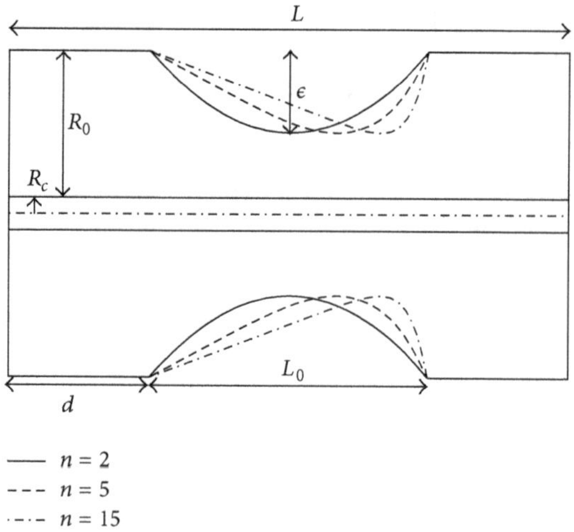

FIGURE 2: 2D view of a catheterized stenosed artery.

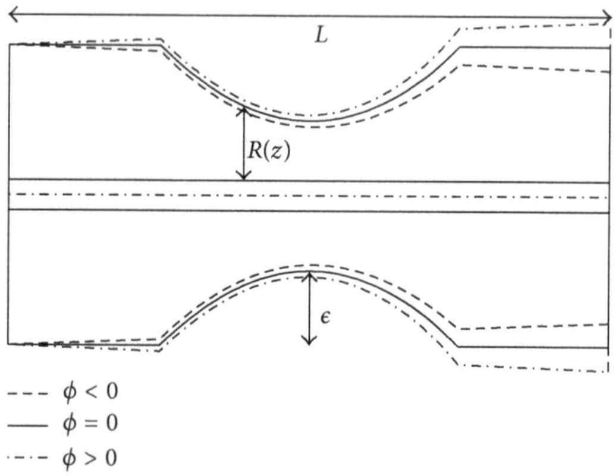

FIGURE 3: 2D view of a tapered stenosed artery.

Equation of axial momentum is

$$\rho\left(u\frac{\partial w}{\partial r}+w\frac{\partial w}{\partial z}\right)=-\frac{\partial p}{\partial z}+\mu\left(\frac{\partial^2 w}{\partial r^2}+\frac{1}{r}\frac{\partial w}{\partial r}+\frac{\partial^2 w}{\partial z^2}\right)-\eta\left(\frac{\partial^4 w}{\partial r^4}+2\frac{\partial^4 w}{\partial r^2\partial z^2}+\frac{\partial^4 w}{\partial z^4}+\frac{2}{r}\frac{\partial^3 w}{\partial r^3}+\frac{2}{r}\frac{\partial^3 w}{\partial r\partial z^2}-\frac{1}{r^2}\frac{\partial^2 u}{\partial r^2}+\frac{1}{r^3}\frac{\partial w}{\partial r}\right)-\sigma B_0^2 w, \quad (10)$$

where B_0 is the external transverse magnetic field. The nondimensional parameters are obtained as follows:

$$z^*=\frac{z}{L_0},\quad r^*=\frac{r}{R_0},\quad w^*=\frac{w}{w_0},\quad \epsilon^*=\frac{\epsilon}{R_0},\quad \zeta^*=\frac{\zeta L_0}{R_0},\quad u^*=\frac{L_0 u}{w_0 R_0},\quad p^*=\frac{r_0^2 p}{w_0 L_0 \mu}, \quad (11)$$

and then the dimensionless form of geometry (1) after dropping the stars is obtained as follows:

$$R(z)=\begin{cases}(1+\zeta z)\left(1-\dfrac{\zeta n^{n/(n-1)}}{(n-1)}\left((z-\gamma)-(z-\gamma)^n\right)\right), & \gamma\le z\le\gamma+1\\ 1+\zeta z, & \text{otherwise,}\end{cases} \quad (12)$$

where $\gamma = L_0/L_1$ and w_0 is a typical axial velocity.

Also, (8)–(10), on dropping the stars, become

$$\frac{\partial u}{\partial r}+\frac{u}{r}+\frac{\partial w}{\partial z}=0,$$

$$\mathrm{Re}\xi^3\left(u\frac{\partial u}{\partial r}+w\frac{\partial u}{\partial z}\right)=-\frac{\partial p}{\partial r}+\xi^2\left(\frac{\partial^2 u}{\partial r^2}+\frac{1}{r}\frac{\partial u}{\partial r}-\frac{u}{r^2}+\xi^2\frac{\partial^2 u}{\partial z^2}\right)-\frac{\xi^2}{\beta^2}\left[\frac{\partial^4 u}{\partial r^4}+\frac{2}{r}\frac{\partial^3 u}{\partial r^3}-\frac{3}{r^2}\frac{\partial^2 u}{\partial r^2}+\frac{3}{r^3}\frac{\partial u}{\partial r}-\frac{3}{r^4}u+\xi^2\left(\frac{2}{r}\frac{\partial^3 u}{\partial r\partial z^2}+2\frac{\partial^4 u}{\partial r^2\partial z^2}+\xi^2\frac{\partial^4 u}{\partial z^4}-\frac{2}{r^2}\frac{\partial^2 u}{\partial z^2}\right)\right],$$

$$\mathrm{Re}\xi\left(u\frac{\partial w}{\partial r}+w\frac{\partial w}{\partial z}\right)=-\frac{\partial p}{\partial z}+\left(\frac{\partial^2 w}{\partial r^2}+\frac{1}{r}\frac{\partial w}{\partial r}+\xi^2\frac{\partial^2 w}{\partial z^2}\right)-\frac{1}{\beta^2}\left[\frac{\partial^4 w}{\partial r^4}+\frac{2}{r}\frac{\partial^3 w}{\partial r^3}-\frac{1}{r^2}\frac{\partial^2 w}{\partial r^2}+\frac{1}{r^3}\frac{\partial w}{\partial r}+\xi^4\frac{\partial^4 w}{\partial z^4}+2\xi^2\left(\frac{\partial^4 w}{\partial r^2\partial z^2}+\frac{1}{r}\frac{\partial^3 w}{\partial r\partial z^2}\right)\right]-H^2 w, \quad (13)$$

where $\xi = R_0/L_0$, $\mathrm{Re} = \rho u_0 R_0/\mu$ is the Reynolds number, $\beta^2 = R_0^2\mu/\eta$ is the couple-stress fluid parameter, and $H^2 = B_0^2R_0^2\sigma/\mu$ is the Hartmann number.

Under the assumption of mild stenosis that is $\epsilon/R_0 \le 1$ and further assuming that $\xi = R_0/L_0 \le 1$, (13) get transformed into

$$\frac{\partial p}{\partial r} = 0, \tag{14}$$

$$\frac{\partial p}{\partial z} = \left(\frac{\partial^2 w}{\partial r^2} + \frac{1}{r}\frac{\partial w}{\partial r}\right) - \frac{1}{\beta^2}\left(\frac{\partial^4 w}{\partial r^4} + \frac{2}{r}\frac{\partial^3 w}{\partial r^3} - \frac{1}{r^2}\frac{\partial^2 w}{\partial r^2} + \frac{1}{r^3}\frac{\partial w}{\partial r}\right) - H^2 w. \tag{15}$$

Equation (15) can be written as

$$\frac{\partial p}{\partial z} = \Delta w - \frac{1}{\beta^2}\Delta^2 w - H^2 w, \tag{16}$$

where

$$\Delta = \frac{\partial^2}{\partial r^2} + \frac{1}{r}\frac{\partial}{\partial r}. \tag{17}$$

It can be seen that the pressure variation depends only on the axial variable. The pressure gradient $\partial p/\partial z$ is produced by the pumping action of the heart.

The corresponding nondimensional boundary conditions are as shown below:

$$\begin{aligned} &w = v \quad \text{at } r = R(z), \\ &w = g \quad \text{at } r = R_c, \\ &\frac{\partial^2 w}{\partial r^2} - \frac{\omega}{r}\frac{\partial w}{\partial r} = 0 \quad \text{at } r = R(z),\ r = R_c, \end{aligned} \tag{18}$$

where v represents the slip velocity at the artery wall, g is the velocity of the moving boundary (catheter), and $\omega = \eta'/\eta$ is the parameter associated with the couple-stress fluid. No couple-stress effects will be present if $\eta' = \eta$, which is equivalent to saying that the couple-stress tensor is symmetric.

3. Solution of the Problem

As a solution of (16) with the pressure gradient being taken as a constant, take

$$-\frac{\partial p}{\partial z} = k_s. \tag{19}$$

Let $n_1^2 + n_2^2 = \beta^2$ and $n_1^2 n_2^2 = \beta^2 H^2$, which are given by

$$n_1^2 = \frac{\beta^2 + \sqrt{\beta^4 - 4\beta^2 H^2}}{2}, \tag{20}$$

$$n_2^2 = \frac{\beta^2 - \sqrt{\beta^4 - 4\beta^2 H^2}}{2}; \tag{21}$$

then, (16) is simplified to the form

$$\left(\Delta - n_1^2\right)\left(\Delta - n_2^2\right) w = \beta^2 k_s. \tag{22}$$

The solution of the above equation is obtained as

$$w = c_1(z) I_0(n_1 r) + c_2(z) K_0(n_1 r) + c_3(z) I_0(n_2 r) + c_4(z) K_0(n_2 r) + \frac{1}{H^2}k_s, \tag{23}$$

where I_0 and K_0 are modified Bessel function of order zero, first and second kinds, respectively. c_i for $i = 1, 2, 3, 4$ are calculated numerically by solving the algebraic system obtained from the boundary conditions.

Volumetric flow rate Q, in steady flow through a tube, is obtained by integrating the velocity profile over a cross section of the tube. The nondimensional volumetric flow rate Q across the radial distance is expressed as

$$Q = \int_{R_c}^{R(z)} 2rw\,dr \tag{24}$$

which is obtained in the form $Q = -2k_s F(R_c, R(z))$, where

$$\begin{aligned} F(R_c, R(z)) = {} & \frac{d_1}{n_1}\left(RI_1(n_1 R) - R_c I_1(n_1 R_c)\right) \\ & + \frac{d_2}{n_1}\left(R_c K_1(n_1 R_c) - RK_1(n_1 R)\right) \\ & + \frac{d_3}{n_2}\left(RI_1(n_2 R) - R_c I_1(n_2 R_c)\right) \\ & + \frac{d_4}{n_2}\left(R_c K_1(n_2 R_c) - RK_1(n_2 R)\right) \\ & + \frac{R^2 - R_c^2}{2H^2}, \end{aligned} \tag{25}$$

where $d_i = -c_i/k_s$, for $i = 1, 2, 3, 4$. The pressure drop Δp across the stenosis between the sections $z = 0$ and $z = L$ is obtained by

$$\Delta p = \frac{Q}{2}\int_0^L \frac{1}{F(R_c, R(z))}dz. \tag{26}$$

The resistance to the flow (impedance) is obtained from

$$\begin{aligned} R_f = \frac{\Delta p}{Q} &= \frac{1}{2}\int_0^L \frac{1}{F(R_c, R(z))}dz \\ &= \frac{1}{2}\left(\int_0^d \frac{1}{F(R_c, R(z))}dz + \int_d^{d+L_0} \frac{1}{F(R_c, R(z))}dz + \int_{d+L_0}^L \frac{1}{F(R_c, R(z))}dz\right). \end{aligned} \tag{27}$$

The dimensionless form of (20) is

$$R_f = \frac{\Gamma}{2}\left(\int_0^\gamma \frac{1}{F(R_c, R(z))}dz + \int_\gamma^{\gamma+1} \frac{1}{F(R_c, R(z))}dz + \int_{\gamma+1}^{1/\gamma} \frac{1}{F(R_c, R(z))}dz\right), \tag{28}$$

where $\Gamma = L_0/d$.

The shear stress τ_{rz} is calculated using the expression

$$\tau_{rz} = \tau_{rz}^{S} + \tau_{rz}^{A}, \tag{29}$$

where the symmetric and skew-symmetric parts of the stress are given by

$$\tau_{rz}^{S} = \mu \frac{\partial w}{\partial r}, \tag{30}$$

$$\tau_{rz}^{A} = \left(\frac{\partial m_{r\theta}}{\partial r} + \frac{m_{r\theta} + m_{\theta r}}{r} \right), \tag{31}$$

respectively. Here,

$$\begin{aligned} m_{r\theta} &= \frac{1}{4}\left(\frac{\eta'}{r}\frac{\partial w}{\partial r} - \eta \frac{\partial^2 w}{\partial r^2} \right), \\ m_{\theta r} &= \frac{1}{4}\left(\frac{\eta}{r}\frac{\partial w}{\partial r} - \eta' \frac{\partial^2 w}{\partial r^2} \right). \end{aligned} \tag{32}$$

Hence, the dimensionless shear stress for the artery is given by

$$\begin{aligned} \tau_{rz} &= n_1 \left(c_1 I_1 (n_1 r) - c_2 K_1 (n_1 r) \right) \left(1 - \frac{n_1^2}{4 (n_1^2 + n_2^2)} \right) \\ &\quad + n_2 \left(c_3 I_1 (n_2 r) - c_4 K_1 (n_2 r) \right) \left(1 - \frac{n_2^2}{4 (n_1^2 + n_2^2)} \right). \end{aligned} \tag{33}$$

Thus, the shear stress at the wall can be computed from (33) by taking $r = R(z)$.

4. Results and Discussion

The study of blood flow through catheterized stenosed tapered artery with the presence of a transverse magnetic field involves the integration of various geometric and fluid variables, which influences the physiological parameters such as the fluid velocity, rate flow, and wall shear stress. Closed form solutions are obtained in terms of modified Bessel's functions. The physiological dimensionless quantities such as the fluid velocity in the stenosis region and the wall shear stress at the maximum height of the stenosis are computed numerically for various values of the fluid and geometric parameters using the program *MATLAB*. The parameters considered are n (shape parameter), ζ (tapered parameter), R_c (catheter radius), Γ (stenosis length), β (couple-stress fluid parameter), H (the Hartmann number), v (slip velocity), and g (catheter wall velocity). The obtained results are analyzed graphically. The results obtained in this study are in good agreement with those reported in the literature.

Figures 4–6 illustrate the variation of axial velocity profile for different values of n, ζ, and β. It is observed from Figure 4 that the velocity profile increases for increasing the shape parameter and that actually this occurred due to the change of the stenosis height. It is observed from Figure 5 that the velocity increases by the increase in the tapered parameter

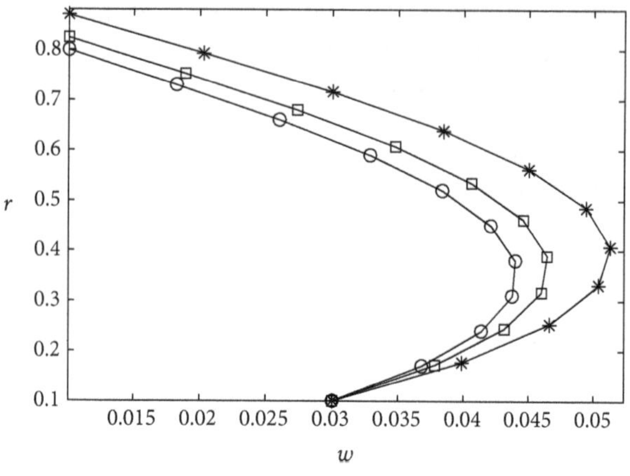

FIGURE 4: Variation of axial velocity w with respect to n when $\zeta = 0$, $\epsilon = 0.2$, $\beta = 5$, $H = 1$, $v = 0.01$, $g = 0.03$, and $R_c = 0.1$.

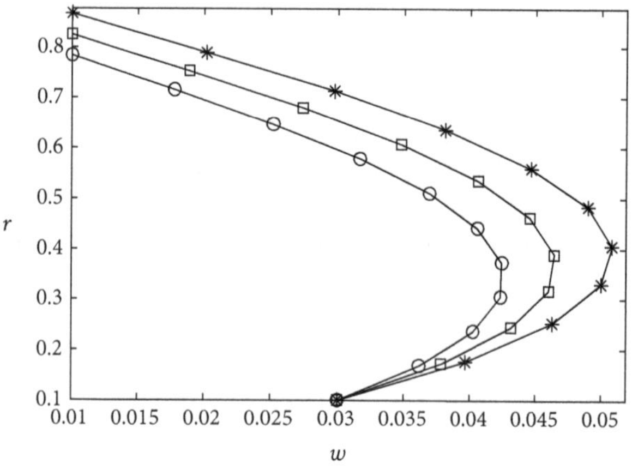

FIGURE 5: Variation of axial velocity w with respect to ζ when $n = 5$, $\epsilon = 0.2$, $\beta = 5$, $H = 1$, $v = 0.01$, $g = 0.03$, and $R_c = 0.1$.

ζ and the same goes for the couple-stress fluid parameter β, Figure 6. It is to be noted as $\beta \to \infty$, the properties of couple-stress in the fluid vanish and hence behave like a Newtonian fluid. Hence it is understood that the velocity is low in couple-stress fluid when compared to that of Newtonian fluid.

In general, from Figure 7, it can be observed that the axial velocity is decreasing. The effect of the Hartmann number H on the axial velocity is shown in Figure 8. By observing Figure 8, the fluid velocity is decreasing as the Hartmann number is increasing. Further, as $H \to 0$, the Hartmann number loses its properties and behaves like a normal blood flow without any magnetic field applied to it. We have concluded that the application of magnetic field reduces the speed of blood flow and that meets the results of Ponalagusamy and Tamil Selvi [25], also indicating that the

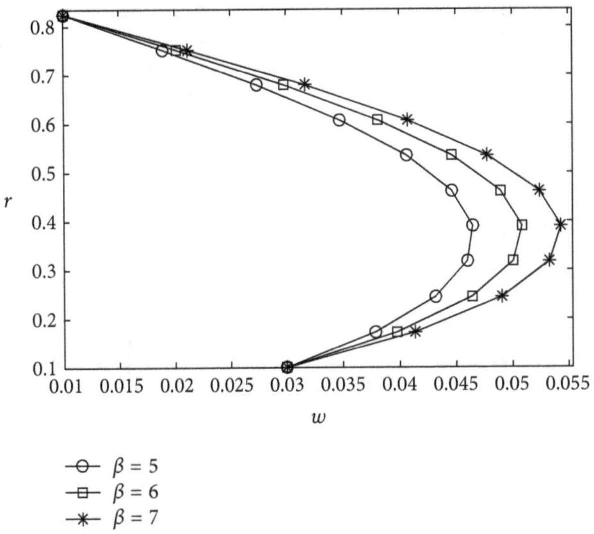

FIGURE 6: Variation of axial velocity w with respect to β when $n = 5$, $\epsilon = 0.2$, $\zeta = 0$, $H = 1$, $\nu = 0.01$, $g = 0.03$, and $R_c = 0.1$.

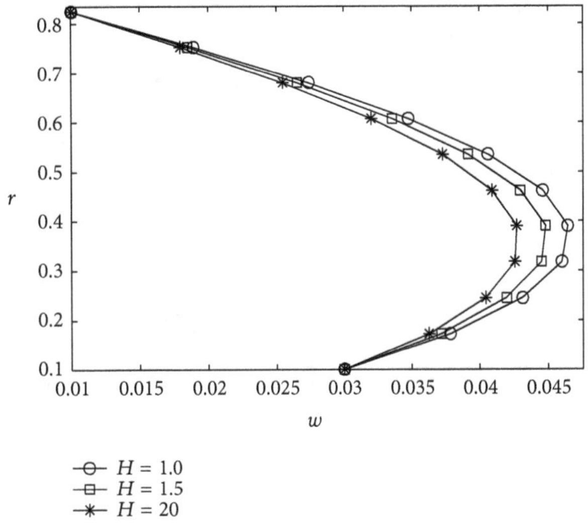

FIGURE 8: Variation of axial velocity w with respect to H when $n = 5$, $\zeta = 0$, $\beta = 5$, $\epsilon = 0.2$, $\nu = 0.01$, $g = 0.03$, and $R_c = 0.1$.

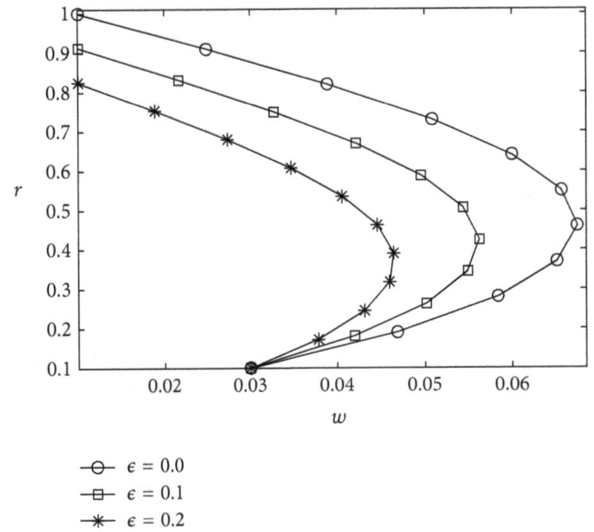

FIGURE 7: Variation of axial velocity w with respect to ϵ when $n = 5$, $\zeta = 0$, $\beta = 5$, $H = 1$, $\nu = 0.01$, $g = 0.03$, and $R_c = 0.1$.

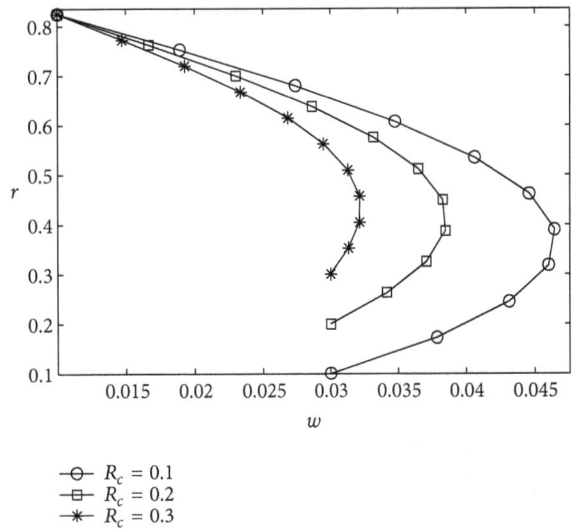

FIGURE 9: Variation of axial velocity w with respect to R_c when $n = 5$, $\zeta = 0$, $\beta = 5$, $\epsilon = 0.2$, $\nu = 0.01$, $g = 0.03$, and $H = 1$.

wall shear stress increases when we increase the Hartmann number. The influence of catheter radius R_c on the axial velocity is directly proportional as shown in Figure 9. From the above result, we understand that as the catheter radius increases, the annular region gets narrowed, which leads to the rise in the obstruction to the flow.

The axial velocity of the blood is high in case of high slip velocity ν or catheter velocity g; it is observed that the slip velocity at the boundary facilitates the fluid flow and the same goes for the moving catheter, Figures 10 and 11.

The shear stress at the wall is a significant physiological parameter to be considered in the blood flow study. Precise predictions of the distribution of the shear stress at the wall are particularly useful in assimilating the effect of blood flow in arteries in general. The shear stress at the wall is calculated at the maximum height of the stenosis. When we increase the height of the stenosis ϵ, the wall shear stress increases, Figure 15. Furthermore, it is observed that the shear stress along the wall is reaching the maximum at the throat of the stenosis.

The effect of the Hartmann number H on the shear stress is shown in Figure 16. As observed from Figure 12, the shear stress at the wall is independent of the shape parameter n as pointed out by Ponalagusamy and Tamil Selvi [25]. The variation of the shear stress at the wall as the tapered parameter increases is depicted in Figure 13. Here, the shear stress at the wall is more significant in the untapered artery compared with the tapered artery which is diverging. Further, the converging tapered artery possesses higher wall

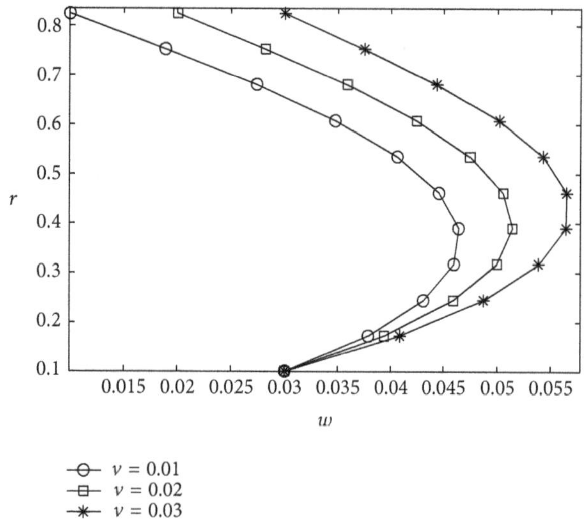

FIGURE 10: Variation of axial velocity w with respect to v when $n = 5$, $\zeta = 0$, $\beta = 5$, $\epsilon = 0.2$, $R_c = 0.1$, $g = 0.03$, and $H = 1$.

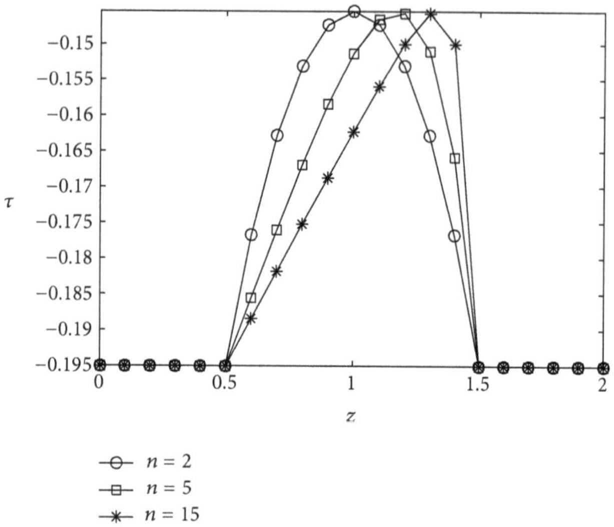

FIGURE 12: Variation of wall shear stress w with respect to n when $\zeta = 0$, $\epsilon = 0.2$, $\beta = 5$, $H = 1$, $v = 0.01$, $g = 0.03$, and $R_c = 0.1$.

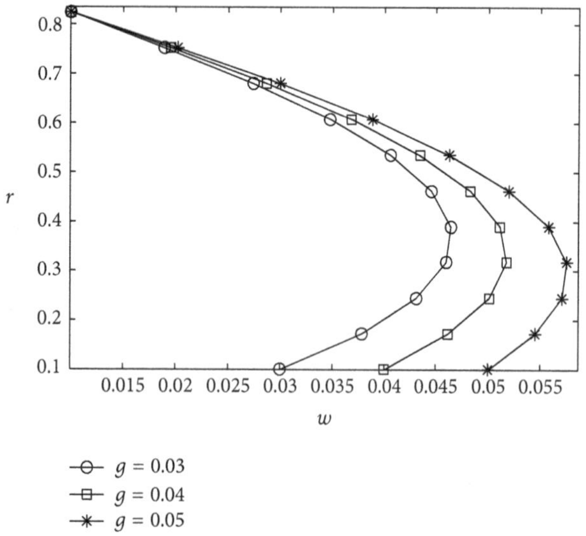

FIGURE 11: Variation of axial velocity w with respect to g when $n = 5$, $\zeta = 0$, $\beta = 5$, $\epsilon = 0.2$, $v = 0.01$, $R_c = 0.1$, and $H = 1$.

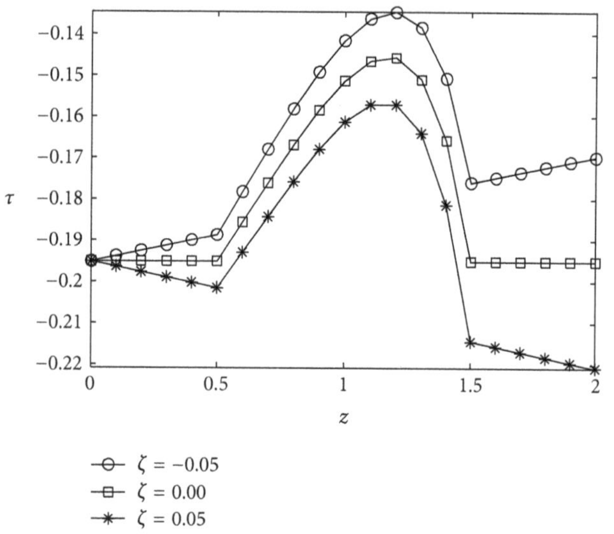

FIGURE 13: Variation of wall shear stress w with respect to ζ when $n = 5$, $\epsilon = 0.2$, $\beta = 5$, $H = 1$, $v = 0.01$, $g = 0.03$, and $R_c = 0.1$.

shear stress than the untapered and tapered artery which is diverging.

The slip velocity at the wall of the stenosed artery and the moving catheter significantly influence the shear stress at the wall, which is noticed from Figures 18 and 19. Here, it is observed that gain in slip velocity (resp., moving catheter) reduces the shear stress at the wall. The influence of catheter radius on shear stress at the wall is shown in Figure 17. The increasing catheter radius narrows down the lumen of the artery thus resulting in the higher values of shear stress at the wall. As couple-stress fluid parameter β increases, the shear stress at the wall decreases. This behavior is shown in Figure 14. It is observed that the theoretical distribution of shear stress along the wall reaches a maximum at throat of the stenosis and then rapidly decreases in the diverging section. This behavior is the same in terms of the change in both the radius of the catheter R_c and the couple-stress fluid parameter β whose behavior is depicted graphically in Figures 17 and 14, respectively.

5. Conclusion

A mathematical model has been built to discuss the flow of blood through a catheterized asymmetric tapered stenosed artery with slip velocity at the stenosed wall and a moving catheter. Closed form solution is obtained and the effects of various geometric, fluid parameters and magnetic field on the axial velocity of the blood and the shear stress at the wall are studied.

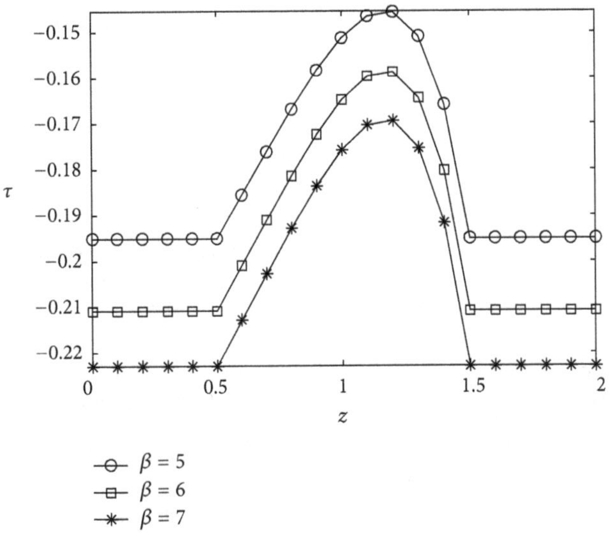

FIGURE 14: Variation of wall shear stress w with respect to β when $n = 5$, $\epsilon = 0.2$, $\zeta = 0$, $H = 1$, $\nu = 0.01$, $g = 0.03$, and $R_c = 0.1$.

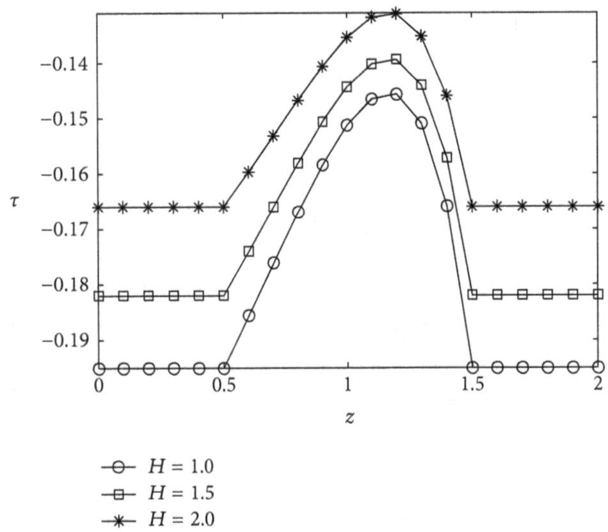

FIGURE 16: Variation of wall shear stress w with respect to H when $n = 5$, $\zeta = 0$, $\beta = 5$, $\epsilon = 0.2$, $\nu = 0.01$, $g = 0.03$, and $R_c = 0.1$.

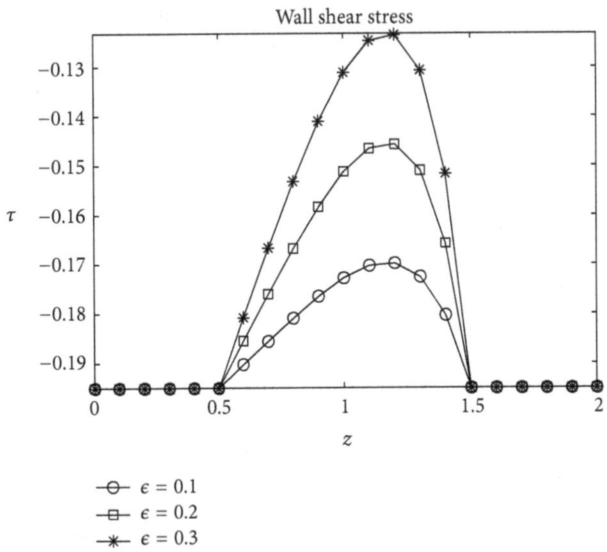

FIGURE 15: Variation of wall shear stress w with respect to ϵ when $n = 5$, $\zeta = 0$, $\beta = 5$, $H = 1$, $\nu = 0.01$, $g = 0.03$, and $R_c = 0.1$.

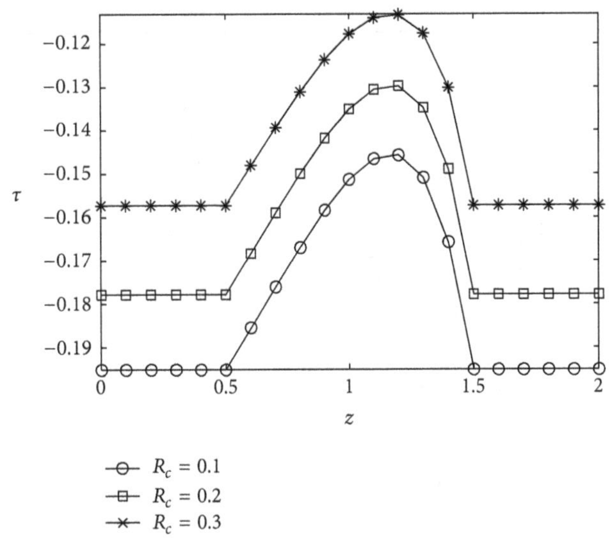

FIGURE 17: Variation of wall shear stress w with respect to R_c when $n = 5$, $\zeta = 0$, $\beta = 5$, $\epsilon = 0.2$, $\nu = 0.01$, $g = 0.03$, and $H = 1$.

There is special importance of couple-stress fluids compared to the Newtonian fluids because of their wide existence such as oil, blood, and polymeric solutions. In view of what is mentioned above, an analytic approach was followed to solve the mathematical model of blood flow through stenosed tapered artery under the assumption of mild stenosis. The resultant observations are summarized as follows:

(i) As the height and the stenosis length are increasing, the obstruction to the flow of blood is increasing.

(ii) Converging tapered artery has more shear stress at the wall than the nontapered and the diverging tapered artery.

(iii) Diverging tapered artery has least shear stress at the wall.

(iv) The increment in the Hartmann number enhances the blood velocity and wall shear stress.

(v) The axial velocity is decreasing while the couple-stress fluid parameters β and the height of the stenosis ϵ are increasing.

(vi) Three different values of the slip velocity and the velocity of the catheter at the arterial boundary and the catheter wall are considered; they are showing a significant influence on the shear stress at the wall and the axial velocity.

The modeling and simulation of the above phenomena are very realistic and are expected to be very useful in

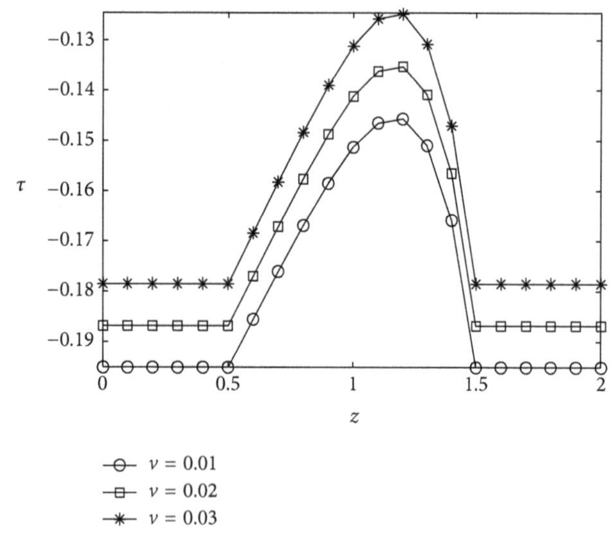

FIGURE 18: Variation of wall shear stress w with respect to ν when $n = 5, \zeta = 0, \beta = 5, \epsilon = 0.2, R_c = 0.1, g = 0.03$, and $H = 1$.

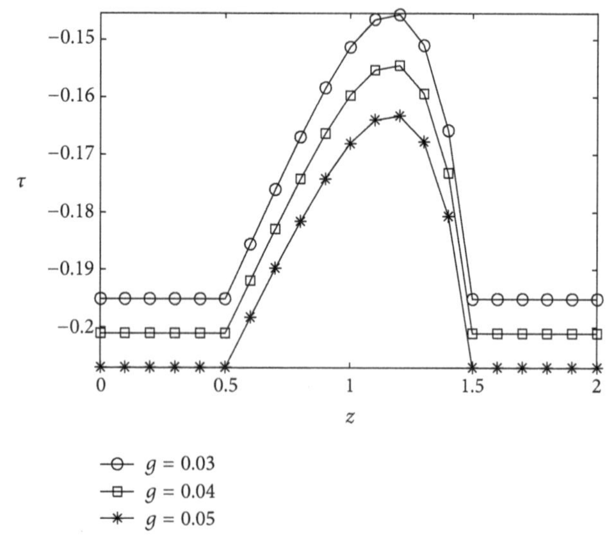

FIGURE 19: Variation of wall shear stress w with respect to g when $n = 5, \zeta = 0, \beta = 5, \epsilon = 0.2, \nu = 0.01, R_c = 0.1$, and $H = 1$.

predicting the behavior of physiological parameters in the diagnosis of various arterial diseases.

Symbols

$\mathbf{B}$: Magnetic flux intensity
B_0: External transverse magnetic field
$c_i,\ i = 1, \ldots, 4$: Constants of integration
d: Location of the stenosis
$d_i = -c_i/k_s,\ i = 1, \ldots, 4$: Defined constants
$\mathbf{E}$: Electric field intensity
$\mathbf{f}$: Body force
$\mathbf{F}_m$: Electromagnetic force
g: Velocity of the moving catheter
$H = B\sqrt{R_0^2\sigma/\mu}$: Hartmann number
I_0: Modified Bessel function of the first kind and order zero
$\mathbf{J}$: Current density
K_0: Modified Bessel function of the second kind and order zero
$\mathbf{l}$: Body couple moment
L_0: Stenosis length
$\mathbf{M}$: Couple-stress tensor
$n_1 = \sqrt{(\beta^2 + \sqrt{\beta^4 - 4\beta^2H^2})/2}$: Defined constant
$n_2 = \sqrt{(\beta^2 - \sqrt{\beta^4 - 4\beta^2H^2})/2}$: Defined constant
p: Pressure
Q: Volumetric flow rate
R_c: Radius of the nonconstricted region
$\text{Re} = \rho u_0 R_0/\mu$: Reynolds number
$R(z)$: Radius of the tube
u: Radial velocity
$\mathbf{V}$: Velocity vector
ν: Slip velocity
w: Axial velocity
w_0: Typical axial velocity
$\beta = R_0\sqrt{\mu/\eta}$: Couple-stress fluid parameter
$\Delta = \partial^2/\partial r^2 + (1/r)(\partial/\partial r)$: Laplacian operator
ϵ: Maximum height of the stenosis
η, η': Couple-stress coefficients
$\Gamma = L_0/d$: Defined constant
$\gamma = L_0/L_1$: Defined constant
R_f: Resistance to the flow
λ, μ: Viscosity coefficients
$\omega = \eta'/\eta$: Parameter associated with the couple-stress fluid
ϕ: Tapering angle
ρ: Density of the fluid
σ: Electrical conductivity
$\boldsymbol{\tau}$: Shear stress tensor
τ_{rz}: Shear stress
τ_{rz}^A: Skew-symmetric part of shear stress
τ_{rz}^S: Symmetric part of shear stress
$\xi = R_0/L_0$: Defined constant
ζ: Tapering parameter.

Conflict of Interests

The authors declare that there is no conflict of interests regarding the publication of this paper.

References

[1] R. C. Diggery and D. T. Grint, *Catheters: Types, Applications and Potential Complications*, Nova Science, 2012.

[2] D. Filgueiras-Rama, A. Estrada, J. Shachar et al., "Remote magnetic navigation for accurate, real-time catheter positioning and ablation in cardiac electrophysiology procedures," *Journal of Visualized Experiments*, no. 74, Article ID e3658, 2013.

[3] B. Mols, "Magnetic-field navigation for catheters," TU Delft Outlook, University Magazine, 2004.

[4] V. K. Stokes, *Theories of Fluids with Microstructure: An Introduction*, Springer, Berlin, Germany, 1984.

[5] A. C. Eringen, *Theory of Micropolar Fluids*, no. RR-27, Purdue University School of Aeronautics and Astronautics, West Lafayette, Ind, USA, 1965.

[6] S. C. Cowin, "The theory of polar fluids," *Advances in Applied Mechanics*, vol. 14, pp. 82–279, 1974.

[7] P. Chaturani and V. S. Upadhya, "Pulsatile flow of a couple stress fluid through circular tubes with applications to blood flow," *Biorheology*, vol. 15, no. 3-4, pp. 193–201, 1978.

[8] P. Chaturani and V. P. Rathod, "A critical study of Poiseuille flow of couple stress fluid with applications to blood flow," *Biorheology*, vol. 18, no. 2, pp. 235–244, 1981.

[9] A. Kollin, "Electromagnetic flowmeter: Principle of method and its application to blood flow measurement," *Proceedings of the Society for Experimental Biology and Medicine*, vol. 35, pp. 235–244, 1936.

[10] E. M. Korchevskii and L. S. Marochnik, "Magnetohydrodynamic version of movement of blood," *Biophysics*, vol. 10, no. 2, pp. 411–413, 1965.

[11] A. R. Rao and K. S. Deshikachar, "Physiological type flow in a circular pipe in the presence of a transverse magnetic field," *Journal of the Indian Institute of Science*, vol. 68, pp. 247–260, 1988.

[12] V. A. Vardanyan, "Effect of a magnetic field on blood flow," *Biofizika*, vol. 18, no. 3, pp. 491–496, 1973.

[13] V. P. Rathod and Gayatri, "Effect of magnetic field on two layer model of pulsatile blood flow with micro-organisms," *Bulletin of Pure and Applied Sciences E. Mathematics and Statistics*, vol. 19, no. 1, pp. 1–13, 2000.

[14] D. S. Sankar and K. Hemalatha, "Pulsatile flow of Herschel-Bulkley fluid through catheterized arteries—a mathematical model," *Applied Mathematical Modelling*, vol. 31, no. 8, pp. 1497–1517, 2007.

[15] D. Srinivasacharya and D. Srikanth, "Effect of couple stresses on the pulsatile flow through a constricted annulus," *Comptes Rendus Mécanique*, vol. 336, no. 11-12, pp. 820–827, 2008.

[16] D. N. Riahi, R. Roy, and S. Cavazos, "On arterial blood flow in the presence of an overlapping stenosis," *Mathematical and Computer Modelling*, vol. 54, no. 11-12, pp. 2999–3006, 2011.

[17] V. P. Srivastava and S. Mishra, "Non-Newtonian arterial blood flow through an overlapping stenosis," *Applications and Applied Mathematics*, vol. 5, no. 1, pp. 225–238, 2010.

[18] S. Chakravarty and P. K. Mandal, "Mathematical modelling of blood flow through an overlapping arterial stenosis," *Mathematical and Computer Modelling*, vol. 19, no. 1, pp. 59–70, 1994.

[19] V. Vand, "Viscosity of solutions and suspensions. I. Theory," *The Journal of Physical Chemistry*, vol. 52, no. 2, pp. 277–299, 1948.

[20] L. Bennett, "Red cell slip at a wall in vitro," *Science*, vol. 155, no. 3769, pp. 1554–1556, 1967.

[21] Y. Nubar, "Blood flow, slip, and viscometry," *Biophysical Journal*, vol. 11, no. 3, pp. 252–264, 1971.

[22] P. Chaturani and D. Biswas, "A comparative study of poiseuille flow of a polar fluid under various boundary conditions with applications to blood flow," *Rheologica Acta*, vol. 23, no. 4, pp. 435–445, 1984.

[23] R. Ponalagusamy, "Blood flow through an artery with mild stenosis: a two-layered model, different shapes of stenoses and slip velocity at the wall," *Journal of Applied Sciences*, vol. 7, no. 7, pp. 1071–1077, 2007.

[24] B. K. Mishra and S. S. Shekhawat, "Magnetic effect on blood flow through a tapered artery with stenosis," *Acta Ciencia Indica*, vol. 1029, pp. 295–301, 2007.

[25] R. Ponalagusamy and R. Tamil Selvi, "Blood flow in stenosed arteries with radially variable viscosity, peripheral plasma layer thickness and magnetic field," *Meccanica*, vol. 48, no. 10, pp. 2427–2438, 2013.

[26] D. Biswas and U. S. Chakraborty, "Pulsatile blood flow through a catheterized artery with an axially nonsymmetrical stenosis," *Applied Mathematical Sciences*, vol. 4, pp. 2865–2880, 2010.

13

Numerical Simulation of Bubble Coalescence and Break-Up in Multinozzle Jet Ejector

Dhanesh Patel,[1] Ashvinkumar Chaudhari,[2] Arto Laari,[3] Matti Heiliö,[4] Jari Hämäläinen,[2] and Kishorilal Agrawal[5]

[1] *Centre for Industrial Mathematics and Department of Applied Mathematics, Faculty of Technology and Engineering, The M. S. University of Baroda, Vadodara, Gujarat 390001, India*
[2] *Centre of Computational Engineering and Integrated Design (CEID), Lappeenranta University of Technology, P.O. Box 20, 53851 Lappeenranta, Finland*
[3] *Department of Chemistry, Lappeenranta University of Technology, P.O. Box 20, 53851 Lappeenranta, Finland*
[4] *Department of Mathematics and Physics, Lappeenranta University of Technology, P.O. Box 20, 53851 Lappeenranta, Finland*
[5] *Department of Chemical Engineering, Faculty of Technology and Engineering, The M. S. University of Baroda, Vadodara, Gujarat 390001, India*

Correspondence should be addressed to Dhanesh Patel; pdhanesh@yahoo.com

Academic Editor: Guan H. Yeoh

Designing the jet ejector optimally is a challenging task and has a great impact on industrial applications. Three different sets of nozzles (namely, 1, 3, and 5) inside the jet ejector are compared in this study by using numerical simulations. More precisely, dynamics of bubble coalescence and breakup in the multinozzle jet ejectors are studied by means of Computational Fluid Dynamics (CFD). The population balance approach is used for the gas phase such that different bubble size groups are included in CFD and the number densities of each of them are predicted in CFD simulations. Here, commercial CFD software *ANSYS Fluent 14.0* is used. The realizable k-ε turbulence model is used in CFD code in three-dimensional computational domains. It is clear that Reynolds-Averaged Navier-Stokes (RANS) models have their limitations, but on the other hand, turbulence modeling is not the key issue in this study and we can assume that the RANS models can predict turbulence of the carrying phase accurately enough. In order to validate our numerical predictions, results of one, three, and five nozzles are compared to laboratory experiments data for Cl_2-NaOH system. Predicted gas volume fractions, bubble size distributions, and resulting number densities of the different bubble size groups as well as the interfacial area concentrations are in good agreement with experimental results.

1. Introduction

There are number of industrial processes in which two-phase flows, that is, gas-liquid mixture, in a jet ejector are encountered. Hence, reactions occurring in gas-liquid systems are of great importance in the chemical as well as in the process industry. Mass transfers in dispersions are directly related to the mass transfer coefficients as well as the interfacial area. The jet ejector is one kind of a venturi scrubber and it is widely used for conducting gas-liquid reactions in practical applications in industry such as pollution control and waste water treatment. Due to their simple construction, low operating cost, high energy efficiency, and good mass transfer characteristics, the jet ejectors have many advantages when used as the gas-liquid contactors. The experimental observations show that dispersed bubbles towards the bottom of the jet ejector cause highly nonuniform volume distribution in the jet ejector. The gas volume fraction, the interfacial area, and the Sauter mean bubble diameter are the three important parameters that characterize the internal flow structure of gas-liquid flows in the jet ejector [1]. The interfacial transport of mass and momentum are proportional to the interfacial area and the driving forces. This is an important parameter required for a two-fluid model formulation [1]. The mean bubble diameter serves as a link between the gas volume fraction and the interfacial area concentration [1]. An accurate knowledge of local distributions of these three parameters is of great importance to eventual understanding and modeling of the interfacial transfer processes [1, 2]. Depending on the gas flow rate, two main flow regimes are observed in the jet ejector, namely, the homogeneous bubbly flow regime and the heterogeneous (churn-turbulent flow) regime [3]. The homogeneous regime is encountered at relatively low gas velocities and characterized by a narrow bubble size distribution and radially uniform gas holdup and it is the most desirable one for practical applications, because it offers a large contact area [2–4].

The bubble size distribution and gas holdup in gas-liquid dispersions depend extensively on the jet ejector geometry, operating conditions, and the physicochemical properties of the two phases. The design of the jet ejector has primarily been carried out by means of empirical or semiempirical correlations based mainly on experimental data. The scale-up of the jet ejector is still poorly understood due to the complexity of flow patterns and their unknown behavior under different sets of design parameters such as area ratio, projection ratio, nozzle diameter, length of free jet, throat, diffuser, convergence angle, divergence angle, and physical properties of the liquid. As a whole, the phenomenon depends strongly on the jet ejector geometry and fluid dynamics involved. It is important to note that the similar kind of study has been developed for bubble column reactor by [3].

The method to gain more knowledge and detailed physical understanding of the hydrodynamics in the jet ejector is Computational Fluid Dynamics (CFD). CFD can be regarded as an effective tool to clarify the importance of physical effects (e.g., gravity, surface tension) on flow by adding or removing them. An increasing number of papers deal with CFD applications of bubble columns [3, 5–7]. To analyze the flow pattern of the jet ejector both in steady and transient state conditions employing CFD, the majority of researchers use either two- or three-dimensional models, when numerical simulations are usually compared to experimental data. As most of the early CFD studies consider monodispersed bubble size distributions ignoring break-up and coalescence mechanisms, their validity is limited [3].

During the last two decades, significant developments have been done in the modeling of two-phase flow processes because of the introduction of the two-fluid models [1]. In the two-fluid models, the interfacial transfer terms are related to the interfacial area concentration and the degree of turbulence near the interfaces [1]. Since the interfacial area concentration represents the key parameter that links the interaction of the phases, significant attention has been paid towards developing a better understanding of the coalescence and breakage effects due to interactions among bubbles and between bubbles and turbulent eddies for gas-liquid bubbly flows [1, 2, 7–11]. The population balance method is a well-known method for tracking the size distribution of the dispersed phase and accounting for the breakage and coalescence dynamics in bubbly flows [1, 12–22]. The population balance method was also used by Hämäläinen et al. [23] and Hämäläinen [24] in paper industry for papermaking suspension flow.

In gas-liquid two-phase systems, bubble break-up and coalescence can greatly influence their overall performance by altering the interfacial area available for the mass transfer between the phases [3]. Therefore, in order to develop reliable predictive tools for designing the jet ejectors, it is essential to obtain some insight into the prevailing phenomena through simulation models based on the bubble formation and distraction mechanisms, that is, bubble coalescence and break-up [3, 12, 15, 25–28], incorporated into CFD simulation making it possible to calculate hydrodynamic variables such as liquid velocity, gas holdup, and bubble size distributions.

In this work, an attempt has been made to demonstrate the possibility of combining the population balance models with Computational Fluid Dynamics (CFD) for the case of a gas-liquid bubbly flow in the jet ejector. In all these processes gas holdup, ε_g, and bubble size distribution are important design parameters, since they define the gas-liquid interfacial area available for interfacial area mass transfer (a), which is given by

$$a = \frac{6\varepsilon_g}{d_{32}}, \tag{1}$$

where d_{32} is the mean Sauter diameter of the bubble size distribution [3]. The MUSIG model implemented in *ANSYS Fluent 14.0*, which accounts for the nonuniform bubble size distribution in a gas-liquid flow [1, 2, 6, 16, 29], is used in this paper. Gas volume fraction, bubble size distribution, number density of bubbles, gas and liquid pressure variation, interfacial area concentration, and gas and liquid velocity variation in jet ejector are predicated. The flow pattern development has been studied at the free jet end, and at the throat end, and at the end of the ejector in detail. In addition, numerical predictions are compared with experiments and the predicated gas interfacial area is in a good agreement with experimental results.

2. Jet Ejector

A large choice of gas-liquid contactors, for example, the falling film column, spray column, packed column, plate column, bubble column, mechanically agitated contactors, spray towers, and venturi scrubbers, are available for understanding the mass transfer process. Among these, the venturi scrubber is a wet type design for gas-liquid contactors. In venturi type of scrubber

(i) liquid is a medium to absorb objectionable gases and particulates from industrial gaseous waste streams;

(ii) a high velocity section of fluid jet is utilized to bring the liquid and gas into intimate contact with each other.

Venturi scrubbers fall into two categories [30, 31].

In the first category, it uses mechanical blower to draw a high velocity gas stream through the system. The liquid was originally at rest but once the gas accelerates, it splits into droplets. Particulates and gases are then confined into the comparatively slower moving droplets. This type is called a "high energy venturi scrubber" (HEVS). The scrubbing liquid can be introduced in two ways:

(i) If the liquid is introduced through nozzles which is usually at the throat it is known as Pearce-Anthony venturi scrubber.

(ii) In this case liquid is introduced as a film which is usually known as the wetted approach type.

Secondly, a mechanical pump or compressor is used to generate a high velocity to the liquid/fluid jet. This liquid/fluid jet creates suction and gas is entrained into it by transfer of momentum. This type of the venturi scrubber is called an ejector venturi scrubber or "jet ejector."

The jet ejectors have some advantages over other types of contactors which are mentioned as follows [31, 32]:

(i) Lower initial capital cost.

(ii) Simple construction and being compact.

(iii) Easy installation and operation.

(iv) No moving parts, so little chances of mechanical failure and hence highly reliable.

(v) Being able to deal with wet, hot, and corrosive gases and thick, aggressive, and inflammable particles.

(vi) Ability to separate gaseous pollutant and fine particulate matter simultaneously.

(vii) Being able to handle large gas flow rates.

(viii) High heat and mass transfer rates and interfacial area.

However, it is not energy efficient equipment as a fluid moving device. But it has been reported that it has high efficiency as a gas-liquid contacting device [33–36].

The jet ejectors are especially effective in chemical and biochemical industries for gas purification and for collaborating gas-liquid reactions like chlorination, oxidation, hydrogenation, and hydroformulation processes. Jet ejectors use high kinetic energy of the operating fluid jet to promote break-up and distribution of the suction fluid into small droplets/bubbles and to pull the gas through the system and push through the connected outlet.

A typical gas-liquid jet ejector is displayed in Figure 1. It consists of a converging section, a throat section, and a diffuser/divergent section. The gas is accelerated to atomize the scrubbing liquid in the convergent section to reach a higher velocity in the throat. Throat is used for interaction of liquid and gases. In the diffuser/divergent section the gas is slowing down allowing some recovery of pressure [37, 38].

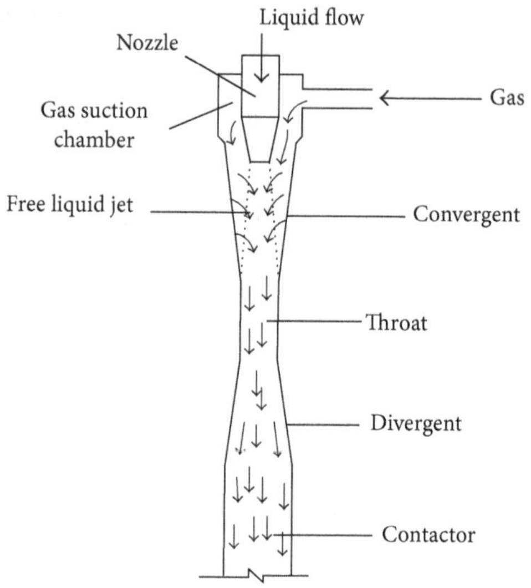

FIGURE 1: Typical gas-liquid jet ejector [31, 113].

Jet ejectors have favorable mixing and mass transfer characteristics so they are more used among other gas-liquid contactors. A jet ejector is a device in which suction, mixing, and dispersal of secondary fluid take place and are based on the principle of utilizing the kinetic energy of a high velocity motive (primary) fluid jet to entrain the secondary phase to create fine dispersion of two phases (Utomo et al. [39]). The secondary fluid may be dispersed by the shearing action of the high velocity motive fluid or motive fluid may get dispersed when it is captured by a secondary fluid [40].

Figure 1 shows the typical ejector system in which the jet of primary fluid, typically liquid, is pumped into the system through high velocity through a nozzle flowing out of a nozzle which creates a low pressure region in the suction chamber, into which secondary fluid, typically gas, moves according to Bernoulli's principle. The driving force for entrainment of the secondary fluid is developed due to the pressure difference between the entry point of the secondary fluid and the nozzle tip. There is a mixing of gas and liquid phases and a gas-liquid dispersion takes place in the mixing tube. In the diffuser the pressure is recovered. Coaxial-flow and froth-flow are the two principal flow regimes in jet ejectors. In the annular region formed between the jet of primary fluid and ejector wall, a central core of primary fluid with secondary fluid flowing was observed. This regime forms coaxial-flow. Froth-flow consists of liquid in which gas phase is completely dispersed in the form of bubble [41]. The phenomenon of change from coaxial-flow to froth-flow is termed as mixing shock [42]. The small bubbles are generated due to mixing shock which turns into creation of large interfacial area (~2000 m^2/m^3). Therefore greater rates of reaction and superior gas-liquid mass transfer rates are obtained in ejectors, in comparison to other common gas-liquid contactors [31].

Depending on application, there may be different objectives for design of an ejector which are as follows [43]:

(a) To achieve greater entrainment of the secondary fluid.

(b) To yield deep mixing between the two fluids.

(c) To inject fluids from a region of low pressure to a region of higher pressure.

Jet ejector may be used as a vacuum producing device as well as jet pump. With the rapid growth of the chemical process industry, their use as entraining and pumping corrosive liquids, slurries, fumes, and dust-laden gases has increased. Their use as mass transfer equipment for liquid-liquid extraction, gas absorption, gas stripping, and slurry reaction, like hydrogenation, oxidation, chlorination, fermentation, and so forth, has increased [31, 44–53].

The numbers of researchers have attempted to optimize performance of jet ejector [40, 51, 54–71] due to its increasing growth rate in usage. Many researchers have studied the mass transfer characteristics and performance of the jet ejectors followed by contactors, draft tube, packed column, or bubble column and they have similar conclusion that there is less mass transfer coefficient in the extended portion in comparison to the ejector [31, 34, 39, 56, 69–87]. Dutta et al. [74] and Gamisans et al. [60] studied the jet ejectors and observe that jet ejector without diffuser or throat is less effective in comparison to ejector with them. Das and Biswas [40] observe that efficiency of ejector depends on the design of the suction chamber, the throat, the diffuser, and the nozzle. Apart from the dimensions of the various sections of the ejector, the factors such as shape of the entrance to the parallel throat, shape of entry of convergent section, throat aspect ratio, angle of divergence and convergence, and the projection ratio are also important factors to be reflected. Various researchers have studied the effect of geometry of jet ejector in the optimal design of jet ejector. Also, many researchers have worked on flow patterns in jet ejector (see [79, 88–92]).

Performance of jet ejector is also depending on the factors like bubble size distributions, correlation for entrainment, bubble diameter, drag force, and gas holdup and factors affecting the bubble size. Measurement of bubble size is critical issue. The researchers in these directions are Lefebvre [93], Liu [94], Schick [95], Xianguo and Tankin [96], Dennis Gary [97], Azad and Syeda [98], Frank [88], Silva et al. [99], Kudzo [100], Ciborowski and Bin [101], Ohkawa et al. [102], Sheng and Irons [103], Kitscha and Ocamustafaogullari [104], Bailer [105], Evans et al. [106], Ceylan et al. [107], Havelka et al. [76], Mandal et al. [81], Zahradník and Fialová [90], Bailer [105], Pawelczyk and Pindur [108], Dutta and Raghavan [74], Biń [109], and Zheng et al. [110].

Figure 2 shows the major parts like primary fluid inlet, secondary fluid inlet, suction chamber, converging section, throat/mixing zone, and diverging section/diffuser of an ejector. The symbols which we have used for describing different components are as follows:

(i) Length of throat (L_T), length of diffuser (L_D), and distance between nozzle and commencement of throat (L_{TN}).

(ii) Diameter of nozzle (D_N), diameter of throat (D_T), and diameter of suction chamber (D_S).

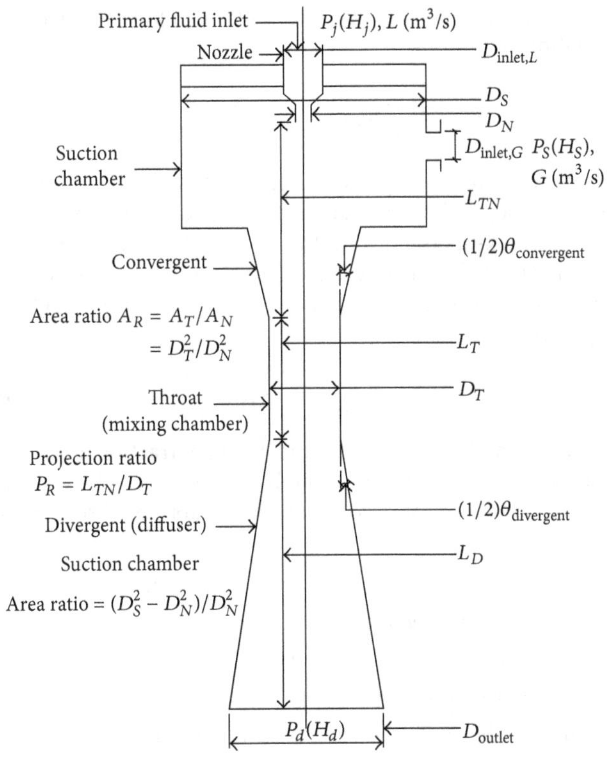

FIGURE 2: Schematic diagram showing geometry of an ejector [31, 111, 112].

(iii) Area of throat (A_T), area of nozzle (A_N), and area of suction (A_S).

(iv) Angle of converging sections ($\theta_{convergent}$) and angle of diverging sections ($\theta_{divergent}$).

Performance of the ejectors has been studied in terms of

(a) area of throat/area of nozzle, that is, area ratio ($A_R = A_T/A_N$),

(b) length of throat/diameter of throat, that is, throat aspect ratio (L_T/D_T),

(c) distance between nozzle tip and the commencement of throat/diameter of throat, that is, projection ratio ($P_R = L_{TN}/D_T$),

(d) suction chamber area ratio ($A_S/A_N = (D_S^2 - D_N^2)/D_N^2$).

3. Mathematical Model

The poly dispersed multiphase flow is observed in jet ejector. It means that the dispersed phase covers a wide range of size groups. One of the characteristics of this flow is that the different sizes of the dispersed phase interact with each other through the processes of break-up and coalescence. A population balance equation is framed to deal with this kind of a flow. It is well known that the population balance model is best fitted method to calculate the size distribution of a dispersed phase which includes break-up and coalescence effect [3]. Our model is in parallel to the model developed by

Mouza et al. [3]. The general form of the population balance equation is

$$\frac{\partial n_i}{\partial t} + \nabla \cdot \left(\vec{u}_g n_i\right) = b_B - d_B + b_C - d_C, \quad (2)$$

where n_i, b_B, d_B, b_C, d_C, and $\vec{u}_g$ represent the number density of size group i, the birth rate due to break-up, the death rate due to break-up, the birth rate due to coalescence, the death rate due to coalescence, and the gas velocity, respectively. The relation between number density n_i and the volume fraction α_g is given by

$$\alpha_g f_i = n_i V_i, \quad (3)$$

where f_i and V_i represent the volume fraction and the corresponding volume of a bubble of group i, respectively. The coalescence of two bubbles occurs in three steps [3, 19]:

(i) First, the bubbles collide trapping a slight amount of liquid between them.

(ii) Secondly, this liquid film drains until it reaches a critical thickness.

(iii) At the end, the film breaks and the bubbles join together.

The process of coalescence depends on the collision rate of the two bubbles and the collision efficiency. Coalescence is a function of t_{ij} and τ_{ij}, where t_{ij} is the time required for coalescence and τ_{ij} is the contact time. Collision is as a result of turbulence ($\theta_{ij}^{\mathrm{Tu}}$), laminar shear ($\theta_{ij}^{\mathrm{Ls}}$), and buoyancy ($\theta_{ij}^{\mathrm{By}}$). The total coalescence rate is

$$Q_{ij} = \left(\theta_{ij}^{\mathrm{Tu}} + \theta_{ij}^{\mathrm{Bu}} + \theta_{ij}^{\mathrm{Ls}}\right) e^{-t_{ij}/\tau_{ij}}. \quad (4)$$

Collision is also developed due to the difference in rise velocities of bubbles with different sizes. The two conditions observed in jet ejector in parallel to the work of Mouza et al. [3] for bubble column are as follows:

(i) Jet ejectors operate at the homogeneous regime. Due to the narrow bubble size distribution in homogenous regime, the relative effect of buoyancy can be neglected which leads to the assumption that bubbles rise with the same velocity regardless of their size.

(ii) Collision occurs as a result of strong circulation pattern in jet ejector. However this happens at gas rates higher than those encountered in the homogeneous regime and hence laminar shear term may be neglected.

Hence, there is only the turbulent contribution in the model which is given by

$$\theta_{ij}^{\;TU} = n_i n_j S_{ij} \left(\overline{u}_{ti}^{\;2} + \overline{u}_{tj}^{\;2}\right), \quad (5)$$

where n_i, n_j, and $\overline{u}_i$ are the concentrations of bubbles of radius r_{bi}, concentrations of bubbles of radius r_{bj}, and the average turbulent fluctuating velocity, respectively. The collision cross-sectional area (S_{ij}) of the bubble, the time required for coalescence, and the contact time are given by (6), (7), and (8), respectively:

$$S_{ij} = \frac{\pi}{4}\left(R_{bi} + R_{bj}\right)^2, \quad (6)$$

$$t_{ij} = \left(\frac{R_{ij}^{\;3}\rho_l}{16\sigma}\right)^{0.5} \ln\frac{h_0}{h_f}, \quad (7)$$

$$\tau_{ij} = \frac{R_{ij}^{\;2/3}}{\varepsilon^{1/3}}, \quad (8)$$

where $R_{ij}, \rho_l, \sigma, \varepsilon, h_0$, and h_f are the equivalent radius, the density of the liquid phase, the surface tension, the turbulent energy dissipation, the film thickness when collision begins, and the film thickness when ruptures of the film occur, respectively. The birth rate of group i due to coalescence of group k and group j bubbles and the death rate of group i due to coalescence with other bubbles are given by (9) and (10), respectively:

$$b_C = \frac{1}{2}\sum_{j=1}^{i}\sum_{k=1}^{i} Q_{jk} n_j n_k, \quad (9)$$

$$d_C = n_i \sum_{j=1}^{N} Q_{ij} n_j. \quad (10)$$

The break-up of bubbles in turbulent dispersions employs the model developed by Luo and Svendsen [28]. The break-up rate of bubbles of size i into bubbles of size j is given by

$$g\left(v_j : v_i\right) = 0.923\left(1 - \alpha_g\right)\sqrt[3]{\left(\frac{\varepsilon}{d_j^2}\right)} \int_{\mu_{\min}}^{1} \frac{(1+\mu)^2}{\mu^{11/3}} e^{-\tau_c} d\mu, \quad (11)$$

where $\varepsilon, d_j, \mu, \tau_c$, and N represent the turbulent energy dissipation rate, the bubble diameter, the dimensionless size of eddies in the inertial subrange of isotropic turbulence, the critical dimensionless energy for break-up, and the total number of groups, respectively.

The birth rate of group i bubbles due to break-up of larger bubbles and the death rate of group i bubbles due to break-up into smaller bubbles are given by (12) and (13), respectively:

$$b_B = \sum_{j=i+1}^{N} g\left(v_j : v_i\right) n_j, \quad (12)$$

$$d_B = g_i n_i. \quad (13)$$

The mass conservation equation for the liquid phase, the mass conservation equation for the gas phase, and the momentum conservation equation in the Eulerian framework are given by (14), (15), and (16), respectively:

$$\frac{\partial(\alpha_l \rho_l)}{\partial t} + \nabla \cdot (\alpha_l \rho_l \vec{u}_l) = 0, \quad (14)$$

$$\frac{\partial(\alpha_g f_i \rho_g)}{\partial t} + \nabla \cdot (\alpha_g f_i \rho_g \vec{u}_g) = S_i, \quad (15)$$

$$\frac{\partial(\alpha_m \rho_m \vec{u}_m)}{\partial t} + \nabla \cdot \left(\alpha_m \rho_m \vec{u}_m \vec{u}_m - \mu_m \alpha_m \left(\nabla \vec{u}_m + (\nabla \vec{u}_m)^T\right)\right) = -\alpha_m \nabla p + M_{mn} + \rho_m g, \quad (16)$$

where $\rho_m, \vec{u}_m, \alpha_m, \mu_m, p, M_{mn}$, and g represent the density, the velocity, the volume fraction, and the viscosity of the mth phase, the pressure, the interphase momentum exchange between phase m and phase n, and the gravitational force, respectively.

The momentum exchange between gas and liquid phase is given by

$$M_{lg} = \frac{3}{4} \rho_m \frac{\alpha_g}{d_{32}} C_D (\vec{u}_g - \vec{u}_l) |\vec{u}_g - \vec{u}_l|, \quad (17)$$

where C_D is the interfacial drag coefficient.

4. Simulation Parameter for Jet Ejector Geometry

Figure 3 shows the details of the geometry of ejector. Agrawal [31, 111, 112] has done the experiments based on the experimental setup as shown in Figure 4. It is important to note that these experiments were conducted on the industrial stage ejector. Data for the geometry of the ejector is shown in Table 1.

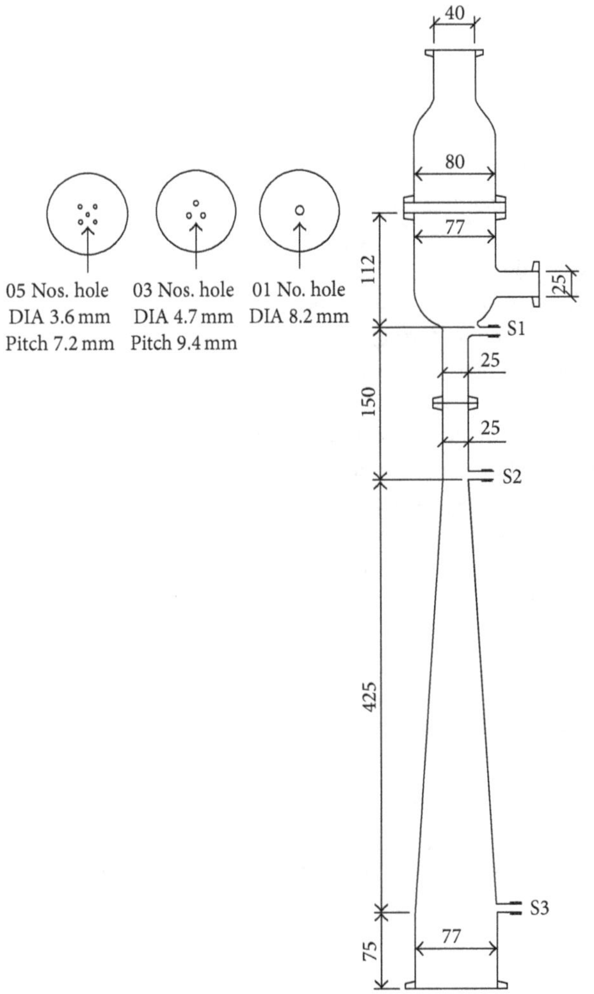

FIGURE 3: Detail of the ejector used in the experimental setup [31, 111, 112].

5. The CFD Simulation

Accurate modeling of the jet ejector requires a Population Balance Modeling (PBM) to be solved simultaneously with a CFD solver because of the presence of bubble-bubble contacts or bubble-liquid contacts. The PBM-CFD model, that is, MUSIG model (*ANSYS Fluent 14.0*), is implemented in this study which combines the population balance method with the break-up [28] and coalescence [19] models in order to predict the bubble size distribution of the gas phase. This model uses the Eulerian-Eulerian model. The MUSIG model has been extensively used for different systems [1–3, 11, 16, 20, 27, 29].

The equations of continuity, momentum, and turbulence for the continuous and dispersed phases give a standard two-phase flow calculation. It can be extended to include number density of bubbles within several size groups using the MUSIG model.

The size range of the bubbles is split into several groups with, for example, groups of equal diameter. Equations are then solved for the number density in each group. These size fractions provide a more accurate measure of the interfacial area density.

In this study, the simulation is done in three dimensions and three geometries of the ejector, namely, with one nozzle, three nozzles, and five nozzles, as described in Figure 3. In the present study, bubbles ranging from 1.4472692×10^{-5} m (bin-20) to 9.7005842×10^{-4} m (bin-0) in diameter are equally divided into 21 classes (see Table 2) as the experimental observation of the gas volume fraction is 100% and therefore we consider 0.005, 0.010, 0.015, 0.018, 0.020, 0.025, 0.032, 0.030, 0.035, 0.040, 0.045, 0.050, 0.055, 0.060, 0.065, 0.070, 0.075, 0.080, 0.085, 0.090, 0.095, and 0.100 which represent the volume fraction of the bubbles of group i, ($i = 0, 1, 2, 3, 4, 5, \ldots, 20$). In view of computational time's resources, we consider the subdivision of the bubble sizes into 21 size groups. All computational results are based on the discretization of the bubble sizes into 21 groups. *ANSYS Fluent 14.0* gives solution to the coupled sets of governing equations for the balances of mass and momentum of each phase. The conservation equations were discretized using the control volume technique. Similar kind of study

Table 1: Dimensions of ejector [31, 111, 112].

			Setup-III	
Nozzle diameter	D_N	8.2 mm	4.7 mm	3.7 mm
Number of nozzles	n	1	3	5
Nozzle number		5	6	7
Pitch*			$2D_N$	
Area ratio (appx.)**	A_R		9.3	
Diameter of throat/mixing tube	D_T		25 mm	
Length of throat/mixing tube***	L_T		150 mm	
Projection ratio#	P_R		4.5	
Angle of convergence	θ_{con}		Well rounded	
Angle of divergence of conical diffuser##	θ_{div}		7°	
Length of the conical diffuser	L_d		425	
Diameter of the diffuser exit	D_C		77	
Diameter of extended contactor	D_C		(—)	
Length of extended contactor	L_C		(—)	
Diameter of the suction chamber	D_S		77 mm	
Length of the suction chamber	L_S		122 mm	
Distance between nozzle & commencement of throat	L_{TN}		112 mm	
Diameter of secondary gas inlet	$D_{G,\text{In}}$		25 mm	
Volume of free jet	V_C		26.32×10^{-6} m^3	
Volume of throat	V_T		73.6×10^{-6} m^3	
Volume of divergence	V_D		1156.78×10^{-6} m^3	
Total ejector volume	V_J		1256.7×10^{-6} m^3	

*Panchal et al. [114], **Acharjee et al. [54], ***Biswas et al. [57], #Yadav and Patwardhan [43], and ##Mukherjee et al. [65].

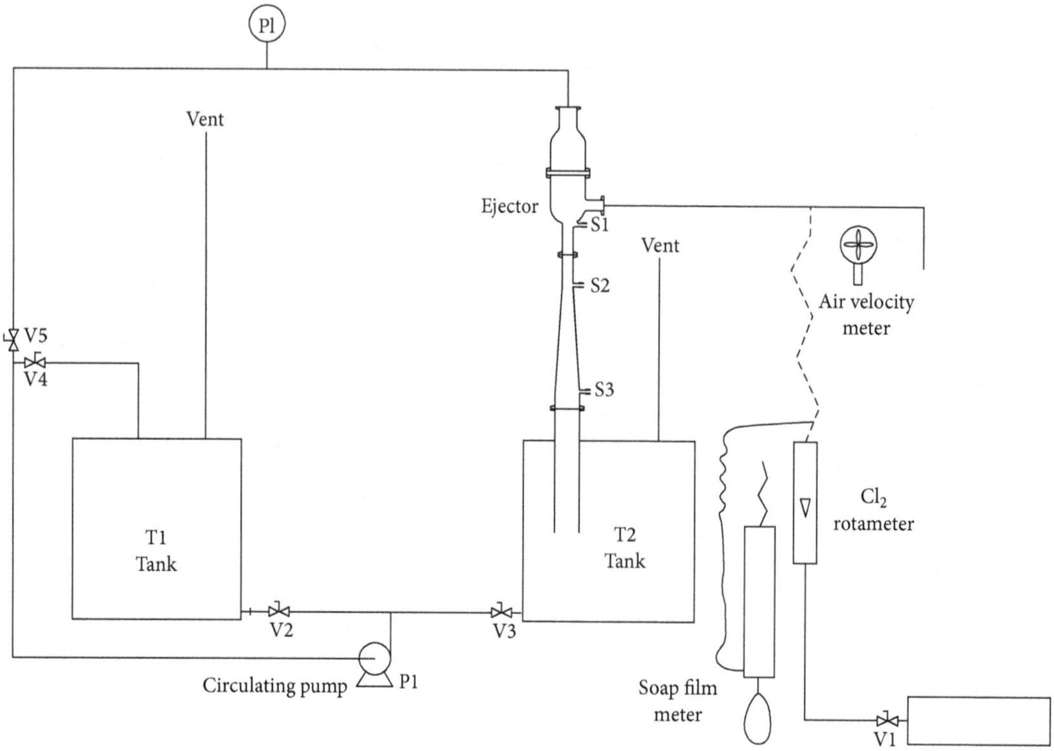

Figure 4: Schematic diagram of the experimental setup [31, 111, 112].

Table 2: Diameter of each bubble class tracked in the simulation.

Bubble diameter, d_i (m) (bin size)	Class index (bin number)
9.7005842×10^{-4}	Bin-0-fraction
7.8611387×10^{-4}	Bin-1-fraction
6.3704928×10^{-4}	Bin-2-fraction
5.1625064×10^{-4}	Bin-3-fraction
4.1835809×10^{-4}	Bin-4-fraction
3.3902814×10^{-4}	Bin-5-fraction
2.747409×10^{-4}	Bin-6-fraction
2.2264395×10^{-4}	Bin-7-fraction
1.802573×10^{-4}	Bin-8-fraction
1.4621301×10^{-4}	Bin-9-fraction
1.1848779×10^{-4}	Bin-10-fraction
9.6019883×10^{-5}	Bin-11-fraction
7.7812388×10^{-5}	Bin-12-fraction
6.3057437×10^{-5}	Bin-13-fraction
5.1100352×10^{-5}	Bin-14-fraction
4.1410594×10^{-5}	Bin-15-fraction
3.3558229×10^{-5}	Bin-16-fraction
2.7194846×10^{-5}	Bin-17-fraction
2.2038101×10^{-5}	Bin-18-fraction
1.7859189×10^{-5}	Bin-19-fraction
1.4472692×10^{-5}	Bin-20-fraction

Table 3: Operating conditions.

Gas phase	Air plus chlorine at 25°C
Liquid phase	Water plus sodium hydroxide at 25°C
Average gas volume fraction	1.0

can be found from Mouza et al. [3] and Ekambara et al. [1].

The simulations have been carried out for nozzle 1 by using 7,22,021 tetrahedral cells, 1,37,3357 triangular interior faces, 1,40,779 triangular wall faces, 430 triangular pressure-outlet faces, 161 triangular velocity-inlet faces, and 156,799 nodes. The simulations have been carried out for nozzle 3 by using 7,23,519 tetrahedral cells, 1,37,6107 triangular interior faces, 1,41,271 triangular wall faces, 430 triangular pressure-outlet faces, 161 triangular velocity-inlet faces, and 157,190 nodes. The simulations have been carried out for nozzle 5 by using 7,15,031 tetrahedral cells, 1,35,9401 triangular interior faces, 1,40,731 triangular wall faces, 430 triangular pressure-outlet faces, 161 triangular velocity-inlet faces, and 155599 nodes.

A second-order discretization scheme was used for the convective terms. At the inlet, gas, liquid, and the average volume fraction have been specified. At the outlet, a relative average static pressure of zero was specified. The operating conditions are summarized in Table 3. The fluid data are taken at room temperature (25°C).

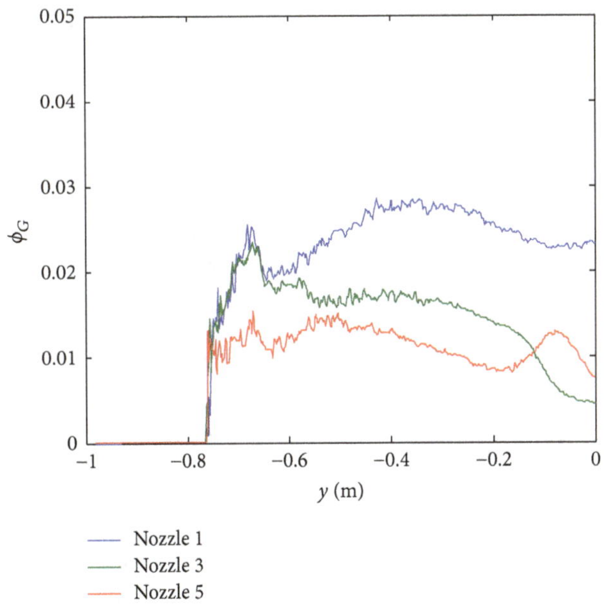

Figure 5: Plot of the volume fraction of gas.

6. Results and Discussion

In this work, we have made an effort to study Chlorine-Aqueous Sodium Hydroxide system. As stated earlier, the bubble sizes were distributed into twenty groups (bins) with diameters between 1.4472692×10^{-5} m and 9.7005842×10^{-4} m.

6.1. Gas Volume Fraction and Liquid Volume Fraction. Figures 5 and 6 show the plot of the predicted gas volume fraction and predicted liquid volume fraction for liquid velocities of 4.6 m/s and gas velocity of 0.2866 m/s for nozzles 1, 3, and 5, respectively. It can be observed that most of the gas volume fraction tends to migrate towards the bottom of the ejector. It is also observed that the gas volume increases in the area of the throat as well as near the gas inlet. As the number of nozzles increases (from nozzle 1, to nozzle 3, to nozzle 5), the gas volume fraction decreases with the same trend in all the nozzles. As we proceed from nozzle to end of jet ejector we are not adding any gas or liquid so it should be parallel to axis. There are three clear cut zones (a) from discharge of liquid nozzle to beginning of the throat, (b) from commencement to the end of the throat, and (c) from end of the throat to end of the diffuser. In zone (a) there is a negative pressure so same quantity of gas has more volume at constant temperature which is clear by peak zone. In the throat there is almost constant pressure which is shown as almost parallel line in this zone. The volume fraction of gas should reduce since the pressure is recovered and higher in zone (c). This fact is evident from Figure 5 as the volume fraction of gas is slanting downward.

6.2. Liquid Velocity and Gas Velocity. Figures 7 and 8 represent the plots of the velocity magnitude of the gas and liquid phases, respectively. These results show that the axial liquid

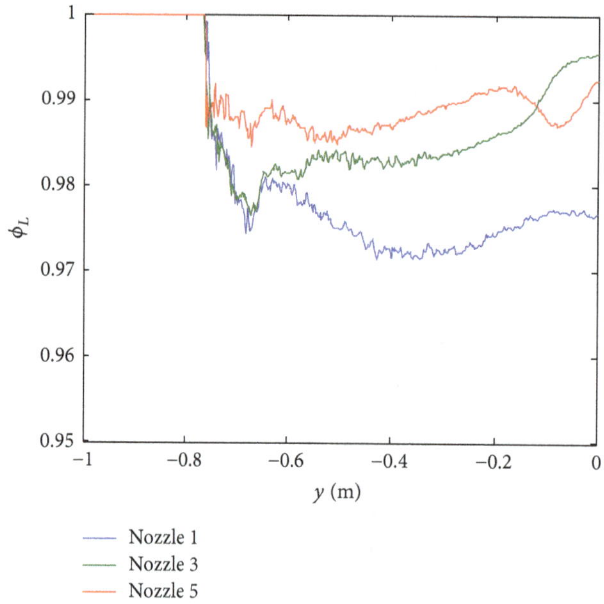

FIGURE 6: Plot of the volume fraction of liquid.

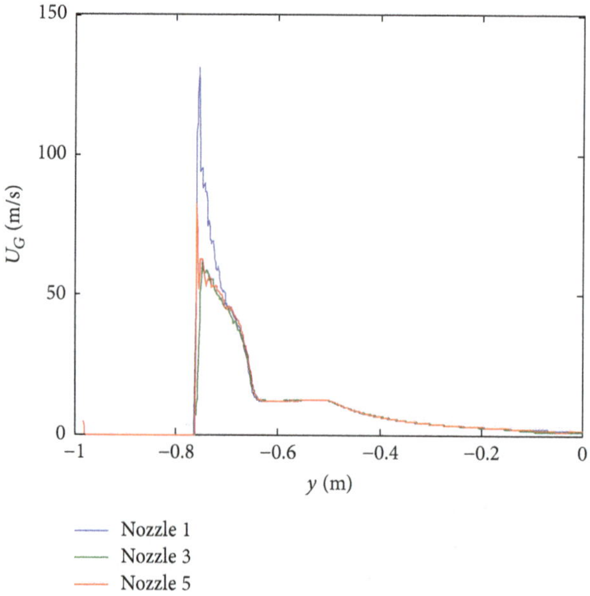

FIGURE 7: Plot of the velocity magnitude of gas.

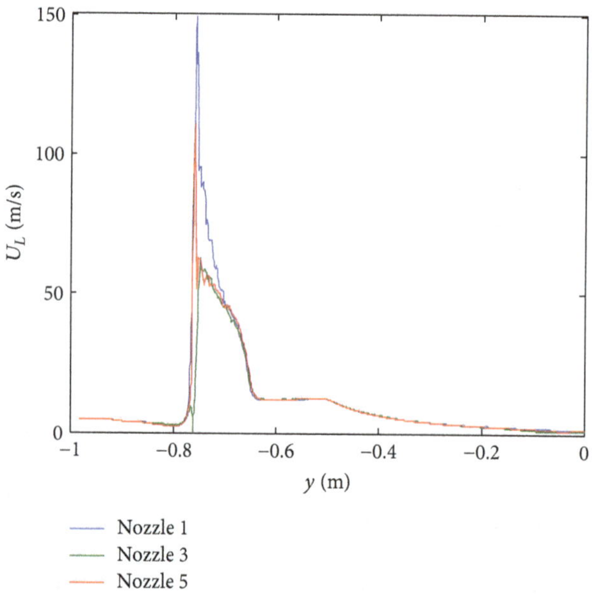

FIGURE 8: Plot of the velocity magnitude of liquid.

velocity profile has a slight degree of asymmetry due to the presence of the gas flow. The liquid velocity in the upper region of the jet ejector is higher than in the lower region. This is due to the increase in cross-sectional area at the bottom of the ejector.

6.3. Bubble Size Distribution. Bubble coalescence and break-up determine the bubble size distribution. Gas volume fraction and the kinetic energy dissipation rate influence the bubble coalescence and break-up in jet ejector. The bubble size distribution varies with the position because of the nonuniform profiles of the gas volume fraction entrapped by liquid stream and dissipation rate (see [1, 3]). The number densities of the bubbles along the vertical direction which is middle line of the jet ejector for bin-0, bin-10, and bin-20 are shown in Figures 10, 12, and 14, respectively, for liquid velocities of 4.6 m/s and gas velocity of 0.2866 m/s. Similarly the number densities of the bubbles in vertical middle plane of the jet ejector for bin-0, bin-10, and bin-20 are shown in Figures 9, 11, and 13, respectively. It is observed that the bubble size having more diameters is at the end. In zones (a) and

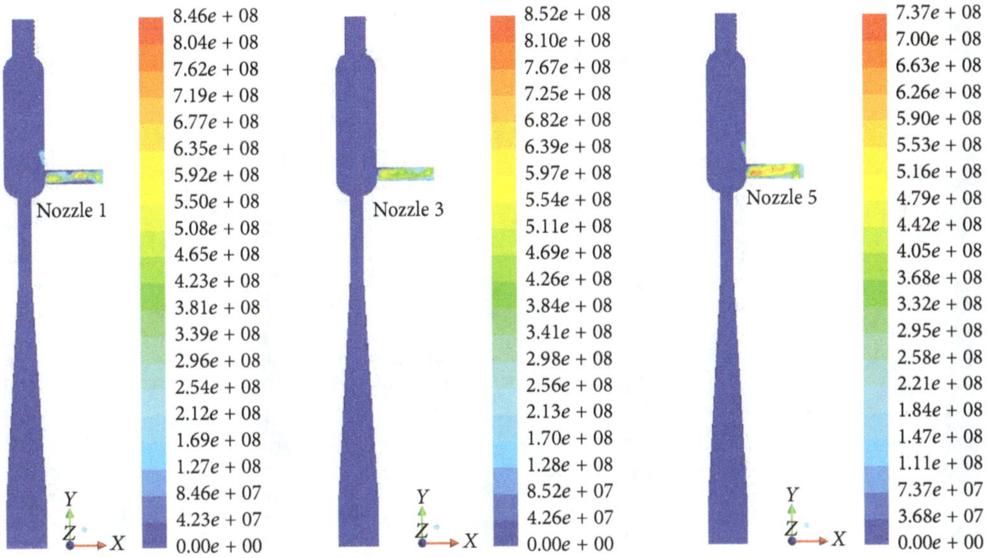

FIGURE 9: Contour of the number density of bin fraction 0 for nozzles 1, 3, and 5.

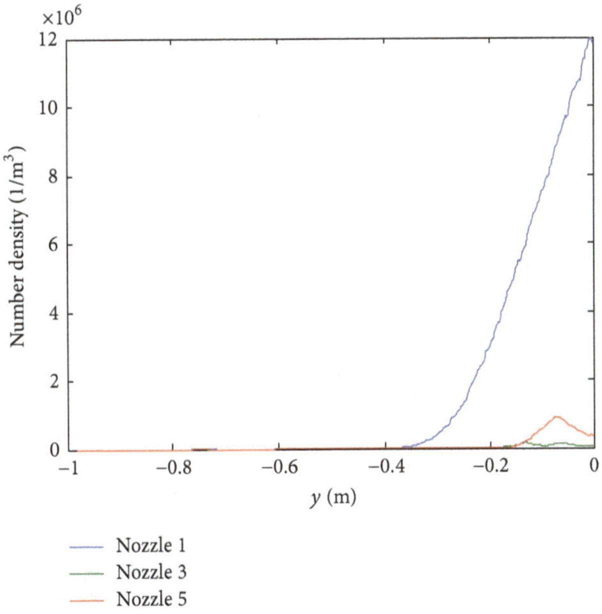

FIGURE 10: Plot of the number density of bin fraction 0.

(b), the bubble size reduces due to the intimate mixing. It can be seen from these figures that the bubble size distribution function reaches an independent state as determined by the balance between birth and death processes that depend on the local flow conditions.

Figures 15–17 represent the discrete number density of the bubbles for nozzles 1, 3, and 5, respectively, for the entire volume. Figures 18–20 represent the discrete number density of the bubbles for nozzles 1, 3, and 5, respectively, for the plane near to gas inlet. It is evident from the figure that as the diameter increases the number density decreases for entire volume as well as the plane near to throat.

6.4. Interfacial Area. The present simulation results show the interfacial area variation along the vertical direction which is middle line of the jet ejector as shown in Figure 22. The variation of interfacial area in vertical middle plane is also shown in Figure 21. From Figures 5 and 22, it can be seen that the appearance of interfacial area concentration looks

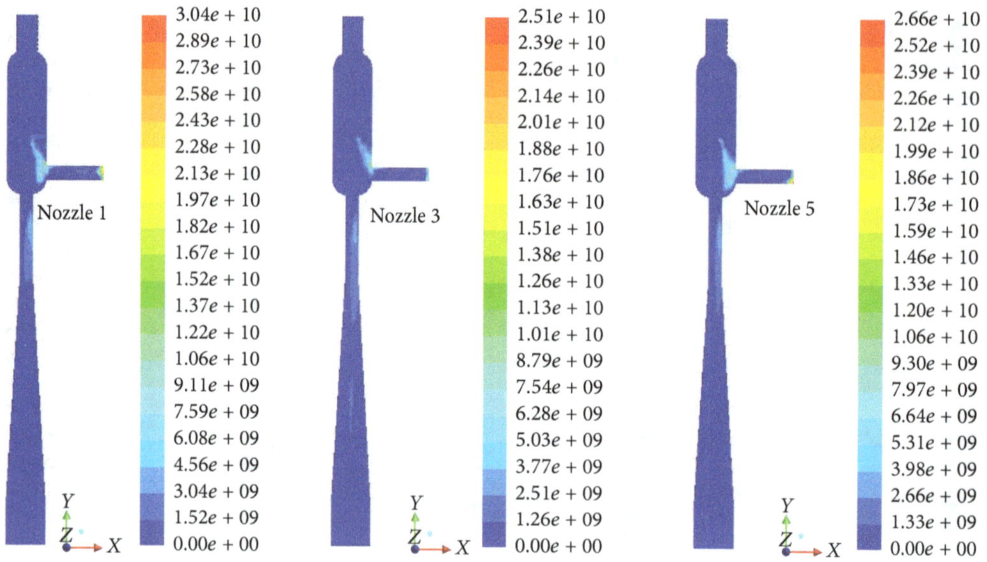

FIGURE 11: Contour of the number density of bin fraction 10 for nozzles 1, 3, and 5.

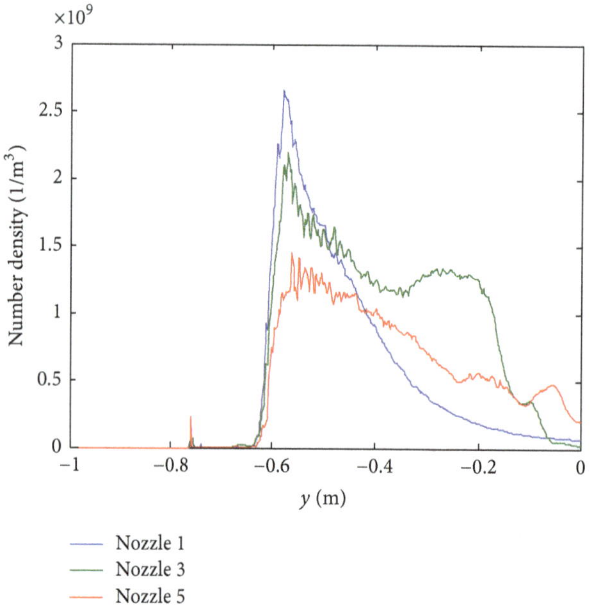

FIGURE 12: Plot of the number density of bin fraction 10.

similar to the gas volume fraction. But the interfacial area is depending not only on the volume fraction of the gas, but also equally on the distribution of the bubble size. Since the measurements of the volume fraction and the interfacial area are independent of each other, the data required for calculating an interfacial area from population balance model provides a valuable test for model prediction [1, 3]. Hence our results show that the birth and the death processes modeled in the population balance model are appropriate to describe the dynamics of bubbles. It can be also seen that the interfacial area concentration reaches the maximum value of 7000 m^2/m^3, 5600 m^2/m^3, and 4500 m^2/m^3 in the mixing zone of the jet ejector for the nozzles 1, 3, and 5, respectively. It is also observed that the interfacial area is decreasing when the number of nozzles is increased. The interfacial area is in good agreement with experimental result [31].

The range of measured values of the interfacial area in the jet ejector is about 3000 to 13000 m^2/m^3 [31]. The experimental values of the interfacial area for nozzle 1, nozzle 3, and nozzle 5 are nearly in the range of 4000–5500 m^2/m^3, 3800–4500 m^2/m^3, and 2500–3500 m^2/m^3, respectively, for different concentrations of gas and liquid [31]. The simulated values of the interfacial area concentration are in the same order of magnitude with the experimental results [31]. More detail on the experimental results can be found in Agrawal [31].

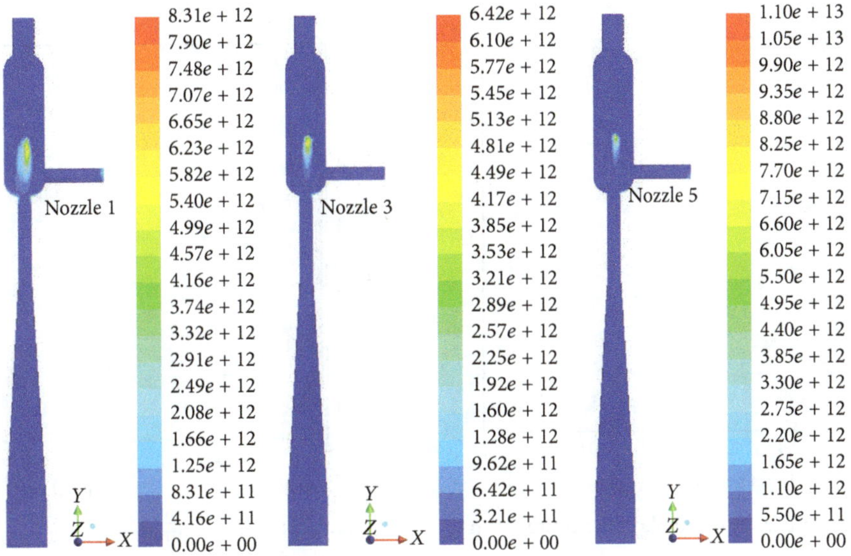

FIGURE 13: Contour of the number density of bin fraction 20 for nozzles 1, 3, and 5.

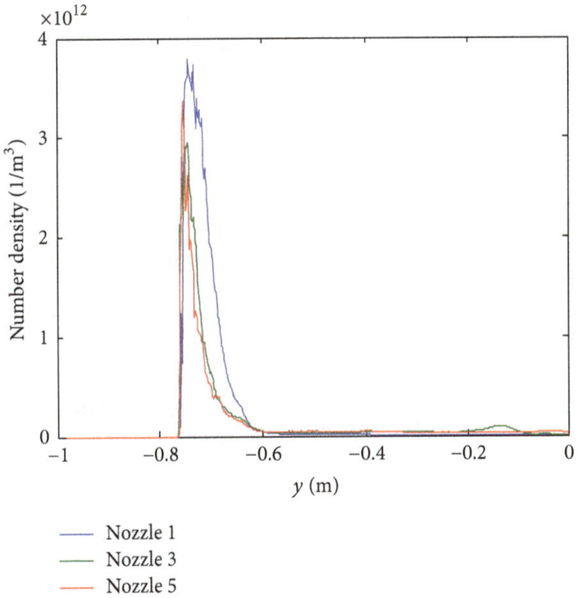

FIGURE 14: Plot of the number density of bin fraction 20.

As shown in Figure 23, the correlation shows that diameter is less in zone (a) but increases in zone (b) and is almost constant in zone (c). In zone (c) there is sudden rise of the diameter in the case of nozzle 1, whereas in the cases of nozzles 3 and 5 the diameter does not show appreciable change. The reason is that in nozzle 1 there is only single jet. Its singularity nature is not disturbed while fluids are passing through the jet ejector. Hence there is fast collision of bubbles and it shows sudden rise in the bubble diameter, while, in the cases of nozzles 3 and 5, the water jet is divided into three and five jets of the same velocity, respectively, so these streams maintain their diversity. While fluid stream enters into the diffuser, the bubbles entrapped in streams are not easily disengaged to form higher size of bubbles and for this reason the two nozzles do not show appreciable increase in their average diameter.

6.5. Path Lines. Figure 24 represents the path lines of liquid for nozzles 1, 3, and 5, respectively. It is evident from the figure

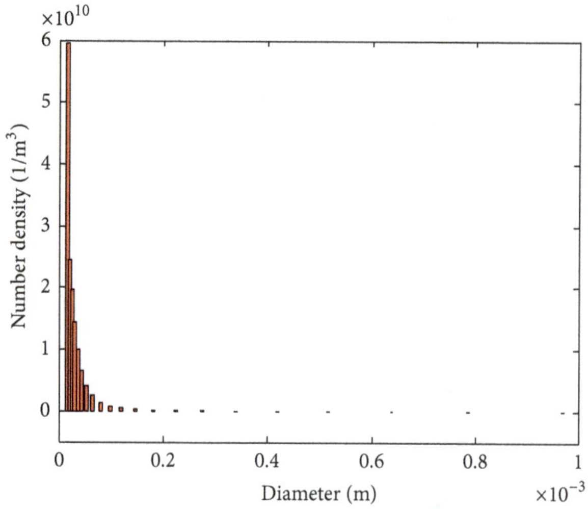

FIGURE 15: Plot of the number density of bubble in the entire volume for nozzle 1.

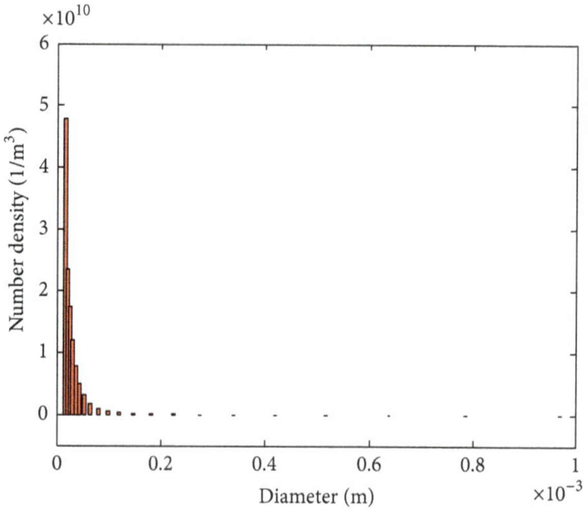

FIGURE 16: Plot of the number density of bubble in the entire volume for nozzle 3.

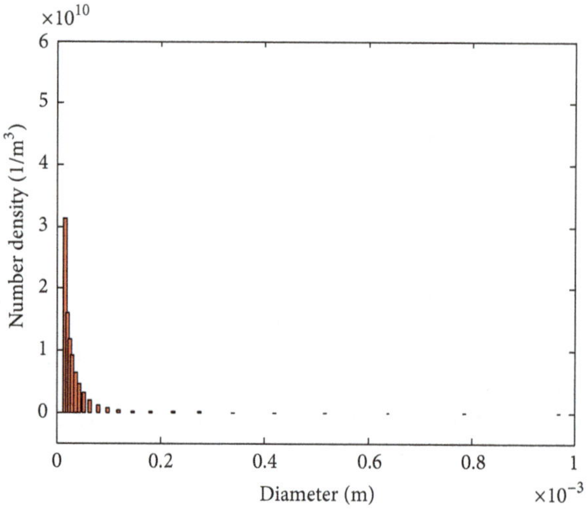

FIGURE 17: Plot of the number density of bubble in the entire volume for nozzle 5.

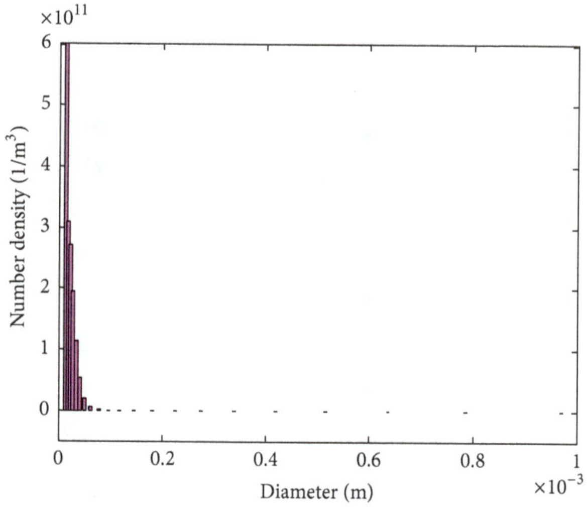

FIGURE 18: Plot of the number density of bubble in a plane near to the gas inlet for nozzle 1.

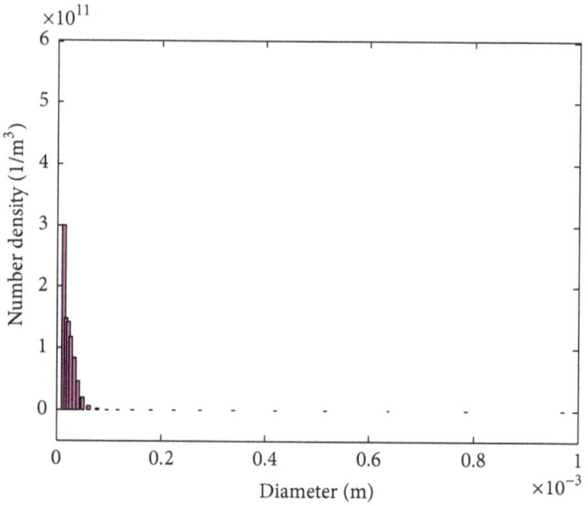

FIGURE 19: Plot of the number density of bubble in a plane near to the gas inlet for nozzle 3.

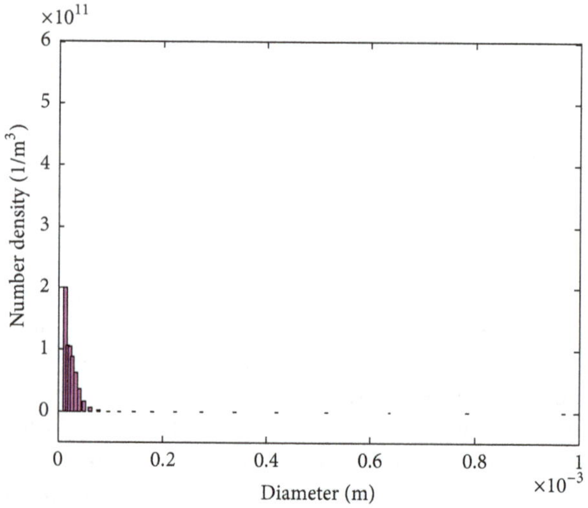

FIGURE 20: Plot of the number density of bubble in a plane near to the gas inlet for nozzle 5.

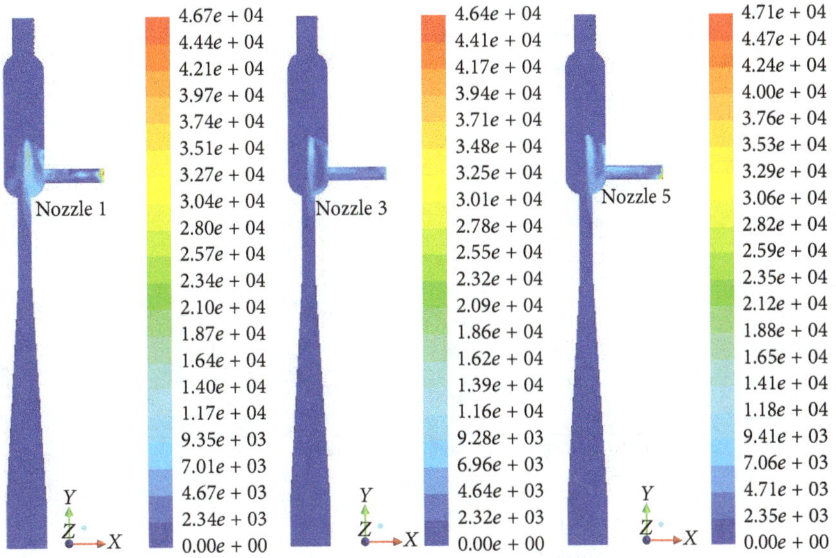

FIGURE 21: Contour of the interfacial area on middle plane for nozzles 1, 3, and 5.

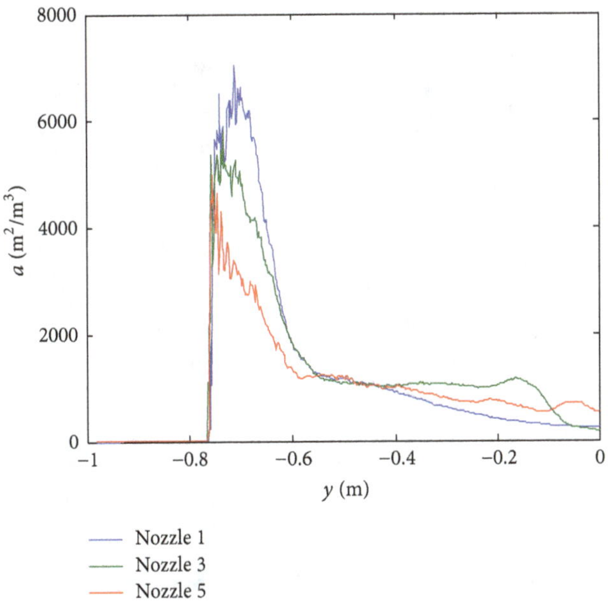

FIGURE 22: Plot of the interfacial area.

that larger recirculation region occurs in the mixing as the number of nozzles increases.

7. Conclusion

Population balance approach, combined with the coalescence and break-up models, is presented to simulate the operation of the jet ejector, using CFD software. The population balance approach was demonstrated by using the equation of bubble number density for gas-liquid flows using *ANSYS Fluent 14.0* to explain the temporal and spatial changes of the gas bubble size distribution. The CFD results have been analyzed to determine bubble size distribution, liquid velocity, gas velocity, and volume fraction in the jet ejector with the Cl_2-NaOH system. Computational results are compared with the experimental results [31]. The simulated interfacial area is in good agreement with the experimental values.

In order to see the validity of the CFD model we require more experimental data on liquid-to-gas ratio, geometry of the jet ejector, properties of gas and liquid, reactivity of fluids, and so forth. The coupling of all these properties gives a good platform to predict the best model and as a consequence we may predict the performance of the jet ejector.

Conflict of Interests

The authors declare that there is no conflict of interests regarding the publication of this paper.

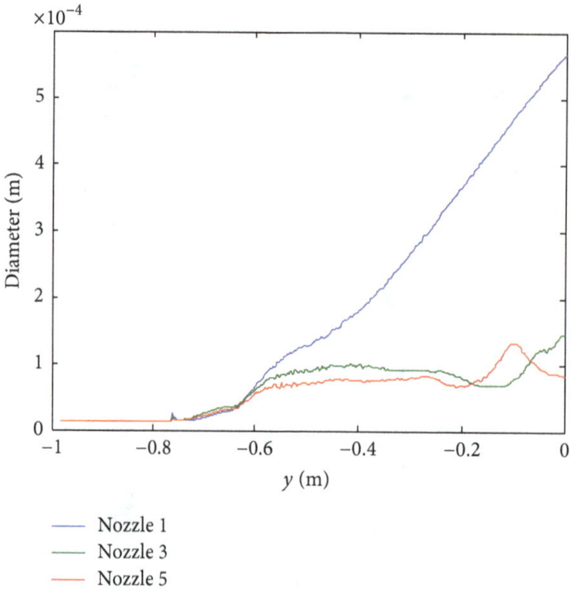

FIGURE 23: Plot of the diameter of bubble.

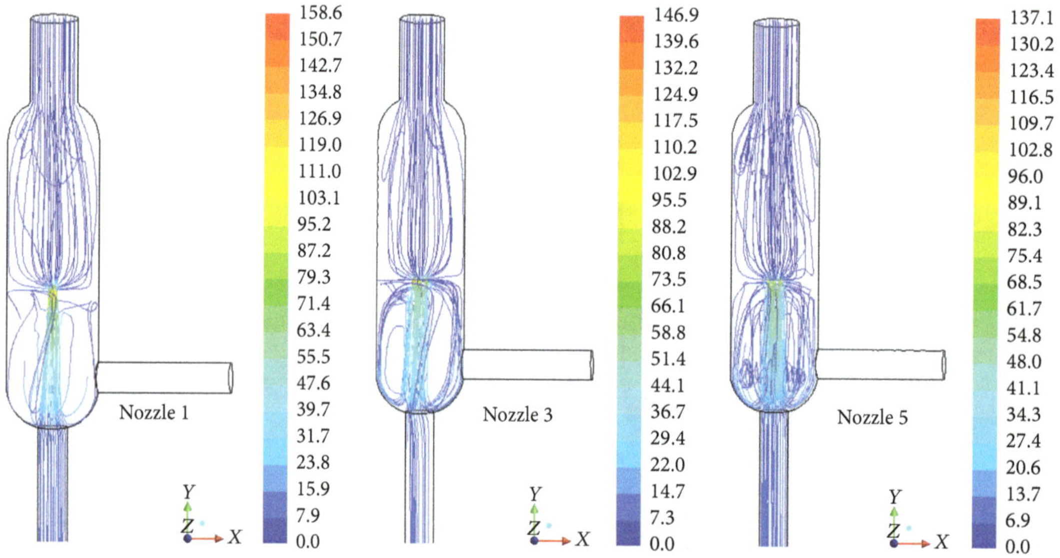

FIGURE 24: Path lines of liquid for nozzles 1, 3, and 5.

Acknowledgment

The first author gratefully acknowledges the financial support received from Department of Mathematics and Physics, Lappeenranta University of Technology, Lappeenranta, Finland, for international mobility support, making this joint research work possible.

References

[1] K. Ekambara, R. Sean Sanders, K. Nandakumar, and J. H. Masliyah, "CFD modeling of gas-liquid bubbly flow in horizontal pipes: influence of bubble coalescence and breakup," *International Journal of Chemical Engineering*, vol. 2012, Article ID 620463, 20 pages, 2012.

[2] E. Olmos, C. Gentric, C. Vial, G. Wild, and N. Midoux, "Numerical simulation of multiphase flow in bubble column reactors. Influence of bubble coalescence and break-up," *Chemical Engineering Science*, vol. 56, no. 21-22, pp. 6359–6365, 2001.

[3] K. A. Mouza, N. A. Kazakis, and S. V. Paras, "Bubble column reactor design using a CFD code," in *Proceedings of the 1st International Conference "From Scientific Computing to Computational Engineering" (IC-SCCE '04)*, Athens, Greece, September 2004.

[4] J. B. Joshi, V. S. Vitankar, A. A. Kulkarni, M. T. Dhotre, and K. Ekambara, "Coherent flow structures in bubble column reactors," *Chemical Engineering Science*, vol. 57, no. 16, pp. 3157–3183, 2002.

[5] G. Kocamustafaogullari and W. D. Huang, "Internal structure and interfacial velocity development for bubbly two-phase

flow," *Nuclear Engineering and Design*, vol. 151, no. 1, pp. 79–101, 1994.

[6] A. Kitagawa, K. Sugiyama, and Y. Murai, "Experimental detection of bubble-bubble interactions in a wall-sliding bubble swarm," *International Journal of Multiphase Flow*, vol. 30, no. 10, pp. 1213–1234, 2004.

[7] G. Wild, S. Poncin, H. Li, and E. Olmos, "Some aspects of the hydrodynamics of bubble columns," *International Journal of Chemical Reactor Engineering*, vol. 1, no. 1, pp. 1–36, 2003.

[8] W.-D. Deckwer, *Bubble Column Reactors*, John Wiley & Sons, Chichester, UK, 1992.

[9] P. Spicka, M. M. Dias, and J. C. B. Lopes, "Gas-liquid flow in a 2D column: comparison between experimental data and CFD modelling," *Chemical Engineering Science*, vol. 56, no. 21-22, pp. 6367–6383, 2001.

[10] J. Cao and R. N. Christensen, "Analysis of moving boundary problem for bubble collapse in binary solutions," *Numerical Heat Transfer, Part A: Applications*, vol. 38, no. 7, pp. 681–699, 2000.

[11] R. D. S. Cavalcanti, S. R. De Farias Neto, and E. O. Vilar, "A computational fluid dynamics study of hydrogen bubbles in an electrochemical reactor," *Brazilian Archives of Biology and Technology*, vol. 48, pp. 219–229, 2005.

[12] R. Krishna and J. M. Van Baten, "Scaling up bubble column reactors with the aid of CFD," *Chemical Engineering Research and Design*, vol. 79, no. 3, pp. 283–309, 2001.

[13] J. M. Van Baten and R. Krishna, "Scale up studies on partitioned bubble column reactors with the aid of CFD simulations," *Catalysis Today*, vol. 79-80, pp. 219–227, 2003.

[14] K. Shimizu, S. Takada, K. Minekawa, and Y. Kawase, "Phenomenological model for bubble column reactors: prediction of gas hold-ups and volumetric mass transfer coefficients," *Chemical Engineering Journal*, vol. 78, no. 1, pp. 21–28, 2000.

[15] V. V. Buwa and V. V. Ranade, "Dynamics of gas-liquid flow in a rectangular bubble column: experiments and single/multi-group CFD simulations," *Chemical Engineering Science*, vol. 57, no. 22-23, pp. 4715–4736, 2002.

[16] S. Lo, "Application of population balance to CFD modelling of gas-liquid reactors," in *Proceedings of the Conference on Trends in Numerical and Physical Modelling for Industrial Multiphase Flows*, Cargèse, France, September 2000.

[17] M. T. Dhotre, K. Ekambara, and J. B. Joshi, "CFD simulation of sparger design and height to diameter ratio on gas hold-up profiles in bubble column reactors," *Experimental Thermal and Fluid Science*, vol. 28, no. 5, pp. 407–421, 2004.

[18] J.-M. Delhaye and J. B. McLaughlin, "Appendix 4: report of study group on microphysics," *International Journal of Multiphase Flow*, vol. 29, no. 7, pp. 1101–1116, 2003.

[19] M. J. Prince and H. W. Blanch, "Bubble coalescence and break-up in air-sparged bubble columns," *AIChE Journal*, vol. 36, no. 10, pp. 1485–1499, 1990.

[20] J. Y. Tu, G. H. Yeoh, G.-C. Park, and M.-O. Kim, "On population balance approach for subcooled boiling flow prediction," *Journal of Heat Transfer*, vol. 127, no. 3, pp. 253–264, 2005.

[21] T. Wang, J. Wang, and Y. Jin, "A novel theoretical breakup kernel function for bubbles/droplets in a turbulent flow," *Chemical Engineering Science*, vol. 58, no. 20, pp. 4629–4637, 2003.

[22] T. Wang, J. Wang, and Y. Jin, "A CFD-PBM coupled model for gas-liquid flows," *AIChE Journal*, vol. 52, no. 1, pp. 125–140, 2006.

[23] J. Hämäläinen, S. B. Lindström, T. Hämäläinen, and H. Niskanen, "Papermaking fibre-suspension flow simulations at multiple scales," *Journal of Engineering Mathematics*, vol. 71, no. 1, pp. 55–79, 2011.

[24] T. Hämäläinen, *Modelling of fibre orientation and fibre flocculation phenomena in paper sheet forming [Ph.D. thesis]*, Tampere University of Technology, Publication 768, Tampere, Finland, 2008.

[25] F. Lehr, M. Millies, and D. Mewes, "Bubble-size distributions and flow fields in bubble columns," *AIChE Journal*, vol. 48, no. 11, pp. 2426–2443, 2002.

[26] F. B. Campos and P. L. C. Lage, "A numerical method for solving the transient multidimensional population balance equation using an Euler-Lagrange formulation," *Chemical Engineering Science*, vol. 58, no. 12, pp. 2725–2744, 2003.

[27] P. Chen, J. Sanyal, and M. P. Dudukovic, "CFD modeling of bubble columns flows: implementation of population balance," *Chemical Engineering Science*, vol. 59, no. 22-23, pp. 5201–5207, 2004.

[28] H. Luo and H. F. Svendsen, "Theoretical model for drop and bubble breakup in turbulent dispersions," *AIChE Journal*, vol. 42, no. 5, pp. 1225–1233, 1996.

[29] G. H. Yeoh and J. Y. Tu, "Population balance modelling for bubbly flows with heat and mass transfer," *Chemical Engineering Science*, vol. 59, no. 15, pp. 3125–3139, 2004.

[30] I. Atay, *Fluid flow and gas absorption, in an ejector venturiscrubber [Dissertation for the Degree of Doctor of Engineering Science]*, New Jersey Institute of Technology, Newark, NJ, USA, 1986.

[31] K. S. Agrawal, *Modeling of multi nozzle jet ejector for absorption with chemical reaction [Doctor of Philosophy Thesis in Chemical Engineering]*, Maharaja Sayajirao University of Baroda, Vadodara, India, 2012.

[32] K. S. Agrawal, "Jet ejector and its importance in context of Indian industry," *International Journal of Applied Environmental Sciences*, vol. 8, no. 3, pp. 267–269, 2013.

[33] A. Laurent and J. C. Charpentier, "Aires interfaciales et coefficients de transfert de matière dans les divers types d'absorbeurs et de réacteurs gaz-liquide," *The Chemical Engineering Journal*, vol. 8, no. 2, pp. 85–101, 1974.

[34] M. Zlokarnik, "Eignung und leistungsfähigkeit von belüftungsvorrichtungen für die biologische abwasserreinigung," *Chemie Ingenieur Technik*, vol. 52, no. 4, pp. 330–331, 1980.

[35] S. Ogawa, H. Yamaguchi, S. Tone, and T. Otake, "Gas-liquid mass transfer in the jet reactor with liquid jet reactor," *Journal of Chemical Engineering of Japan*, vol. 16, no. 5, pp. 419–425, 1983.

[36] J. C. Charpentier, "What's new in absorption with chemical reaction?" *Transactions of the Institution of Chemical Engineers*, vol. 60, pp. 131–156, 1982.

[37] M. A. M. Costa, A. P. R. A. Ribeiro, É. R. Tognetti, M. L. Aguiar, J. A. S. Gonçalves, and J. R. Coury, "Performance of a Venturi scrubber in the removal of fine powder from a confined gas stream," *Materials Research*, vol. 8, no. 2, pp. 177–179, 2005.

[38] A. Majid, Y. C. Qi, and K. Mehboob, "A review of performance of a venturi scrubber," *Research Journal of Applied Sciences, Engineering and Technology*, vol. 4, no. 19, pp. 3811–3818, 2012.

[39] T. Utomo, Z. Jin, M. Rahman, H. Jeong, and H. Chung, "Investigation on hydrodynamics and mass transfer characteristics of a gas-liquid ejector using three-dimensional CFD modeling," *Journal of Mechanical Science and Technology*, vol. 22, no. 9, pp. 1821–1829, 2008.

[40] S. K. Das and M. N. Biswas, "Studies on ejector-venturi fume scrubber," *Chemical Engineering Journal*, vol. 119, no. 2-3, pp. 153–160, 2006.

[41] D. Li, *Investigation of an ejector-expansion device in a transcritical carbon dioxide cycle for military ECU applications [Ph.D. thesis]*, Purdue University, West Lafayette, Ind, USA, 2006.

[42] J. H. Witte, "Mixing shocks in two-phase flow," *Journal of Fluid Mechanics*, vol. 36, no. 4, pp. 639–655, 1969.

[43] R. L. Yadav and A. W. Patwardhan, "Design aspects of ejectors: effects of suction chamber geometry," *Chemical Engineering Science*, vol. 63, no. 15, pp. 3886–3897, 2008.

[44] H. Blenke, K. Bohner, and E. Vollmerhaus, "Untersuchungen zur berechnung des betriebsverhaltens von treibstrahlförderern," *Chemie Ingenieur Technik*, vol. 35, no. 3, pp. 201–208, 1963.

[45] S. T. Bonnington, "Water jet ejectors," BHRA Publication RR540, 1956.

[46] S. T. Bonnington, "Water driven air ejectors," Tech. Rep. SP664, BHRA Publication, 1960.

[47] S. T. Bonnington, "A guide to jet pump design," *British Chemical Engineering*, vol. 9, no. 3, pp. 150–154, 1964.

[48] S. T. Bonnington and A. L. King, *Jet Pumps and Ejectors: A State of the Art Review and Bibliography*, BHRA Fluid Engineering, Bedford, UK, 1972.

[49] R. G. Cunningham, "Gas compression with the liquid jet pump," *Journal of Fluids Engineering*, vol. 96, no. 3, pp. 203–215, 1974.

[50] R. G. Cunningham and R. J. Dopkin, "Jet breakup and mixing throat lengths for the liquid jet gas pump," *Transactions of the ASME, Journal of Fluids Engineering*, vol. 96, no. 3, pp. 216–226, 1974.

[51] X. Gamisans, M. Sarrà, and F. J. Lafuente, "The role of the liquid film on the mass transfer in venturi-based scrubbers," *Chemical Engineering Research and Design*, vol. 82, no. 3, pp. 372–380, 2004.

[52] G. C. Govatos, "The slurry pump," *Journal of Pipelines*, vol. 1, pp. 145–157, 1981.

[53] A. E. Kroll, "The design of jet pump," *Chemical Engineering Progress*, vol. 43, no. 2, pp. 21–24, 1947.

[54] D. K. Acharjee, P. A. Bhat, A. K. Mitra, and N. A. Roy, "Studies on momentum transfer in vertical liquid jet ejectors," *Indian Journal of Technology*, vol. 13, pp. 205–210, 1975.

[55] P. A. Bhat, A. K. Mitra, and A. N. Roy, "Momentum transfer in a horizontal liquid-jet ejector," *The Canadian Journal of Chemical Engineering*, vol. 50, no. 3, pp. 313–317, 1972.

[56] M. N. Biswas, A. K. Mitra, and A. N. Roy, "Effective interfacial area in a liquid-jet induced horizontal gas-liquid pipeline contactor," *Indian Chemical Engineer*, vol. 19, no. 2, pp. 15–21, 1977.

[57] M. N. Biswas, A. K. Mitra, and A. N. Roy, "Studies on gas dispersion in a horizontal liquid jet ejector," in *Proceedings of the 2nd Symposium on Jet Pumps and Ejectors and Gas Lift Techniques*, E3-27-42, BHRA, Cambridge, UK, March 1975.

[58] W. J. Davis, "The effect of the Froude number in estimating vertical 2-phase gas-liquid friction losses," *British Chemical Engineering*, vol. 8, pp. 462–465, 1963.

[59] G. S. Davles, A. K. Mitra, and A. N. Roy, "Momentum transfer studies in ejectors. Correlalations for single-phase and two-phase systems," *Industrial & Engineering Chemistry Process Design and Development*, vol. 6, no. 3, pp. 293–299, 1967.

[60] X. Gamisans, M. Sarrà, and F. J. Lafuente, "Gas pollutants removal in a single- and two-stage ejector-venturi scrubber," *Journal of Hazardous Materials*, vol. 90, no. 3, pp. 251–266, 2002.

[61] S. Dasappa, P. J. Paul, and H. J. Mukunda, "Fluid dynamic studies on ejectors for thermal applications of gasifiers," in *Proceedings of the 4th National Meet on Recent Advances in Biomass Gasification Technology*, Mysore, India, 1993.

[62] A. K. Mitra and A. N. Roy, "Studies on the performance of ejector: correlation for air-air system," *Indian Journal of Technology*, vol. 2, no. 9, pp. 315–316, 1964.

[63] A. K. Mitra, D. K. Guha, and A. N. Roy, "Studies on the performance of ejector Part-I, air-air system," *Indian Chemical Engineering Transactions*, vol. 5, article 59, 1963.

[64] D. Mukherjee, M. N. Biswas, and A. K. Mitra, *Holdup Studies in Liquid; Liquid Ejector System*, vol. 1, Institute of Chemical Engineers, 1981.

[65] D. Mukherjee, M. N. Biswas, and A. K. Mitra, "Hydrodynamics of liquid-liquid dispersion in ejectors and vertical two phase flow," *Canadian Journal of Chemical Engineering*, vol. 66, no. 6, pp. 896–907, 1988.

[66] V. R. Radhakrishnan and A. K. Mitra, "Pressure drop, holdup and interfacial area in vertical two-phase flow of multi-jet ejector induced dispersions," *Canadian Journal of Chemical Engineering*, vol. 62, no. 2, pp. 170–178, 1984.

[67] S. S. Pal, A. K. Mitra, and A. N. Roy, "Pressure drop and holdup in a vertical two-phase counter current flow with improved gas mixing liquid," *Industrial & Engineering Chemistry Process Design and Development*, vol. 19, no. 1, pp. 67–75, 1979.

[68] K. P. Singh, N. K. Purohit, and A. K. Mitra, "Performance characteristics of a horizontal ejector: water-water system," *Indian Chemical Engineer*, vol. 16, no. 3, pp. 1–6, 1974.

[69] K. S. Agrawal, "Performance of venturi scrubber," *International Journal of Engineering Research and Development*, vol. 7, no. 11, pp. 53–69, 2013.

[70] K. S. Agrawal, "Interfacial area and mass transfer characteristics in multi nozzle jet ejector," *International Journal of Emerging Technologies in Computational and Applied Sciences*, vol. 3, no. 5, pp. 270–280, 2013.

[71] S. K. Agrawal, "Bubble dynamics and interface phenomenon," *Journal of Engineering and Technology Research*, vol. 5, no. 3, pp. 42–50, 2013.

[72] S. Balamurugan, M. D. Lad, V. G. Gaikar, and A. W. Patwardhan, "Effect of geometry on mass transfer characteristics of ejectors," *Industrial and Engineering Chemistry Research*, vol. 46, no. 25, pp. 8505–8517, 2007.

[73] S. Balamurugan, V. G. Gaikar, and A. W. Patwardhan, "Effect of ejector configuration on hydrodynamic characteristics of gas-liquid ejectors," *Chemical Engineering Science*, vol. 63, no. 3, pp. 721–731, 2008.

[74] N. N. Dutta and K. V. Raghavan, "Mass transfer and hydrodynamic characteristics of loop reactors with downflow liquid jet ejector," *The Chemical Engineering Journal*, vol. 36, no. 2, pp. 111–121, 1987.

[75] P. Havelka, V. Linek, J. Sinkule, J. Zahradník, and M. Fialová, "Hydrodynamic and mass transfer characteristics of ejector loop reactors," *Chemical Engineering Science*, vol. 55, no. 3, pp. 535–549, 2000.

[76] P. Havelka, V. Linek, J. Sinkule, J. Zahradnik, and M. Fialova, "Effect of the ejector configuration on the gas suction rate and gas hold-up in ejector loop reactors," *Chemical Engineering Science*, vol. 52, no. 11, pp. 1701–1713, 1997.

[77] C. Li and Y. Z. Li, "Investigation of entrainment behavior and characteristics of gas-liquid ejectors based on CFD simulation," *Chemical Engineering Science*, vol. 66, no. 3, pp. 405–416, 2011.

[78] A. Mandal, "Characterization of gas-liquid parameters in a down-flow jet loop bubble column," *Brazilian Journal of Chemical Engineering*, vol. 27, no. 2, pp. 253–264, 2010.

[79] A. Mandal, G. Kundu, and D. Mukherjee, "Gas-holdup distribution and energy dissipation in an ejector-induced downflow bubble column: the case of non-Newtonian liquid," *Chemical Engineering Science*, vol. 59, no. 13, pp. 2705–2713, 2004.

[80] A. Mandal, G. Kundu, and D. Mukherjee, "Interfacial area and liquid-side volumetric mass transfer coefficient in a downflow bubble column," *Canadian Journal of Chemical Engineering*, vol. 81, no. 2, pp. 212–219, 2003.

[81] A. Mandal, G. Kundu, and D. Mukherjee, "Gas holdup and entrainment characteristics in a modified downflow bubble column with Newtonian and non-Newtonian liquid," *Chemical Engineering and Processing: Process Intensification*, vol. 42, no. 10, pp. 777–787, 2003.

[82] A. Mandal, G. Kundu, and D. Mukherjee, "A comparative study of gas holdup, bubble size distribution and interfacial area in a downflow bubble column," *Chemical Engineering Research and Design*, vol. 83, no. 4, pp. 423–428, 2005.

[83] A. Mandal, G. Kundu, and D. Mukherjee, "Energy analysis and air entrainment in an ejector induced downflow bubble column with non-Newtonian motive fluid," *Chemical Engineering & Technology*, vol. 28, no. 2, pp. 210–218, 2005.

[84] D. A. Mitchell, "Improving the efficiency of free-jet scrubbers," *Environment International*, vol. 6, no. 1–6, pp. 21–24, 1981.

[85] F. Rahman, D. B. Umesh, D. Subbarao, and M. Ramasamy, "Enhancement of entrainment rates in liquid-gas ejectors," *Chemical Engineering and Processing: Process Intensification*, vol. 49, no. 10, pp. 1128–1135, 2010.

[86] W. Yaici, A. Laurent, N. Midoux, and J.-C. Charpentier, "Determination of gas-side mass transfer coefficients in trickle-bed reactors in the presence of an aqueous or an organic liquid phase," *International Chemical Engineering*, vol. 28, no. 2, pp. 299–305, 1988.

[87] K. Yamagiwa, D. Kusabiraki, and A. Ohkawa, "Gas holdup and gas entrainment rate in downflow bubble column with gas entrainment by a liquid jet operating at high liquid throughput," *Journal of Chemical Engineering of Japan*, vol. 23, no. 3, pp. 343–348, 1990.

[88] T. Frank, "Advances in computational fluid dynamics (CFD) of 3-dimensional gas-liquid multiphase flows," in *Proceedings of the NAFEMS Seminar: Simulation of Complex Flows (CFD)*, Niedernhausen/Wiesbaden, Germany, April 2005.

[89] A. A. Kendoush and S. A. W. Al-khatab, "Flow regimes characterization in vertical downward two phase flow," in *Proceedings of the 2nd International Symposium on Multiphase Flow and Heat Transfer*, X. Z. Chen, T. N. Veziroglu, and C. L. Tien, Eds., pp. 215–220, 1989.

[90] J. Zahradník and M. Fialová, "The effect of bubbling regime on gas and liquid phase mixing in bubble column reactors," *Chemical Engineering Science*, vol. 51, no. 10, pp. 2491–2500, 1996.

[91] R. G. Rice and M. A. Littlefield, "Dispersion coefficients for ideal bubbly flow in truly vertical bubble columns," *Chemical Engineering Science*, vol. 42, no. 8, pp. 2045–2053, 1987.

[92] B. R. Bakshi, P. Jiang, and L. S. Fan, "Analysis of flow in gas-liquid bubble columns using multi-resolution methods," in *Proceedings of the 2nd International Conference on Gas-Liquid-Solid Reactor Engineering*, pp. A1–A8, Cambridge, Mass, USA, March 1995.

[93] H. Lefebvre, *Atomization and Sprays*, Hemisphere Publishing Corporation, New York, NY, USA, 1989.

[94] H. Liu, *Science and Engineering of Droplets: Fundamentals and Applications*, Noyes Publications, New York, NY, USA, 2000.

[95] R. J. Schick, "Spray technology reference guide: understanding drop size," Bulletne 4598, Spraying Systems, Wheaton, Ill, USA, 2006.

[96] L. Xianguo and R. S. Tankin, "Droplet size distribution: a derivative of a Nukiyama—Tanasawa type distribution function," *Combustion Science and Technology*, vol. 56, no. 1, pp. 65–76, 1987.

[97] A. Dennis Gary, "A study of injector spray characteristics in simulated rocket combustion chamber including longitudinal mode pressure oscillation," Tech. Rep. 730, NASA, 1966.

[98] M. Azad and S. R. Syeda, "A numerical model for bubble size distribution in turbulent gas-liquid dispersion," *Journal of Chemical Engineering*, vol. 24, no. 1, pp. 25–34, 2006.

[99] M. K. Silva, M. A. D. Ávila, and M. Mori, "CFD modelling of a bubble column with an external loop in the heterogeneous regime," *The Canadian Journal of Chemical Engineering*, vol. 89, no. 4, pp. 671–681, 2011.

[100] A. N. Kudzo, *Visualization and characterization of ultrasonic cavitating atomizer and other automotive paint sprayers using infrared thermography [A Dissertation Submitted in Partial Fulfillment of the Requirements for the Degree of Doctor of Philosophy]*, College of Engineering, University of Kentucky, Lexington, Ky, USA, 2009.

[101] J. Ciborowski and A. Bin, "Investigation of the aeration effect of plunging liquid jets," *Inzynieria Chemiczna*, vol. 2, pp. 557–577, 1972.

[102] A. Ohkawa, D. Kusabiraki, Y. Kawai, N. Sakai, and K. Endoh, "Some flow characteristics of a vertical liquid jet system having downcomers," *Chemical Engineering Science*, vol. 41, no. 9, pp. 2347–2361, 1986.

[103] Y. Y. Sheng and G. A. Irons, "The impact of bubble dynamics on the flow in plumes of ladle water models," *Metallurgical and Materials Transactions B*, vol. 26, no. 3, pp. 625–635, 1995.

[104] J. Kitscha and G. Kocamustafaogullari, "Breakup criteria for fluid particles," *International Journal of Multiphase Flow*, vol. 15, no. 4, pp. 573–588, 1989.

[105] F. O. Bailer, *Mass transfer characteristics of a novel gas-liquid contactor, the advanced buss loop reactor [A Dissertation Submitted to the Swiss Federal Institute of Technology for Degree of Doctor of Technical Sciences]*, Swiss Federal Institute of Technology, Zürich, Switzerland, 2001.

[106] G. M. Evans, G. J. Jameson, and B. W. Atkinson, "Prediction of the bubble size generated by a plunging liquid jet bubble column," *Chemical Engineering Science*, vol. 47, no. 13-14, pp. 3265–3272, 1992.

[107] K. Ceylan, A. Altunbaş, and G. Kelbaliyev, "A new model for estimation of drag force in the flow of Newtonian fluids around rigid or deformable particles," *Powder Technology*, vol. 119, no. 2-3, pp. 250–256, 2001.

[108] R. Pawelczyk and K. Pindur, "A dynamic method for dispersing gases in liquids," *Chemical Engineering and Processing: Process Intensification*, vol. 38, no. 2, pp. 95–107, 1999.

[109] A. K. Biń, "Gas entrainment by plunging liquid jets," *Chemical Engineering Science*, vol. 48, no. 21, pp. 3585–3630, 1993.

[110] S.-Q. Zheng, Y. Yao, F.-F. Guo, R.-S. Bi, and J.-Y. Li, "Local bubble size distribution, gas-liquid interfacial areas and gas holdups in an up-flow ejector," *Chemical Engineering Science*, vol. 65, no. 18, pp. 5264–5271, 2010.

[111] K. S. Agrawal, "Removal efficiency in industrial scale liquid jet ejector for chlorine-aqueous caustic soda system," *International Journal of Engineering Trends and Technology*, vol. 4, no. 7, pp. 2931–2940, 2013.

[112] K. S. Agrawal, "Experimental observation for liquid jet ejector for chlorine-aqueous caustic soda system at laboratory scale," *American International Journal of Research in Science, Technology, Engineering and Mathematics*, vol. 2, no. 2, pp. 177–181, 2013.

[113] K. S. Agrawal, "Performance of venturi scrubber," *International Journal of Engineering Research and Development*, vol. 17, no. 11, pp. 53–69, 2013.

[114] N. A. Panchal, S. R. Bhutada, and V. G. Pangarkar, "Gas induction and hold up characteristics of liquid jet loop reactor using multi orifice nozzles," *Chemical Engineering Communications*, vol. 102, no. 1, pp. 59–68, 1991.

14

A Comparative Study on Stabilized Finite Element Methods for the Convection-Diffusion-Reaction Problems

Ali Sendur

Alanya Alaaddin Keykubat University, 07490 Antalya, Turkey

Correspondence should be addressed to Ali Sendur; ali.sendur@alanya.edu.tr

Academic Editor: Srinivasan Natesan

The disproportionality in the problem parameters of the convection-diffusion-reaction equation may lead to the formation of layer structures in some parts of the problem domain which are difficult to resolve by the standard numerical algorithms. Therefore the use of a stabilized numerical method is inevitable. In this work, we employ and compare three classical stabilized finite element formulations, namely, the Streamline-Upwind Petrov-Galerkin (SUPG), Galerkin/Least-Squares (GLS), and Subgrid Scale (SGS) methods, and a recent Link-Cutting Bubble (LCB) strategy proposed by Brezzi and his coworkers for the numerical solution of the convection-diffusion-reaction equation, especially in the case of small diffusion. On the other hand, we also consider the pseudo residual-free bubble (PRFB) method as another alternative that is based on enlarging the finite element space by a set of appropriate enriching functions. We compare the performances of these stabilized methods on several benchmark problems. Numerical experiments show that the proposed methods are comparable and display good performance, especially in the convection-dominated regime. However, as the problem turns into reaction-dominated case, the PRFB method is slightly better than the other well-known and extensively used stabilized finite element formulations as they start to exhibit oscillations.

1. Introduction

The modeling of real world problems to predict physical quantities is an important subject in science and engineering. Many physical, chemical, and biological phenomena contain interactions of convection, diffusion, and reaction processes, which can be modeled in terms of the convection-diffusion-reaction (CDR) problems [1–3]. Explicit solutions to those problems are only known under specific circumstances; therefore numerical simulation of the problem is required. However, the numerical approximation of those equations is a challenging problem, in general, and it is not possible to get reasonable approximations with the standard numerical algorithms. Therefore developing efficient and effective computational methods for solving those problems has been drawing attention of many researchers for several decades [4–7].

The exact solutions of the convection-diffusion-reaction (CDR) problems exhibit very fine structures, so-called layers that are small subregions where the derivatives of the solution are very large. In this case, it is well-known that the Galerkin finite element method produces nonphysical oscillations polluting all over the domain. To improve the numerical approximations, a new class of the finite element methods, the so-called stabilized FEMs, have been introduced. The main idea underlying those methods is to augment the variational formulation by mesh-dependent terms in order to gain control over the derivatives of the solution. The Streamline-Upwind Petrov-Galerkin (SUPG) method proposed by Brooks and Hughes in [8] is one of the most well-known stabilized formulations in the literature. It is extensively used in the simulation of many real world problems. There are also several variants of the SUPG method in the finite element literature which are also popular [9, 10]. Not only can those methods be applied to many physical problems, but also their computer implementation is quite practical. However the lack of formalism in derivation of the stability parameters used in stabilized formulations has been seen as a major drawback of those strategies.

Another important and a more recent class of the stable discretization for the CDR problem is based on augmenting the finite element space by a set of suitable enriching functions. As an important and well-known example to that class, the residual-free bubbles (RFBs) method could be mentioned that is first proposed by Brezzi and Russo in [11]. The RFB method is based on enlarging the polynomial finite element space by special type of enriching functions, the so-called residual-free bubbles, and then eliminating those functions from the formulation to get an improved approximation for the polynomial part of the solution. Although the RFBs form an effective platform to derive improved discretizations, they are defined by a set of local differential equations posed inside each element which may not be easier to solve than the original one, except that the problem domains are simple element geometries [12]. That observation has motivated the introduction of a further option, the so-called pseudo residual-free bubble (PRFB) method. In that approach, the RFB functions are basically approximated by piecewise linear functions on an appropriately chosen simple subgrid established inside each element and then they are replaced by their approximate counterparts, the so-called pseudo residual-free bubbles, in the numerical computations. Here, the locations of the subgrid nodes are of critical importance and therefore they should be chosen specially so that the fine scale-effect of the exact solution can accurately be represented in the coarse scale numerical approximation [13–15]. In particular, their location is determined by minimizing the residual of a local differential problem defining the bubbles, with respect to the L_1-norm. It seems that the PRFB method is quite robust and effective algorithm in the numerical approximation of the convection-diffusion-reaction equations.

In this work, we compare the performances of the Streamline-Upwind Petrov-Galerkin (SUPG), Galerkin/Least-Squares (GLS), Subgrid Scale (SGS), Link-Cutting Bubble (LCB) [16], and pseudo residual-free bubble (PRFB) methods for the numerical solution of the convection-diffusion-reaction equation on several benchmark problems. A wide range of problem parameters have been examined and the corresponding numerical results are presented. As it can be seen from the numerical results, the results are comparable especially in the convection-dominated regimes. However the approximations generated by the PRFB method are slightly better than the others in the reaction-dominated cases.

The outline of this paper is as follows. In Section 2, we state the model problem, the variational form, and its standard Galerkin discretization. This is followed by a brief survey to the basic ideas of the stabilizations through augmented variational forms in Section 3. In Section 4, we describe the idea of augmenting finite element space which is the starting point for deriving the pseudo residual-free bubbles. We display the computation of the PRFBs and the derivation of the subgrid points in Section 5. Finally, we perform the numerical tests for several benchmark problems in both 1D and 2D in Section 6.

2. Statement of the Problem

The convection-diffusion-reaction problem in two space dimensions consists of finding the scalar-valued function u in a polygonal domain Ω with a smooth boundary $\Gamma = \partial\Omega$ and Dirichlet boundary conditions, such that

$$\begin{aligned} \mathscr{L}u &= -\epsilon\Delta u + \beta\cdot\nabla u + \sigma u = f \quad \text{in } \Omega, \\ u &= 0 \quad \text{on } \Gamma, \end{aligned} \tag{1}$$

where $\epsilon > 0$ is the diffusion coefficient, β is the velocity field, $\sigma > 0$ is the reaction coefficient, and f is a smooth source function in $L_2(\Omega)$. This model may describe how the concentration of the substance distributed in the medium changes under the influence of interactions of convection, diffusion, and reaction processes. The variational formulation corresponding to problem (1) reads as follows: Find $u \in H_0^1(\Omega)$ such that

$$a(u,v) = (f,v) \quad \forall v \in H_0^1(\Omega), \tag{2}$$

where

$$a(u,v) = \epsilon\int_\Omega \nabla u\cdot\nabla v + \int_\Omega (\beta\cdot\nabla u)\,v + \int_\Omega \sigma u v \tag{3}$$

is a continuous and coercive bilinear form on the Hilbert space $H_0^1(\Omega)$. Let us consider a partition $\mathscr{T}_h$ of the computational domain Ω into triangular elements K that do not overlap and let $h_K = \operatorname{diam}(K)$ with $h = \max_{K\in\mathscr{T}_h} h_K$. If we restrict ourselves to the case of continuous, piecewise linear elements which we denote by

$$V_L = \left\{v \in H_0^1(\Omega) : v|_K \text{ is linear for } K \in \mathscr{T}_h\right\}, \tag{4}$$

then the standard Galerkin finite element method reads as follows: Find $u_L \in V_L$ such that

$$a(u_L, v_L) = (f, v_L) \quad \forall v_L \in V_L. \tag{5}$$

It is well-known that if the problem is convection or reaction-dominated, then, unless h is of the same or smaller size of ϵ, the solution of (5) possesses spurious oscillations spreading all over the domain. Therefore a stabilization is needed.

3. Stabilizations through Augmented Variational Forms

A major class of the stabilized finite element methods is based on enlarging the discrete variational formulation by a mesh-dependent stabilization term. The general form of those methods for problem (5) reads as follows: Find $u_L \in V_L$ such that

$$a(u_L, v_L) + S(u_L, v_L) = (f, v_L) \quad \forall v_L \in V_L, \tag{6}$$

where $S(\cdot,\cdot)$ indicates a consistent stabilizing term added to the standard Galerkin formulation (5). The general form of the operator $S(\cdot,\cdot)$ is given by

$$S(u_L, v_L) = \sum_K \tau_K \left(\mathscr{L}u_L - f, \widetilde{\mathscr{L}}v_L\right)_K, \tag{7}$$

where τ_K is the stabilization parameter that needs to be tuned up by user and $(\cdot,\cdot)_K$ denotes the restriction of the inner product over K. Each choice of the operator $\widetilde{\mathscr{L}}$ gives rise to a different stabilized method and some popular examples to the choice of $\widetilde{\mathscr{L}}$ appearing in the literature are

$$\begin{aligned} \text{SUPG} &: \widetilde{\mathscr{L}}v = \beta \cdot \nabla v, \\ \text{GLS} &: \widetilde{\mathscr{L}}v = -\epsilon \Delta v + \beta \cdot \nabla v + \sigma v, \\ \text{SGS} &: \widetilde{\mathscr{L}}v = \epsilon \Delta v + \beta \cdot \nabla v - \sigma v. \end{aligned} \tag{8}$$

The SUPG method was designed by Brooks and Hughes [8] and was later generalized for multidimensional problems by Hughes and Mallet [17]. This method adds up additional diffusion to the Galerkin problem (5) in the direction of streamlines. A generalization of the SUPG method is the GLS method whose essential feature is to augment the Galerkin problem (5) with a least-squares form of the residuals that are based on the corresponding differential equations. The SGS method was introduced in [18] which is similar to the GLS method yet is derived from a more theoretical setting. There are also other stabilized methods used in the literature and a thorough comparison of those methods can be found in [19–23].

One of the important points in the implementation of the stabilized methods is the choice of the stabilization parameter τ_K. The parameter τ_K is constructed by user based on the problem and/or algorithm properties (the discrete maximum principle, convergence and stability analysis, etc.). Some of the possibilities that have been proposed for τ_K in the literature [21, 24, 25] are as follows:

$$\begin{aligned} \tau_K^c &= \frac{1}{4\epsilon/h^2 + 2|\beta|/h + \sigma}, \\ \tau_K^s &= \left(9\left(\frac{4\epsilon}{h^2}\right)^2 + \left(\frac{2|\beta|}{h}\right)^2 + \sigma^2\right)^{-1/2}, \\ \tau_K^a &= \frac{1}{12\epsilon/h^2 + 2|\beta|/h + 2\sigma}, \\ \tau_K^{fv} &= \frac{1}{(6\epsilon/h_K^2)\,\zeta(Pe_2) + \sigma\zeta(Pe_1)}, \end{aligned} \tag{9}$$

where

$$\begin{aligned} \zeta(x) &= \begin{cases} 1, & 0 \le x \le 1; \\ x, & x \ge 1, \end{cases} \\ Pe_1 &= \frac{6\epsilon}{\sigma h_K^2} \\ Pe_2 &= \frac{|\beta|\, h_K}{3\epsilon}. \end{aligned} \tag{10}$$

An appropriate choice from the stability parameters (9) can be used for the selected method through (8) in computations. In the numerical calculations herein, we will employ the classical stabilized finite element formulations with the parameter τ_K^c in (9) for all test problems under investigation, as the others produce comparable results.

The fact that the stabilized methods can be applied to many physical problems and their computer implementation is quite practical made those methods favorable to many researchers. However the lack of general method in derivation of the stability parameters has been seen as a major drawback of those strategies. Therefore alternative algorithms have still been of interest to researchers.

4. Stabilization through Augmented Spaces

Another successful class of the stabilized finite element methods is based on enlarging the polynomial finite element space V_L with more rich functions inspired by the original differential problem. Employing special type of enriching functions, the so-called bubble functions, it is possible to get improved approximations without increasing the size of the discrete problem. The method basically consists of defining the enriched finite element space $V_h = V_L \oplus V_B$ where the enriching space V_B is defined by

$$V_B = \bigoplus_{K \in \mathscr{T}_h} B_K \quad \text{with } B_K = H_0^1(K) \tag{11}$$

and solving problem (5) on V_h. In this approach, it is required that the bubble component u_B of u_h satisfies the original differential equation in each K strongly; that is,

$$\begin{aligned} &\mathscr{L}u_B = -\mathscr{L}u_L + f \\ &\text{in } K \text{ subject to the boundary condition } u_B = 0 \text{ on } \partial K. \end{aligned} \tag{12}$$

With that choice, it is possible to eliminate the bubbles from the formulation, keeping the size of the computation matrices the same, yet getting an improved approximation. This approach is known as the residual-free bubbles (RFBs) method in the finite element literature [11, 12], which, for the present problem, reads as follows: Find $u_h = u_L + u_B$ in V_h such that

$$a(u_L, v_L) + a(u_B, v_L) = (f, v_L) \quad \forall v_L \in V_L. \tag{13}$$

The term $a(u_B, v_L)$ is responsible for the stabilization of the numerical method and the bubble component u_B should be computed before we solve (13) for its linear part u_L. However, since u_B is identified by the linear part u_L and the source function f through (12), its solution is not an easy task, in general.

On the other hand, regarding the configuration of the problem and the simple element geometry, it is possible to compute a reasonable approximation to u_B that provides a similar stabilizing effect into the numerical method (13). Indeed, in a recent approach based on the RFB method, the bubble functions are replaced by their suitable approximate counterparts, the so-called pseudo bubbles [14, 15]. The development and the details of that approach are given in the following section.

5. The Pseudo Residual-Free Bubbles and Stabilizing Subgrids

In this section, we shall display the computation of the pseudo residual-free bubbles and the construction of the associated

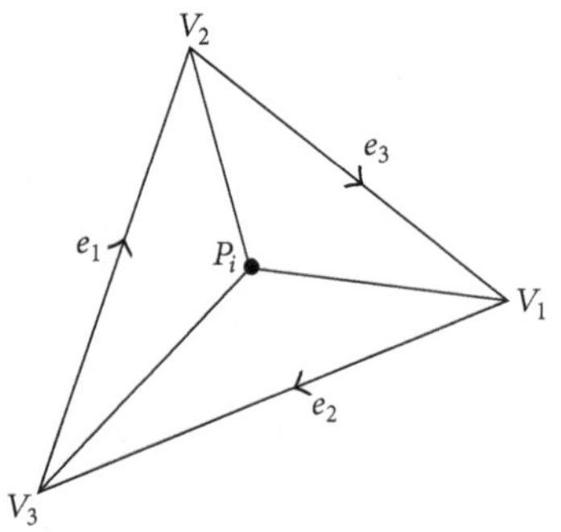

FIGURE 1: Splitting K into three subregions by P_i (figure reproduced from [15]).

subgrid nodes on which the bubble functions are approximated (see [15] for details). We note that the configuration of the subgrid nodes is of crucial importance and therefore should be determined in a special manner. Before we outline the algorithm, we introduce the notations used throughout the presentation.

Denote the vertices of K by V_i, the edges of K by e_i opposite to V_i, the length of e_i by $|e_i|$, the midpoint of edge e_i by M_i, the outward unit normal to e_i by n^i, $\nu_i = |e_i|n^i$, $\beta_{\nu_i} = (\beta, \nu_i)$, and the area of K by $|K|$ (see Figure 1). The actual numbering of the vertices will be chosen according to the direction of β.

Now let us assume that the location of P_i is along the median from V_i; that is,

$$P_i = (1 - t_i) M_i + t_i V_i, \quad 0 < t_i < 1, \; i = 1, 2, 3. \tag{14}$$

To specify the problem regimes, we further introduce

$$\epsilon_i^* = \frac{2|K|\left(-3\beta_{\nu_i} + \sigma|K|\right)}{9\left(|e_1|^2 + |e_2|^2 + |e_3|^2\right)} \quad i = 1, 2, 3. \tag{15}$$

Consider the bubble functions B_i, $i = 1, 2, 3$, defined by

$$\begin{aligned} \mathscr{L}B_i &= -\mathscr{L}\psi_i \quad \text{in } K, \\ B_i &= 0 \quad \text{on } \partial K, \\ &\quad i = 1, 2, 3, \end{aligned} \tag{16}$$

where ψ_i are the restrictions of the piecewise linear basis functions for V_L to K and

$$\begin{aligned} \mathscr{L}B_f &= f \quad \text{in } K, \\ B_f &= 0 \quad \text{on } \partial K. \end{aligned} \tag{17}$$

Now if

$$u_L|_K = \sum_{i=1}^{3} c_i\psi_i \quad \text{then} \quad u_B|_K = \sum_{i=1}^{3} c_i B_i + B_f, \tag{18}$$

where $c_i \in \mathbb{R}$, $i = 1, 2, 3$. Thus, (12) is automatically satisfied with the present choice of bubble functions. Using the element geometry and the problem properties, it is possible to construct the approximate bubbles, say B_i^*, that have the same qualitative behavior with their continuous counterpart. To start with, let b_i be a piecewise linear function with $b_i(V_j) = 0$ and $b_i(P_i) = 1 \; \forall i, j = 1, 2, 3$ and $B_i^*(x) = \alpha_i b_i(x)$ be the classical Galerkin approximation of B_i through (16); that is,

$$a\left(B_i^*, b_i\right)_K = \left(-\mathscr{L}\psi_i, b_i\right)_K, \quad i = 1, 2, 3. \tag{19}$$

Here the choice of the internal point P_i is critical and the main criteria used to determine its location are to minimize L_1 norm of the residual coming out from the bubble equation (16) in the critical case where a layer structure exists. In other words, we choose P_i such that

$$J_i = \int_K \left|\mathscr{L}B_i^* + \mathscr{L}\psi_i\right|, \quad i = 1, 2, 3 \tag{20}$$

is minimum. The criteria (20) are employed to determine the locations of subgrid nodes P_i in convection and reaction-dominated cases separately.

5.1. Convection-Dominated Flows. In order to present the explicit descriptions of P_i, we have to distinguish among the following two cases, in which the convection-dominated regime is considered with respect to the number of inflow edges (see Figure 2).

5.1.1. One Inflow Edge. The problem is assumed to be convection-dominated if

$$\epsilon \le \epsilon_1^* \quad \text{with } \sigma|K| < 3\min\left\{\beta_{\nu_2}, \beta_{\nu_3}\right\}. \tag{21}$$

From a previous calculation [15], the explicit location of P_1 along the median from V_1 is given by

$$t_1^* = 1 - \frac{-\rho_1 + \sqrt{\rho_1^2 + \lambda_1}}{2\sigma|K|^2}, \tag{22}$$

where $\rho_1 = -2\beta_{\nu_1}|K| + 3\epsilon|e_2 - e_3|^2$ and $\lambda_1 = 24\epsilon\sigma|K|^2(|e_2|^2 + |e_3|^2)$. The choice of other points P_2 and P_3 should be consistent with the physics of the problem. Thus we take $t_1 = t_1^*, t_2 = t_3 = \min\{1/3, 1 - t_1^*\}$.

5.1.2. Two Inflow Edges. The problem is assumed to be convection-dominated if

$$\epsilon \le \min\left\{\epsilon_2^*, \epsilon_3^*\right\} \quad \text{with } \sigma|K| < 3\beta_{\nu_1}. \tag{23}$$

From [15], the position of P_2 along the median from V_2 is given by

$$t_2^* = 1 - \frac{-\rho_2 + \sqrt{\rho_2^2 + \lambda_2}}{2\sigma|K|^2}, \tag{24}$$

where $\rho_2 = -2\beta_{\nu_2}|K| + 3\epsilon|e_1 - e_3|^2$ and $\lambda_2 = 24\epsilon\sigma|K|^2(|e_1|^2 + |e_3|^2)$, and the position of P_3 along the median from V_3 is given by

$$t_3^* = 1 - \frac{-\rho_3 + \sqrt{\rho_3^2 + \lambda_3}}{2\sigma|K|^2}, \tag{25}$$

where $\rho_3 = -2\beta_{\nu_3}|K| + 3\epsilon|e_1 - e_2|^2$ and $\lambda_3 = 24\epsilon\sigma|K|^2(|e_1|^2 + |e_2|^2)$. The choice of P_1 should be consistent with the physics

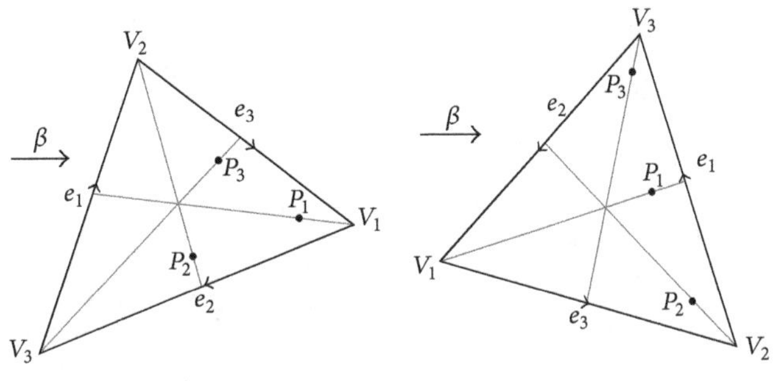

FIGURE 2: Configuration of internal nodes for convection-dominated regime: one inflow edge (a) and two inflow edges (b) (figure reproduced from [15]).

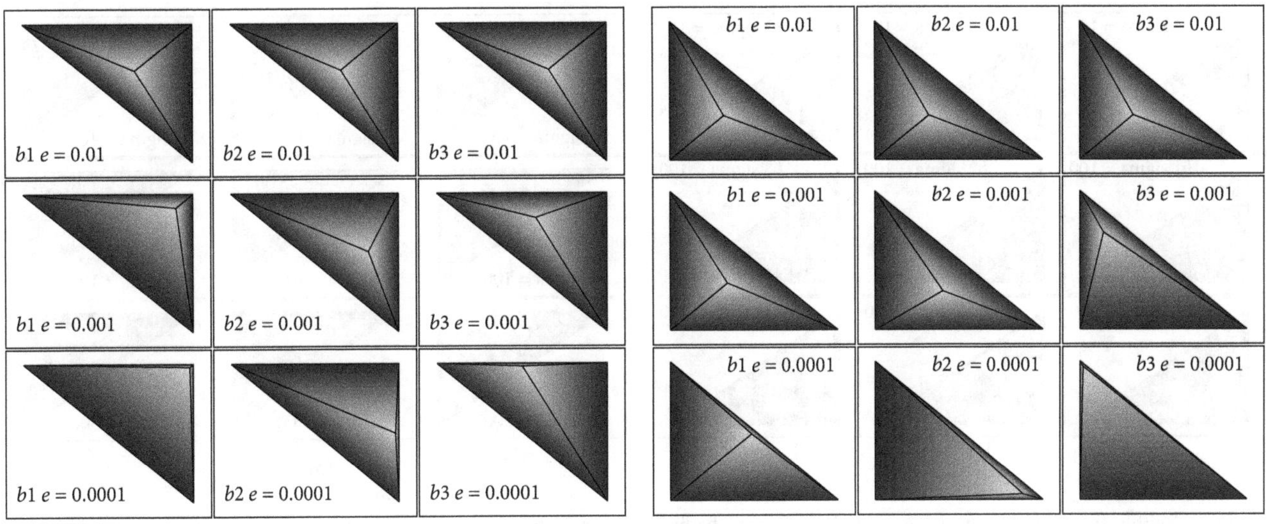

FIGURE 3: Bubble functions, b_1, b_2, b_3, in a typical element having (a) one inflow edge and (b) two inflow edges, when $\beta = (\sin 72^\circ, \cos 72^\circ)$, $\sigma = 0.001, \epsilon = 10^{-2}, 10^{-3}, 10^{-4}$, and $N = 10$ (figure reproduced from [15]).

of the problem. Thus we take $t_2 = t_2^*, t_3 = t_3^*, t_1 = \min\{1/3, 1 - t_2^*, 1 - t_3^*\}$.

With the present choice, it is noteworthy to observe the self-adjustment of subgrid nodes as the problem evolves from the diffusion-dominated regime into the convection-dominated one, so that the pseudo bubbles contribute to the stability of the numerical method at their most (see Figure 3).

5.2. Reaction-Dominated Flows. In order to present the explicit descriptions of P_i in the reaction-dominated regime, we have to distinguish among the following two cases with respect to the number of inflow edges (see Figure 4).

5.2.1. One Inflow Edge. The problem is assumed to be reaction-dominated if

$$\epsilon \leq \epsilon_1^* \quad \text{with } \sigma |K| > 3 \min \left\{ \beta_{\nu_2}, \beta_{\nu_3} \right\}. \tag{26}$$

The location of P_1 is taken from (22) and the position of P_2 along the median from V_2 is given by

$$t_2^{**} = 1 - \frac{\rho_2 + \sqrt{\rho_2^2 + \lambda_2}}{2\sigma |K|^2}, \tag{27}$$

where $\rho_2 = 2\beta_{\nu_2}|K| - 3\epsilon|e_1 - e_3|^2$ and $\lambda_2 = 24\epsilon\sigma|K|^2(|e_1|^2 + |e_3|^2)$, and the position of P_3 along the median from V_3 is given by

$$t_3^{**} = 1 - \frac{\rho_3 + \sqrt{\rho_3^2 + \lambda_3}}{2\sigma |K|^2}, \tag{28}$$

where $\rho_3 = 2\beta_{\nu_3}|K| - 3\epsilon|e_1 - e_2|^2$ and $\lambda_3 = 24\epsilon\sigma|K|^2(|e_1|^2 + |e_2|^2)$ (see [15]). Thus the positions of P_i, $i = 1, 2, 3$, are determined by taking $t_1 = t_1^*, t_2 = \max\{1 - t_1, t_2^{**}\}$ and $t_3 = \max\{1 - t_1, t_3^{**}\}$.

5.2.2. Two Inflow Edges. In this case, the problem is assumed to be reaction-dominated if

$$\epsilon \leq \min \left\{ \epsilon_2^*, \epsilon_3^* \right\} \quad \text{with } \sigma |K| > 3\beta_{\nu_1} \tag{29}$$

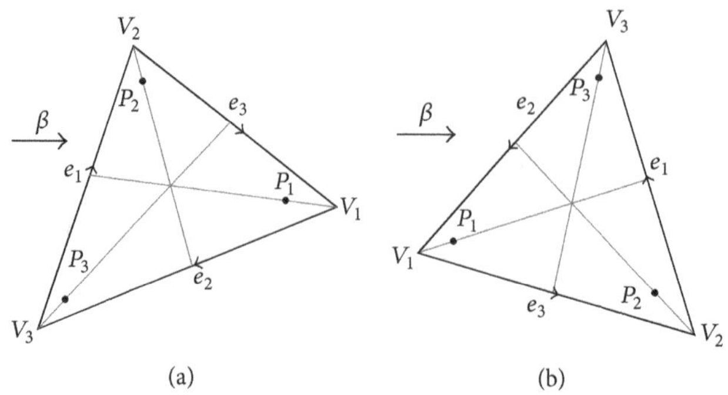

FIGURE 4: Configuration of internal nodes for reaction-dominated regime: one inflow edge (a) and two inflow edges (b) (figure reproduced from [15]).

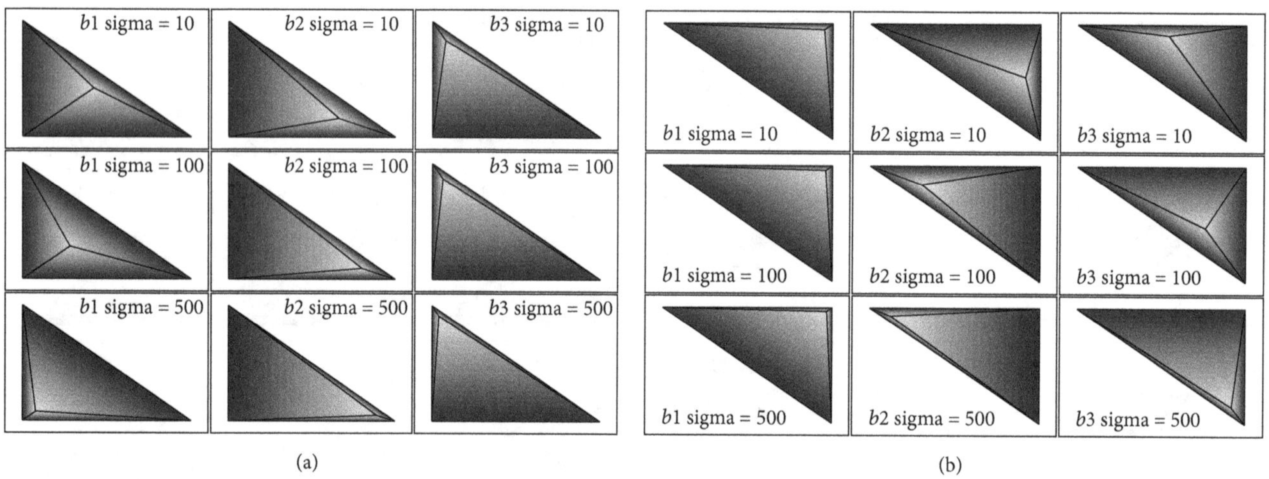

FIGURE 5: Bubble functions, b_1, b_2, b_3, in a typical element having (a) two inflow edges and (b) one inflow edge, when $\epsilon = 10^{-3}$, $\beta = (\sin 72^\circ, \cos 72^\circ)$, $\sigma = 10, 100, 500$, and $N = 10$ (figure reproduced from [15]).

and the locations of P_2 and P_3 are taken from (24)-(25). Now, the position of P_1 along the median from V_1 is given by

$$t_1^{**} = 1 - \frac{\rho_1 + \sqrt{\rho_1^2 + \lambda_1}}{2\sigma |K|^2}, \tag{30}$$

where $\rho_1 = 2\beta_{\nu_1}|K| - 3\epsilon|e_2 - e_3|^2$ and $\lambda_1 = 24\epsilon\sigma|K|^2(|e_2|^2 + |e_3|^2)$. Thus the positions of P_i, $i = 1, 2, 3$, are determined by taking $t_1 = \max\{\min\{1/3, 1 - t_2, 1 - t_3\}, t_1^{**}\}$, $t_2 = t_2^*$, $t_3 = t_3^*$.

With the present choice, it is noteworthy to observe the self-adjustment of subgrid nodes as the problem evolves from the convection-dominated regime into the reaction-dominated one, so that the pseudo bubbles contribute to the stability of the numerical method at their most (see Figure 5).

Finally, we recall that the pseudo bubble functions B_i^* ($i = 1, 2, 3$) are approximations to B_i on the subgrid specified above, through (19), and they are used in place of B_i to represent u_B. The approximate representation of u_B by bubble functions B_i^* ($i = 1, 2, 3$) is eventually used to solve (13) for its linear part.

6. Numerical Results

The methods discussed above and Link-Cutting Bubble (LCB) strategy proposed in [16] are compared for one- and two-dimensional convection-diffusion-reaction problems. This comparison is carried out by computing the numerical solutions for a wide range of problem parameters especially in the interesting case of small diffusion which corresponds to the convection-dominated or reaction-dominated regimes. We also remark that the LCB method is employed only for one-dimensional problems as it is not designed for higher dimensions.

6.1. Numerical Results for 1D Convection-Diffusion-Reaction Problems

6.1.1. Experiment 1. We first consider an equation with constant coefficients and compute the approximate solutions on uniform mesh. The uniform mesh is generated by dividing the unit interval $[0, 1]$ into ten elements; that is, the mesh size $h = 1/10$. In Figure 6, we present the linear parts of the numerical solutions together with the exact solution for

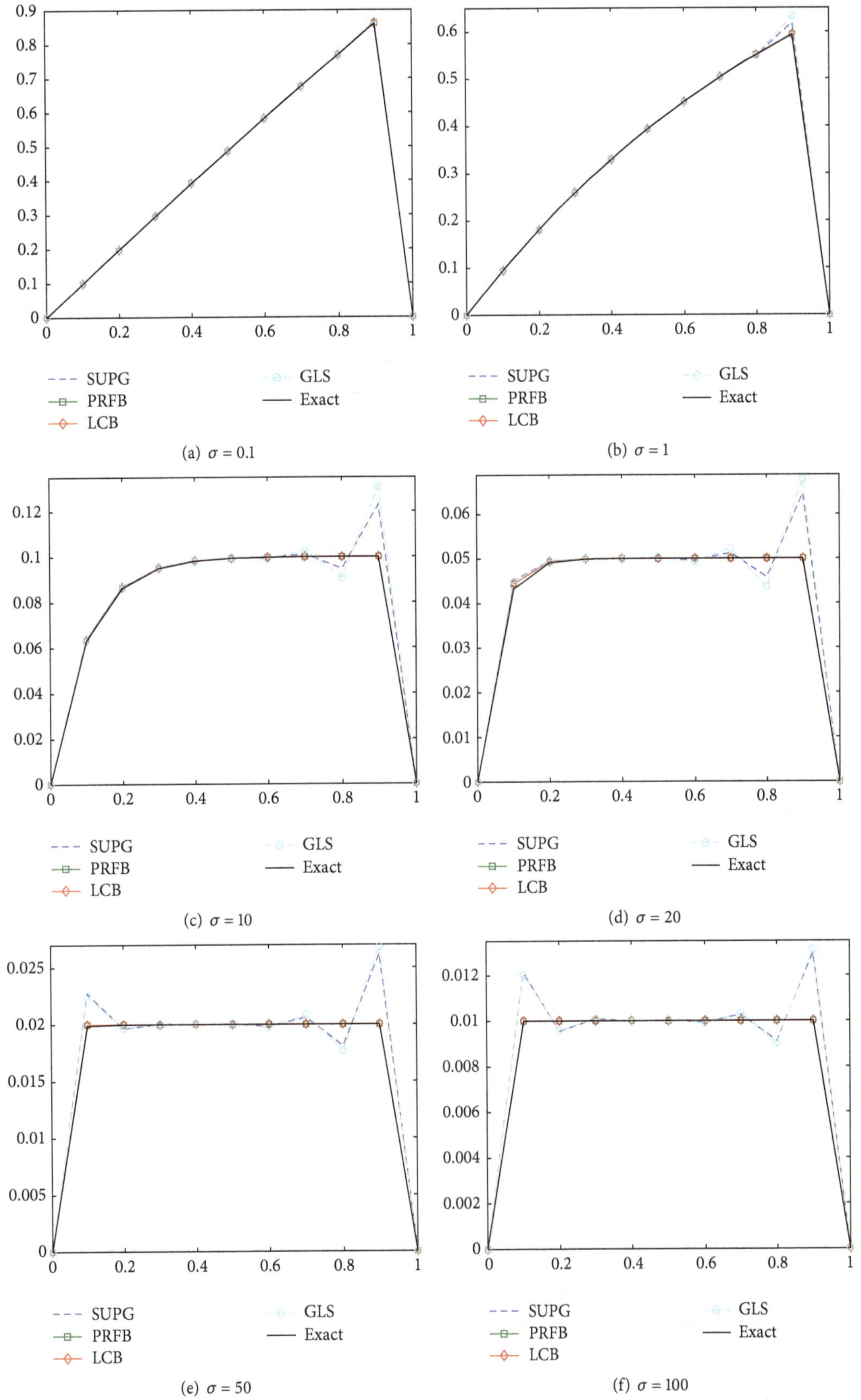

(a) $\sigma = 0.1$

(b) $\sigma = 1$

(c) $\sigma = 10$

(d) $\sigma = 20$

(e) $\sigma = 50$

(f) $\sigma = 100$

FIGURE 6: The numerical solutions and the exact solution for several values of σ when $f(x) = 1$ and $\epsilon = 10^{-5}$.

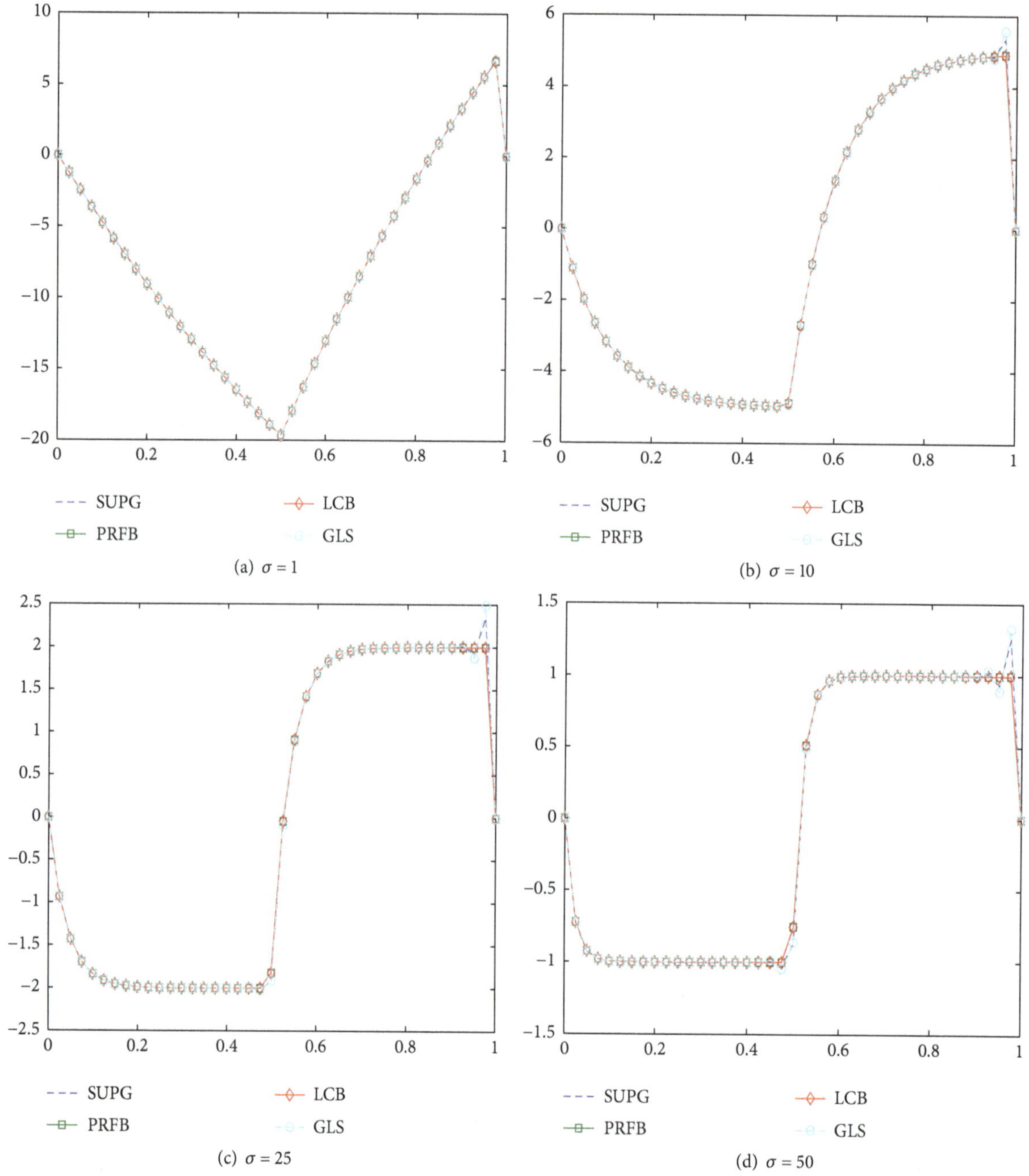

FIGURE 7: The numerical solutions for several values of σ when $\epsilon = 10^{-5}$ and $\beta = 1$.

various intensities of reaction ($\sigma = 0.1, 1, 10, 20, 50, 100$) when $\epsilon = 10^{-5}$, $\beta = 1$, and $f(x) = 1$. The numerical solutions obtained with the PRFB and LCB methods are in excellent agreement with the exact solution for the whole problem parameters while the approximations generated by the SUPG and GLS methods possess spurious oscillations in layer regions, especially in reaction-dominated cases. We also note that there is no major difference between SUPG and GLS methods; however, the instabilities introduced by Galerkin are a little more amplified in GLS compared with SUPG.

6.1.2. Experiment 2. Next, we consider a test problem which exhibits both internal and boundary layers. External force is defined as

$$f(x) = \begin{cases} 50, & x \leq 0.5, \\ -50, & x > 0.5. \end{cases} \tag{31}$$

In Figure 7, we present the linear parts of the numerical solutions for various intensities of reaction ($\sigma = 1, 10, 25, 50$) when $\epsilon = 10^{-5}$, $\beta = 1$, and $h = 1/40$. The PRFB and

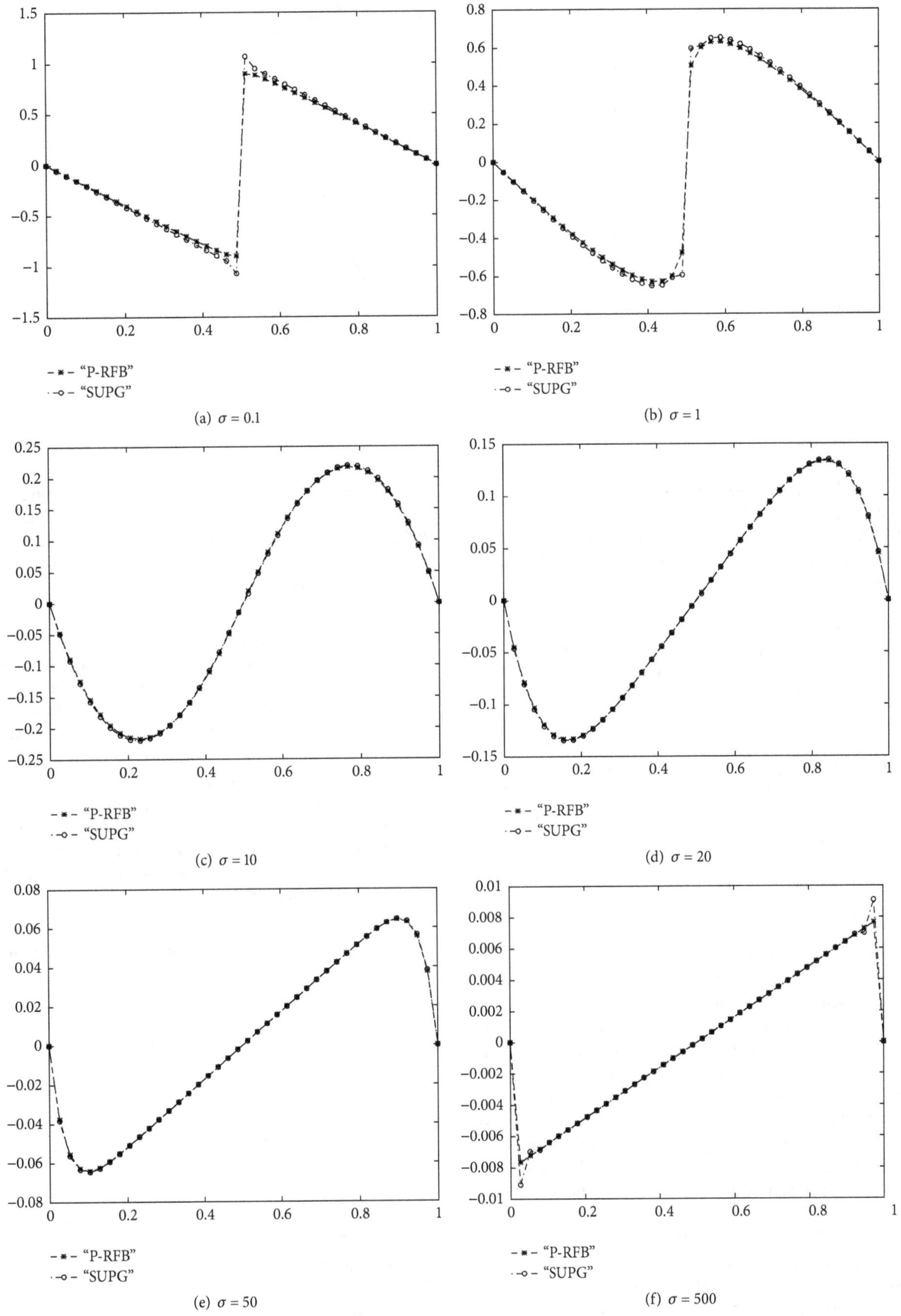

FIGURE 8: The numerical solutions when $f(x) = 4(2x - 1)$, $\epsilon = 10^{-5}$, and $\beta = -2(2x - 1)$.

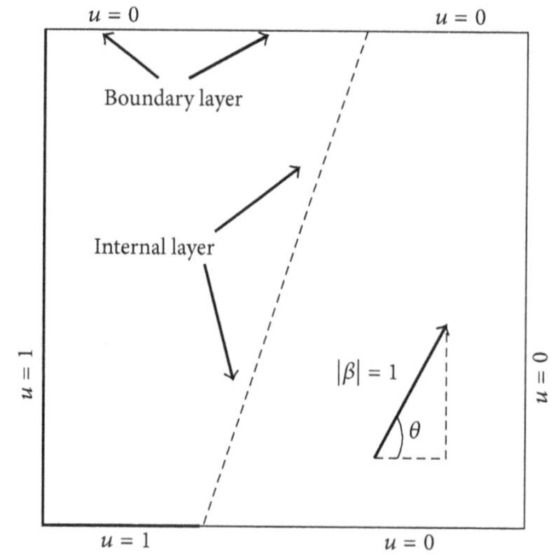

FIGURE 9: Configuration of Experiment 1.

LCB methods are able to capture both the internal and boundary layers and produce approximations consistent with the physical solution while the approximations generated by the SUPG and GLS methods exhibit spurious oscillations in the reaction-dominated regime.

6.1.3. Experiment 3. Now, we consider a variable coefficient case with a turning point. We set $\epsilon = 10^{-5}, \beta = -2(2x-1)$, and $f(x) = 4(2x-1)$ and decompose the domain into uniform discretization of 40 elements. Since there is no major difference between SUPG/GLS methods and PRFB/LCB methods, respectively, we only display the results with the SUPG and PRFB methods in Figure 8. The comparative study shows that the PRFB method works well even for turning point problems while the SUPG method exhibits slight oscillations around both the internal and the boundary layer.

6.2. Numerical Results for 2D Convection-Diffusion-Reaction Problems

6.2.1. Experiment 1. We consider a test problem taken from [26]. This problem is a benchmark problem, which exhibits numerical oscillations in both interior and boundary layer regions. The discontinuity in the boundary data causes interior layer that is propagated across unit square domain and there will also be layers at the outflow boundaries. Boundary conditions are displayed in Figure 9. We first take a set of uniform triangular meshes which is made up of $N = 10$ elements in x and y directions, respectively. We set $\epsilon = 10^{-4}$ and $\theta = 72^\circ$ and present elevation and contour plots of the solutions obtained with the PRFB and SUPG methods for various values of reaction ($\sigma = f = 0.001, 1, 10, 20, 50, 10^2, 10^3, 10^4, 10^5$) in Figures 10 and 11. The performances of the SUPG, GLS, SGS, and the PRFB methods are displayed in Figures 12 and 13 when $N = 20$. The stabilized methods deliver similar results unless the reaction term is dominant. We also remark that, in reaction-dominated regimes, there is no significant difference between SUPG and GLS methods; in contrast, the approximations generated by the SGS method are a little more oscillatory. Although the PRFB method could not eliminate the oscillations at all, it captures the layers well like in the RFB method (see [11, 12, 26, 27] and references therein) in any regimes.

6.2.2. Experiment 2. Now, we consider a problem with convection skew to the discretization. We display the boundary conditions and the triangular elements used in discretization of the problem domain in Figure 14. In Figure 15, we present the solutions obtained with the SUPG, GLS, SGS, and PRFB methods when $\epsilon = 10^{-4}$, $\beta = (0.15, 0.1)$, and $f = 0$ for various values of reaction ($\sigma = 10, 10^2, 10^3$). The numerical solutions in Figure 15 show that the PRFB method is more robust as both results are consistent with the physical configuration of the problem even on nonuniform meshes while the classical stabilization techniques, SUPG, GLS, and SGS, present similar qualitative behavior.

6.2.3. Experiment 3: Thermal Boundary Layer. Finally, we consider the problem statement shown in Figure 16 with the following boundary conditions:

$$u(x,y) = \begin{cases} 1, & \text{if } x = 0,\ 0 \le y \le 0.5, \\ 1, & \text{if } y = 0.5,\ 0 \le x \le 1, \\ 0, & \text{if } y = 0,\ 0 \le x \le 1, \\ 2y, & \text{if } x = 1,\ 0 \le y \le 0.5. \end{cases} \quad (32)$$

The flow is taken as $\beta = (2y, 0)$. This problem may be viewed as the simulation of the development of a thermal boundary layer on a fully developed flow between two parallel plates, where the top plate is moving with velocity equal to one and the bottom plate is fixed. We take a set of uniform triangular meshes which is made up of $N = 10$ elements in x and y directions, respectively (see Figure 16). We set the

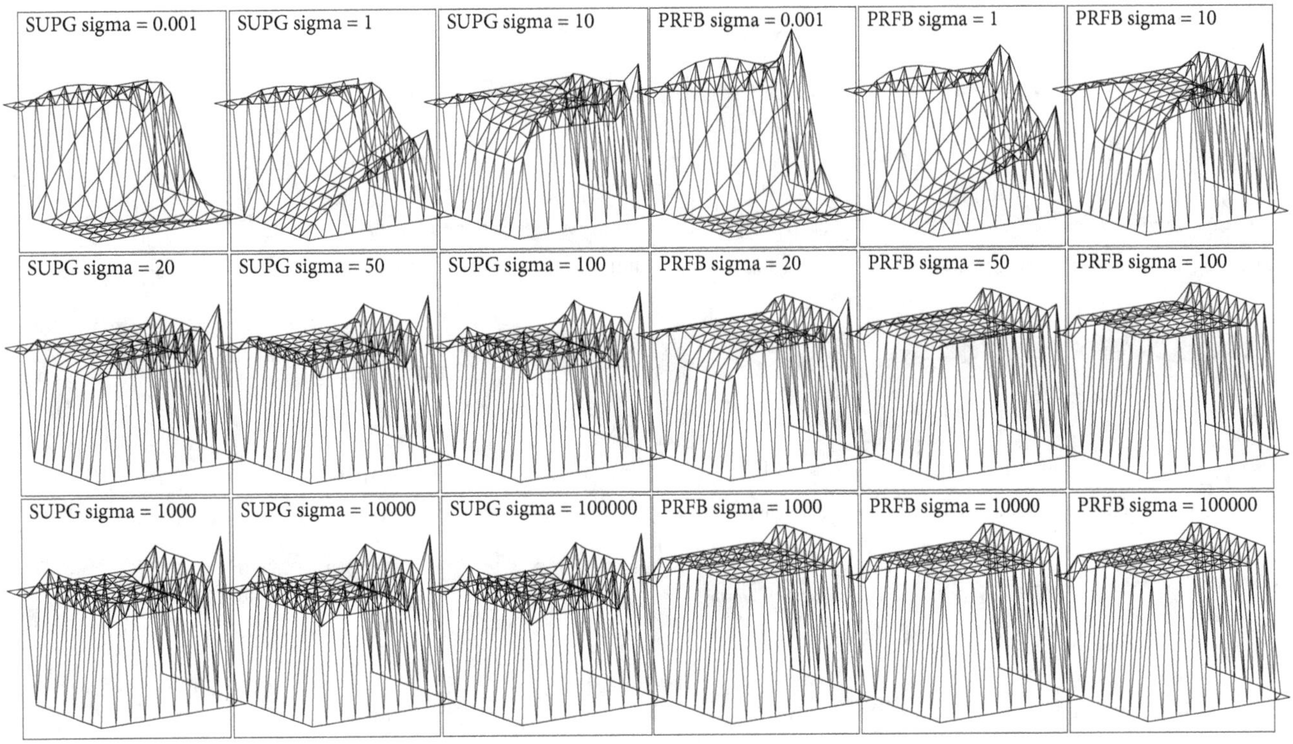

FIGURE 10: Numerical solutions for $\theta = 72^{\circ}$, $\epsilon = 10^{-4}$, and $N = 10$.

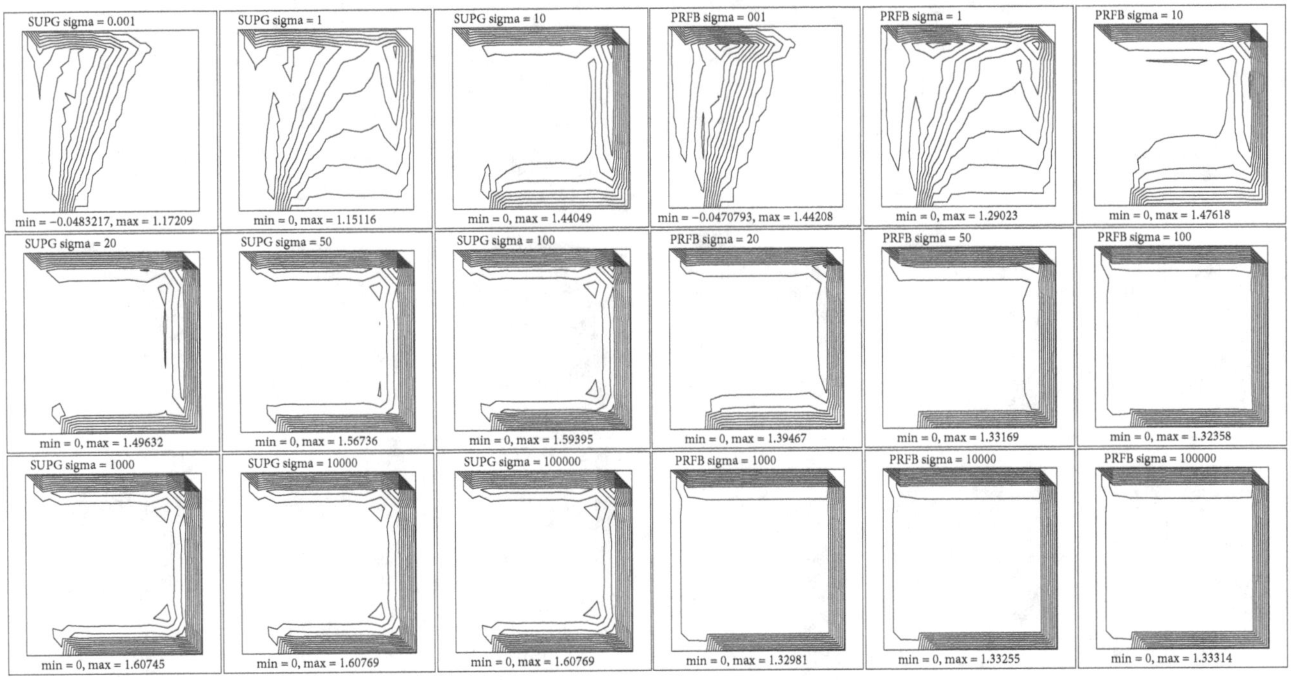

FIGURE 11: Corresponding contour plots.

diffusion $\epsilon = 10^{-5}$ and the reaction term $\sigma = 10^{-3}, 10^{-1}, 5 \times 10^{-1}, 1, 5, 10, 10^2, 10^3, 10^4$. The numerical results and the corresponding contour plots are reported in Figures 17 and 18. The plots show that the PRFB solution captures the characteristic features of the exact solution even on coarse meshes. The SUPG method is again more oscillatory than the PRFB method. The results for SGS and GLS methods present a similar qualitative behavior as in Experiment 1 and thus are not shown.

7. Conclusion

In this paper, several benchmark problems exhibiting boundary/internal layers are used to test the performance and the

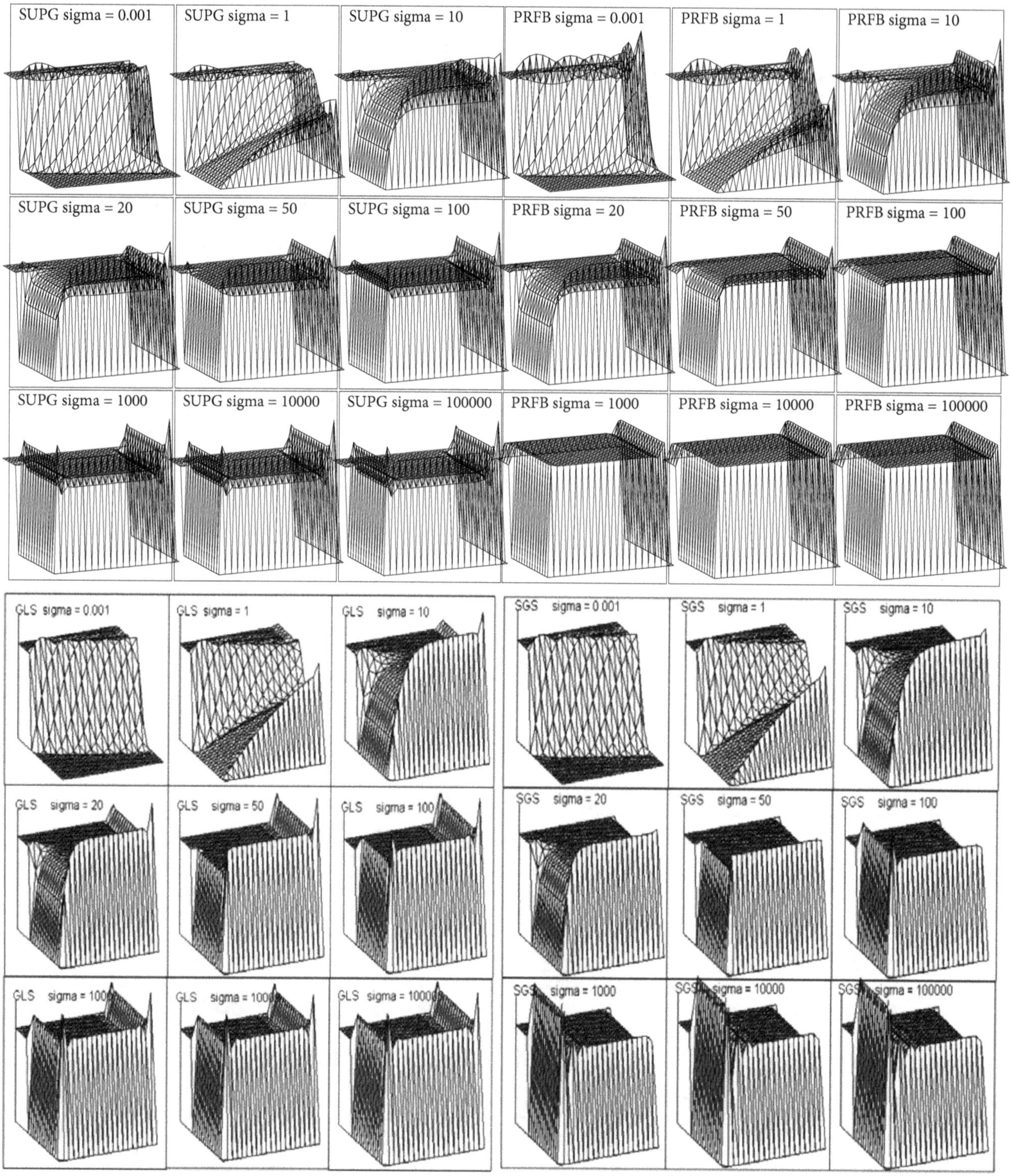

Figure 12: Numerical solutions for $\theta = 72^{\circ}$, $\epsilon = 10^{-4}$, and $N = 20$.

robustness of the stabilized FEMs for the numerical solution of the convection-diffusion-reaction problems. We observe that the classical SUPG method is very similar to the GLS method, whereas the PRFB method has a stabilization effect similar to LCB for one-dimensional convection-diffusion-reaction problems. We also note that the numerical results obtained with the SUPG, GLS, SGS, and the PRFB methods are comparable for the convection-dominated regimes. However, as the problem turns into reaction-dominated case, the PRFB approach achieves a better performance than the SUPG, GLS, and SGS methods. Therefore the PRFB strategy can be viewed as a promising tool to solve both

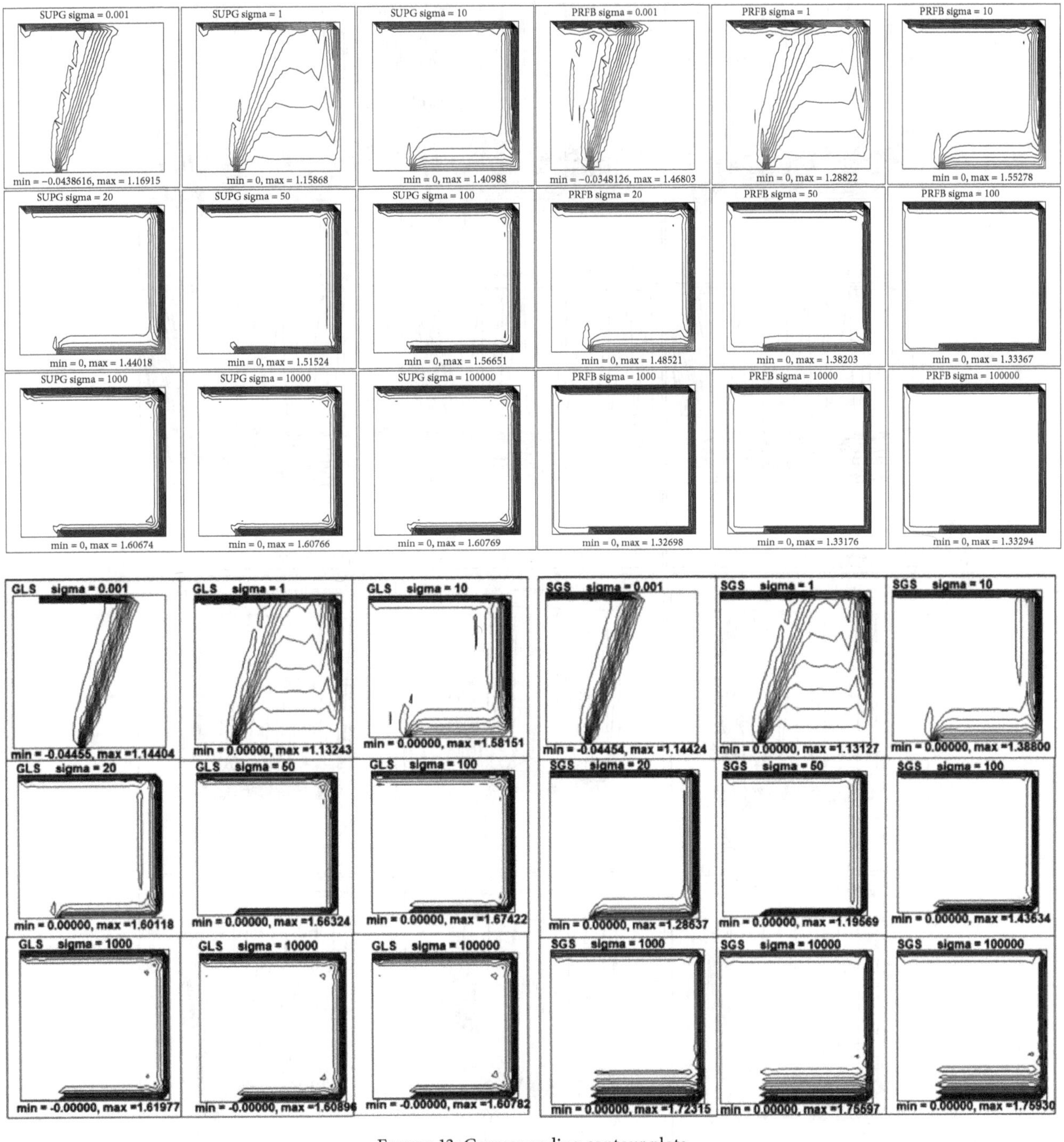

FIGURE 13: Corresponding contour plots.

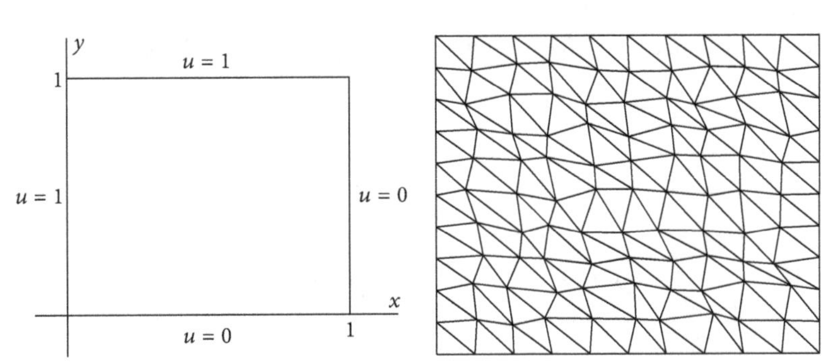

(a) (b)

FIGURE 14: (a) Configuration of Experiment 2. (b) Triangular elements used in discretization of the problem domain.

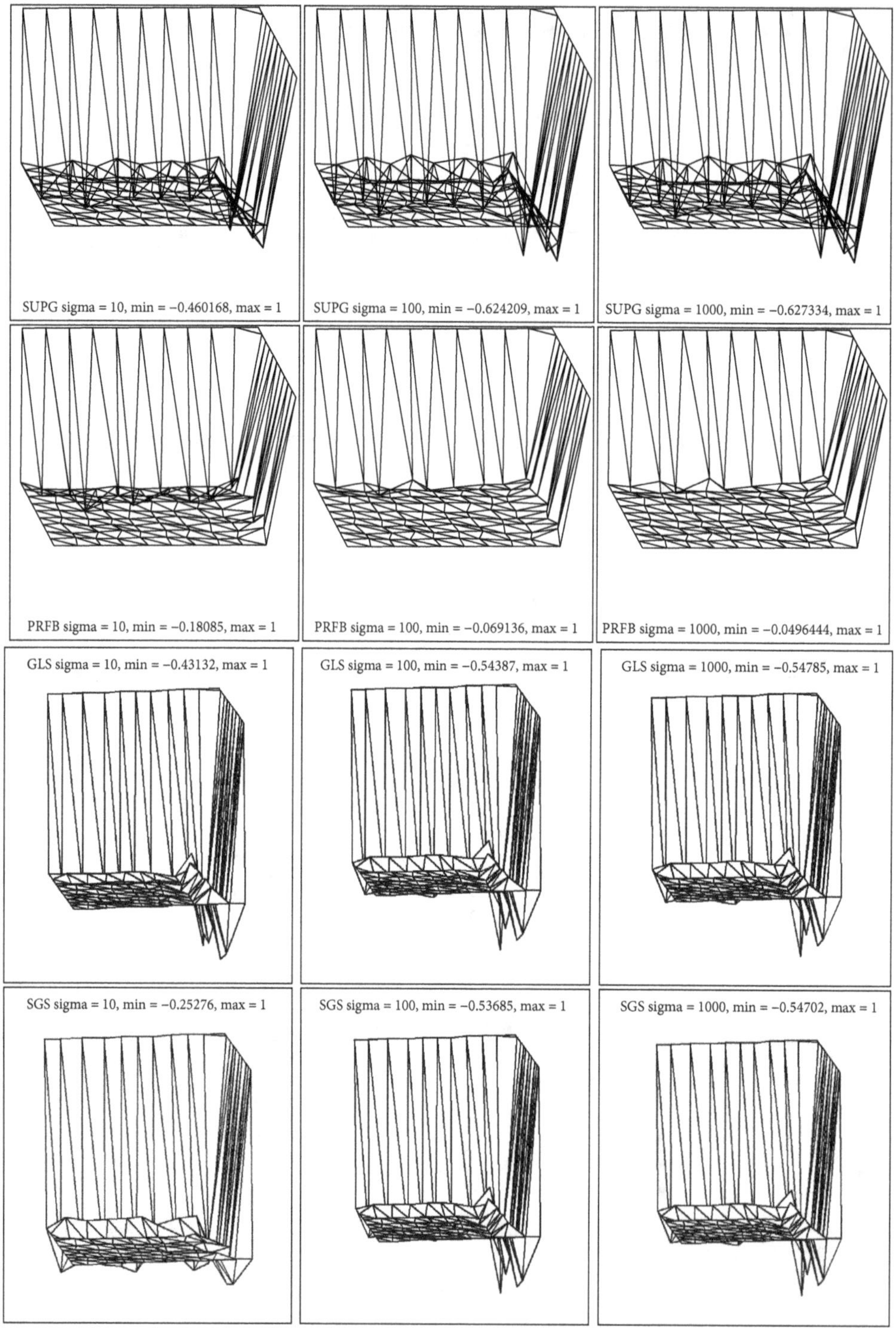

FIGURE 15: Numerical solution for $\epsilon = 10^{-4}$ (top to bottom: SUPG, P-RFB, GLS, and SGS; left to right: $\sigma = 10, 10^2, 10^3$).

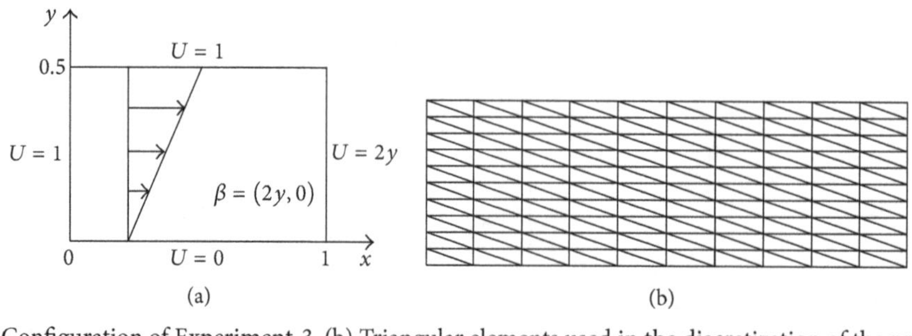

FIGURE 16: (a) Configuration of Experiment 3. (b) Triangular elements used in the discretization of the problem domain.

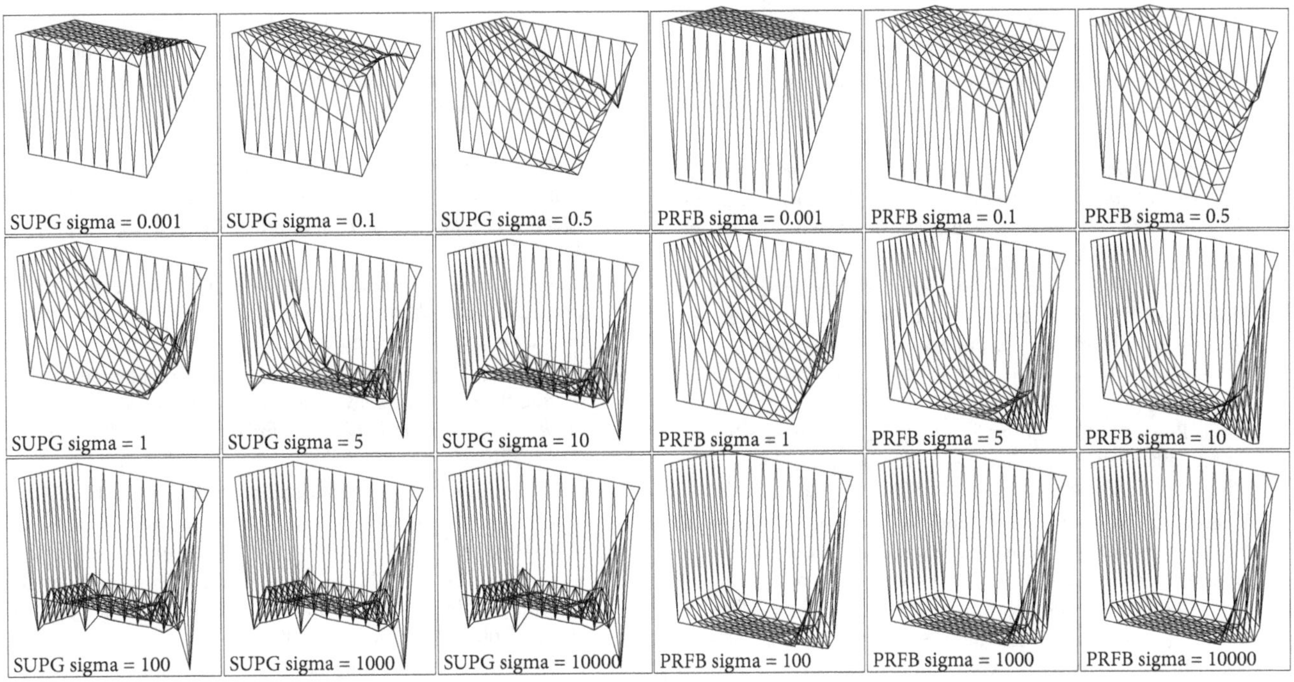

FIGURE 17: Numerical approximations for Experiment 3 when $\epsilon = 10^{-5}$, $\beta = (2y, 0)$, and $N = 10$.

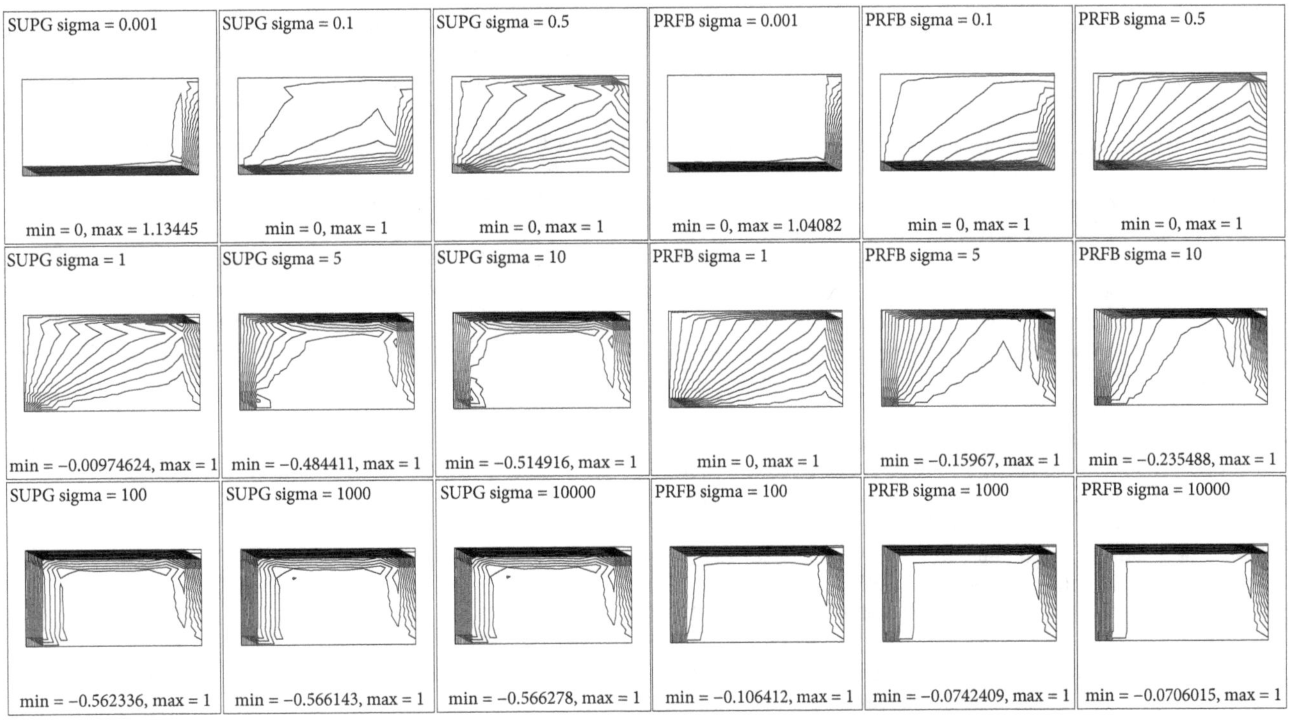

FIGURE 18: The corresponding contour plots.

one- and two-dimensional convection-diffusion-reaction problems.

Conflicts of Interest

The author declares that there are no conflicts of interest regarding the publication of this paper.

References

[1] M. I. Asensio, B. Ayuso, L. Ferragut, and G. Sangalli, "Numerical methods for modelling leaching of pollutants in soils," *Advances in Engineering Software*, vol. 38, pp. 429–438, 2007.

[2] V. John, T. Mitkova, M. Roland, K. Sundmacher, L. Tobiska, and A. Voigt, "Simulations of population balance systems with one

internal coordinate using finite element methods," *Chemical Engineering Science*, vol. 64, no. 4, pp. 733–741, 2009.

[3] J. D. Murray, *Mathematical Biology. II*, vol. 18 of *Interdisciplinary Applied Mathematics*, Springer, New York, NY, USA, 3rd edition, 2003.

[4] S. Natesan, *Efficient Numerical Methods for Singularly Perturbed Differential Equations, in Mathematical Modeling with Multidisciplinary Applications*, X.-S. Yang, Ed., John Wiley and Sons, Hoboken, NJ, USA, 2013.

[5] H. Ramos, J. Vigo-Aguiar, S. Natesan, R. Garcia-Rubio, and M. A. Queiruga, "Numerical solution of nonlinear singularly perturbed problems on nonuniform meshes by using a non-standard algorithm," *Journal of Mathematical Chemistry*, vol. 48, no. 1, pp. 38–54, 2010.

[6] H.-G. Roos, "Robust numerical methods for singularly perturbed differential equations: a survey covering 2008–2012," *ISRN Applied Mathematics*, Art. ID 379547, 30 pages, 2012.

[7] V. Shanthi, N. Ramanujam, and S. Natesan, "Fitted mesh method for singularly perturbed reaction-convection-diffusion problems with boundary and interior layers," *Journal of Applied Mathematics & Computing*, vol. 22, no. 1-2, pp. 49–65, 2006.

[8] A. N. Brooks and T. J. Hughes, "Streamline upwind/Petrov-Galerkin formulations for convection dominated flows with particular emphasis on the incompressible Navier-Stokes equations," *Computer Methods Applied Mechanics and Engineering*, vol. 32, no. 1-3, pp. 199–259, 1982.

[9] T. J. R. Hughes and M. Mizukami, "A new finite element formulation for computational fluid dynamics: II. Beyond SUPG," *Computer Methods in Applied Mechanics and Engineering*, vol. 54, pp. 341–355, 1986.

[10] M. Stynes, "Steady-state convection-diffusion problems," *Acta Numerica*, vol. 14, pp. 445–508, 2005.

[11] F. Brezzi and A. Russo, "Choosing bubbles for advection-diffusion problems," *Mathematical Models and Methods in Applied Sciences*, vol. 4, pp. 571–587, 1994.

[12] L. P. Franca, A. Nesliturk, and M. Stynes, "On the stability of residual-free bubbles for convection-diffusion problems and their approximation by a two-level finite element method," *Computer Methods Applied Mechanics and Engineering*, vol. 166, no. 1-2, pp. 35–49, 1998.

[13] F. Brezzi, D. Marini, and A. Russo, "Applications of pseudo residual-free bubbles to the stabilization of convection-diffusion problems," *Computer Methods in Applied Mechanics and Engineering*, vol. 166, pp. 51–63, 1998.

[14] A. Sendur and A. I. Nesliturk, "Applications of the pseudo residual-free bubbles to the stabilization of convection-diffusion-reaction problems," *Calcolo*, vol. 49, no. 1, pp. 1–19, 2012.

[15] A. Sendur, A. I. Nesliturk, and A. Kaya, "Applications of the pseudo residual-free bubbles to the stabilization of the convection-diffusion-reaction problems in 2D," *Computer Methods in Applied Mechanics and Engineering*, vol. 277, pp. 154–179, 2014.

[16] F. Brezzi, G. Hauke, D. Marini, and G. Sangalli, "Link-cutting bubbles for the stabilization of convection-diffusion-reaction problems," *Mathematical Models and Methods in Applied Sciences*, vol. 13, pp. 445–461, 2003.

[17] T. J. R. Hughes and M. Mallet, "A new finite element formulation for computational fluid dynamics: III. The generalized streamline operator for multidimensional advective-diffusive systems," *Computer Methods in Applied Mechanics and Engineering*, vol. 58, pp. 305–328, 1986.

[18] T. J. Hughes, "Multiscale phenomena: Green's functions, the Dirichlet-to-Neumann formulation, subgrid scale models, bubbles and the origins of stabilized methods," *Computer Methods Applied Mechanics and Engineering*, vol. 127, no. 1-4, pp. 387–401, 1995.

[19] R. Araya, E. Behrens, and R. Rodriguez, "An adaptive stabilized finite element scheme for the advection-reaction-diffusion equation," *Applied Numerical Mathematics*, vol. 54, no. 3-4, pp. 491–503, 2005.

[20] R. P. B. Chaple, "Numerical stabilization of convection-diffusion-reaction problems," *Reports of the Department of Applied Mathematical Analysis*, 2006.

[21] R. Codina, "Comparison of some finite element methods for solving the diffusion-convection-reaction equation," *Computer Methods Applied Mechanics and Engineering*, vol. 156, no. 1-4, pp. 185–210, 1998.

[22] X. Liu and J. Zhang, "Comparison of SUPG with bubble stabilization parameters and the standard supg," *Abstract and Applied Analysis*, vol. 2014, Article ID 364675, 2014.

[23] T. E. Tezduyar, "Stabilized Finite Element Formulations for Incompressible Flow Computations," *Advances in Applied Mechanics*, vol. 28, no. C, pp. 1–44, 1991.

[24] L. P. Franca and F. Valentin, "On an improved unusual stabilized finite element method for the advective-reactive-diffusive equation," *Computer Methods in Applied Mechanics and Engineering*, vol. 190, pp. 1785–1800, 2000.

[25] G. Hauke and A. García-Olivares, "Variational subgrid scale formulations for the advection-diffusion-reaction equation," *Computer Methods Applied Mechanics and Engineering*, vol. 190, no. 51-52, pp. 6847–6865, 2001.

[26] M. I. Asensio, A. Russo, and G. Sangalli, "The Residual-Free Bubble numerical method with quadratic elements," *Mathematical Models and Methods in Applied Sciences*, vol. 14, pp. 641–661, 2004.

[27] F. Brezzi, M. O. Bristeau, L. P. Franca, M. Mallet, and G. Roge, "A relationship between stabilized finite element methods and the Galerkin method with bubble functions," *Computer Methods in Applied Mechanics and Engineering*, vol. 96, pp. 117–129, 1992.

15

Urban Lead: Modeling Its Distribution and Effects on Children

Zhixiong Chen,[1] Yi Ding,[1] Andrew Getz,[1] and Bernard Lipat[2]

[1]*Department of Mathematics, New Jersey City University, Jersey City, NJ 07305-1597, USA*
[2]*Applied Physics and Applied Mathematics Department, Columbia University, New York, NY 10027, USA*

Correspondence should be addressed to Zhixiong Chen; zchen@njcu.edu

Academic Editor: Keshlan S. Govinder

We model the transportation of lead from the atmosphere and from the surface of the soil simultaneously at the macroscale and mesoscale to study its health effects on children in Jersey City, NJ. We conceptualize Jersey City as an open system where lead is continuously emitted from a local smelting plant and a local power plant, deposited onto the surface soil of playgrounds, and ingested by children. The model is constructed using the diffusion-advection partial differential equation in three spatial dimensions and one temporal dimension with an initial condition and boundary conditions. The model is solved using the Crank-Nicolson numerical method at the macroscale to determine the deposition of lead from the smelting plant and the local power plant and at the mesoscale to refine the amount of lead deposition for the areas considered. We then determine the health consequences for the average child using the bioaccessibility of lead from soil to children, the bioavailability of ingested lead to the circulatory system, and the biological half-life of lead isotopes in the blood. The health effects on children from lead are directly proportional to the blood lead concentration.

1. Conceptualization of the System

On June 22, 2012, officials of the City of Jersey City announced the closure of downtown-area Mary Benson Park [1]. The reason was lead. The problem of lead in the urban environment so epitomized by Jersey City has been a long one and a challenging one. Although the second largest city in the State of New Jersey, Jersey City, is a preeminent commercial, distribution, and transportation hub for the Tristate Area [2–4], Jersey City has not been immune to the decline in manufacturing so characteristic of postindustrial, globalized America. Once home to the production of such popular brands as American Can, Emerson Radio, Lorillard, Colgate, Dixon Ticonderoga, and Chloro, Jersey City now claims only four industrial centers: Greenville Yards, Claremont Industrial Park, Montgomery Industrial Park, and Liberty Industrial Park [2–4]. It is unfortunate, however, that this reduction in manufacturing coupled with the ban on leaded gasoline has not resulted in the elimination of the threat of lead poisoning: the accumulation of lead in the soil of industrial areas has created a problem that persists.

In the present study, we construct a mathematical model to determine the spatial distribution of lead in the City of Jersey City so as to determine the resultant effects on the health of children. We begin with the foundational assumptions about the system. First we assume that the City of Jersey City is an open cube, wherein lead is free to enter and exit. We assume that the two major sources of lead in the vicinity are the only sources of lead and that the effects of the other sources are negligible. To make calculations simpler, we assume that all isotopes of lead have the same atomic radius. We assume, by the insolubility of lead at reasonably balanced acidity levels, that lead will not be transported by rain or surface water, nor will it diffuse through the soil. Further, we assume that lead, because of its particulate form, will diffuse through the air, assumed to be homogeneous, and be transported by wind advection. Our initial and boundary conditions are approximately chosen to simulate data for Jersey City, keeping the simplistic 3D-City model. We focus specifically on four public parks in Jersey City.

2. Diffusion-Advection

The transportation of lead through the air is governed by the diffusion-advection equation:

$$\frac{\partial C}{\partial t} = D\left(\frac{\partial^2 C}{\partial x^2} + \frac{\partial^2 C}{\partial y^2} + \frac{\partial^2 C}{\partial z^2}\right) - v_x \frac{\partial C}{\partial x} - v_y \frac{\partial C}{\partial y} - (v_z + v_T)\frac{\partial C}{\partial z}, \quad (1)$$

where $C = C(x, y, z, t)$ is the concentration of lead at a point (x, y, z) in space and at time t, D is the diffusion coefficient for lead particles in the air, v_x is the velocity of the wind in the x, or eastwest, direction, v_y is the velocity of the wind in the y, or northsouth, direction, v_z is the velocity of the wind in the z, or vertical, direction, and v_T is the terminal velocity of a falling particle of lead.

The diffusion coefficient for lead in air is calculated using the mean velocity and the mean free path [5]; that is,

$$D = \frac{B_k T}{3\pi\sqrt{2}P\left(d_{\text{air}} + d_{Pb}\right)^2}\sqrt{\frac{8B_k T}{\pi m_N}}, \quad (2)$$

where $B_k = 1.380650 \times 10^{-23}$ J/K is Boltzmann's constant, $T = 298.15$ K is the temperature, $m_N = 3.44063610352 \times 10^{-25}$ kg is the mass of a particle of lead, $P = 100000\,\text{kg/m}^3$ is the pressure, $d_{\text{air}} = 3.66 \times 10^{-10}$ m is the average diameter of a particle in air, and $d_{Pb} = 1.4193 \times 10^{-7}$ m is the diameter of a particle of lead [5–7].

The advective coefficients, that is, velocity of the wind in the x- and y-directions, respectively, v_x and v_y are taken as functions of season, as in Table 1 [8], where $\alpha = \pi/4$ and $\beta = \pi/8$.

The directions of the velocities reflect the choice of origin to be the bottommost southeast corner of the cube. During the autumn, winter, and spring, the direction of the wind is southeasterly, that is, in the direction of the origin. However, during the summer, the direction of the wind is northeasterly. To maintain the strict diagonal dominance of the coefficient matrix **A**, which will be detailed in Section 3, the cube is reflected so that the origin becomes the bottommost northeast corner of the cube.

The advective coefficient, or velocity of the wind in the z-direction, $v_z = dz/dt$ is calculated by solving the hydrostatic ordinary differential equation, which relates the vertical velocity of the wind with the ω-vertical velocity, which is the rate of change of pressure over time. The hydrostatic equation is as follows:

$$\begin{aligned} \frac{dP}{dz} &= -d \cdot g \Longrightarrow \\ \frac{dP}{dt}\frac{dt}{dz} &= -d \cdot g \Longrightarrow \\ \frac{dz}{dt} &= \frac{1}{-d \cdot g}\frac{dP}{dt}, \end{aligned} \quad (3)$$

where dP/dt is the ω-vertical velocity in Pa/s, d is the air density in kg/m^3, and g is gravitational acceleration [9–16].

Table 1

	Autumn	Winter	Spring	Summer
v_x	$3 \cdot \cos\alpha$	$4 \cdot \cos\alpha$	$3 \cdot \cos\beta$	$2 \cdot \cos\alpha$
v_y	$3 \cdot \sin\alpha$	$4 \cdot \sin\alpha$	$3 \cdot \sin\beta$	$2 \cdot \sin\alpha$

Air density d is calculated according to the ideal gas law for mixtures of ideal gases

$$d = \frac{p_a}{287.05T} + \frac{p_v}{461.495T} \quad (4)$$

p_a being the partial pressure of dry air in Pa, p_v being the partial pressure of water vapor also in Pa, and T being the air temperature in K and 287.05 and 461.495 being the specific gas constants for dry air and water vapor, respectively [9–16]. The partial pressure of water vapor is determined by

$$p_v = \frac{\text{RH}}{100p_{\text{vsat}}}, \quad (5)$$

where RH is the relative humidity and p_{vsat} is the partial pressure of water vapor at saturation, which since p_{vsat} is [17]

$$p_{\text{vsat}} = \text{RH} \cdot 6.1078 \times 10^{((7.5T-2048.625)/(T-35.85))} \quad (6)$$

solves for p_v. Again, using standard atmospheric temperature and pressure, and since $dP/dt = 0.02$ Pa/s, we have $dz/dt = v_z = -0.00158$ m/s [16, 17].

The terminal velocity v_T is calculated using the definition

$$v_T = \sqrt{\frac{2m_N g}{dC_D A}}, \quad (7)$$

where $m_N = 3.44063610352 \times 10^{-25}$ kg, $C_D = 0.47$ is the dimensionless drag coefficient of a spherical particle of lead, and $A = 3.16423 \times 10^{-14}$ m^2/s is the cross-sectional area of a lead particle [5–7]. Thus $v_T = 4.9267 \times 10^{-6}$ m/s.

2.1. Initial Conditions. Data available on the mixing ratios of lead, measured in parts per million (ppm) in 2007 at each of the parks under consideration, was used in construction the initial conditions of the system [18]. We have taken the data of lead found in the surface of soil, which we will use to show the effects on human life. We do not consider the lead below the surface. They are summarized in Table 2.

To avoid the unreasonable and distorting assumption that the concentration of lead at points in the system with unavailable data is nonexistent, the following interpolation method for the initial conditions is used:

$$\begin{aligned} \overline{C} &= \frac{\sum_{i=1}^{n^2-p} C + \sum_{i=1}^{p} f(x_i, y_i)}{n^2} \Longrightarrow \\ C &= \frac{n^2\overline{C} - \sum_{i=1}^{p} f(x_i, y_i)}{n^2 - p} \end{aligned} \quad (8)$$

to agree with the known concentration data and to preserve the average concentration over the surface soil.

Once the data values were uncovered, two-dimensional cubic spline interpolation was used to refine the surface soil data.

TABLE 2

Park	Concentration (ppm)
Hamilton Park	500
Mary Benson Park	383.34
E. F. Jones Park	840
Lincoln Park	512.5

TABLE 3

	Year					
Source concentration	2007	2008	2009	2010	2011	2012
Kearny point (ppm)	2325	2194	1515	680	2259.75	2259.75
Hudson generating (ppm)	210	581	407	55	313.25	313.25

2.2. Boundary Conditions. The concentrations at the two sources of lead from 2007 to the present were available from the Environmental Protection Agency [19] and are summarized in Table 3.

Consistent with the assumption of openness, Neumann boundary conditions were imposed at the six faces of the cube:

$$\begin{aligned} \left.\frac{\partial C}{\partial x}\right|_{x=0} &= 0 \\ \left.\frac{\partial C}{\partial y}\right|_{y=0} &= 0 \\ \left.\frac{\partial C}{\partial z}\right|_{z=0} &= 0, \\ \left.\frac{\partial C}{\partial x}\right|_{x=L} &= 0 \\ \left.\frac{\partial C}{\partial y}\right|_{y=L} &= 0 \\ \left.\frac{\partial C}{\partial z}\right|_{z=L} &= 0, \end{aligned} \tag{9}$$

where L is the distance to the opposite face of the cube.

The boundary conditions are incorporated into the numerical scheme at each of the boundaries. The emissions from a local smelting factory and a local power plant are taken into account.

3. Numerical Method

To solve the diffusion-advection PDE, the Crank-Nicolson numerical scheme is employed, and we used MATLAB to assist us. It is the average of the forward finite difference method and of the backward finite difference method with the index shifted to agree with the forward difference method. Essentially, the scheme uses the relation, as set out by the two discretizations of the diffusion-advection PDE, between the values of the concentration at the six surrounding points in 3-dimesional space at one time step and the values of the concentration at the same six surrounding points in the next time step, to approximate the value of the concentration at the central point over time. The resulting finite difference scheme is desirable over either of its constituents because of its unconditional stability [20].

The $n \times n \times n$ cube is reshaped into an $n^3 \times 1$ column vector by iterating first over x, then over y, and finally over z. The following matrix equation results:

$$\begin{aligned} \mathbf{A} \cdot \mathbf{C}_{i,j,k}^{l+1} &= \mathbf{B} \cdot \mathbf{C}_{i,j,k}^{l} \Longrightarrow \\ \mathbf{C}_{i,j,k}^{l+1} &= \mathbf{A}^{-1}\mathbf{B} \cdot \mathbf{C}_{i,j,k}^{l}, \end{aligned} \tag{10}$$

where l is the time step, i is the space step in the x-direction, j is the space step in the y-direction, k is the space step in the z-direction, C is the $n^3 \times 1$ column vector of concentration values, and $\mathbf{A}$ and $\mathbf{B}$ are $n^3 \times n^3$ square matrix of coefficients [21]. The invertibility of $\mathbf{A}$, a strictly diagonally dominant matrix, is guaranteed by the Levy-Desplanques Theorem [22] and the reflection described in Section 2.

The boundary conditions are incorporated into the matrix using the centered difference method at each boundary, that is, at each of the six faces of the cube. For example,

$$\begin{aligned} \left.\frac{\partial C}{\partial x}\right|_{x=0} &\approx \frac{C_{0,j,k}^{l} - C_{2,j,k}^{l}}{2h} = 0 \Longrightarrow \\ C_{0,j,k}^{l} &= C_{2,j,k}^{l}, \\ \left.\frac{\partial C}{\partial x}\right|_{x=L} &\approx \frac{C_{n-1,j,k}^{l} - C_{n,j,k}^{l}}{2h} = 0 \Longrightarrow \\ C_{n,j,k}^{l} &= C_{n-1,j,k}^{l}, \end{aligned} \tag{11}$$

where the space step is $h = 1300$ m at the macroscale and $h = 100$ m at the mesoscale. Let $\{a_i\}_1^{n^3}$ be the sequence of row of the column vector C. Given the set-up of the cube, the $x = 0$ boundary occurs where $a_i \bmod n = 1$, $x = L$ occurs, where $a_i \bmod n = 0$, $y = 0$ occurs, where $1 \leq a_i \bmod n^2 \leq n$, $y = L$ occurs, where $n^2 - n \leq a_i \bmod n^2 \leq 0$, $z = 0$ occurs, where $1 \leq a_i \bmod n^3 \leq n^2$, and $z = L$ occurs where $n^3 - n^2 \leq a_i \bmod n^3 \leq 0$.

The matrix equation is programmed into MATLAB and iterated with a time step consistent with the stability requirement [8]

$$D \cdot \frac{l}{h^2} \leq \frac{1}{2} \tag{12}$$

from autumn 2007 to autumn of 2012, that is, twenty-four seasons.

The results are depicted in Figures 1 and 2. Although the model takes into consideration the seasonal variations in the directions and the velocities of the wind, there is, as expected, a southeasterly tendency for lead transportation across the surface soil. In fact, even though this lead is blown across the surface soil, the increase in the concentration of lead, particularly at the parks, indicates that the rate of lead deposition from the two sources of lead is greater than that of lead transportation out of the parks. As in Figure 1, if we calculate in macroscale, we can only see the lead deposition of Lincoln Park clearly; after we used mesoscale as shown in Figure 2, it is clear to see the lead deposition of E. F. Jones Park, Mary Benson Park, and Hamilton Park.

Further, a sensitivity analysis was conducted on the macroscale numerical solution. A ten percent (±10%) change

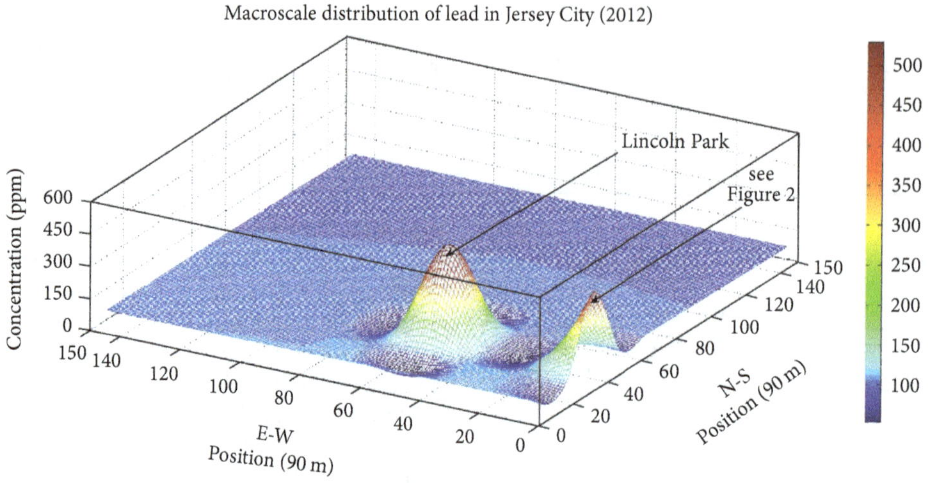

FIGURE 1

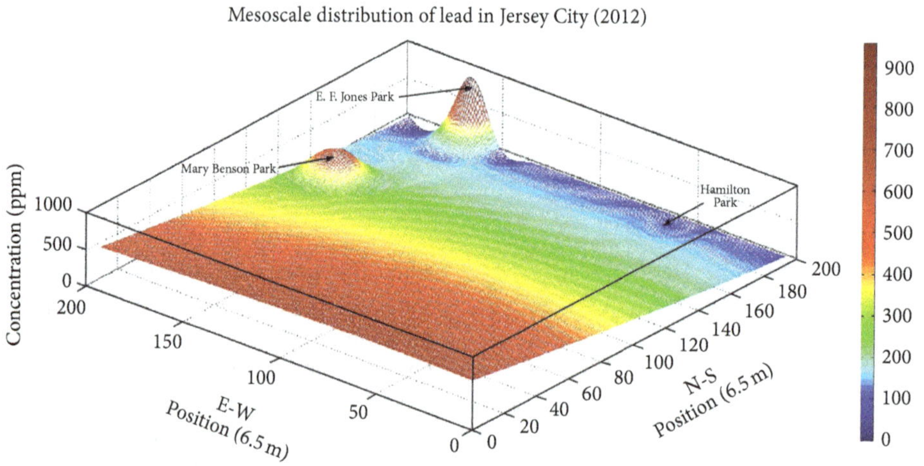

FIGURE 2

in each parameter (e.g., source concentrations, initial concentrations, vertical velocity, and angle of advection) yielded only slight percent differences in the final concentrations of lead. It was found that the model was relatively insensitive to most parameters. However, it should be noted that the model was moderately sensitive to a change in the angle of advection. This is depicted in Figure 3. The absolute differences are so small, that is, on the order of 10^{-74}, that the significance of a moderately large percent change (at the very maximum about −70% and only in the upper atmosphere 13 km above the surface) should not be conflated.

Of note, the model reflects the fact that in 2010 Hamilton Park underwent renovations that included removal and replacement of the surface soil by setting the concentration at that point to 100 ppm, which the authors thought reasonable in the absence of data to reflect a significant decrease in the concentration of lead but so as not to distort the lead deficiency and therefore the rate of lead deposition over time. The continued increase of the concentration of lead at Hamilton Park after 2010 at a rate commensurate with the rate of increase before 2010 implies, through something of a natural experiment, that remediation—the removal and replacement of surface soil—is not an effective long-term strategy because it will only be a matter of years before the concentration of lead at, say, Hamilton Park, returns to elevated and, perhaps, prohibited levels.

Figure 4 demonstrates the comparison of the data generated by the numerical solution of the model with data published by the City of Jersey City [1], and these are seasonal averages. Here we use our numerical solutions to compare with sample 2 (410 mg/Kg) and sample 4 (563 mg/Kg) from [1]. "Sample 2 and sample 4 from [1] were taken by an outside geotechnical and environmental materials testing consultant and were taken in four distinct locations within the playground area" [1]. The numerical solution of the model herein developed agrees with the data available from the aforementioned closure of Mary Benson Park in June 2012.

4. Health Effects

Lead's neurotoxic effects have been widely studied [14, 23–32]. They result mainly from lead's similarity to and consequent

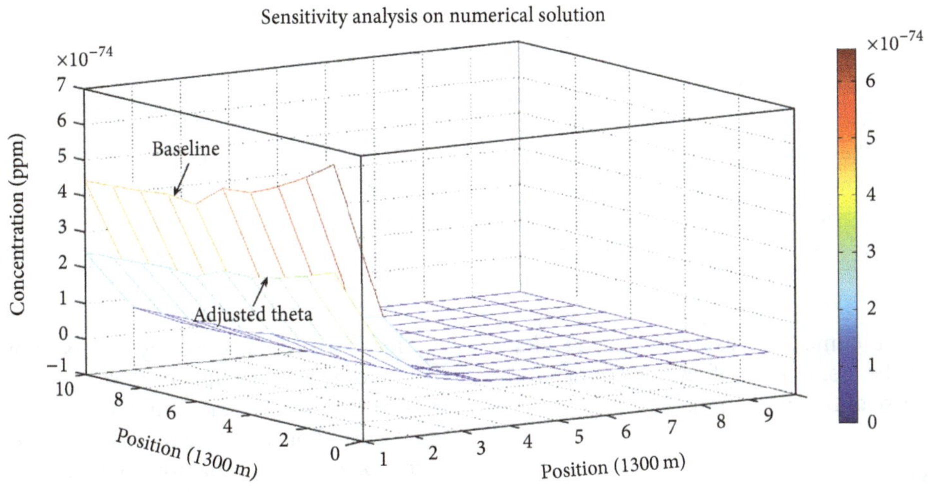

FIGURE 3

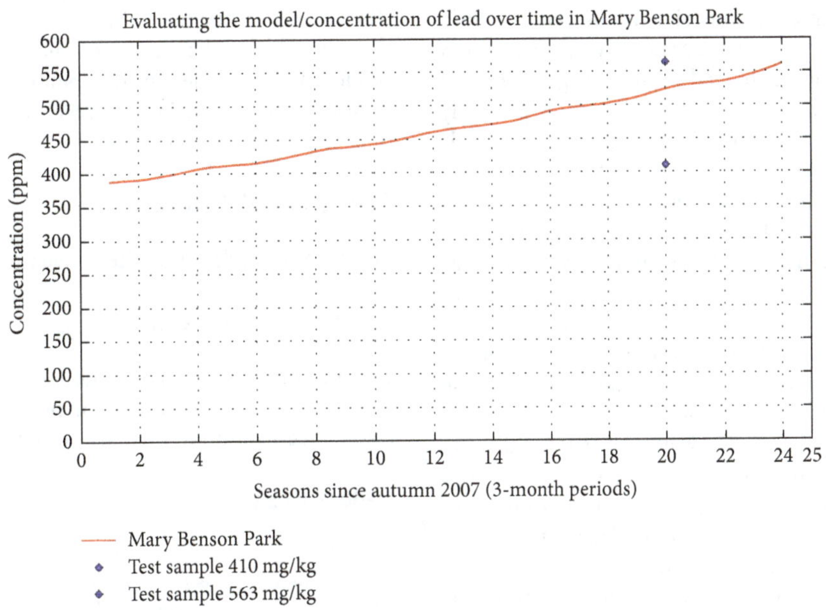

FIGURE 4

competition with calcium, which is vital for the effective delivery of action potentials along axons. Replacement of calcium ions in the brain with lead results in retardation of this process and can have severe neurological effects.

The following integral is used to calculate the cumulative exposure, that is, blood lead level (BLL), to lead at a point (x_1, y_1, z_1), for a time period $[t_1, t_2]$:

$$\mathrm{BLL} = B_{\mathrm{acc}} \cdot B_{\mathrm{av}} \cdot \int_{t_1}^{t_2} C(x_1, y_1, z_1, t) \left(\frac{1}{2}\right)^{t/30} dt, \quad (13)$$

where $B_{\mathrm{acc}} = 0.154897$ is the bioaccessibility factor and $B_{\mathrm{av}} = 0.108$ is the bioavailability factor [30, 32]. Bioaccessibility is the measure of the percentage of lead deposited onto the surface soil accessible for absorption into a biological system, and bioavailability is the percentage of lead absorbed into a biological system available for uptake into the circulatory system [33].

The health effects of lead considered in this paper are those directly proportional to the blood lead level (BLL), with the proportionality summarized in Table 4 [8, 15–17, 19, 23, 27, 33–36].

TABLE 4

Health effects	Proportionality
Fine motor composite score	−4.6
IQ	−0.25
Mortality	0.0035
Average lifetime earning	0.005975

Table 5

Park	Blood lead level (μg/dL)	Summary of health effects Fine motor decrease	IQ decrease	Excess relative risk (%)	Earnings decrease (%)
Downtown	6.0428	0.1314	1.5107	2.115	0.9026
Hamilton	1.6447	0.0358	0.4112	0.5756	0.2457
Lincoln	7.0903	0.1541	1.7726	2.4816	1.0591
Mary Benson	7.136	0.1551	1.784	2.4976	1.0659
E. F. Jones	12.6348	0.2747	3.1587	4.4222	1.8873

Table 5 depicts the cumulative exposure and resultant health effects under the assumption of an hour of exposure per day after the course of the simulation.

5. Conclusions

This model is tuned to the conditions presented in the City of Jersey City, but it is not unreasonable to generalize these results to similar postindustrial urban centers. Still, future research may wish to refine the simplifying assumption regarding the bioaccessibility of lead by modeling saltation of granular particles near the surface of the soil to more precisely quantify the probability of uptake. A more accurate bioavailability factor may be achieved by modeling the competition between lead and calcium, for example, for neural receptors. The initial conditions are derived from samples taken on location and so, while necessarily accurate, are limited as any samples are limited by questions of representativeness and quality of equipment. The interpolation thereof, however, while preserving the invariant of average concentration, assumes uniform distribution. Further research may choose to assume that concentrations of lead are higher near highways and factories than they are away from such. Still, the diffusive and advective processes, while possibly weak in the short-term, may have resulted in a redistribution of lead in, say, the four decades since the ban of leaded gasoline.

Nevertheless, the qualitative predictions of the solution to the model are no less worrisome. Of local interest, our model predicts especially elevated levels of lead concentration at many highly trafficked and popular parks, such as E. F. Jones Park. Public health officials may wish to take action to protect all who benefit from public recreation areas, especially children, by collecting and testing soil samples to validate experimentally the results predicted and to take the actions necessary to safeguard the public from adverse health effects. These actions may include public notices about the potential threats of lead ingestion and inhalation, covering exposed topsoil, prohibition of access to the parks, and soil remediation. Although none of the aforementioned are likely to be effective long-term solutions, they may be effective in the short run. Cost-benefit analyses may be done to determine which options are best.

However, the results of the model's predictions for Hamilton Park, while a single data point, reveal that even the most drastic of the short-term solutions, that is, soil remediation, is ineffective in the long run because more lead is being deposited by the emissions than is transported away by advection. Long-term solutions may include inspections of the local refineries, smelting factors, power plants, and so forth, to ensure compliance with EPA emissions regulations, reduction in lead emissions, and elimination of lead emissions altogether. There may be a steady-state solution with nonzero lead emissions, but, considering the current concentrations of lead in local parks, the health consequences may outweigh any potential economic benefit.

How prescient it was of Dr. Clair C. Patterson, who wrote: "Sometime in the near future it probably will be shown that the older urban areas of the United States have been rendered more or less uninhabitable by the millions of tons of poisonous industrial lead residues that have accumulated in cities during the past century." [37]

Conflicts of Interest

The authors declare that they have no conflicts of interest.

Acknowledgments

The authors wish to thank the Geo Science Department of New Jersey City University for providing the initial data of lead concentration in the parks of Jersey City.

References

[1] "City to close playground area at Mary Benson Park as a precautionary measure. Soil testing results show elevated lead levels, City of Jersey City," 2012, http://www.cityofjerseycity.com/uploadedFiles/Public_Notices/Press_Releases/Mary%20Benson%20Park.pdf.

[2] "Living in Jersey City, History of a Working City, Ruby Press, Inc, and Antonicello and Company, Inc," http://www.jerseycity-online.com/history/jc_history.htm.

[3] History of Jersey City, http://www.jerseycityonline.com/history/history.htm.

[4] Jersey City: America's Golden Door, http://www.jerseycityonline.com/history/jc/_golden/_door.htm.

[5] L. B. Loeb, *The Kinetic Theory of Gases*, Dover Publications, 5th edition, 1961.

[6] W. Benenson, J. W. Harris, H. Stocker, and H. Lutz, *Handbook of Physics*, Springer, New York, NY, USA, 2002.

[7] R. W. Haynes, Ed., *Handbook of Chemistry and Physics*, CRC Press Taylor and Francis Group, 9th edition, 2010.

[8] R. Dunk, Offshore Wind Analysis for New Jersey and Delaware, https://marine.rutgers.edu/cool/weather/wind_analysis/.

[9] A. E. Gill, *Atmosphere-Ocean Dynamics*, Academic Press, 1982.

[10] D. Byun and K. L. Schere, "Review of the governing equations, computational algorithms, and other components of

the models-3 Community Multiscale Air Quality (CMAQ) modeling system," *Applied Mechanics Reviews*, vol. 59, no. 1–6, pp. 51–76, 2006.

[11] J. A. Dutton, *Dynamics of Atmospheric Motion*, Dover Publications, 1995.

[12] J. B. Klemp and R. B. Wilhelmson, "The simulation of three-dimensional convective storm dynamics," *Journal of the Atmospheric Sciences*, vol. 35, no. 6, pp. 1070–1096, 1978.

[13] J. F. Kok, E. J. R. Parteli, T. I. Michaels, and D. B. Karam, "The physics of wind-blown sand and dust," *Reports on Progress in Physics*, vol. 75, no. 10, Article ID 106901, 2012.

[14] Public Health Statement., Lead, Agency for Toxic Substances and Disease Registry, 2007.

[15] R. S. Anderson and B. Hallet, "Sediment transport by wind: toward a general model," *Geological Society of America Bulletin*, vol. 97, no. 5, pp. 523–535, 1986.

[16] W. G. Nickling, "The initiation of particle movement by wind," *Sedimentology*, vol. 35, no. 3, pp. 499–511, 1988.

[17] K. Madanipour, S. Fatehi, and P. Parvin, "Measurement of temperature, refractive index, density distribution and convective heat transfer coefficient around a vertical wire by michelson interferometer," in *Proceedings of the Optical Measurement Systems for Industrial Inspection VI*, Munich, Germany, June 2009.

[18] A. Rosiello, J. Raia, D. Freile, and W. Montgomery, "Spatial distribution of lead in soils near the New Jersey City University campus, Jersey City, NJ and their potential sources," *Geological Society of America (Abstracts with Programs)*, vol. 41, p. 247, 2008.

[19] Enforcement and Compliance History Online (ECHO), United States Environmental Protection Agency, 2013, http://epa-sites .findthedata.org/l/380248/Kearny-Smelting-And-Refining-Corp.

[20] R. L. Burden and J. D. Faires, *Cengage Learning*, Brooks/Cole, Cengage Learning, 9th edition, 2011.

[21] Yanenko, 1971, https://www.abebooks.com/servlet/BookDetailsPL?bi=12793660665&searchurl=isbn%3D0387052720%26sort-by%3D17.

[22] R. A. Horn and C. R. Johnson, *Topics in Matrix Analysis*, Cambridge University Press, New York, NY, USA, 1994.

[23] A. L. Mendelsohn, B. P. Dreyer, A. H. Fierman et al., "Low-level lead exposure and behavior in early childhood," *Pediatrics*, vol. 101, no. 3, pp. 1–7, 1998.

[24] J. Hsiang and E. Díaz, "Lead and developmental neurotoxicity of the central nervous system," *Current Neurobiology*, vol. 2, no. 1, pp. 35–42, 2011.

[25] K. M. Cecil, C. J. Brubaker, C. M. Adler et al., "Decreased brain volume in adults with childhood lead exposure," *PLoS Medicine*, vol. 5, no. 5, Article ID e112, 2008.

[26] P. J. Landrigan, C. B. Schechter, J. M. Lipton, M. C. Fahs, and J. Schwartz, "Environmental pollutants and disease in American children: estimates of morbidity, mortality, and costs for lead poisoning, asthma, cancer, and developmental disabilities," *Environmental Health Perspectives*, vol. 110, no. 7, pp. 721–728, 2002.

[27] S. J. Pocock, M. Smith, and P. Baghurst, "Environmental lead and children's intelligence: a systematic review of the epidemiological evidence," *BMJ*, vol. 309, no. 6963, p. 1189, 1994.

[28] S. Tong, P. A. Baghurst, M. G. Sawyer, J. Burns, and A. J. McMichael, "Declining blood lead levels and changes in cognitive function during childhood. The port pirie cohort study," *Journal of the American Medical Association*, vol. 280, no. 22, pp. 1915–1919, 1998.

[29] T. I. Lidsky and J. S. Schneider, "Lead neurotoxicity in children: basic mechanisms and clinical correlates," *Brain*, vol. 126, part 1, pp. 5–19, 2003.

[30] T. Sanders, Y. Liu, V. Buchner, and P. B. Tchounwou, "Neurotoxic effects and biomarkers of lead exposure: a review," *Reviews on Environmental Health*, vol. 24, no. 1, pp. 15–45, 2009.

[31] ToxGuide for Lead, Agency for Toxic Substances and Disease Registry, 2007.

[32] Y. Finkelstein, M. E. Markowitz, and J. F. Rosen, "Low-level lead-induced neurotoxicity in children: an update on central nervous system effects," *Brain Research Reviews*, vol. 27, no. 2, pp. 168–176, 1998.

[33] Standard operating procedure for an in vitro bioaccessibility assay for lead in soil, U.S. EPA, 2008.

[34] M. P. Elless, C. A. Bray, and M. J. Blaylock, "Chemical behavior of residential lead in urban yards in the United States," *Environmental Pollution*, vol. 148, no. 1, pp. 291–300, 2007.

[35] K. T. Semple, K. J. Doick, K. C. Jones, P. Burauel, A. Craven, and H. Harms, "Defining bioavailability and bioaccessibility of contaminated soil and sediment is complicated," *Environmental Science and Technology*, vol. 38, no. 12, pp. 229A–231A, 2004.

[36] E. Rossi, "Low level environmental lead exposure-A continuing challenge," *The Clinical Biochemist Reviews*, vol. 29, pp. 63–70, 2008.

[37] Committee on Lead in the Human Environment Environmental Studies Board Commission on Natural Resources National Research Council: Lead in the Human Environment, 271, 1980.

16

Analysis of Economic Burden of Seasonal Influenza: An Actuarial Based Conceptual Model

S. S. N. Perera

Research and Development Center for Mathematical Modeling, Faculty of Science, University of Colombo, 00300 Colombo, Sri Lanka

Correspondence should be addressed to S. S. N. Perera; ssnp@maths.cmb.ac.lk

Academic Editor: Winston Garira

Analysing the economic burden of the seasonal influenza is highly essential due to the large number of outbreaks in recent years. Mathematical and actuarial models can be considered as management tools to understand the dynamical behavior, predict the risk, and compute it. This study is an attempt to develop conceptual model to investigate the economic burden due to seasonal influenza. The compartment *SIS* (susceptible-infected-susceptible) model is used to capture the dynamical behavior of influenza. Considering the current investment and future medical care expenditure as premium payment and benefit (claim), respectively, the insurance and actuarial based conceptual model is proposed to model the present economic burden due to the spread of influenza. Simulation is carried out to demonstrate the variation of the present economic burden with respect to model parameters. The sensitivity of the present economic burden is studied with respect to the risk of disease spread. The basic reproduction is used to identify the risk of disease spread. Impact of the seasonality is studied by introducing the seasonally varying infection rate. The proposed model provides theoretical background to investigate the economic burden of seasonal influenza.

1. Introduction

Infectious diseases such as Dengue, Ebola, SARS, Zika, and various types of influenza are viral infections that are of major public health concern in tropical and subtropical countries [1]. Influenza, generally known as "the flu," is a sessional infectious disease caused by virus. High fever, runny nose, sore throat, muscle pains, headache, cough, and feeling lethargic can be considered as common symptoms. Seasonal influenza can cause mild to severe illness and even death, particularly in some high-risk individuals such as pregnant women, children, very old people, and people with chronic underlying medical conditions such as cancer, diabetes mellitus, and heart diseases. Three different types of influenza viruses generally affect humans, namely, Type A, Type B, and Type C. In recent years, the influenza A virus subtypes H1N1 and H3N2 have been in circulation. In addition, there are two types of B viruses that are also circulating as seasonal influenza viruses. However, influenza C may be considered as less of a disease burden [2].

Normally, the influenza virus spreads over a short distance through the air from coughs or sneezes. Human mobility is the main responsible factor for the spread over long distance. Annual influenza outbreaks occur around the world and cause significant morbidity and mortality. Since the virus survives in cold environments with low relative humidity, normally the influenza outbreak can be seen in winter season in the Northern and Southern Hemispheres. Generally, in the tropical and subtropical region outbreaks may occur at any time of the year. However, in most of the tropical countries these peaks of infection are seen mainly during the rainy season [3]. In Sri Lanka, for the last few years, influenza epidemic outbreak has been generally observed during April to June and again in November to January.

From a public health and community point of view, seasonal influenza epidemics spread rapidly and are very difficult to control. Seasonal influenza spreads all over the world as an annual outbreak, making about three to five million severe illness cases and about 250,000 to 500,000 death cases [4]. In the United States approximately, annually, from 5% to 20% of the individuals are infected with influenza. Further, the United States health report indicates that about 3,000 to 49,000 influenza-associated deaths occur every year. According to the WHO and health report, in Sri Lanka, for the last few years, it has been generally observed that there is high trend concerning the spread of influenza. This epidemics risk causes enormous economic loss and serious adverse events leading to hospitalization. In USA, annual economic burden due to seasonal influenza is approximately more than 1 billion US $ [3–5]. However, in Sri Lanka the annual economic losses due to this epidemic risk have not yet been addressed.

Due to significant level of outbreaks in and around the world including Sri Lanka during recent years, investigating the economic burden of annual seasonal influenza is motivated. The economic burden can be classified in two ways, namely, individual level which accounts for medical and death expenditures and public level which accounts for the burden due to loss of human days. This study is an attempt to develop the conceptual model to capture economic losses in an individual level due to influenza. However, the proposed tool is only responsible for identifying losses for personal medical care. Classical compartment model is used to describe the spread of disease; model parameters are used to capture the seasonal/climatic impacts. Epidemiological insurance is designed to represent expenditures due to personal medical care and actuarial based computation tool is proposed to compute the economic burden.

This paper is organized in the following way. Classical compartment model, insurance based model, and actuarial tools are presented in Section 2. Mathematical analysis of the model is also presented in Section 2. In Section 3, numerical results and discussion are presented. Finally, conclusion remarks and further steps are highlighted in Section 4.

2. Mathematical Models

In this section, first, we introduce the classical compartment deterministic model to study the dynamics of influenza. Secondly, mathematical and actuarial analysis of the epidemic model is discussed and a basic insurance model is constructed.

2.1. Classical Compartment Model. Since influenza virus is constantly changing, people can get infected multiple times throughout their lives. This concept is mathematically modeled by using classical *SIS* (susceptible (S), infected (I), susceptible) compartments [6, 7]. Entire population is divided into two compartments, namely, susceptible as previously unexposed to the disease and infected as currently colonized by the virus. Since the objective is to investigate the annual seasonal economic losses due to individual medical care, impacts of the population demography, births, deaths, and migration, are not significant. Hence, omitting such effects, model is developed by considering transition only from compartment S to I and I to S. Let $S(t)$ and $I(t)$ denote the number of susceptible and infected ones at time t, respectively. The increment of rate of change of infected cases is proportional to the number of susceptible individuals and the number of individuals previously infected. It also noted that such a rate is also proportional to number of transitions from infected to susceptible cases. Taking $b > 0$, as a potentially infective contact rate, $p > 0$ as the probability of infection per contact, $D > 0$ as the average disease duration, and N as a total population, simple *SIS* model is given by (1a), (1b), and (1c).

$$\frac{dS}{dt} = -\frac{pb}{N}S(t)I(t) + \frac{1}{D}I(t), \tag{1a}$$

$$\frac{dI}{dt} = \frac{pb}{N}S(t)I(t) - \frac{1}{D}I(t), \tag{1b}$$

with

$$\begin{gathered} S(0) > 0, \\ I(0) > 0, \\ S(t) + I(t) = N, \quad \forall t \geq 0. \end{gathered} \tag{1c}$$

Taking population fraction, $S^*(t) = S(t)/N$, $I^*(t) = I(t)/N$, introducing rate of infection, β, as product of p and b and the rate of recovery, $\gamma = 1/D$, and considering close environment (i.e., $S(t)+I(t) = N$, $\forall t \geq 0$) and dropping $*$, the system (1a), (1b), and (1c) reads as [8, 9]

$$\frac{dI}{dt} = \beta S(t)I(t) - \gamma I(t), \quad \text{with initial condition } S(0) > 0,\ I(0) > 0. \tag{2}$$

2.2. Mathematical Analysis of the Model. Equation (2) can be further simplified and one can obtain an analytical solution given by

$$\frac{dI}{dt} = (\beta(1 - I(t)) - \gamma)I(t) \tag{3a}$$

$$I(t) = I(0)\int_0^T (\beta(1 - I(t)) - \gamma)I(t)\,dt \tag{3b}$$

$$I(t) = \frac{(\beta - \gamma)}{\beta + (((\beta - \gamma)/I(0)) - \beta)\exp(-(\beta - \gamma)t)}. \tag{3c}$$

Equation (2) has two equilibrium points, namely, disease-free equilibrium point ($I_{de} = 0$) and the endemic equilibrium point ($I_{ee} = (\beta - \gamma)/\beta$). One can see, using linear stability analysis, disease-free equilibrium point, I_{de}, is locally asymptotically stable if $\beta < \gamma$ and unstable if $\beta > \gamma$, whereas the endemic equilibrium point, I_{ee}, is locally asymptotically stable if $\beta > \gamma$ and unstable if $\beta < \gamma$ [9–11]. The product of the rate of infection, β, and the average duration of the infectious period, $1/\gamma$, is the expected number of new infections from one infected individual in a fully susceptible population through the entire duration of the infectious period. This parameter is known as a basic reproduction number, R_0, and if $R_0 > 1$, an introduced infectious individual leads to more than one infection so the disease spreads in the population. One can easily see that if $R_0 < 1$, the infection in one individual cannot replace itself so the disease dies out. Figure 1 displays variation of infected population fraction with respect to different rate of infection, β.

The infection rate, β, depends on the season, demography, and other external factors such as socioeconomic factors,

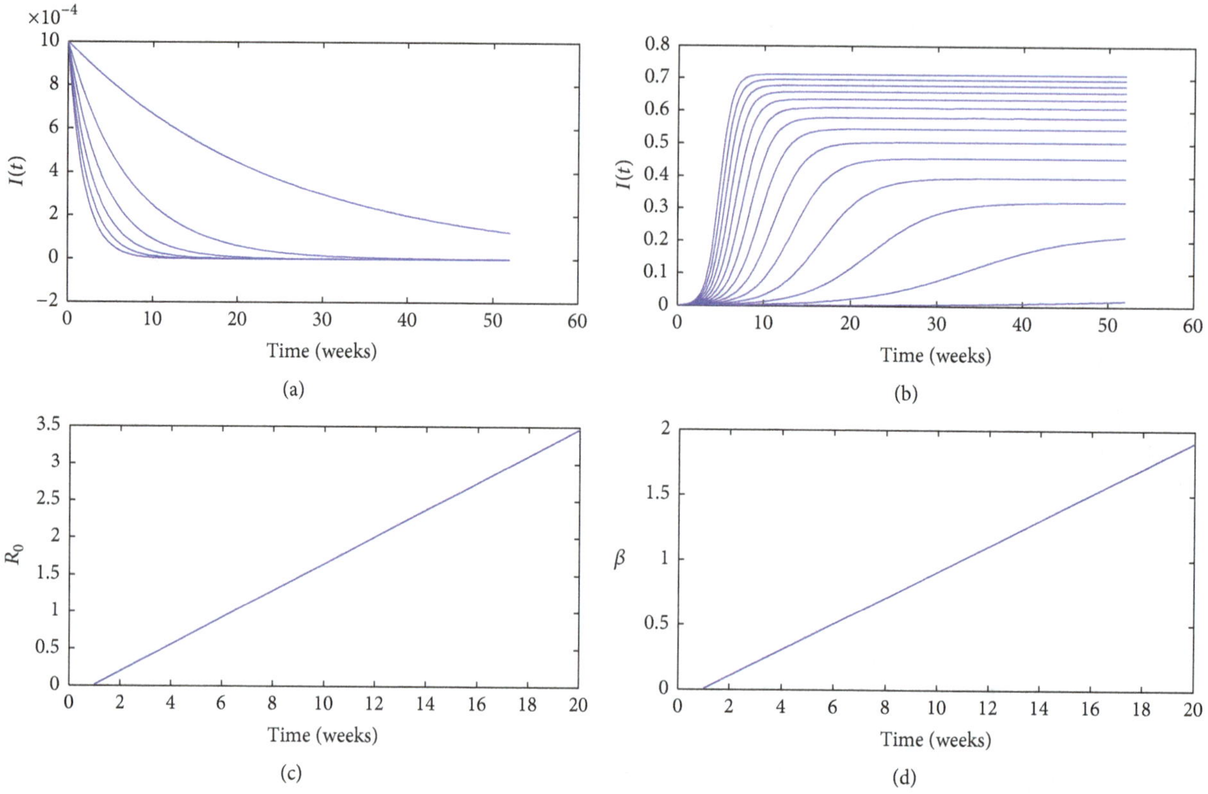

Figure 1: (a) Variation of infected population fraction when $R_0 < 1$. (b) Variation of infected population fraction when $R_0 > 1$. (c) Corresponding R_0 variation. (d) Corresponding β: this simulation is done, fixing $\gamma = 0.55$.

living conditions and life style, and the recovery rate, γ, which highly depends on the type of population, age group, and other biological factors. Since our objective is to include the seasonality impact, β can be chosen to represent seasonal variation. Since the basic reproduction number, R_0, is defined as the ratio between infection and recovery rates, R_0 can be considered as the representor of the spread risk of influenza due to external factors. Now β may vary periodically due to climatic variations. Mathematically, the seasonal contact rate is generally defined as sinusoidal, such that

$$\beta(t) = \beta_0 (1 + \omega \cos(2\pi t)), \quad (4)$$

where ω is the amplitude of seasonal variation in transmission and β_0 is the constant infection rate without seasonal impact [12, 13]. It is understood that sinusoidal forcing of transmission is a crude assumption for influenza. However for simplicity and since our objective is to develop a conceptual model, in this research we incorporate sinusoidal functions.

2.3. Insurance Based Actuarial/Probabilistic Model. In classical *SIS* model defined in (2), one can see these compartments play similar roles in an insurance model. Susceptible group facing the risk of being infected by influenza form the investment or insurance fund by depositing or investing or paying premiums, to get benefit or return to cover the medical care expenses incurred if infected [14]. During the outbreak of the influenza, the infected policyholders/investors would benefit from the claim (return) payments provided by the insurance/investment fund. Simply speaking the susceptible group make investment to cover future medical expenses. Present investment amount or premium can also be considered as the current economic burden due to the future medical expenditure. Now considering $S(t)$ and $I(t)$ as the probability of an individual being susceptible and infected, respectively, at the time t, one can redefine (2) as a probabilistic (an actuarial) model to describe the dynamics of investments versus expenditures.

Considering the expenditure side of the investment (insurance), expected (actuarial) present value of t period unit benefit (claim) payment, $E[B]$, is defined in (5a) [15]:

$$E[B(t)] = \int_0^t \exp(-\delta t) \text{ Prob. of an individual being infected at time } t \, dt \quad (5a)$$

$$= \int_0^t \exp(-\delta t) I(t) \, dt, \quad (5b)$$

where $\delta > 0$ is known as the discounting force of interest. Similarly, one can define the income side of the investment package as the expected (actuarial) present value of t period premium payment (investment), $E[P]$, as in (5c) [15].

$$E[P(t)] = \int_0^t \exp(-\delta t) \text{ Prob. of an individual being susceptible at time } t \, dt \quad (5c)$$

$$= \int_0^t \exp(-\delta t) S(t) \, dt. \quad (5d)$$

Using the equivalence principle technique, premium payment/current economic burden, π, can be obtained [15].

$$\pi E[\text{present value of premium}] = E[\text{present value of future benefit}]. \quad (6)$$

Therefore, π can be given in

$$\pi(t) = \frac{E[B(t)]}{E[P(t)]} = \frac{\int_0^t \exp(-\delta t) I(t) \, dt}{\int_0^t \exp(-\delta t) S(t) \, dt}. \quad (7)$$

Expected present premium value given in (5c) can be further modified.

$$\begin{aligned} E[P(t)] &= \int_0^t \exp(-\delta t) S(t) \, dt \\ &= \int_0^t S(t) \frac{d}{dt}\left(\frac{\exp(-\delta t)}{-\delta}\right) dt \\ &= \frac{1}{\delta}(S(0) - \exp(-\delta t) S(t)) - \frac{1}{\delta} \\ &\cdot \int_0^t \exp(-\delta t) \frac{dI(t)}{dt} dt = \frac{1}{\delta}\Big(S(0) + I(0) \\ &- \exp(-\delta t)(S(t) + I(t)) - \delta \int_0^t \exp(-\delta t) I(t) \, dt\Big). \end{aligned} \quad (8)$$

Since $S(t) + I(t) = 1$, $\forall t$, $E[P(t)]$ can be expressed as in

$$E[P(t)] = \frac{1}{\delta}(1 - \exp(-\delta t) - \delta E[B(t)]). \quad (9)$$

Now the premium payment/current economic burden, π, is given in

$$\pi(t) = \frac{\delta E[B(t)]}{1 - \exp(-\delta t) - \delta E[B(t)]}. \quad (10)$$

The conceptual model to compute the economic burden due to seasonal influenza is given in

$$\pi(t) = \frac{\delta E[B(t)]}{1 - \exp(-\delta t) - \delta E[B(t)]}, \quad (11a)$$

$$E[B(t)] = \int_0^t \exp(-\delta t) I(t) \, dt, \quad (11b)$$

$$\frac{dI}{dt} = \beta(1 - I(t)) I(t) - \gamma I(t), \quad \text{with initial condition } I(0) > 0. \quad (11c)$$

2.4. Algorithm for Computing the Economic Burden. The system (11a), (11b), and (11c) can be evaluated taking analytical solution given in (3a). By keeping the record of infected population, it is possible to set up the reported infected table and hence the approximated method can be proposed. From the infected table, we obtain a piecewise constant approximation of the continuous function, $I(t)$, and current economic burden, π, is now given in (12a), (12b), (12c), and (12d).

$$\tilde{I}(t) = \begin{cases} I_k; & k - 1 < t \le k \\ 0; & \text{otherwise.} \end{cases} \quad (12a)$$

$$\begin{aligned} E[B(t)] &= \int_0^t \exp(-\delta t) I(t) \, dt \\ &\approx \int_0^t \exp(-\delta t) \tilde{I}(t) \, dt \end{aligned} \quad (12b)$$

$$E[B(t)] \approx \sum_{k=1}^{m} \left(\frac{\exp(-\delta(k-1)) - \exp(-\delta k)}{\delta}\right) I_k. \quad (12c)$$

Now π can be expressed as

$$\pi(m) = \frac{\vartheta}{1 - \exp(-\delta m) - \vartheta}, \quad (12d)$$

where

$$\vartheta = \sum_{k=1}^{m} (\exp(-\delta(k-1)) - \exp(-\delta k)). \quad (13)$$

3. Results and Discussion

Figure 2 displays the variation of the current economic burden (premium) with respect to the contact rate. It is observed that the economic burden rises up when the infection rate, β, increases. It is clear that, with higher infection rate, the disease spreads out quickly and hence infected fraction is dominant. One can see that current premium is higher than 1 when β exceeds 1.8. It means that when infected rate exceeds this threshold value, we need to pay more than one unit in order to have a unit benefit (claim) in future.

Figure 3 shows the variation of the current economic burden (premium) with respect to the infected rate, β, and recovery rate, γ. The corresponding variation of basic reproduction number, R_0, under same domain of infected

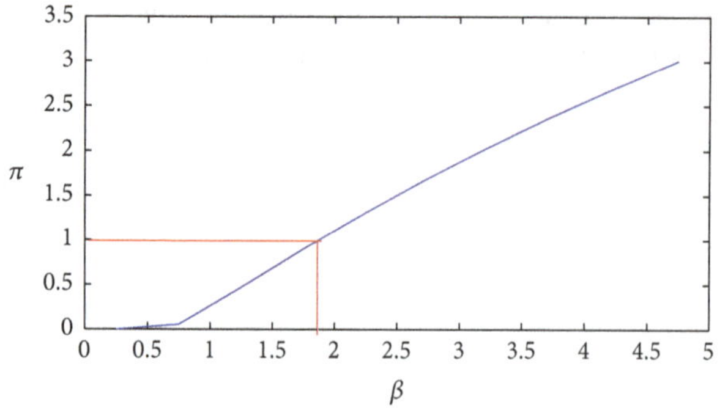

FIGURE 2: Variation of the present economic burden/premium (π) with respect to rate of infection (β). Here $\gamma = 0.55$ and $\delta = 0.05$.

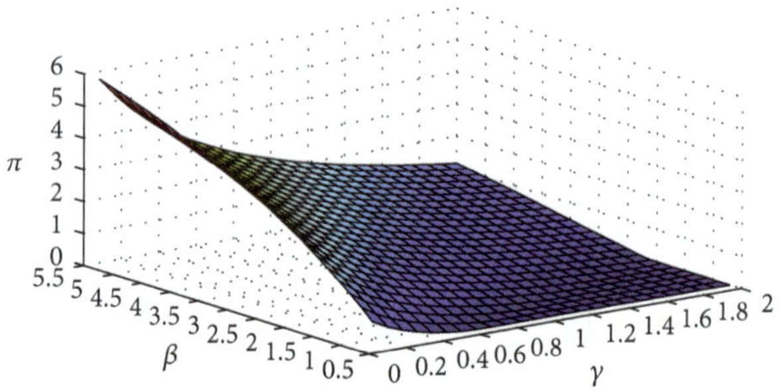

FIGURE 3: Variation of the present economic burden/premium (π) with respect to rate of contact (β) and γ. Here $\delta = 0.05$.

rate and recovery rate, is given by Figure 4. Since the basic reproduction number increases significantly, it is understood that the current economic burden also rises. Further, since R_0 accounts for the spread risk due to various external factors, it is possible to identify the sensitivity of the premium with respect to particular external factor if R_0 is defined accordingly.

Figure 5 shows the variation of infected population fraction with the constant infection rate and the seasonal infection rate with $\beta_0 = 1.66$. Seasonal variation can be captured via the seasonal infection rate. However, it is noted that, to get realistic situation, we need to identify appropriate parameters in (4).

Figure 6 displays the variation of the current economic burden with respect to time under the infection contact rate and seasonal infection rate. The dynamics of the present economic burden now depends on the amplitude (ω) of the seasonal variation.

Figure 7 demonstrates the variation of the current economic burden with respect to the discounting force interest, δ, and the basic reproduction number, R_0. The discounting force of interest is responsible factor for computing the equivalent present amount of future sums. Further, Figure 7 displays the variability of the current economic burden with respect to various external factors as well as future uncertainty too. If R_0 and δ are defined appropriately, it is possible to capture the variability of economic burden due to uncertainty of external parameters.

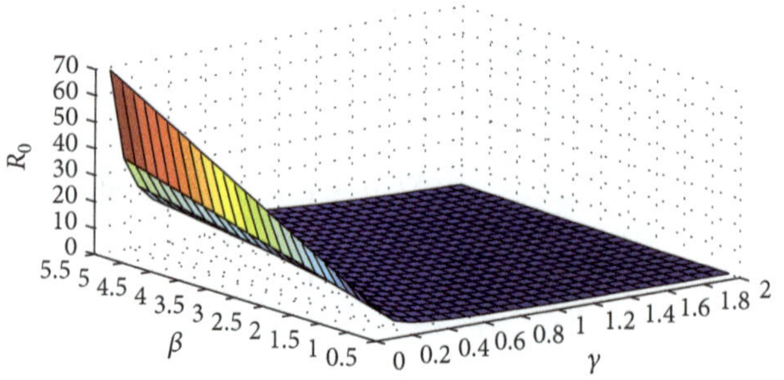

FIGURE 4: Variation of the basic reproduction number (R_0) with respect to rate of contact (β) and γ.

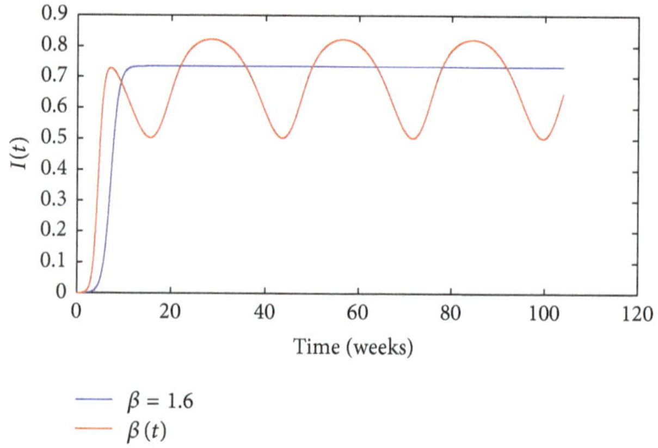

FIGURE 5: Variation of the infected fraction with the constant infection rate, β (in blue), and the seasonal infection rate, $\beta(t)$ (in red), defined in (4). Here $\beta_0 = 1.66$ and $\omega = 0.28$.

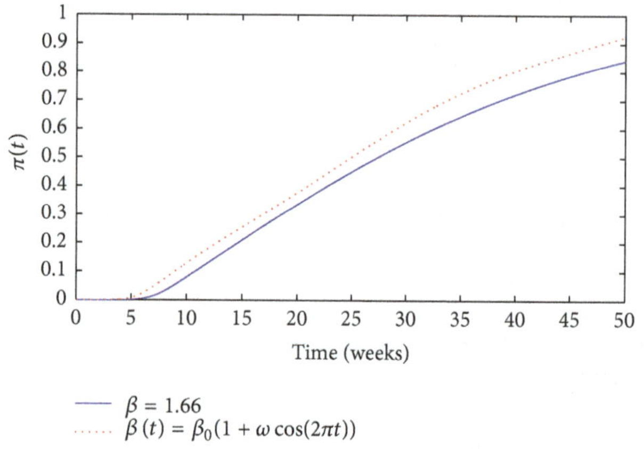

FIGURE 6: Variation of the current economic burden with respect to time under constant infection rate, β (in blue), and the seasonal infection rate, $\beta(t)$ (in red), defined in (4). Here $\beta_0 = 1.66$ and $\omega = 0.28$.

4. Conclusions

The economic burden due to seasonal influenza outbreaks is a critical budgetary problem in many countries due to loss of human hours as well as individual medical care expenditures. Using classical compartment *SIS* model, the dynamics of influenza is studied. Considering the current

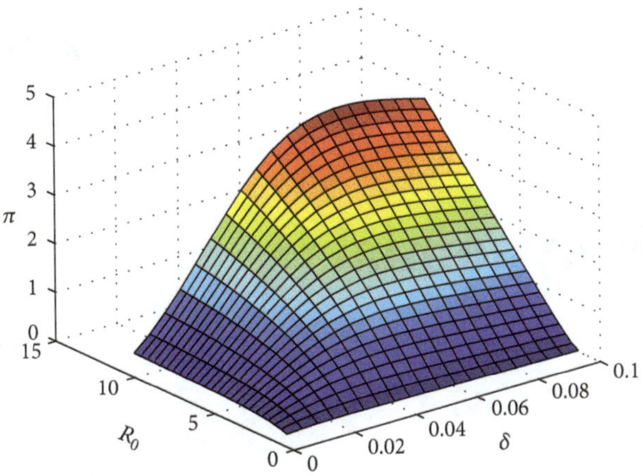

FIGURE 7: Variation of the current economic burden with respect to discounting force interest, δ, and basic reproduction number, R_0.

investment and future medical care expenditure as premium payment and benefit (claim), respectively, the insurance based conceptual model is proposed to investigate the present economic burden due to the spread of disease. The expected present economic burden is defined as the present investment to cover the future unit medical expenditure. This value is computed using the actuarial based model via equivalence principle. It is observed that the present economic burden depends on model parameters such as the infection rate, β, and recovery rate, γ. Since the infection rate highly depends on climatic variation, hence seasonality can be captured by introducing appropriate seasonally dependent function for it. It is also noted that dynamics of the economic burden depends on the amplitude of the seasonality.

This study can be extended to develop feasible premium schemes depending on different seasonal conditions by incorporating with reported infected data. Further, the model can be extended by including indirect cost such as expenditure due to loss of human hours, death, and community expenditures.

Conflicts of Interest

The author declares that there are no conflicts of interest regarding the publication of this paper.

References

[1] H. J. Wearing, P. Rohani, and M. J. Keeling, "Appropriate models for the management of infectious diseases," *PLOS Medicine*, vol. 2, no. 7, pp. 621–627, 2005.

[2] N. J. Cox and K. Subbarao, "Global epidemiology of influenza: past and present," *Annual Review of Medicine*, vol. 51, pp. 407–421, 2000.

[3] Centers for Disease Control and Prevention, "Estimates of deaths associated with seasonal influenza - United States 1976–2007," *Morbidity and Mortality Weekly Report*, vol. 59, no. 33, pp. 1057–1062, 2010.

[4] K. M. Sullivan, A. S. Monto, and I. M. Longini Jr., "Estimates of the US health impact of influenza," *American Journal of Public Health*, vol. 83, no. 12, pp. 1712–1716, 1993.

[5] L. Mao, Y. Yang, Y. Qiu, and Y. Yang, "Annual economic impacts of seasonal influenza on US counties: Spatial heterogeneity and patterns," *International Journal of Health Geographics*, vol. 11, pp. 1-8, 2012.

[6] J. D. Murray, *Mathematical Biology, I. Introduction*, Springer, 3rd edition, 2002.

[7] D. J. Daley and J. Gani, *Epidemic Modeling: An Introduction*, Cambridge University Press, 1999.

[8] J. A. Jacquez, *Compartmental Analysis in Biology and Medicine*, BioMedware, Ann Arbor, MI, USA, 3rd edition, 1996.

[9] H. W. Hethcote, "The mathematics of infectious diseases," *SIAM Review*, vol. 42, no. 4, pp. 599–653, 2000.

[10] R. M. Anderson and R. M. May, *Infectious Diseases of Humans*, Oxford Science Publications, 1991.

[11] J. A. Heesterbeek and M. G. Roberts, "Threshold quantities for infectious diseases in periodic environments," *Journal of Biological Systems*, vol. 3, no. 3, pp. 779–787, 1995.

[12] M. J. Keeling, P. Rohani, and B. T. Grenfell, "Seasonally forced disease dynamics explored as switching between attractors," *Physica D: Nonlinear Phenomena*, vol. 148, no. 3-4, pp. 317–335, 2001.

[13] N. C. Grassly and C. Fraser, "Seasonal infectious disease epidemiology," *Proceedings of the Royal Society B: Biological Sciences*, vol. 273, no. 1600, pp. 2541–2550, 2006.

[14] N. Ishimura, D. Komadel, and Y. Yoshizawa, "Risk estimation model on epidemic outbreaks for an insurer," *Astin Colloquium, LISBOA*, 2016.

[15] N. L. Bowers, H. U. Gerber, J. C. Hickman, D. A. Jones, and C. J. Nesbitt, *Actuarial Mathematics*, Society of Actuaries, Schaumburg, IL, USA, 1997.

17

Pricing Basket Options by Polynomial Approximations

Pablo Olivares and Alexander Alvarez

Department of Mathematics, Ryerson University, 350 Victoria St., Toronto, ON, Canada M5B 2K3

Correspondence should be addressed to Pablo Olivares; pablo.olivares@ryerson.ca

Academic Editor: Turgut Öziş

We propose a closed-form approximation for the price of basket options under a multivariate Black-Scholes model. The method is based on Taylor and Chebyshev expansions and involves mixed exponential-power moments of a Gaussian distribution. Our numerical results show that both approaches are comparable in accuracy to a standard Monte Carlo method, with a lesser computational effort.

1. Introduction

The objective of the paper is the study of the pricing of basket options under a multivariate Black-Scholes model, using two different polynomial approximations of a related conditional contract.

Our main contribution consists in the proposal of an efficient methodology that combines two approaches previously considered. First, we compute the conditional expected value of the payoff with respect to some of the underlying assets in order to reduce the problem to one-dimensional pricing; see [1]. Then, we use Taylor and Chebyshev polynomials to approximate the resulting conditional price. Numerical findings show that, within an equivalent accuracy, both approximations considerably reduce the computational time necessary to obtain the price, when compared with a standard Monte Carlo method.

While a Taylor expansion of the second order for spread pricing has been considered in [2], an approach based on Chebyshev expansion remains less explored. Recently, it has been considered in [3] for a single-asset option contract and in a multiasset setting in [4]. The latter combines with a Fourier series development, offering an interesting analysis of the error in the approximation. Our method conceptually differs from the works cited above in the way the expansion is carried on.

In addition, our expansions are presented in a general setting with the purpose of illustrating applications in higher dimensions, to other contracts and models, at the expense of somehow more complicated notations.

Basket options are multivariate extensions of univariate European calls or puts. A call basket option takes the weighted average of a group of d stocks (the basket) as the underlying asset and produces a payoff equal to the maximum of zero and the difference between the weighted average and the strike (or the opposite difference for the case of a put). Financial indices such as the S&P 500 and real options based on the difference between gas and oil prices are examples of such contracts.

In the case of European spread options, whose payoffs are given in terms of the spread of both prices at maturity, several approximations have been considered in the works of [2, 5–10], among others, where different ad hoc approaches are studied.

The approach to pricing by Taylor expansions can be traced back to [11], where a method to the price of one-dimensional derivatives is proposed. See also [12] for the use of Taylor in valuation of basket and Asian options, where the expansion is made about the characteristic function. Furthermore, in [10] the Taylor expansion is compared with other pricing techniques, proving to be effective and accurate for most values in the parametric space. In our case, the expansion is done on the function resulting from the conditional price, as opposed to a development based on the conditional strike price, as previously considered. Moreover, our methodology hinges on the calculation of mixed exponential-power moments of a Gaussian distribution and is extended to expansions about any point and order. Our point of view may allow for a better control on the approximation.

Taylor expansions produce reasonable approximations in terms of a simple closed-formula. Nonetheless, they are very sensible to the point around which the development is done. Moreover, as this expansion is local, it may introduce significant errors, albeit infrequent, at values far from the point where the expansion is considered, typically the conditional average price under the risk neutral measure. Significant errors for out-of-the-money contracts or extreme values in the parametric set are reported.

In order to overcome this potential problem, we study developments in terms of Chebyshev polynomials, which offer a uniform convergence of the conditional price on a predetermined closed interval. Interesting alternative approximations in terms of Fourier series can be found in [13, 14].

The organization of the paper is the following: in Section 2, we introduce the model and main notations and derive a Taylor approximation for d-dimensional basket options; in Section 3, we implement Taylor method to spread contracts while in Section 4 we study the case of a Chebyshev approximation and the sensitivity with respect to the spot prices; in Section 4, we discuss the numerical implementation and results; finally, in Section 5 we conclude.

2. Basket Derivatives and Taylor Expansions

We introduce some notations. Let $(\Omega, \mathscr{F}, (\mathscr{F}_t)_{t\geq 0}, \mathbb{P})$ be a filtered probability space. We define the filtration $\mathscr{F}^{X_t} := \sigma(X_s,\ 0 \leq s \leq t) \subset \mathscr{F}_t$ as the σ-algebra generated by the random variables $\{X_s,\ 0 \leq s \leq t\}$ completed in the usual way. Denote by $\mathbb{Q}$ an equivalent martingale risk neutral measure (EMM) and by $E_{\mathbb{Q}}$ the expectation under $\mathbb{Q}$. Quantities $\mu_{a,b}(k)$, $M_X(u)$ and $M_X(u,a,b)$ represent, respectively, the truncated kth moment and the moment generating function (m.g.f.) truncated on $[a,b]$, while the function $N(\cdot)$ is the cumulative distribution function (c.d.f.) of a standard normal distribution.

The matrix A' represents the transpose of matrix $A = (a_{ij})$, while $\mathrm{diag}(A)$ is a column vector with components from the diagonal of the matrix A. On the other hand, $A^{1/2}$ denotes a matrix such that $A^{1/2}A^{1/2} = A$. For a d-dimensional vector Y, denote by $\tilde{Y}$ the same vector, excluding the first component.

By r, we denote the (constant) interest rate or a vector with components equal to r. The matrix I_d is the $d \times d$ identity matrix.

For l-times differentiable function f on $\mathbb{R}^d$ and a vector $L = (l_1, l_2, \ldots, l_d)$ with $l_k \in \mathbb{N}$ such that $\sum_{k=1}^{d} l_k = l$, $D^L f$ represents its mixed partial derivative of order l differentiated l_k times w.r.t. y_k.

The process of spot prices is denoted by $S = (S_t^{(1)}, S_t^{(2)}, \ldots, S_t^{(d)})_{t\geq 0}$ for $0 \leq t \leq T$, while $Y = (Y_t^{(1)}, Y_t^{(2)}, \ldots, Y_t^{(d)})_{t\geq 0}$ is the asset log-return process; they are related by

$$S_t^{(j)} = S_0^{(j)} \exp\left(Y_t^{(j)}\right) \quad \text{for } j = 1, 2, \ldots, d. \tag{1}$$

The log-return process follows the dynamic

$$dY_t = \left(r - \frac{1}{2}\,\mathrm{diag}\,(\Sigma)\right) dt + \Sigma^{1/2} dB_t, \tag{2}$$

where $(B_t)_{t\geq 0}$ is a multivariate standard Brownian motion and Σ is a positive definite symmetric matrix with components $(\sigma_{ij})_{i,j=1,2,\ldots,d}$, with $\sigma_{ii} = \sigma_i^2$.

We analyze European Basket options whose payoff at maturity T, for a strike price K, is given by

$$h(S_T) = \left(\sum_{j=1}^{d} w_j S_T^{(j)} - K\right)_+, \tag{3}$$

where $(w_j)_{1\leq j\leq d}$ are some deterministic weights and $x_+ = \max(x, 0)$.

The main examples are spread options, defined for $d = 2$ by a payoff:

$$h(S_T) = \left(S_T^{(1)} - S_T^{(2)} - K\right)_+. \tag{4}$$

Other related contracts are the so-called 3 : 2 : 1 crack spreads with payoff:

$$h(S_T) = \left(\frac{2}{3}S_T^{(1)} - \frac{1}{3}S_T^{(2)} - S_T^{(3)} - K\right)_+, \tag{5}$$

where $S_T^{(1)}$, $S_T^{(2)}$, and $S_T^{(3)}$ are, respectively, the spot prices of gasoline, heating oil, and crude oil.

Exchange options are spread options with $K = 0$; see [15].

We start writing the price of the basket option, denoted by C_B, in terms of a conditional price via the following elementary proposition.

Proposition 1. *Let C_B be the price of a European Basket contract with maturity at T, strike price K, and payoff $h(S_T)$ under the model given by (1) and (2). Then,*

$$C_B = w_1 E_{\mathbb{Q}}\left[\exp\left(-\left(r - \frac{1}{2}\sigma^2_{Y_T^{(1)}/\tilde{Y}_T}\right)T + \mu_{Y_T^{(1)}/\tilde{Y}_T}\right) \cdot C\left(\tilde{Y}_T\right)\right], \tag{6}$$

where, for $y \in \mathbb{R}^{d-1}$,

$$C(y) := C_{BS}\left(S_0^{(1)}, K(y), r, \sigma_{Y_T^{(1)}/\tilde{Y}_T=y}, T\right) \tag{7}$$

is the Black-Scholes price of a call option with maturity at $T > 0$, volatility $\sigma_{Y_T^{(1)}/\tilde{Y}_T=y}$, spot price $S_0^{(1)}$, and strike price

$$K(y) = \frac{1}{w_1}\exp\left(\left(r - \frac{1}{2}\sigma^2_{Y_T^{(1)}/\tilde{Y}_T=y}\right)T - \mu_{Y_T^{(1)}/\tilde{Y}_T=y}\right) \cdot \left(K - \sum_{j=2}^{d} w_j S_0^{(j)} e^{y^{(j)}}\right) \tag{8}$$

with

$$\mu_{Y_T^{(1)}/\widetilde{Y}_T} = \left(r - \frac{1}{2}\sigma_{11}\right)T + \Sigma_{1\widetilde{Y}}\Sigma_{\widetilde{Y}}^{-1}\left(\widetilde{Y}_T - r + \frac{1}{2}\operatorname{diag}\left(\Sigma_{\widetilde{Y}}\right)\right)T, \qquad (9)$$
$$\sigma_{Y_T^{(1)}/\widetilde{Y}_T} = \sigma_{11} - \Sigma_{1\widetilde{Y}}\Sigma_{\widetilde{Y}}^{-1}\Sigma'_{1\widetilde{Y}},$$

where

$$\Sigma_{1\widetilde{Y}} = \left(\sigma_{12}, \sigma_{13}, \ldots, \sigma_{1,d-1}\right) \qquad (10)$$

and $\Sigma_{\widetilde{Y}}$ *is the covariance matrix of the vector* $\widetilde{Y}_T$.

Proof. From (2),

$$Y_T = \left(r - \frac{1}{2}\operatorname{diag}(\Sigma)\right)T + \sqrt{T}\Sigma^{1/2}Z_d \qquad (11)$$

in law, where Z_d is a random variable with a multivariate standard normal probability distribution in $\mathbb{R}^d$. Moreover, independently of $\widetilde{Y}_T$, the random variable $Y_T^{(1)}$ has a univariate normal distribution. Thus, we can write

$$Y_T^{(1)} = \mu_{Y_T^{(1)}/\widetilde{Y}_T} + \sqrt{T}\sigma_{Y_T^{(1)}/\widetilde{Y}_T}Z_1 \qquad (12)$$

in law, where Z_1 is independent of Y_T and it has, conditionally on $\widetilde{Y}_T$, a standard univariate normal distribution. Moreover, it is well known (see, e.g., [16]) that $\mu_{Y_T^{(1)}/\widetilde{Y}_T}$ and $\sigma_{Y_T^{(1)}/\widetilde{Y}_T}$ are, respectively, given by (9).

Next, by the iterated property of the conditional expected value, we have

$$\begin{aligned} C_B &:= e^{-rT}E_{\mathbb{Q}}\left(h\left(S_T\right)\right) = e^{-rT}E_{\mathbb{Q}}\left(E_{\mathbb{Q}}\left(h\left(S_T\right) \mid \mathcal{F}^{\widetilde{Y}_T}\right)\right) \\ &= w_1 e^{-rT}E_{\mathbb{Q}}\left(E_{\mathbb{Q}}\left[\left(S_0^{(1)}\exp\left(Y_T^{(1)}\right) - \left(\frac{K}{w_1} - \sum_{j=2}^{d}\frac{w_j}{w_1}S_0^{(j)}\exp\left(Y_T^{(j)}\right)\right)\right)_+ \mid \mathcal{F}^{\widetilde{Y}_T}\right]\right) \\ &= w_1 e^{-rT}E_{\mathbb{Q}}\left(E_{\mathbb{Q}}\left[\left(S_0^{(1)}\exp\left(Y_T^{(1)}\right) - K_1\left(\widetilde{Y}_T\right)\right)_+ \mid \mathcal{F}^{\widetilde{Y}_T}\right]\right), \end{aligned} \qquad (13)$$

where $K_1(y) = K/w_1 - \sum_{j=2}^{d}(w_j/w_1)S_0^{(j)}e^{y^{(j)}}$.

Moreover, substituting (12) into (13), we have

$$\begin{aligned} C_B &= w_1 e^{-rT}E_{\mathbb{Q}}\left[E_{\mathbb{Q}}\left(\left(S_0^{(1)} \cdot \exp\left(\mu_{Y_T^{(1)}/\widetilde{Y}_T} + \sigma_{Y_T^{(1)}/\widetilde{Y}_T}\sqrt{T}Z^{(1)}\right) - K'\left(\widetilde{Y}_T\right)\right)_+ \mid \mathcal{F}^{\widetilde{Y}_T}\right)\right] = w_1 e^{-rT}E_{\mathbb{Q}}\left[\exp\left(-rT + \frac{1}{2}\sigma^2_{Y_T^{(1)}/\widetilde{Y}_T}T \right.\right. \\ &\left.\left. + \mu_{Y_T^{(1)}/\widetilde{Y}_T}\right)E_{\mathbb{Q}}\left(\left(S_0^{(1)} \cdot \exp\left(rT - \frac{1}{2}\sigma^2_{Y_T^{(1)}/\widetilde{Y}_T}T + \sigma_{Y_T^{(1)}/\widetilde{Y}_T}\sqrt{T}Z^{(1)}\right) - K_1\left(\widetilde{Y}_T\right)\right)_+ \mid \mathcal{F}^{\widetilde{Y}_T}\right)\right], \end{aligned} \qquad (14)$$

where

$$C\left(\widetilde{Y}_T\right) = e^{-rT}E_{\mathbb{Q}}\left[\left(S_0^{(1)} \cdot \exp\left(\left(r - \frac{1}{2}\sigma^2_{Y_T^{(1)}/\widetilde{Y}_T}\right)T + \sigma_{Y_T^{(1)}/\widetilde{Y}_T}\sqrt{T}Z^{(1)}\right) - K\left(\widetilde{Y}_T\right)\right)_+ \mid \mathcal{F}^{\widetilde{Y}_T}\right]. \qquad (15)$$

Hence, we can recognize in $C(\widetilde{Y}_T)$ the expression of a Black-Scholes price with the parameters mentioned above. □

Remark 2. Notice that $K(y)$ may be negative. In this case, $C(y)$ is not the Black-Scholes price, but the formula remains valid. Condition $w_j \le 0$, $j = 2, \ldots, d$, guarantees its positiveness, as it is the case of spreads and 3 : 2 : 1 crack spread contracts defined above.

Notice that $C(y)$ is smooth enough to apply a kth-order Taylor development around any point $y^* \in \mathbb{R}^{d-1}$. After conditioning on the last $d-1$ underlying assets, it leads to the approximated price:

$$C^{(k)}(y) = \sum_{l=0}^{k}\sum_{R_l}\frac{D^L C\left(y^*\right)}{\prod_{k=1}^{d} l_k!}\prod_{k=1}^{d-1}\left(y^{(k+1)} - y^{*(k)}\right)^{l_k}, \qquad (16)$$

where $L = (l_1, l_2, \ldots, l_{d-1})$ and $R_l = \{L \in \mathbb{N}^{d-1}/l_1 + l_2 + \cdots + l_{d-1} = l,\ 0 \le l_k \le l\}$.

Taking into account Proposition 1, a natural kth-order Taylor approximation around $y^* = (y_2^*, y_3^*, \ldots, y_{d-1}^*)$ of the price C_B of a basket option with payoff $h(S_T)$, defined as $C_B^{(k)} := e^{-rT} E_{\mathbb{Q}}[C^{(k)}(\tilde{Y}_T)]$, is obtained by

$$C_B^{(k)}(y^*) = w_1 \sum_{l=0}^{k} \sum_{R_l} \frac{D^L C(y^*)}{l_1! l_2! \cdots l_{d-1}!} \cdot E_{\mathbb{Q}}\left[\exp\left(-\left(r - \frac{1}{2}\sigma^2_{Y_T^{(1)}/\tilde{Y}_T}\right)T + \mu_{Y_T^{(1)}/\tilde{Y}_T}\right) \cdot \prod_{k=1}^{d-1} \left(Y_T^{(k+1)} - y^{*(k)}\right)^{l_k}\right] \quad (17)$$

after replacing (16) into the expression for C_B in the proposition above.

Remark 3. Notice that the approximation $C_B^{(k)}$ depends only on the derivatives of the function $C(y)$ with respect to y, which turns out to be the Black-Scholes price composed with the function $K(y)$, and the mixed exponential-power moments of a Gaussian multivariate distribution.

Remark 4. Sensitivities to the parameters can be computed by a similar approximation, as Greeks for a Black-Scholes option model are known. For example, the Delta with respect to the jth asset can be approximated by

$$\Delta_k^{(j)} = w_1 \sum_{l=0}^{k} \sum_{R_l} \frac{D^L\left(\partial C(y^*)/\partial s^{(j)}\right)}{l_1! l_2! \cdots l_{d-1}!} E_{\mathbb{Q}}\left[e \cdot \exp\left(-\left(r - \frac{1}{2}\sigma^2_{Y_T^{(1)}/\tilde{Y}_T}\right)T + \mu_{Y_T^{(1)}/\tilde{Y}_T}\right) \cdot \prod_{k=2}^{d} \left(Y_T^{(k)} - y_k^*\right)^{l_k}\right]. \quad (18)$$

3. Pricing Spreads Options by Taylor Approximations

In order to illustrate the method studied in the previous section, we consider the case of a bidimensional spread option under the model given by (1) and (2) with covariance matrix:

$$\Sigma = \begin{pmatrix} \sigma_1^2 & \rho\sigma_1\sigma_2 \\ \rho\sigma_1\sigma_2 & \sigma_2^2 \end{pmatrix}. \quad (19)$$

Notice that, for convenience, we have slightly changed some notations.

From (11), the conditional distribution of $Y_T^{(1)}$ given $Y_T^{(2)}$ is normal with mean and variance given by

$$\begin{aligned} \mu\left(Y_T^{(2)}\right) &:= \mu_{Y_T^{(1)}/\tilde{Y}_T} \\ &= r\left(1 - \frac{\sigma_1}{\sigma_2}\rho\right)T + \frac{1}{2}\sigma_1(\sigma_2\rho - \sigma_1)T \\ &\quad + \frac{\sigma_1}{\sigma_2}\rho Y_T^{(2)}, \\ \sigma &:= \sigma_{Y_T^{(1)}/\tilde{Y}_T} = \sqrt{(1-\rho^2)}\sigma_1. \end{aligned} \quad (20)$$

Thus, we can write

$$Y_T^{(1)} = \mu\left(Y_T^{(2)}\right) + \sqrt{T}\sigma Z \quad \text{in law}, \quad (21)$$

where $Z \sim N(0,1)$ independent of Y_T.

The kth-order Taylor approximation in this case simplifies to

$$C_B^{(k)}(y^*) = \sum_{l=0}^{k} \frac{D^l C(y^*)}{l!} E_{\mathbb{Q}}\left[\exp\left(-\left(r - \frac{1}{2}\sigma^2\right)T + \mu\left(Y_T^{(2)}\right)\right) \cdot \left(Y_T^{(2)} - y^*\right)^l\right]. \quad (22)$$

Moreover,

$$\begin{aligned} &E_{\mathbb{Q}}\left[\exp\left(\left(-r + \frac{1}{2}\sigma^2\right)T + \mu\left(Y_T^{(2)}\right)\right)\left(Y_T^{(2)} - y^*\right)^l\right] \\ &= e^A E_{\mathbb{Q}}\left[\exp\left(\frac{\sigma_1}{\sigma_2}\rho Y_T^{(2)}\right)\left(Y_T^{(2)} - y^*\right)^l\right], \end{aligned} \quad (23)$$

where

$$A = -\left(\frac{1}{2}\rho^2\sigma_1^2 + r\frac{\sigma_1}{\sigma_2}\rho - \frac{1}{2}\sigma_1\sigma_2\rho\right)T. \quad (24)$$

Now, marginals of a multivariate normal distribution are normal too. Hence, we have that $Y_T^{(2)} \sim N((r - (1/2)\sigma_2^2)T, T\sigma_2^2)$. The exponential-power moments can be calculated as

$$\begin{aligned} &E_{\mathbb{Q}}\left[\exp\left(\frac{\sigma_1}{\sigma_2}\rho Y_T^{(2)}\right)\left(Y_T^{(2)} - y^*\right)^l\right] \\ &= \exp\left(\frac{\sigma_1}{\sigma_2}\rho\left(r - \frac{1}{2}\sigma_2^2\right)T\right) \\ &\quad \cdot \sum_{m=0}^{l} \binom{l}{m} \left(\sqrt{T}\sigma_2\right)^m B(y^*)^{l-m} D^m M_Z\left(\sqrt{T}\sigma_1\rho\right) \end{aligned} \quad (25)$$

with

$$B(y^*) = \left(r - \frac{1}{2}\sigma_2^2\right)T - y^*. \quad (26)$$

Next, integrating by parts,

$$D^m M_Z\left(\sqrt{T}\sigma_1\rho\right) = \exp\left(\frac{\sigma_1^2\rho^2 T}{2}\right) \cdot \sum_{\nu=0}^{[m/2]} \binom{m}{2\nu} \left(\sigma_1\rho\sqrt{T}\right)^{m-2\nu} (2\nu-1)!!, \quad (27)$$

where $n!!$ is the double factorial defined as the product of all odd numbers between 1 and n including both. When the set is empty, by convention, the product is equal to one.

As a consequence, for $y^* = E_{\mathbb{Q}}(Y_T^{(2)})$, we have

$$\begin{aligned} E_{\mathbb{Q}}\left[\exp\left(\frac{\sigma_1}{\sigma_2}\rho Y_T^{(2)}\right)\left(Y_T^{(2)} - y^*\right)^l\right] &= T^{l/2}\sigma_2^l \cdot \exp\left(\frac{\sigma_1}{\sigma_2}\rho\left(r - \frac{1}{2}\sigma_2^2\right)T\right)\exp\left(\frac{\sigma_1^2\rho^2 T}{2}\right) \cdot \sum_{\nu=0}^{l} \binom{l}{\nu}\left(\sigma_1\rho\sqrt{T}\right)^{l-\nu} E\left(Z^\nu\right) \\ &= T^{l/2}\sigma_2^l e^{-A}\sum_{\nu=0}^{[l/2]}\binom{l}{2\nu}\left(\sigma_1\rho\sqrt{T}\right)^{l-2\nu}(2\nu-1)!!. \end{aligned} \quad (28)$$

After gathering all pieces and substituting in (22), we have the following result.

Proposition 5. *The kth-order Taylor approximation of a spread contract with maturity at T and strike price K, under the model described by (1) and (2), is given by*

$$C_B^{(k)}\left(y^*\right) = \sum_{l=0}^{k}\sum_{m=0}^{l} \frac{D^l C\left(y^*\right)}{l!}\binom{l}{m}\left(\sqrt{T}\sigma_2\right)^m B\left(y^*\right)^{l-m} E(m) \quad (29)$$

with

$$E(m) = \sum_{\nu=0}^{m}\binom{m}{\nu}\left(\sigma_1\rho\sqrt{T}\right)^{m-\nu} E_{\mathbb{Q}}\left(Z^\nu\right) \quad (30)$$

for $m = 1, 2, \ldots, k$ and $E(0) = 1$, where $E_{\mathbb{Q}}Z^\nu = (\nu-1)!!$ if ν is even or zero if it is odd, and

$$K(y) = e^{-A}\left(K\exp\left(-\frac{\sigma_1}{\sigma_2}\rho y\right) + S_0^{(2)}\exp\left(\left(1-\frac{\sigma_1}{\sigma_2}\rho\right)y\right)\right). \quad (31)$$

We now need to compute the derivatives of the function $C(y)$ with respect to y. By a straightforward calculation from Black-Scholes formula,

$$\begin{aligned} D^1 K(y) &= e^{-A}\left(-\frac{\sigma_1}{\sigma_2}\rho K\exp\left(-\frac{\sigma_1}{\sigma_2}\rho y\right) + S_0^{(2)}\left(1-\frac{\sigma_1}{\sigma_2}\rho\right)\exp\left(\left(1-\frac{\sigma_1}{\sigma_2}\rho y\right)\right)\right), \\ D^2 K(y) &= e^{-A}\left(\left(\frac{\sigma_1}{\sigma_2}\rho\right)^2 K\exp\left(-\frac{\sigma_1}{\sigma_2}\rho y\right) + S_0^{(2)}\left(1-\frac{\sigma_1}{\sigma_2}\rho\right)^2\exp\left(\left(1-\frac{\sigma_1}{\sigma_2}\rho\right)y\right)\right). \end{aligned} \quad (32)$$

Also,

$$\begin{aligned} D^1 C(y) &= S_0^{(1)} f_Z\left(d_1(K(y))\right) D^{(1)} d_1(K(y)) - e^{-rT} D^1 K(y) N\left(d_2(K(y))\right) - e^{-rT} K(y) f_Z\left(d_2(K(y))\right) D^1 d_1(K(y)) \\ &= -\frac{D^{(1)}K(y)}{K(y)\sigma\sqrt{T}} A_2(y), \end{aligned} \quad (33)$$

where f_Z is the density function of a standard normal random variable and

$$A_2(y) = S_0^{(1)} f_Z\left(d_1(K(y))\right) + \sigma\sqrt{T}e^{-rT}K(y)N\left(d_2(K(y))\right) - e^{-rT}K(y) f_Z\left(d_2(K(y))\right). \quad (34)$$

Similarly, the second derivative is obtained as

$$D^2 C(y) = -\frac{1}{\sigma\sqrt{T}}\left[A_2(y)\frac{K(y)D^{(2)}K(y) - \left(D^1 K(y)\right)^2}{K^2(y)} + D^{(1)}A_2(y)\frac{D^{(1)}K(y)}{K(y)}\right] \quad (35)$$

with

$$\begin{aligned} D^1 A_2(y) = \frac{D^1 K(y)}{K(y)\sigma\sqrt{T}}\Big[& S_0^{(1)} f_Z\left(d_1(K(y))\right) d_1(K(y)) + \sigma^2 T e^{-rT} K(y) N\left(d_2(K(y))\right) - 2\sigma\sqrt{T}e^{-rT}K(y) f_Z\left(d_2(K(y))\right) \\ & - e^{-rT}K(y) f_Z\left(d_2(K(y))\right) d_2(K(y)) + e^{-rT}\sigma\sqrt{T}K^2(y) f_Z\left(d_2(K(y))\right)\Big]. \end{aligned} \quad (36)$$

In particular, when we expand the price around the point $y_{\text{mean}} = E_{\mathbb{Q}}(Y_T^{(2)}) = (r - (1/2)\sigma_2^2)T$ we have the first and second approximations given, respectively, by

$$C^{(1)}(y_{\text{mean}}) = C(y_{\text{mean}}) + \sigma_1\sigma_2\rho T D^1 C(y_{\text{mean}}),$$
$$C^{(2)}(y_{\text{mean}}) = \widehat{p}_1 + \frac{1}{2}\left[T\sigma_2^2\left(1+\sigma_1^2\rho^2 T\right)\right] D^2 C(y_{\text{mean}}). \quad (37)$$

More generally, expanding around $y^* \in \mathbb{R}$, the first two approximations denoted by $\widehat{p}_1(y^*)$ and $\widehat{p}_2(y^*)$, respectively, are given by

$$C_B^{(1)}(y^*) = C(y^*) + D^1C(y^*) \cdot \left(B(y^*) + \sqrt{T}\sigma_2 E(1)\right) = C(y^*) + D^1C(y^*) \cdot \left(B(y^*) + T\sigma_1\sigma_2\rho\right),$$
$$C^{(2)}(y^*) = C_B^{(1)}(y^*) + \frac{1}{2}D^2C(y^*) \cdot \left[B^2(y^*) + 2T\sigma_1\sigma_2\rho B(y^*) + T\sigma_2^2\left(1+T\sigma_1^2\rho^2\right)\right]. \quad (38)$$

4. Approximation by Chebyshev Polynomials

We study an alternative approximation of the price via Chebyshev polynomials. For definition and their basic properties see, for example, [17].

Denoting by $(T_k(y))_{k\in\mathbb{N}}$ the sequence of Chebyshev polynomials of first type on $[-1,1]$, we consider the nth approximation of the function $C(y)$ on the interval $[a,b]$ described by equation in terms of Chebyshev polynomials, the one given by

$$C_{\text{CH}}(y,a,b,n) = \frac{1}{2}\widehat{c}_0 1_{[a,b]}(y) + \sum_{k=1}^{n}\widehat{c}_k T_k^{a,b}(y) 1_{[a,b]}(y), \quad (39)$$

where $(T_k^{a,b}(x))_{k\in\mathbb{N}}$ is the sequence of Chebyshev polynomials of first type on $[a,b]$ defined by

$$T_k^{a,b}(y) = T_k\left(-1 + \frac{2}{(b-a)}(y-a)\right), \quad a \le y \le b, \quad (40)$$

and the values $(\widehat{c}_k)_{0\le k\le N}$ are estimators of the corresponding coefficients in the Chebyshev expansion.

Chebyshev polynomials on $[a,b]$ are orthogonal with respect to the scalar product defined as

$$\langle f, g\rangle = \int_a^b f(x)\, g(x)\, w_{a,b}(x)\, dx \quad (41)$$

with weight function $w_{a,b}(x) = (1-(2(x-a)/(b-a)-1)^2)^{-1/2}$.

Notice that

$$\left\|T_k^{a,b}\right\|^2 = \begin{cases} \dfrac{(b-a)\pi}{4} & \text{for } k \neq 0 \\ \dfrac{(b-a)\pi}{2} & \text{for } k = 0. \end{cases} \quad (42)$$

Then, for $k \neq 0$ the coefficients in the expansion can be calculated as

$$c_k = \frac{\langle C, T_k^{a,b}\rangle}{\left\|T_k^{a,b}\right\|^2} = \frac{4}{(b-a)\pi} \cdot \int_a^b C(y) T_k^{a,b}(y) w_{a,b}(y)\, dy = \frac{4}{(b-a)\pi} \cdot \int_a^b C(y) T_k\left(-1+\frac{2}{(b-a)}(y-a)\right) w_{a,b}(y)\, dy$$
$$= \frac{2}{\pi}\int_{-1}^{1} C\left(a + \frac{b-a}{2}(x+1)\right) T_k(x)\, w_{-1,1}(x)\, dx = \frac{2}{\pi}\int_0^{\pi} C\left(a+\frac{b-a}{2}(\cos\theta+1)\right)\cos(k\theta)\, d\theta \quad (43)$$

after changes of variables $x = -1 + (2/(b-a))(y-a)$ and $x = \cos\theta$.

Also,

$$c_0 = \frac{1}{\pi}\int_0^{\pi} C\left(a+\frac{b-a}{2}(\cos\theta+1)\right) d\theta. \quad (44)$$

From the trapezoidal rule to approximate Riemann integrals, the coefficients $(c_k)_{0\le k\le N}$ can be estimated by an equidistant partition of N points on $[0,\pi]$.

Chebyshev polynomials of first type can be written in terms of powers of the variable. From [18],

$$T_k^{a,b}(x) = \sum_{l=0}^{[k/2]} b_l^{(k)}\left(-1+\frac{2}{b-a}(x-a)\right)^{k-2l}, \quad (45)$$

where

$$b_l^{(k)} = \begin{cases} (-1)^{k/2} & \text{for } l = \dfrac{k}{2} \text{ and } k \text{ even} \\ (-1)^l\, 2^{k-2l-1} \dfrac{k}{k-l} \dbinom{k-l}{l} & \text{for } l < \dfrac{k}{2}. \end{cases} \quad (46)$$

In a similar way to the case of Taylor polynomials, define the nth order Chebyshev approximation for the basket option price as

$$C_{\mathrm{CH}}(a,b,n) := w_1 E_{\mathbb{Q}}\left[\exp\left(-\left(r-\frac{1}{2}\sigma^2\right)T + \mu\left(Y_T^{(2)}\right)\right) \cdot \widetilde{C}_{\mathrm{CH}}\left(Y_T^{(2)},a,b,n\right)\right]. \quad (47)$$

The next theorem provides the Chebyshev approximation for the price of a basket option.

Theorem 6. *The nth order Chebyshev's approximation of the price C_B of a European Basket option under the model given by (1) and (2) is given by*

$$C_{CH}(a,b,n) = \frac{w_1}{2} \cdot \widehat{c}_0\left[N\left(\widetilde{b}-\sigma_1\rho\sqrt{T}\right) - N\left(\widetilde{a}-\sigma_1\rho\sqrt{T}\right)\right] + w_1 \cdot \exp\left(-\frac{1}{2}\rho^2\sigma_1^2 T\right) \cdot \sum_{k=1}^{n}\sum_{l=0}^{[k/2]} \widehat{c}_k b_l^{(k)} \left(\frac{2\sigma_2\sqrt{T}}{b-a}\right)^{k-2l} \widehat{G}(k-2l), \quad (48)$$

where

$$\widehat{G}(k) = \sum_{m=0}^{k} \binom{k}{m} \left(\frac{2\left(r-(1/2)\sigma_2^2\right)T - a - b}{2\sigma_2\sqrt{T}}\right)^m \cdot \left.\frac{d^{k-m} M_Z\left(u,\widetilde{a},\widetilde{b}\right)}{du^{k-m}}\right|_{u=\sigma_1\rho\sqrt{T}} \quad (49)$$

for $k = 0,1,2,\ldots,n$,

$$\widetilde{b} = \frac{b-\left(r-(1/2)\sigma_2^2\right)T}{\sigma_2 T}, \qquad \widetilde{a} = \frac{a-\left(r-(1/2)\sigma_2^2\right)T}{\sigma_2 T}. \quad (50)$$

Proof. Notice that

$$e^A E_{\mathbb{Q}}\left[\exp\left(\frac{\sigma_1}{\sigma_2}\rho Y_T^{(2)}\right) 1_{(-\infty,b)}\left(Y_T^{(2)}\right)\right] = e^A \cdot \exp\left(\left(r-\frac{1}{2}\sigma_2^2\right)\frac{\sigma_1}{\sigma_2}\rho T\right) \cdot E_{\mathbb{Q}}\left[\exp\left(\sigma_1\rho\sqrt{T}Z\right) 1_{(-\infty,\widetilde{b})}(Z)\right] = e^A \cdot \exp\left(\left(r-\frac{1}{2}\sigma_2^2\right)\frac{\sigma_1}{\sigma_2}\rho T + \frac{1}{2}\sigma_1^2\rho^2 T\right)\frac{1}{\sqrt{2\pi}} \cdot \int_{-\infty}^{\widetilde{b}-\sigma_1\rho\sqrt{T}} \exp\left(-\frac{1}{2}z^2\right)dz = N\left(\widetilde{b}-\sigma_1\rho\sqrt{T}\right). \quad (51)$$

Hence,

$$e^A E_{\mathbb{Q}}\left[\exp\left(\frac{\sigma_1}{\sigma_2}\rho Y_T^{(2)}\right) 1_{[a,b]}\left(Y_T^{(2)}\right)\right] = N\left(\widetilde{b}-\sigma_1\rho\sqrt{T}\right) - N\left(\widetilde{a}-\sigma_1\rho\sqrt{T}\right). \quad (52)$$

From Proposition 1, we take into account (39) and (45) to get

$$C_{\mathrm{CH}}(a,b,n) = w_1 e^A E_{\mathbb{Q}}\left(\exp\left(\frac{\sigma_1}{\sigma_2}\rho Y_T^{(2)}\right) \cdot C_{\mathrm{CH}}\left(Y_T^{(2)},a,b,n\right)\right) = \frac{w_1}{2}\widehat{c}_0\left[N\left(\widetilde{b}-\sigma_1\rho\sqrt{T}\right) - N\left(\widetilde{a}-\sigma_1\rho\sqrt{T}\right)\right] + w_1 e^A \sum_{k=1}^{n} \widehat{c}_k E_{\mathbb{Q}}\left[\exp\left(\frac{\sigma_1}{\sigma_2}\rho Y_T^{(2)}\right) T_k^{a,b}\left(Y_T^{(2)}\right) \cdot 1_{[a,b]}\left(Y_T^{(2)}\right)\right] = \frac{w_1}{2}\widehat{c}_0\left[N\left(\widetilde{b}-\sigma_1\rho\sqrt{T}\right) - N\left(\widetilde{a} - \sigma_1\rho\sqrt{T}\right)\right] + w_1 e^A \sum_{k=1}^{n}\sum_{l=0}^{[k/2]} \frac{\widehat{c}_k b_l^{(k)}}{(b-a)^{k-2l}} \cdot E_{\mathbb{Q}}\left[\exp\left(\frac{\sigma_1}{\sigma_2}\rho Y_T^{(2)}\right)\left(2Y_T^{(2)}-a-b\right)^{k-2l} \cdot 1_{[a,b]}\left(Y_T^{(2)}\right)\right]. \quad (53)$$

Now,

$$
\begin{aligned}
&E_{\mathbb{Q}}\left[\exp\left(\frac{\sigma_1}{\sigma_2}\rho Y_T^{(2)}\right)\left(2Y_T^{(2)}-a-b\right)^{k-2l}1_{[a,b]}\left(Y_T^{(2)}\right)\right] \\
&=\sum_{m=0}^{k-2l}\binom{k-2l}{m}\left(2E_{\mathbb{Q}}\left[Y_T^{(2)}\right]-a-b\right)^{m} \\
&\cdot 2^{k-2l-m}E_{\mathbb{Q}}\left[\exp\left(\frac{\sigma_1}{\sigma_2}\rho Y_T^{(2)}\right)\right. \\
&\left.\cdot\left(Y_T^{(2)}-E_{\mathbb{Q}}\left[Y_T^{(2)}\right]\right)^{k-2l-m}1_{[a,b]}\left(Y_T^{(2)}\right)\right]=\exp\left(\frac{\sigma_1}{\sigma_2}\right. \\
&\left.\cdot\rho\left(r-\frac{1}{2}\sigma_2^2\right)T\right)\left(2\sigma_2\sqrt{T}\right)^{k-2l} \\
&\cdot\sum_{m=0}^{k-2l}\binom{k-2l}{m}\left(\frac{2\left(r-(1/2)\sigma_2^2\right)T-a-b}{2\sigma_2\sqrt{T}}\right)^{m} \qquad (54)\\
&\cdot E_{\mathbb{Q}}\left[\exp\left(\sigma_1\rho\sqrt{T}Z\right)Z^{k-2l-m}1_{[\tilde{a},\tilde{b}]}(Z)\right]=\exp\left(\frac{\sigma_1}{\sigma_2}\right. \\
&\left.\cdot\rho\left(r-\frac{1}{2}\sigma_2^2\right)T\right)\left(2\sigma_2\sqrt{T}\right)^{k-2l} \\
&\cdot\sum_{m=0}^{k-2l}\binom{k-2l}{m}\left(\frac{2\left(r-(1/2)\sigma_2^2\right)T-a-b}{2\sigma_2\sqrt{T}}\right)^{m} \\
&\cdot\left.\frac{d^{k-2l-m}M_Z\left(u,\tilde{a},\tilde{b}\right)}{du^{k-2l-m}}\right|_{u=\sigma_1\rho\sqrt{T}}.
\end{aligned}
$$

Substituting the last expression into (53), we get (48). □

5. Numerical Results

As a benchmark setting, we consider initial prices of $S_0^{(1)} = 100$ and $S_0^{(2)} = 96$ dollars, strike price of $K = 1$ dollar, maturity at $T = 1$ year, and an annual interest rate of $r = 3\%$ under a bivariate Black-Scholes model, with a negative correlation $\rho = -0.3$ and respective volatilities equal to $\sigma_1 = 0.3$ and $\sigma_2 = 0.1$.

In Figure 1, we show the conditional price function on $[-1, 1]$ for the benchmark setting (blue line), together with the first and second Taylor expansions around the mean value of $Y_T^{(2)}$; that is, $E_{\mathbb{Q}}(Y_T^{(2)}) = (r - (1/2)\sigma_2^2)T$ (green and red lines, resp.). While the second-order approximation offers a reasonable local fit, significant errors may be found for values far from the mean. These errors may impact the price given by its conditional expected value under the risk neutral probability when, for example, volatilities are high.

Notice that, due to the concavity of function C, the first approximation underestimates the price. Not surprisingly, the second approximation estimates the price fairly well for values close to the point mean value while it is less accurate for values far from the mean. Although it seems as a drawback of the method, it does not constitute a serious problem as values far from the mean are infrequent, and thus the error in calculating the expected value from a Taylor approximation is fairly small.

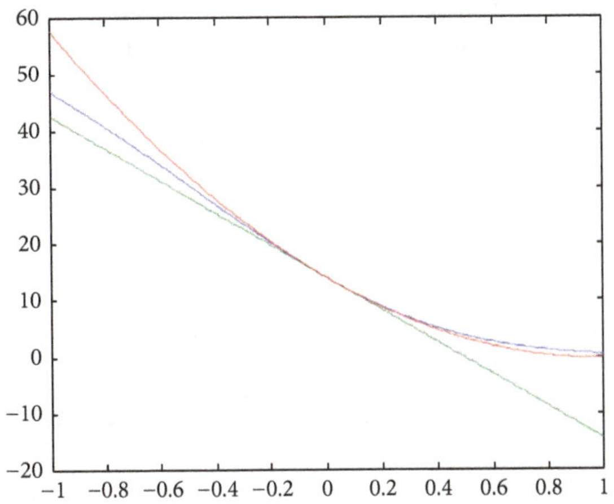

FIGURE 1: Approximations of the conditional price based on Taylor polynomials using the benchmark set of parameters. The blue line describes the conditional price on $[-1, 1]$ while the green and red lines show first- and second-order expansions, respectively.

A more promising result is obtained when an approximation of the conditional price is done via Chebyshev polynomials. Figure 2 represents, clockwise from top left, approximations of 4th, 10th, and 15th orders, respectively. Expansions of the 10th and 15th orders are practically indistinguishable from the original function. Coefficients in the expansion are calculated following a trapezoidal rule with 100 points on the interval $[-1.5, 1.5]$.

In Table 1, we show the impact of correlation on the prices of spread contracts obtained under the benchmark parameter set. Prices are computed by a Monte Carlo approach based on 10 million simulations of the asset prices following a geometric bivariate Black-Scholes model with correlated Brownian motions (column 2). In addition, we implement a second-order Taylor expansion and an approximation by Chebyshev polynomials of order $n = 15$. We consider positive, negative, large, moderated, and weak correlations. In all cases, the Chebyshev approximation shows a notable agreement with Monte Carlo prices at a lesser computational cost.

Notice that Chebyshev expansion requires extra quadrature to compute the polynomial coefficients as opposed to Taylor where we simply need to evaluate the corresponding derivatives. Hence, the computational complexity in the former multiplies by the number of points used in the trapezoid rule.

In Table 2, we show average routine times for the price calculation following Monte Carlo, Taylor, and Chebyshev expansions. The average is taken upon runs for different parameters in the parametric set. Routines were implemented on a standard PC using MATLAB language. Chebyshev approximation works about 200 times faster than the standard Monte Carlo approach. Taylor approach has an even

TABLE 1: Spread prices for the benchmark parameters and several values of ρ, using Monte Carlo, a Taylor second-order approximation around $y^* = 0$, and a Chebyshev approximation with 15 terms, $a = -4$ and $b = 0.25$.

Correlation	Monte Carlo	Taylor second approx.	Chebyshev, $n = 15$
$\rho = -0.1$	14.2921	13.8709	14.2906
$\rho = 0.1$	13.56278	14.78882	13.5649
$\rho = -0.3$	14.9734	15.0065	14.9629
$\rho = 0.3$	12.8085	12.7901	12.7903
$\rho = -0.5$	15.6273	15.9238	15.6316
$\rho = 0.5$	11.9525	11.9646	11.9566
$\rho = -0.7$	16.2421	17.5217	16.2521
$\rho = 0.7$	11.0315	11.1947	11.0529

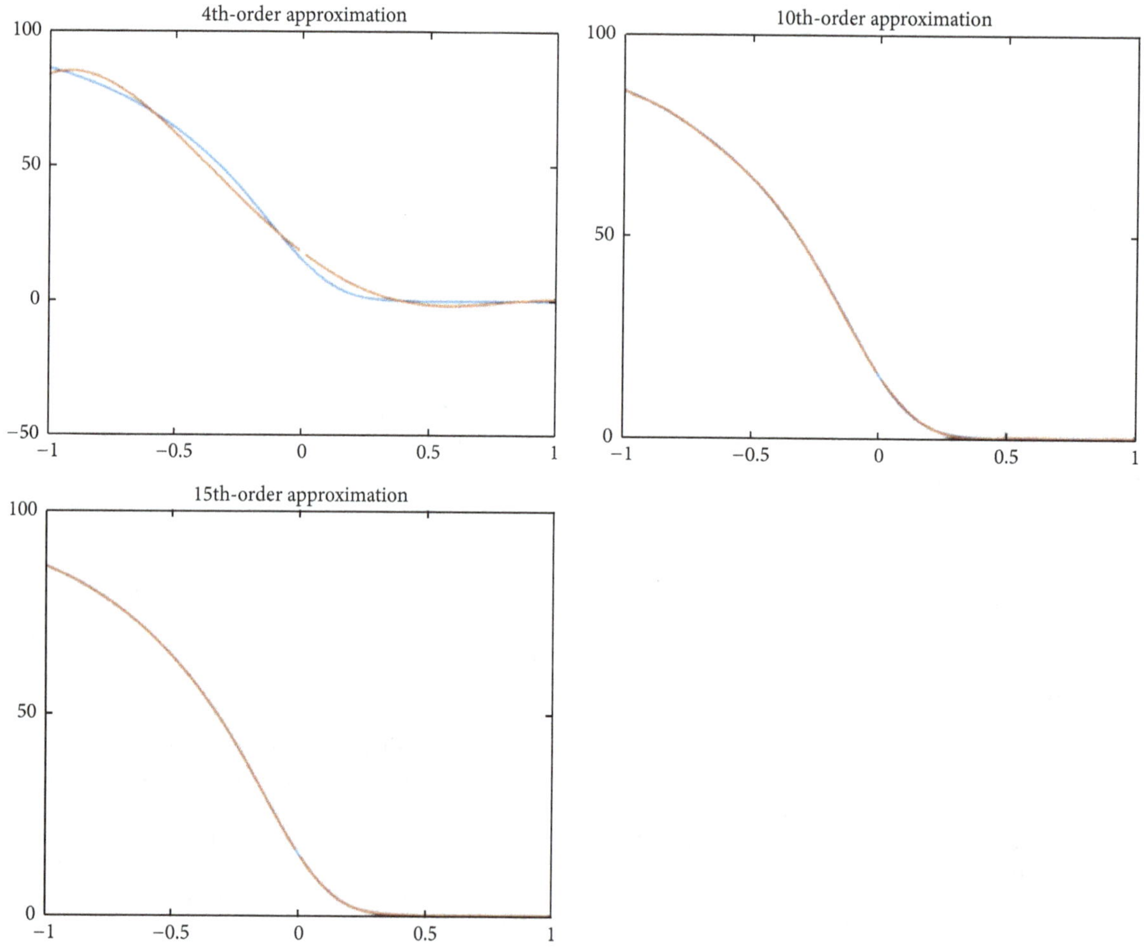

FIGURE 2: Approximations of the conditional price based on Chebyshev polynomials using the benchmark set of parameters. Clockwise, from top left are 4th, 10th, and 15th orders, respectively. The truncation interval is $[-1.5, 1.5]$.

lesser computational time, but for large asset correlations it is not as accurate as the former.

Due to the steepness of the function $C(y)$, the Chebyshev approximation is sensible to the truncation interval $[a, b]$. In our numerical computations, we have used $a = -4$ and $b = 0.25$. Within the range of parameter considered, most values of $Y_T^{(2)}$ lie on the selected interval $[a, b]$; hence, truncation does not affect the mixed exponential-power function by much. Otherwise, the approximation might be improved by taking $C(a)$ if $y < a$ and by $C(b)$ if $y > b$.

The selection of a and b is a delicate point. A priori, there is no theoretical limit in the length of the interval. Moreover, a large interval will result in a better approximation. On the flip side, the effort in the estimation of coefficients c_k in

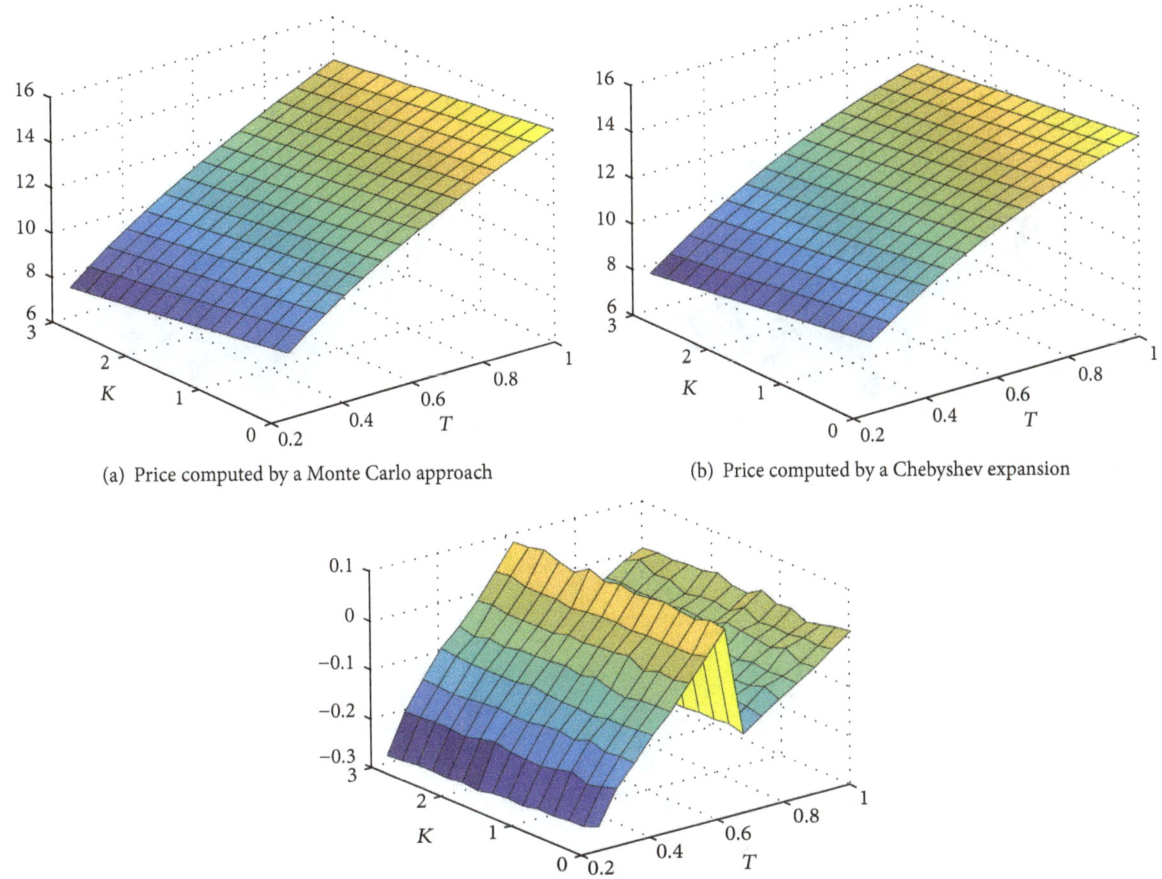

(a) Price computed by a Monte Carlo approach

(b) Price computed by a Chebyshev expansion

(c) Difference of prices between Monte Carlo and Chebyshev

FIGURE 3: Prices of a spread contract with maturity between one month and one year and strike price between 0 and 3 dollars under the benchmark parametric set. (c) Difference between prices obtained for both methods.

TABLE 2: Average routine times for Monte Carlo, Taylor, and Chebyshev approaches on a PC using MATLAB language.

Monte Carlo	Taylor second approx.	Chebyshev, $n = 15$
2.048 sec.	0.0076 sec.	0.012 sec.

the polynomial expansion will increase for larger intervals. An intuitive criterion to select the truncation values might consist in looking at points where the function $C(y)$ flattens out.

On the other hand, if the second asset has a high volatility, the length of the truncating interval should increase to avoid having too many points outside the interval.

The method is stable for the number of points considered in the trapezoidal rule. Also, the approximation gets close to the actual price after a fairly moderate number of polynomials. For $n = 10$, the method shows a good approximation within an error in the order of a penny. For $n = 15$ and $n = 20$ the approximation improves even more. For approximations of larger orders, the gain in precision does not compensate the increase in computational time.

Figure 3 shows prices of a spread contract based on a Chebyshev approximation of order $n = 15$. Maturity times range from one month to one year, while strike prices go from zero (exchange option) to 3 dollars. Results are consistent with an increase in the contract prices with higher maturity and their decrease with the increase of the strike price. Figure 3(c) at the bottom, conveniently scaled, shows the difference between both prices. The average relative error is 0.0075%.

Figure 4 shows the price of the contract as a function of volatilities of both underlying assets. For symmetry, we only have considered $\sigma_1 \geq \sigma_2$ with volatilities ranging within 10%–50%. Again, we can observe a remarkable agreement between prices obtained by Monte Carlo and Chebyshev approximation. As expected, the price of the spread increases with the increment in the volatility. The average relative error is 0.0023%.

6. Conclusions

We compare three methods to price spreads options under a bivariate Black-Scholes model with correlated Brownian motions versus a standard Monte Carlo approach. Our results

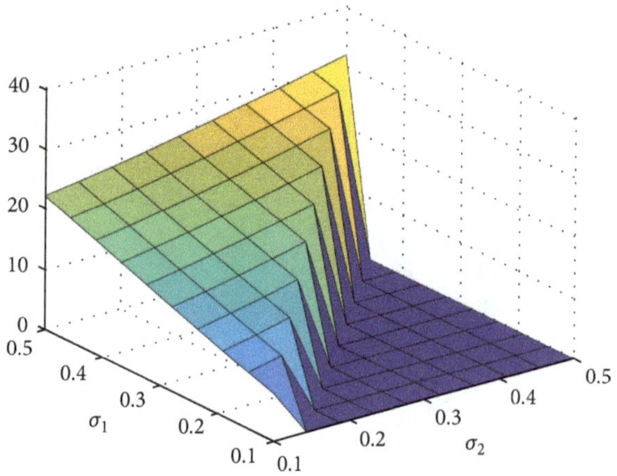

(a) Price computed by a Monte Carlo approach, $n = 10^7$

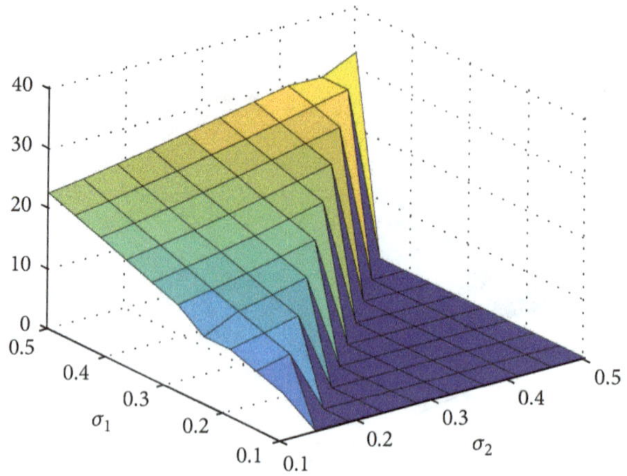

(b) Price computed by a Chebyshev expansion, order 15

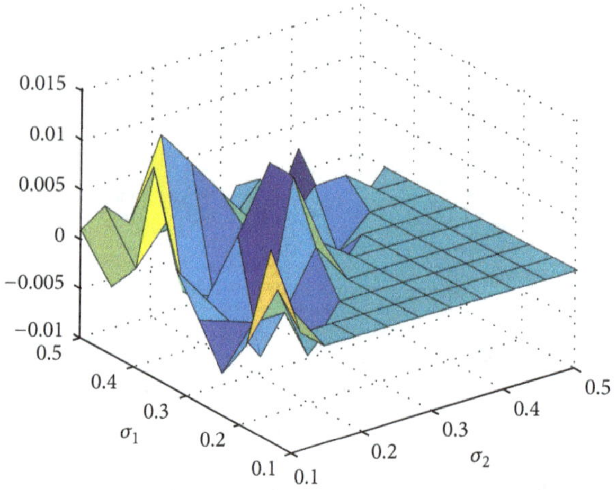

(c) Difference of prices between Monte Carlo and Chebyshev

FIGURE 4: Prices of a spread contract with log-return volatilities between 10% and 50% under the benchmark parametric set. (c) Difference between prices obtained for both methods. We consider, without loss of generality, the case $\sigma_1 \geq \sigma_2$.

show that Taylor approximation does not offer a uniform convergence, and hence a poor result when values are far from the point around the expansion is taken. For some values in the parameter set, it may affect the corresponding expected value.

The approximation based on Chebyshev polynomials seems to be appropriate in terms of the balance offered between accuracy and computational cost. Moreover, the method is suitable to be implemented in more general models provided the conditional distribution is available.

Competing Interests

The authors declare that there is no conflict of interests regarding the publication of this manuscript.

Acknowledgments

Both authors would like to recognize NSERC for its support.

References

[1] N. D. Pearson, "An Efficient Approach for Pricing Spread Options," *The Journal of Derivatives*, vol. 3, no. 1, pp. 76–91, 1995.

[2] M. Li, S.-J. Deng, and J. Zhou, "Closed-form approximations for spread option prices and Greeks," *Journal of Derivatives*, vol. 15, no. 3, pp. 58–80, 2008.

[3] R. Pachon, "Numerical pricing of european options with arbitrary payoffs," *SSRN Electronic Journal*, 2016.

[4] M. Gass, K. Glau, M. Mahlstedt, and M. Mair, "Chebyshev interpolation for parametric option pricing," https://arxiv.org/abs/1505.04648.

[5] E. Kirk, *Correlation in the Energy Markets, in Managing Energy Price Risk*, Risk Publications and Enron, London, UK, 1995.

[6] R. Carmona and V. Durrleman, "Pricing and hedging spread options," *SIAM Review*, vol. 45, no. 4, pp. 627–685, 2003.

[7] S. Borokova, F. J. Permana, and H. Weide, "A closed form approach to the valuation and hedging of basket and spread options," *The Journal of Derivatives*, vol. 14, no. 4, pp. 8–24, 2009.

[8] A. Alvarez, M. Escobar, and P. Olivares, "Spread Options under Stochastic Covariance and Jumps. Pricing two dimensional

derivatives under stochastic correlation," *International Journal of Financial Markets and Derivatives*, vol. 2, no. 4, pp. 265–287, 2011.

[9] A. Venkatramanan and C. Alexander, "Closed form approximations for spread options," *Applied Mathematical Finance*, vol. 18, no. 5, pp. 447–472, 2011.

[10] M. Li, J. Zhou, and S.-J. Deng, "Multi-asset spread option pricing and hedging," *Quantitative Finance*, vol. 10, no. 3, pp. 305–324, 2010.

[11] J. Hull and A. White, "The pricing of options on assets with stochastic volatilities," *The Journal of Finance*, vol. 42, no. 2, pp. 281–300, 1987.

[12] N. Ju, "Pricing Asian and basket options via Taylor expansion," *The Journal of Computational Finance*, vol. 5, no. 3, pp. 79–103, 2002.

[13] Q.-J. Meng and D. Ding, "An efficient pricing method for rainbow options based on two-dimensional modified sine-sine series expansions," *International Journal of Computer Mathematics*, vol. 90, no. 5, pp. 1096–1113, 2013.

[14] F. Fang and C. W. Oosterlee, "A novel pricing method for European options based on Fourier-cosine series expansions," *SIAM Journal on Scientific Computing*, vol. 31, no. 2, pp. 826–848, 2008.

[15] W. Margrabe, "The value of an option to exchange one asset for another," *The Journal of Finance*, vol. 33, no. 1, pp. 177–186, 1978.

[16] Y. L. Tong, *The Multivariate Normal Distribution*, Springer, Berlin, Germany, 1989.

[17] J. C. Mason and D. C. Handscomb, *Chebyshev Polynomials*, CRC Press Company, Boca Raton, Fla, USA, 2003.

[18] A. Gil, J. Segura, and N. Temme, *Numerical Methods for Special Functions*, SIAM, Philadelphia, Pa, USA, 2007.

Permissions

All chapters in this book were first published in JAM, by Hindawi Publishing Corporation; hereby published with permission under the Creative Commons Attribution License or equivalent. Every chapter published in this book has been scrutinized by our experts. Their significance has been extensively debated. The topics covered herein carry significant findings which will fuel the growth of the discipline. They may even be implemented as practical applications or may be referred to as a beginning point for another development.

The contributors of this book come from diverse backgrounds, making this book a truly international effort. This book will bring forth new frontiers with its revolutionizing research information and detailed analysis of the nascent developments around the world.

We would like to thank all the contributing authors for lending their expertise to make the book truly unique. They have played a crucial role in the development of this book. Without their invaluable contributions this book wouldn't have been possible. They have made vital efforts to compile up to date information on the varied aspects of this subject to make this book a valuable addition to the collection of many professionals and students.

This book was conceptualized with the vision of imparting up-to-date information and advanced data in this field. To ensure the same, a matchless editorial board was set up. Every individual on the board went through rigorous rounds of assessment to prove their worth. After which they invested a large part of their time researching and compiling the most relevant data for our readers.

The editorial board has been involved in producing this book since its inception. They have spent rigorous hours researching and exploring the diverse topics which have resulted in the successful publishing of this book. They have passed on their knowledge of decades through this book. To expedite this challenging task, the publisher supported the team at every step. A small team of assistant editors was also appointed to further simplify the editing procedure and attain best results for the readers.

Apart from the editorial board, the designing team has also invested a significant amount of their time in understanding the subject and creating the most relevant covers. They scrutinized every image to scout for the most suitable representation of the subject and create an appropriate cover for the book.

The publishing team has been an ardent support to the editorial, designing and production team. Their endless efforts to recruit the best for this project, has resulted in the accomplishment of this book. They are a veteran in the field of academics and their pool of knowledge is as vast as their experience in printing. Their expertise and guidance has proved useful at every step. Their uncompromising quality standards have made this book an exceptional effort. Their encouragement from time to time has been an inspiration for everyone.

The publisher and the editorial board hope that this book will prove to be a valuable piece of knowledge for researchers, students, practitioners and scholars across the globe.

List of Contributors

T. M. Ajayi, A. J. Omowaye and I. L. Animasaun
Department of Mathematical Sciences, Federal University of Technology, Akure, Ondo State, Nigeria

Inês Simões, António Tadeu and Nuno Simões
ITeCons, Rua Pedro Hispano, s/n, 3030-289 Coimbra, Portugal
Department of Civil Engineering, University of Coimbra, Rua Luís Reis Santos, Pólo II, 3030-788 Coimbra, Portugal

Siyuan Li
School of Mathematics and Applied Statistics, Faculty of Engineering and Information Sciences, University of Wollongong, Wollongong, NSW 2522, Australia

Lihua You, Yafei Chen and Pingzhi Yuan
School of Mathematical Sciences, South China Normal University, Guangzhou 510631, China

Sandile S. Motsa and Zodwa G. Makukula
University of Swaziland, Private Bag 4, Kwaluseni, Matsapha M201, Swaziland

Mawardi Bahri and Muh. Saleh Arif Fatimah
Department of Mathematics, Hasanuddin University, Makassar 90245, Indonesia

A. Srinivasan and G. Geetharamani
Department of Mathematics, Anna University, Bharathidasan Institute of Technology (BIT) Campus, Tiruchirappalli, Tamilnadu 620 024, India

A. Sakabekov
Research Laboratory "Applied Modeling Oil and Gas Fields", Kazakh-British Technical University, 59 Tole Bi Street, Almaty 050000, Kazakhstan

Y. Auzhani
Al-Farabi Kazakh National University, 71 Al-Farabi Avenue, Almaty 050040, Kazakhstan

Arild Wikan
School of Business and Economics, The Arctic University of Norway, Campus Harstad, Havnegata 5, 9480 Harstad, Norway

Samsul Ariffin Abdul Karim
Fundamental and Applied Sciences Department, Universiti Teknologi PETRONAS, Bandar Seri Iskandar, 32610 Seri Iskandar, Perak Darul Ridzuan, Malaysia

Kong Voon Pang
School of Mathematical Sciences, Universiti Sains Malaysia (USM), 11800 Minden, Penang, Malaysia

Chriscella Jalius and Zanariah Abdul Majid
Institute for Mathematical Research, Universiti Putra Malaysia (UPM), 43400 Serdang, Selangor, Malaysia

Hamzah Bakhti
MMC Team, Department of Mathematics, Faculty of Sciences & Techniques, AEU, BP 416 Tangier, Morocco

Lahcen Azrar
MMC Team, Department of Mathematics, Faculty of Sciences & Techniques, AEU, BP 416 Tangier, Morocco
MMC Team, Department of Mechanical Engineering, Faculty of Engineering, KAU, Jeddah, Saudi Arabia

Dhanesh Patel
Centre for Industrial Mathematics and Department of Applied Mathematics, Faculty of Technology and Engineering, The M. S. University of Baroda, Vadodara, Gujarat 390001, India

Ashvinkumar Chaudhari and Jari Hämäläinen
Centre of Computational Engineering and Integrated Design (CEID), Lappeenranta University of Technology, Lappeenranta, Finland

Arto Laari
Department of Chemistry, Lappeenranta University of Technology, Lappeenranta, Finland

Matti Heiliö
Department of Mathematics and Physics, Lappeenranta University of Technology, Lappeenranta, Finland

Kishorilal Agrawal
Department of Chemical Engineering, Faculty of Technology and Engineering, The M. S. University of Baroda, Vadodara, Gujarat 390001, India

Ali Sendur
Alanya Alaaddin Keykubat University, 07490 Antalya, Turkey

Zhixiong Chen, Yi Ding and Andrew Getz
Department of Mathematics, New Jersey City University, Jersey City, NJ 07305-1597, USA

Bernard Lipat
Applied Physics and Applied Mathematics Department, Columbia University, New York, NY 10027, USA

S. S. N. Perera
Research and Development Center for Mathematical Modeling, Faculty of Science, University of Colombo, 00300 Colombo, Sri Lanka

Pablo Olivares and Alexander Alvarez
Department of Mathematics, Ryerson University, 350 Victoria St., Toronto, ON, CanadaM5B 2K3

Index

www.ingramcontent.com/pod-product-compliance
Lightning Source LLC
LaVergne TN
LVHW060607210926
848485LV00005B/1383